EXPLORATION OF THE SOLAR SYSTEM BY INFRARED REMOTE SENSING

This book describes all aspects of the theory, instrumental techniques, and observational results of the remote sensing of objects in our Solar System through studies of infrared radiation. Fully revised since publication of the first edition in 1992, it now incorporates the latest technologies, new mission results, and scientific discoveries. It also includes a fully up-dated list of references to reflect the advances made in this field during the past ten years.

The theories of radiative transfer, molecular spectroscopy, and atmospheric physics are first combined to show how it is possible to calculate the infrared spectra of model planetary atmospheres. Next the authors describe the instrumental techniques, in order to assess the effect of real instruments on the measurement of the emerging radiation field. Finally, techniques that allow the retrieval of atmospheric and surface parameters from observations are examined. There are plenty of examples from ground-based and space observations that demonstrate the methods of finding temperatures, gas compositions, and certain parameters of the solid surface. All planets from Mercury to Pluto, many of their satellites, asteroids, and comets are discussed.

The presentation will appeal to advanced students and professional planetary science researchers, although some chapters are of wider interest. The authors have drawn on their extensive experience at the NASA–Goddard Space Flight Center to produce a definitive account of what can be learned from infrared studies of our planetary system.

RUDOLF HANEL worked at the Goddard Space Flight Center for 31 years where he served as Principal Investigator on missions around Earth (Tiros, Nimbus), Mars (Mariner 9), and the outer planets (Voyagers 1 and 2). During this time he published over 100 scientific papers and book chapters, and was the recipient of numerous prestigious awards including the G. P. Kuiper Award of the Planetary Division of the American Astronomical Society.

BARNEY CONRATH was affiliated to the Goddard Space Flight Center from 1960 until 1995 and is currently a visiting faculty member in the Cornell University Center for Radiophysics and Space Research. His research interests include the study of the thermal structure, composition, and dynamics of planetary atmospheres. He has participated in spacecraft missions to the Earth, Mars, and the outer planets, and is currently a member of the infrared spectrometer teams on the Mars Global Surveyor and Cassini missions.

DONALD JENNINGS has worked at the Goddard Space Flight Center since 1977 in the area of infrared spectroscopy; developing and using a variety of spectrometers to study atmospheres of planets, the Sun, and stars. His research has ranged from examining molecules in the laboratory to measuring the infrared glow of the Space Shuttle and he is presently Instrument Scientist for the Composite Infrared Spectrometer on the Cassini mission to Saturn.

ROBERT SAMUELSON was a research scientist at the Goddard Space Flight Center for 39 years and is presently a research associate with the Astronomy Department at the University of Maryland. His specialities include radiative transfer in scattering atmospheres and the interpretation of radiometric and spectroscopic data from ground-based and space-borne infrared instruments. He is a co-investigator for the Cassini Orbiter infrared spectrometer and the Huygens Probe aerosol collector/pyrolizer experiment.

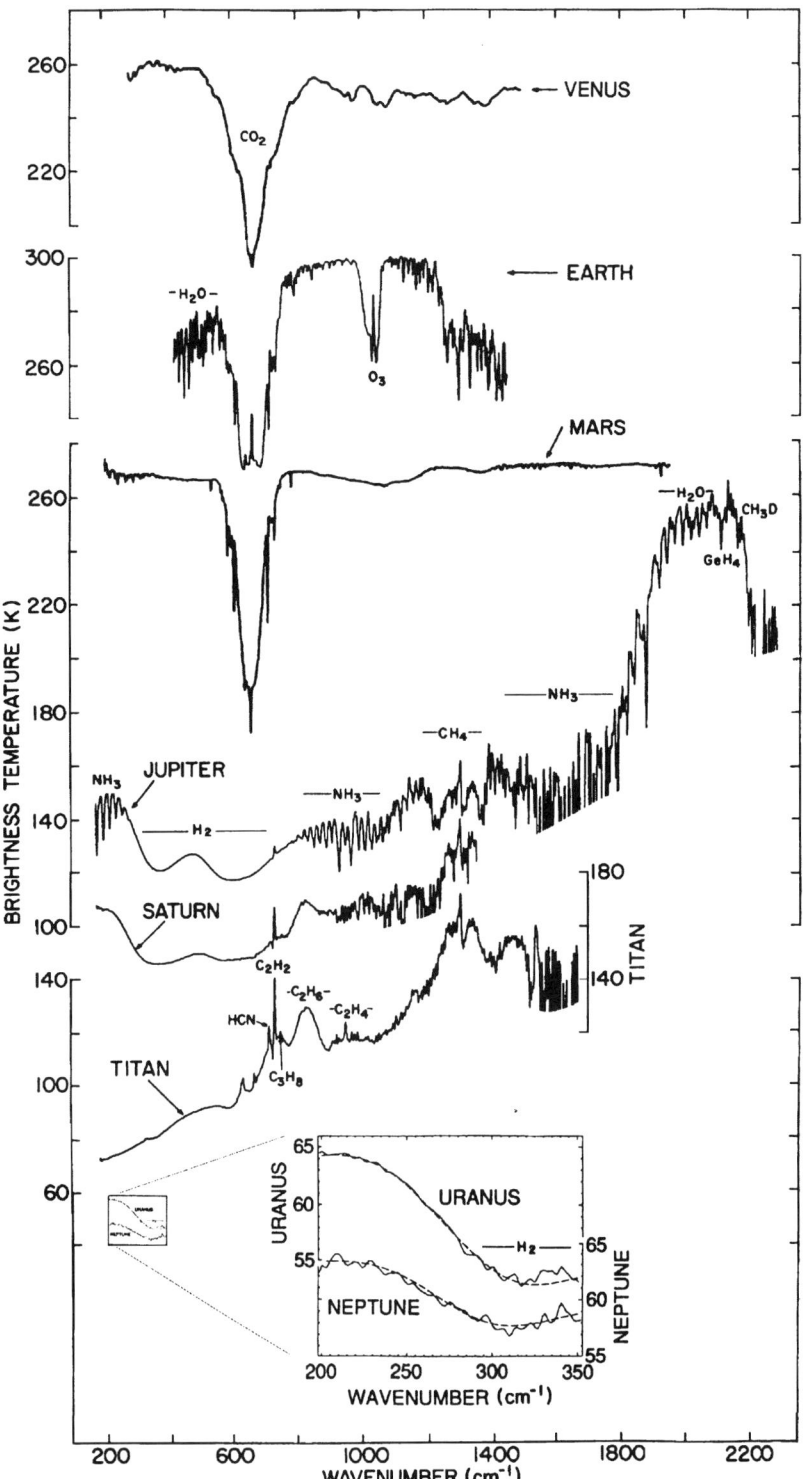

Examples of infrared spectra of the planets (except Mercury and Pluto) and of Titan recorded by Michelson interferometers on Venera, Nimbus, Mariner 9, and Voyager.

EXPLORATION OF THE SOLAR SYSTEM BY INFRARED REMOTE SENSING

Second edition

R. A. HANEL
B. J. CONRATH
D. E. JENNINGS
and
R. E. SAMUELSON

PUBLISHED BY THE PRESS SYNDICATE OF THE UNIVERSITY OF CAMBRIDGE
The Pitt Building, Trumpington Street, Cambridge, United Kingdom

CAMBRIDGE UNIVERSITY PRESS
The Edinburgh Building, Cambridge CB2 2RU, UK
40 West 20th Street, New York, NY 10011-4211, USA
477 Williamstown Road, Port Melbourne, VIC 3207, Australia
Ruiz de Alarcón 13, 28014 Madrid, Spain
Dock House, The Waterfront, Cape Town 8001, South Africa

http://www.cambridge.org

First edition © Cambridge University Press 1992, except in the jurisdictional territory of the
United States of America. The United States Government has a licence under any copyright
for Governmental purposes.
Second edition © Cambridge University Press 2003, except in the jurisdictional territory of the
United States of America. The United States Government has a licence under any copyright
for Governmental purposes.

First published 1992

Second edition 2003

Printed in the United Kingdom at the University Press, Cambridge

Typeface Times 11/14 pt   *System* LaTeX 2_ε [TB]

A catalogue record for this book is available from the British Library

Library of Congress Cataloguing in Publication data

Exploration of the solar system by infrared remote sensing / R. A. Hanel . . . [et al.].–2nd ed.
p.   cm.
Includes bibliographical references and index.
ISBN 0 521 81897 4
1. Planets–Remote sensing.   2. Infrared astronomy.   3. Outer space–Exploration.   I. Hanel, R. A.
QB603.I52 E88   2002

523.2–dc21   2002025665

ISBN 0 521 81897 4 hardback

Contents

Introduction to first edition		*page* xi
Introduction to second edition		xv
1	Foundation of radiation theory	1
	1.1 Maxwell's equations	2
	1.2 Conservation of energy and the Poynting vector	4
	1.3 Wave propagation	5
	1.4 Polarization	13
	1.5 Boundary conditions	14
	1.6 Reflection, refraction, and the Fresnel equations	17
	1.7 The Planck function	21
	1.8 The Poynting vector, specific intensity, and net flux	25
2	Radiative transfer	27
	2.1 The equation of transfer	28
	a. Definitions and geometry	28
	b. Microscopic processes	29
	c. The total field	37
	d. The diffuse field	40
	2.2 Formal solutions	42
	2.3 Invariance principles	44
	a. Definitions	44
	b. The stacking of layers	45
	c. Composite scattering and transmission functions	47
	d. Starting solutions	49
	2.4 Special cases	50
	a. Nonscattering atmospheres	50
	b. Optically thin atmospheres	51
	2.5 Scattering atmospheres; the two-stream approximation	52

			a.	Single scattering phase function	52
			b.	Separation of variables	53
			c.	Discrete streams	54
			d.	Homogeneous solution	55
			e.	Outside point source	56
3	Interaction of radiation with matter				58
	3.1	Absorption and emission in gases			59
			a.	The old quantum theory	59
			b.	The Schrödinger equation	60
			c.	Energy levels and radiative transitions	62
	3.2	Vibration and rotation of molecules			64
	3.3	Diatomic molecules			66
			a.	Vibration	67
			b.	Rotation	72
			c.	Vibration–rotation interaction	75
			d.	Collision-induced transitions	78
	3.4	Polyatomic molecules			80
			a.	Vibration	80
			b.	Rotation	86
			c.	Vibration–rotation transitions	90
	3.5	Line strength			93
	3.6	Line shape			99
	3.7	Solid and liquid surfaces			103
			a.	Solid and liquid phases	103
			b.	Complex refractive indices	105
	3.8	Cloud and aerosol particles			110
			a.	Asymptotic scattering functions	110
			b.	Rigorous scattering theory; general solution	113
			c.	Particular solutions and boundary conditions	117
			d.	The far field; phase function and efficiency factors	122
4	The emerging radiation field				129
	4.1	Models with one isothermal layer			129
			a.	Without scattering	129
			b.	With scattering	132
	4.2	Models with a vertical temperature structure			140
			a.	Single lapse rate	141
			b.	Multiple lapse rates	144
	4.3	Model with realistic molecular parameters			148
5	Instruments to measure the radiation field				152
	5.1	Introduction to infrared radiometry			154

5.2	Optical elements	155
5.3	Diffraction limit	166
5.4	Chopping, scanning, and image motion compensation	170
	a. D.C. radiometers	170
	b. Chopped or a.c. radiometers	179
	c. Image motion compensation	185
5.5	Intrinsic material properties	188
	a. Absorbing and reflecting filters	188
	b. Prism spectrometers	190
	c. Gas filter, selective chopper, and the pressure modulated radiometer	192
5.6	Interference phenomena in thin films	194
	a. Outline of thin film theory	195
	b. Antireflection coatings	198
	c. Beam dividers	200
	d. Interference filters and Fabry–Perot interferometers	204
5.7	Grating spectrometers	209
5.8	Fourier transform spectrometers	220
	a. Michelson interferometer	220
	b. Post-dispersion	240
	c. Martin–Puplett interferometer	243
	d. Lamellar grating interferometer	247
5.9	Heterodyne detection	249
5.10	Infrared detectors in general	253
5.11	Thermal detectors	255
	a. Temperature change	255
	b. Noise in thermal detectors	260
	c. Temperature to voltage conversion	264
5.12	Photon detectors	272
	a. Intrinsic and extrinsic semiconductors	273
	b. Photoconductors and photodiodes	274
	c. Responsivities	275
	d. Noise in photon detectors	277
	e. Circuits for photon detectors	278
	f. Detector arrays	280
5.13	Calibration	281
	a. Concepts	281
	b. Middle and far infrared calibration	284
	c. Near infrared calibration	291
	d. Wavenumber calibration	293

		5.14	Choice of measurement techniques	294
			a. Scientific objectives	294
			b. Instrument parameters	296
6	Measured radiation from planetary objects up to Neptune			301
	6.1	Instrument effects		301
	6.2	The terrestrial planets		305
	6.3	The giant planets		317
	6.4	Titan		325
	6.5	Objects without substantial atmospheres		333
			a. Tenuous atmospheres	333
			b. Surfaces	334
7	Trans-Neptunian objects and asteroids			342
	7.1	Pluto and Charon		342
	7.2	Comets		346
	7.3	Asteroids		349
8	Retrieval of physical parameters from measurements			352
	8.1	Retrieval of atmospheric parameters		352
	8.2	Temperature profile retrieval		355
			a. General consideration	355
			b. Constrained linear inversion	356
			c. Relaxation algorithms	360
			d. Backus–Gilbert formulation	361
			e. Statistical estimation	365
			f. Limb-tangent geometry	367
	8.3	Atmospheric composition		368
			a. Principles	369
			b. Feature identification	369
			c. Correlation analysis	370
			d. Abundance determination	371
			e. Profile retrieval	372
			f. Simultaneous retrieval of temperature and gas abundance	376
			g. Limb-tangent observations	378
	8.4	Clouds and aerosols		380
			a. Small absorbing particles	381
			b. Titan's stratospheric aerosol	382
	8.5	Solid surface parameters		385
			a. Surface temperature	385
			b. Thermal inertia	388
			c. Refractive index and texture	392

	8.6	Photometric investigations	394
		a. Introduction	394
		b. The Bond albedo	396
		c. Thermal emission	402
9	Interpretation of results		405
	9.1	Radiative equilibrium	405
		a. Governing principles	406
		b. The solar radiation field	407
		c. Thermal radiation and the temperature profile	410
		d. General atmospheric properties	413
	9.2	Atmospheric motion	420
		a. Governing equations	421
		b. Mars	428
		c. The outer planets	436
		d. Venus	442
	9.3	Evolution and composition of the Solar System	444
		a. Formation of the Solar System	445
		b. Evolution of the terrestrial planets	450
		c. Evolution of the giant planets	452
	9.4	Energy balance	457
		a. The terrestrial planets	459
		b. The giant planets	459
Closing remarks			465
Appendices			
1	Mathematical formulas		467
2	Physical constants		471
3	Planetary and satellite parameters		472
References			475
Abbreviations			511
Index			513

Introduction to first edition

The advent of spaceflight has ushered in a new era of Solar System exploration. Man has walked on the Moon and returned with soil samples. Instrumented probes have descended through the atmospheres of Venus and Mars. The Mariner, Pioneer, Venera, Viking, and Voyager space flight programs have provided opportunities to study the planets from Mercury to Neptune and most of the satellites. Remote sensing investigations have been conducted with unprecedented spatial and spectral resolutions, permitting detailed examinations of atmospheres and surfaces. Even for the Earth, space-borne observations, obtained with global coverage and high spatial, spectral, and temporal resolutions, have revolutionized weather forcasting, climate research, and the exploration of natural resources.

The collective study of the various atmospheres and surfaces in the Solar System constitutes the field of comparative planetology. Wide ranges in surface gravity, solar flux, internal heat, obliquity, rotation rate, mass, and composition provide a broad spectrum of boundary conditions for atmospheric systems. Analyses of data within this context lead to an understanding of physical processes applicable to all planets. Once the general physical principles are identified, the evolution of planetary systems can be explored.

Some of the data needed to address the broader questions have already been collected. Infrared spectra, images, and many other types of data are available in varying amounts for Mercury, Venus, Earth, Mars, Jupiter, Saturn, Uranus, Neptune, and many of their satellites. It is now appropriate to review and assess the techniques used in obtaining the existing information. This will not only provide a summary of our present capabilities, but will also suggest ways of extending our knowledge to better address the issues of comparative planetology and Solar System evolution.

Remote sensing is an interdisciplinary task. Theories of radiative transfer, molecular quantum mechanics, atmospheric physics, photochemistry, and planetary geology overlap with the design of advanced instrumentation, complex data processing, and a wide range of analysis methods. The purpose of this book is to bring many of these disciplines together with emphasis on the acquisition and interpretation of thermal infrared data. We address the advanced student and active researcher in the field. It is our intent to examine the basic principles in some depth. To meet this goal we strive to develop a consistent and essentially self-contained review. It is necessary to be highly selective in choosing illustrative cases because the development of each is fairly complex.

Although some *in situ* measurements have been made, planetary investigations have largely been restricted to remote sensing of emitted and reflected radiation. Planets emit most of their thermal radiation in the middle and far infrared while reflected sunlight dominates their visible and near infrared spectra. Planetary spectra, recorded from orbiting or fly-by spacecraft, make it possible to simultaneously obtain good horizontal and vertical resolutions of both atmospheric composition and thermal structure. These quantities and their gradients lead to a description of energetic and dynamical processes characteristic of each atmosphere. High resolution images at visible and infrared wavelengths display cloud patterns, which manifest this dynamical activity and provide highly complementary information to the spectral data. Ground-based astronomy has contributed additional information, with the significant advantage of providing observations over relatively long time spans.

Emitted and reflected radiation fields can be regarded as coded descriptions of planetary atmospheres and surfaces. Radiative transfer theory provides a means of transforming the codes into intelligible terms. This approach requires an understanding of electromagnetic radiation and its interaction with matter. Chapters 1 through 3 are directed towards these ends. A review of Maxwell's equations, wave propagation, polarization, reflection, refraction, and the Planck function is undertaken in Chapter 1. In Chapter 2 the equation of radiative transfer is derived in a form suitable for remote sensing from space, and various solutions of the transfer equation are obtained. In Chapter 3 we examine the interaction of radiation with matter. Quantum mechanical concepts, the principles of vibrational and rotational spectra, and other tools necessary to understand planetary spectra are developed. Investigation of matter in condensed phases – solid surfaces, ice crystals, and liquid droplets – requires an understanding of the emission and reflection of radiation at surfaces characterized by a complex index of refraction and such topics as the Mie theory.

With the tools developed in Chapters 1 through 3, it is possible to construct models of the emission and reflection of gas layers over a solid surface. Such models, with increasing complexity, including scattering, are the subject of Chapter 4. However, it is impossible to separate a study of planetary systems by remote sensing from the instruments which record the data. Inferences of atmospheric and surface parameters require the analysis of observed spectra, which have been subjected to modifications characteristic of the instruments used. In Chapter 5 we consider concepts of remote sensing instruments. The discussion of certain principles and techniques is supplemented with specific examples of instruments, such as the Thematic Mapper and the Voyager infrared spectrometer. Special attention is given to radiometric calibration. Examination of scientific objectives and instrumental techniques leads to a discussion of trade-offs between spatial and spectral resolution, signal-to-noise ratio, data rate, and other parameters.

In Chapter 6 we consider instrumental effects, such as spectral resolution and signal-to-noise ratio, and discuss data from the terrestrial and the giant planets in a qualitative manner. In Chapter 7 we examine methods for interpreting spectroscopic and radiometric data produced by real instruments in terms of physical properties of atmospheres and surfaces. Emphasis is placed on the retrieval of thermal structure, gas composition and cloud properties of the atmospheres, and thermal properties and texture of surfaces. Limitations on the information content inherent in measured quantities are assessed.

In Chapter 8 we associate measured quantities with the underlying physical processes. The connection between thermal equilibrium and the vertical temperature profile is investigated. Dynamical regimes are explored with emphasis on wind fields and circulations. Certain aspects of Solar System composition, internal heat sources, and the concept of global energy balance are discussed in the context of planetary evolution.

In Appendix 1 we list some of the properties of vectors and mathematical functions used in the text. Important physical constants are listed in Appendix 2. The most important planetary and satellite parameters, such as dimensions and composition, are summarized in Appendix 3.

Throughout the book we adopt the International System (SI), with the basic units of meter, kilogram, second, ampere, mole, and kinetic temperature (kelvin). However, we make exceptions in deference to common usage. For example, in atmospheric physics and specifically in meteorology the bar and millibar are firmly entrenched in the literature as units of pressure; we retain these here. The corresponding SI unit, the pascal (newton per square meter, or $N\,m^{-2}$), which equals 10^{-5} bar, is only slowly gaining acceptance in the planetary literature.

The SI unit of intensity, the candela, is defined (1985) as the luminous intensity in a given direction of a source that emits at 540×10^{12} hertz (Hz) and has a radiant intensity in that direction of 1/683 watt per steradian ($W\,sr^{-1}$). Although the candela should be a convenient unit in the discussion of radiative processes, it is not used in planetary astronomy or in the field of remote sensing. Hence we follow tradition and express the spectrally integrated intensity in $W\,cm^{-2}\,s^{-1}$; the spectral intensity itself is then expressed in $W\,cm^{-2}\,sr^{-1}/cm^{-1}$ (we prefer to retain this explicit expression rather than use the equivalent term $W\,cm^{-1}\,sr^{-1}$). The term spectral radiance is synonymous with specific or spectral intensity.

Another exception concerns the units of wavenumber and wavelength. Radio astronomy is a rather modern branch of science and has easily adopted the SI (e.g., flux in W m^{-2}), while spectroscopy is an old discipline of physics. The roots of spectroscopy lie deep in the nineteenth century, when the Gaussian system ruled with the centimeter as the unit of length. The common spectroscopic unit of wavenumber is, therefore, cm^{-1}; wavelength is usually measured in μm. We follow that tradition.

In writing this book the authors gained from numerous discussions with many colleagues and friends. Several have made specific comments on the manuscript. We would like to acknowledge contributions from W. Bandeen, G. Birnbaum, R. Born, M. Flasar, P. Gierasch, G. Hunt, T. Kostiuk, V. Kunde, J. Mangus, J. Mather, J. Pearl, and D. Reuter. J. Guerber and L. Mayo helped with computer programming. We also appreciate the encouragement and patience of the editor S. Mitton and his staff at Cambridge University Press.

The following journals and publishers have given permission to reproduce figures:

Applied Optics. Optical Society of America, Washington DC: Figs. 5.2.10, 5.3.2, 5.8.2, 5.8.3, 5.8.9, 5.8.10, and 5.8.12.

The Astrophysical Journal. The University of Chicago Press, Chicago IL: Figs. 3.8.2, 7.3.4, and 7.3.5.

Icarus. Academic Press, Orlando FL: Figs. 5.9.1, 6.2.2, 6.2.9, 7.3.3, 7.5.1, 8.2.2, and 8.2.3.

Journal of Atmospheric Sciences. American Metereological Society, Boston MA: Fig. 8.2.4.

Journal of Geophysical Research. American Geophysical Union, Washington DC: Figs. 6.2.5, 6.2.6, 6.2.7, and 6.4.1.

Nature. Macmillan Magazines Ltd, London: Figs. 5.12.5, 6.4.2, 6.4.3, and 6.5.1.

Proceedings of the Twenty-first Astronautical Congress. North Holland Publishing Co.: Fig. 6.2.4.

Canadian Journal of Physics. National Research Council of Canada: Fig. 3.3.6.

Science. American Association for the Advancement of Science. Washington DC: Figs. 6.2.8, 6.2.11, 6.3.3, and 7.5.2.

Spectrometric Techniques III. Academic Press, Orlando FL: Fig. 5.8.5.

Satellites of Jupiter. University of Arizona Press, Tucson AZ: Fig. 6.5.4.

We also thank the authors for making this material available.

Introduction to second edition

Since the first edition of this book appeared in print, infrared observations have been responsible for a number of significant new results from many objects in the Solar System. Besides highly sophisticated ground-based measurements, instruments on space probes such as Galileo, Mars Global Surveyor, Vega, Giotto, Phobos-2, the Infrared Space Observatory, and others have produced new data leading to interesting conclusions. Even the spectacular impact of comet Shoemaker–Levy 9 yielded unique information on the atmosphere of Jupiter as well as on the structure of comets. More refined analyses of older data sets have also contributed new insight.

Clearly, an identical reprint of the first edition would have been out of date. To bring the book up to the present state of the art it was necessary to incorporate the latest results. Although discussion of the Solar System bodies has been broadened by including Pluto, comets, and asteroids, the basic format and structure of the book has been preserved. The first four chapters, dealing primarily with fundamental aspects, radiative transfer theory, molecular physics, and modeling of atmospheric spectra, have not been affected by new information. Only minor changes have been made to the text, in some cases to correct errors, in others to clarify certain points. The latest results have been added primarily to Chapters 5 through 9. Some new instrumental techniques needed to be included. More recent information on atmospheric composition and structure had to be compared to older results. Although the Galileo probe data are *in situ* measurements, the composition of the Jovian atmosphere cannot be treated without referring to them. Therefore, we made an exception and included results from the helium-to-hydrogen detector and the mass spectrometer along with the remote sensing information. A new chapter (7) dealing with trans-Neptunian objects and asteroids has been inserted. In some cases, the treatment of earlier work was shortened to make room for interesting newer findings.

We are grateful to Dr Heidi Hammel and her colleagues at the Massachusetts Institute of Technology, who have used the first edition as part of a course. They have pointed out errors, misprints, and several areas where changes might benefit the reader. We thank also other reviewers for their suggestions. Again, we would like to thank authors, editors, and publishers for permission to use recently published figures. We also appreciate the support and patience of S. Milton and his staff at Cambridge University Press.

The following journals and publishers have given permission to reproduce figures:

Applied Optics. Optical Society of America, Washington DC: Figs. 5.2.10; 5.3.2; 5.8.3; 5.8.9; 5.8.10; 5.8.12; 5.8.13; and 6.2.1.

Astronomy and Astrophysics. Springer Verlag, New York: Figs. 5.7.5; 5.7.6; and 6.3.4b.

The Astrophysical Journal. The University of Chicago Press, Chicago, IL: Figs. 3.8.2; 8.3.2; and 8.3.3.

Canadian Journal of Physics. National Research Council of Canada: Fig. 3.3.6.

Geophysical Research Letters. American Geophysical Union, Washington, DC: Figs. 9.2.10 and 9.2.11.

Icarus. Academic Press, New York, NY: Figs. 5.9.1; 6.2.2; 8.3.1; 8.4.2; 8.5.1; 9.2.3; 9.2.6; 9.2.7; 9.2.8; and 9.2.9.

Journal of Atmospheric Sciences. American Meteorological Society, Boston, MA: Figs. 9.2.4 and 9.2.5.

Journal of Geophysical Research. American Geophysical Union, Washington, DC: Figs. 6.2.4; 6.2.5; 6.2.6; 6.2.10; 6.4.1; 8.2.1; 9.2.1; and 9.2.2.

Nature. Macmillan Magazines Ltd, London, UK: Figs. 6.4.2; 6.4.3; 6.4.4; 6.4.6; and 6.5.1.

Proceedings of the First Workshop on Analytical Spectroscopy. (ESA SP419): Fig. 6.3.6.

Science. American Association for the Advancement of Science, Washington, DC: Figs. 6.2.9; 6.3.4a; 7.1.2; and 8.5.2.

Spectrometric Techniques III. Academic Press, New York NY: Fig. 5.8.5.

Satellites of Jupiter. University of Arizona Press, Tucson, AZ: Figs. 6.5.2; 6.5.3; and 6.5.4.

Aspen International Conference on Fourier Spectroscopy. Air Force Cambridge Research Laboratories, Hanscom Field, Bedford, MA: Fig. 3.7.3.

The Atmospheres of Saturn and Titan (ESA: SP241): Fig. 8.4.1.

Lunar and Planetary Science Conference. Pergamon Press: Figs. 6.5.5 and 6.5.6.

Pluto and Charon. Editors: S. A. Stern & D. A. Tholen. University of Arizona Press, Tucson, AZ: Fig. 7.1.1.

Titan: The Earth-Like Moon. World Scientific Publishing Co. Pte. Ltd: Fig. 6.4.5.

SPIE Proceedings, IEEE Publications, Piscataway, NJ: Fig. 5.8.11.

1
Foundation of radiation theory

In this chapter we review the physical foundation of remote sensing. Except for possible gravitational effects, information accessible to a distant observer must be sensed as electromagnetic radiation, either in the form of reflected or refracted solar or stellar radiation, or in the form of thermal or nonthermal emission. We restrict the discussion to passive techniques. Active methods, involving the generation of electromagnetic radiation (radar, lidar), are not explicitly treated. However, the physical principles discussed in this text are equally applicable to passive and active methods. In either case a discussion of the measurement and interpretation of remotely sensed data must be based on electromagnetic theory. In Section 1.1 we begin with that theory by reviewing Maxwell's equations. The application of the principle of energy conservation to Maxwell's equations leads to the Poynting theorem with the Poynting vector describing radiative energy transport; this is discussed in Section 1.2. However, the Poynting vector does not characterize more complex phenomena, such as reflection, refraction, polarization, or interference; all of these phenomena play significant roles in many aspects of remote sensing. Their study requires, first, a derivation of the wave equation from Maxwell's formulas, and second, finding appropriate solutions for the electric and magnetic field vectors; this is the subject of Section 1.3. Polarization is briefly reviewed in Section 1.4. Effects of electromagnetic waves striking an interface between two media and the conditions that must be satisfied at the boundary are treated in Section 1.5. The derived conditions are then applied to the boundary to find expressions for reflected and refracted waves. These expressions, the Fresnel equations, are discussed in Section 1.6. The same boundary conditions are used again in Section 5.6 to describe the behavior of thin films employed in many ways in remote sensing instruments. The Planck function is introduced in Section 1.7. In Section 1.8, we return to the Poynting vector in a discussion of quantities used in the theory of radiative transfer, such as spectral intensity and radiative flux.

1.1 Maxwell's equations

Electromagnetic radiation between the red limit of the visible spectrum and the microwave region is called the infrared. In round numbers the infrared covers the spectral range from 1 to 1000 μm. Although only the range from 0.35 to 0.75 μm is truly visible to the human eye, the region between 0.75 and 1 μm is often considered as a part of the 'visible' spectrum because many detectors common to that spectral domain, such as conventional photomultipliers, photographic film, and charge-coupled silicon devices, work well up to about 1 μm. At the far end of the infrared spectrum, tuned circuits, waveguides, and other elements associated with radio and microwave technology become the commonly employed detection tools.

Whatever the wavelength, electromagnetic radiation obeys the laws expressed by Maxwell's equations. These equations describe the interrelationship of electric and magnetic quantities by field action, in contrast to action at a distance, which up to Maxwell's time (1873) was the generally accepted point of view. The field concept goes back to Michael Faraday. In all likelihood, the concept suggested itself to him in experiments with magnets and iron filings in which lines of force become almost an observable reality. However, it was left to James Clerk Maxwell to give the field concept a far-reaching and elegant mathematical formulation. Fifteen years after the publication of Maxwell's treatise (1873), Heinrich Hertz (1888) discovered electromagnetic waves, an experimental verification of Maxwell's theory.

In differential form, using the rationalized system and vector notation, the first pair of Maxwell's equations is (e.g. Sommerfeld, 1952):

$$\dot{\mathbf{D}} + \mathbf{J} = \nabla \times \mathbf{H} \qquad (1.1.1)$$

and

$$\dot{\mathbf{B}} = -\nabla \times \mathbf{E}, \qquad (1.1.2)$$

where $\mathbf{D}$ and $\mathbf{B}$ are the electric displacement and magnetic induction, and $\mathbf{E}$ and $\mathbf{H}$ the electric and magnetic field strengths, respectively; $\mathbf{J}$ is the current density. The dot symbolizes differentiation with respect to time. Definitions of the curl ($\nabla \times$) and the divergence ($\nabla \cdot$) operators are given in Appendix 1. The concept of the electric displacement was introduced by Maxwell. The first equation includes Ampère's law and the second represents Faraday's law of induction.

Besides the main equations (1.1.1) and (1.1.2), two more expressions are traditionally considered part of Maxwell's equations,

$$\nabla \cdot \mathbf{D} = \rho \qquad (1.1.3)$$

and

$$\nabla \cdot \mathbf{B} = 0. \tag{1.1.4}$$

Equation (1.1.3) defines the electric charge density, ρ, while Eq. (1.1.4) states the nonexistence of magnetic charges or monopoles. Strictly from symmetry considerations of Maxwell's equations one may be led to postulate the existence of magnetic charges, but despite many attempts none has been found.

By applying the divergence operator to Eq. (1.1.1) and substituting ρ for $\nabla \cdot \mathbf{D}$, one arrives at the electric continuity equation,

$$\dot{\rho} + \nabla \cdot \mathbf{J} = 0, \tag{1.1.5}$$

which states the conservation of electric charge: a change in the charge density of a volume element must be associated with a current flow across the boundary of that arbitrarily chosen element. The continuity equation in fluid dynamics is an analogous expression of the conservation of mass.

In order to study the interaction of matter with electric and magnetic fields, three material constants are introduced: the electric conductivity, σ,

$$\mathbf{J} = \sigma \mathbf{E}, \tag{1.1.6}$$

the dielectric constant, ε,

$$\mathbf{D} = \varepsilon \mathbf{E}, \tag{1.1.7}$$

and the magnetic permeability, μ,

$$\mathbf{B} = \mu \mathbf{H}. \tag{1.1.8}$$

Equation (1.1.6) is a form of Ohm's law. Since $\mathbf{J}$ is the current density (A m^{-2}) and $\mathbf{E}$ the electric field strength (V m^{-1}), σ is expressed in Ω^{-1} m^{-1}. The inverse conductivity is the resistivity. In the rationalized system the dielectric constant is conveniently written

$$\varepsilon = \varepsilon_0 \varepsilon_{\text{rel}}, \tag{1.1.9}$$

where ε_0 is the dielectric constant of free space (see Appendix 2 for numerical values) and ε_{rel} is a dimensionless quantity, which is unity for free space and which has the same value as the dielectric constant in the Gaussian system of units. The permeability is

$$\mu = \mu_0 \mu_{\text{rel}}, \tag{1.1.10}$$

where μ_0 represents the permeability of free space. The relative permeability is unity for free space, larger than unity for paramagnetic materials, and less than unity for diamagnetic substances.

Maxwell's equations are linear. However, the parameters that describe material properties may become nonlinear in exceptionally strong fields, such as in powerful lasers. In these cases nonlinear terms have to be included. The linear material equations, Eqs. (1.1.6) to (1.1.8), are not applicable to ferroelectric or ferromagnetic substances where the relationship between the electric field strength, **E**, and the electric displacement, **D**, or between the magnetic field strength, **H**, and the magnetic induction, **B**, are not only nonlinear, but show hysteresis effects as well. In any case, Maxwell's equations are the foundation of electromagnetism, which includes optics and infrared physics.

1.2 Conservation of energy and the Poynting vector

The Poynting theorem expresses the conservation of energy in electromagnetism. If one takes the scalar product of Eq. (1.1.1) with **E** and of Eq. (1.1.2) with **H**, and adds the results one finds

$$\mathbf{H} \cdot \dot{\mathbf{B}} + \mathbf{E} \cdot \dot{\mathbf{D}} + \mathbf{E} \cdot \mathbf{J} = \mathbf{E} \cdot (\nabla \times \mathbf{H}) - \mathbf{H} \cdot (\nabla \times \mathbf{E}). \tag{1.2.1}$$

With the vector identity

$$\mathbf{E} \cdot (\nabla \times \mathbf{H}) - \mathbf{H} \cdot (\nabla \times \mathbf{E}) \equiv -\nabla \cdot (\mathbf{E} \times \mathbf{H}) \tag{1.2.2}$$

and the definition

$$\mathbf{S} = \mathbf{E} \times \mathbf{H} \tag{1.2.3}$$

one obtains

$$\mathbf{H} \cdot \dot{\mathbf{B}} + \mathbf{E} \cdot \dot{\mathbf{D}} + \mathbf{E} \cdot \mathbf{J} + \nabla \cdot \mathbf{S} = 0. \tag{1.2.4}$$

This is the Poynting theorem; **S** is the Poynting vector. The first two terms in Eq. (1.2.4) represent rate of change of the magnetic and electric energy densities in the field. The third term, $\mathbf{E} \cdot \mathbf{J}$, describes the energy dissipated by the motion of electric charges. Generally, this motion results in Joule heating and, therefore, in losses to the energy stored in the field. The last term, $\nabla \cdot \mathbf{S}$, represents the net flow of electromagnetic energy across the boundaries of the chosen volume. All terms of Eq. (1.2.4) are measured in J m^{-3} s^{-1}, which is energy per unit volume and unit time. Since the divergence operator corresponds to a differentiation with respect to space coordinates, the units of $\mathbf{S}$ are J m^{-2} s^{-1} or W m^{-2}, thus $\mathbf{S}$ is an energy flux through a surface element.

The definition of the Poynting vector, Eq. (1.2.3), requires that $\mathbf{S}$ be orthogonal to both $\mathbf{E}$ and $\mathbf{H}$. In order to better visualize the relative orientation of these three vectors, we align a Cartesian coordinate system so that the x-axis coincides with the direction of the Poynting vector. The components of $\mathbf{S}$ along the y- and z-axes, as well as the components of $\mathbf{E}$ and $\mathbf{H}$ in the direction of the x-axis, must then be zero: $S_y = S_z = E_x = H_x = 0$. The vectors $\mathbf{E}$ and $\mathbf{H}$ do not have components in the direction of energy transport represented by $\mathbf{S}$. Electromagnetic waves are transverse, in contrast to sound waves, which are longitudinal. To investigate the relative orientation between $\mathbf{E}$ and $\mathbf{H}$, we use the second of Maxwell's equations (Eq. 1.1.2) and the explicit expression of the curl operator (see Appendix 1). With the assumption that μ is constant and E_x and H_x equal zero, one obtains one scalar equation for each of the $\hat{\mathbf{j}}$- and $\hat{\mathbf{k}}$-directions ($\hat{\mathbf{i}}$, $\hat{\mathbf{j}}$, and $\hat{\mathbf{k}}$ are the unit vectors in the x-, y-, and z-directions):

$$\mu \frac{\partial H_y}{\partial t} = \frac{\partial E_z}{\partial x}; \qquad \mu \frac{\partial H_z}{\partial t} = -\frac{\partial E_y}{\partial x}. \tag{1.2.5}$$

Except for a static field, which is not of interest in this context, Eq. (1.2.5) indicates that H_y must be zero if E_z vanishes and, conversely, H_z must disappear when E_y is zero. These conditions require $\mathbf{E}$ and $\mathbf{H}$ to be at right angles to each other; $\mathbf{E}$, $\mathbf{H}$, and $\mathbf{S}$ form a right-handed, orthogonal system of vectors.

1.3 Wave propagation

In an isotropic, stationary medium, the material constants σ, ε, and μ are uniform and constant scalars. The first pair of Maxwell's equations may then be stated:

$$\varepsilon \dot{\mathbf{E}} + \sigma \mathbf{E} = \nabla \times \mathbf{H} \tag{1.3.1}$$

and

$$\mu \dot{\mathbf{H}} = -\nabla \times \mathbf{E}. \tag{1.3.2}$$

If one differentiates Eq. (1.3.1) with respect to time and multiplies by μ, one obtains

$$\varepsilon\mu\ddot{\mathbf{E}} + \sigma\mu\dot{\mathbf{E}} = \mu\frac{\partial}{\partial t}(\nabla \times \mathbf{H}). \tag{1.3.3}$$

Application of the curl operator to Eq. (1.3.2) yields

$$\mu\nabla \times \dot{\mathbf{H}} = -\nabla \times (\nabla \times \mathbf{E}). \tag{1.3.4}$$

For a medium at rest the order of differentiation with respect to space and time may be interchanged. Applying the vector identity

$$\nabla \times (\nabla \times \mathbf{E}) \equiv \nabla(\nabla \cdot \mathbf{E}) - \nabla^2\mathbf{E} \tag{1.3.5}$$

and assuming the medium to be free of electric charges $[(\nabla \cdot \mathbf{E}) = 0]$ leads to

$$\varepsilon\mu\ddot{\mathbf{E}} + \sigma\mu\dot{\mathbf{E}} = \nabla^2\mathbf{E}. \tag{1.3.6}$$

The Laplace operator, ∇^2, is defined in Appendix 1. This partial differential equation characterizes wave and relaxation phenomena. Again, we assume the x-axis to be aligned with the Poynting vector, so that $E_x = 0$. To simplify matters further, we rotate the coordinate system around the x-axis until the y-axis coincides with the direction of the electric field strength, so that $E_z = 0$ also. Only the y-component of $\mathbf{E}$ remains and Eq. (1.3.6) becomes a scalar equation for the unknown $E_y(x, t)$,

$$\varepsilon\mu\ddot{E}_y + \sigma\mu\dot{E}_y = E_y''. \tag{1.3.7}$$

We denote differentiation with respect to time by a dot and with respect to a space coordinate (in this case with respect to x) by a prime. The assumption $E_y = T(t)X(x)$ separates the variables,

$$\varepsilon\mu\frac{\ddot{T}}{T} + \sigma\mu\frac{\dot{T}}{T} = \frac{X''}{X} = -k^2. \tag{1.3.8}$$

Since the left side depends only on the variable t and the middle part only on the variable x, Eq. (1.3.8) can only be satisfied if the left and the middle part equal a constant, $-k^2$. The reason for choosing a negative square and the physical meaning

1.3 Wave propagation

of k will become apparent later. With the introduction of k, Eq. (1.3.8) yields two ordinary differential equations:

$$\varepsilon\mu\ddot{T} + \sigma\mu\dot{T} + k^2 T = 0 \tag{1.3.9}$$

and

$$X'' + k^2 X = 0. \tag{1.3.10}$$

A solution of Eq. (1.3.10) is readily shown to be

$$X = A\,\mathrm{e}^{\pm i k x}. \tag{1.3.11}$$

The amplitude A is not defined by Eq. (1.3.10); it is determined by boundary conditions. For convenience we use notation with complex arguments in the treatment of wave phenomena. To simplify notation we omit the amplitudes but reintroduce them when needed. To solve Eq. (1.3.9) one may assume a solution of exponential form,

$$T = \mathrm{e}^{pt}, \tag{1.3.12}$$

which yields a characteristic equation for p,

$$\varepsilon\mu p^2 + \sigma\mu p + k^2 = 0. \tag{1.3.13}$$

We make two choices for p. In the first case we find the roots of Eq. (1.3.13) for p, assuming the coefficients ε, μ, σ, and k to be real quantities. Later, we will be interested in periodic solutions of Eq. (1.3.12), which imply $p = \pm i\omega$. In that case, if $\sigma \neq 0$, at least one of the coefficients must be complex. The roots of Eq. (1.3.13) for p are

$$p = -\frac{\sigma}{2\varepsilon} \pm \left(\frac{\sigma^2}{4\varepsilon^2} - \frac{k^2}{\varepsilon\mu}\right)^{\frac{1}{2}}. \tag{1.3.14}$$

The parameter p is complex because the term with σ^2 in the parentheses is generally smaller than the term containing k^2,

$$E_y = \exp\left[-\frac{\sigma t}{2\varepsilon} \pm i\left(\frac{k^2}{\varepsilon\mu} - \frac{\sigma^2}{4\varepsilon^2}\right)^{\frac{1}{2}} t\right] \exp(\pm i k x). \tag{1.3.15}$$

E_y is an oscillating function of t and x. Before we discuss the physical content of Eq. (1.3.15) we consider the meaning of some of the quantities involved. It is convenient to introduce new terms pertinent to the description of optical phenomena in the infrared. Consider the inverse product $\varepsilon^{-1}\mu^{-1}$, which has the dimension of the square of a velocity, m² s⁻². This is the propagation velocity, v, of electromagnetic waves in a medium with dielectric constant ε and permeability μ. For free space this velocity is the velocity of light, c. We have

$$v = (\varepsilon\mu)^{-\frac{1}{2}}; \qquad c = (\varepsilon_0\mu_0)^{-\frac{1}{2}}. \qquad (1.3.16)$$

Consequently

$$\frac{c}{v} = \left(\frac{\varepsilon\mu}{\varepsilon_0\mu_0}\right)^{\frac{1}{2}} = (\varepsilon_{\text{rel}}\mu_{\text{rel}})^{\frac{1}{2}} = n. \qquad (1.3.17)$$

The ratio of the propagation velocity of free space to that of a medium is the refractive index, n, of the medium. In this case both n and k are real quantities. Since μ_{rel} is nearly unity for most materials of importance in the infrared, the refractive index can often be approximated by $n \sim (\varepsilon_{\text{rel}})^{\frac{1}{2}}$.

The constant k has the dimension of inverse length; it is the number of radians per meter, the angular wavenumber. Therefore,

$$k\lambda = 2\pi, \qquad (1.3.18)$$

where λ is the wavelength in meters. The angular frequency, ω, measured in radians per second, is then

$$\omega = kv. \qquad (1.3.19)$$

The frequency, f, in hertz (cycles per second), and the wavenumber, ν, in m⁻¹, are

$$f = \frac{\omega}{2\pi}; \qquad \nu = \frac{k}{2\pi}. \qquad (1.3.20)$$

Even for a wavelength of 1000 μm the frequency is approximately 3×10^{11} Hz, a very high frequency compared with radio waves. The FM broadcast band is about 100 MHz or 10^8 Hz, for comparison. The term $k^2/\varepsilon\mu$ in Eq. (1.3.15) is simply ω^2

and the solution for E_y becomes:

$$E_y = \exp\left(-\frac{\sigma t}{2\varepsilon}\right) \exp\left\{\pm i\omega\left[1 - \left(\frac{\sigma}{2\varepsilon\omega}\right)^2\right]^{\frac{1}{2}} t\right\} \exp(\pm ikx). \qquad (1.3.21)$$

As required for a second order differential equation, Eq. (1.3.21) represents two solutions, indicated by the $\pm$ signs. One solution describes a wave traveling in the direction of x (outgoing wave, opposite signs, $+ -$ or $- +$), and the other, a wave traveling in the opposite direction (incoming wave, equal signs, $+ +$ or $- -$). If the amplitudes of these waves are equal, only a standing wave exists. For a nonconductive medium, where σ is zero, the solution for the outgoing wave simplifies to

$$E_y(\sigma = 0) = e^{\pm i(kx - \omega t)}, \qquad (1.3.22)$$

which is a plane, unattenuated wave traveling in the x-direction. This case is shown in Fig. 1.3.1 by the periodic curve marked '0'.

For a weakly conducting material – dry soil or rocks, for example – two effects may be noted. First, due to the factor $\exp(-\sigma t/2\varepsilon)$ in Eq. (1.3.21), the amplitudes of the waves diminish exponentially with time. Materials with good optical transmission properties must, therefore, be electrical insulators, but not all insulators are transparent. For many substances the frequency dependence of the refractive index is due to quantum mechanical resonances. Equation (1.3.17) is valid for low frequencies where v and n can be determined from the static values of ε and μ, but not necessarily at infrared or visible wavelengths. The second effect to be noted in Eq. (1.3.21) concerns a frequency shift by the factor $[1 - (\sigma/2\varepsilon\omega)^2]^{\frac{1}{2}}$. As long as σ is small compared with $2\varepsilon\omega$, as in the case marked 0.05 in Fig. 1.3.1, the frequency shift is negligible, but it becomes noticeable for the case $\sigma/2\varepsilon\omega = 0.2$. If σ is equal to or larger than $2\varepsilon\omega$ – that is, if the conduction current is comparable to or larger than the displacement current, as in metals – then the square root in Eq. (1.3.21) becomes zero or imaginary; in either case periodic solutions disappear and only an exponential decay exists, shown by curve 1 of Fig. 1.3.1.

Now we return to the choice of p in Eq. (1.3.12). With the assumption $p = \pm i\omega$ the solution for T becomes

$$T = e^{\pm i\omega t}, \qquad (1.3.23)$$

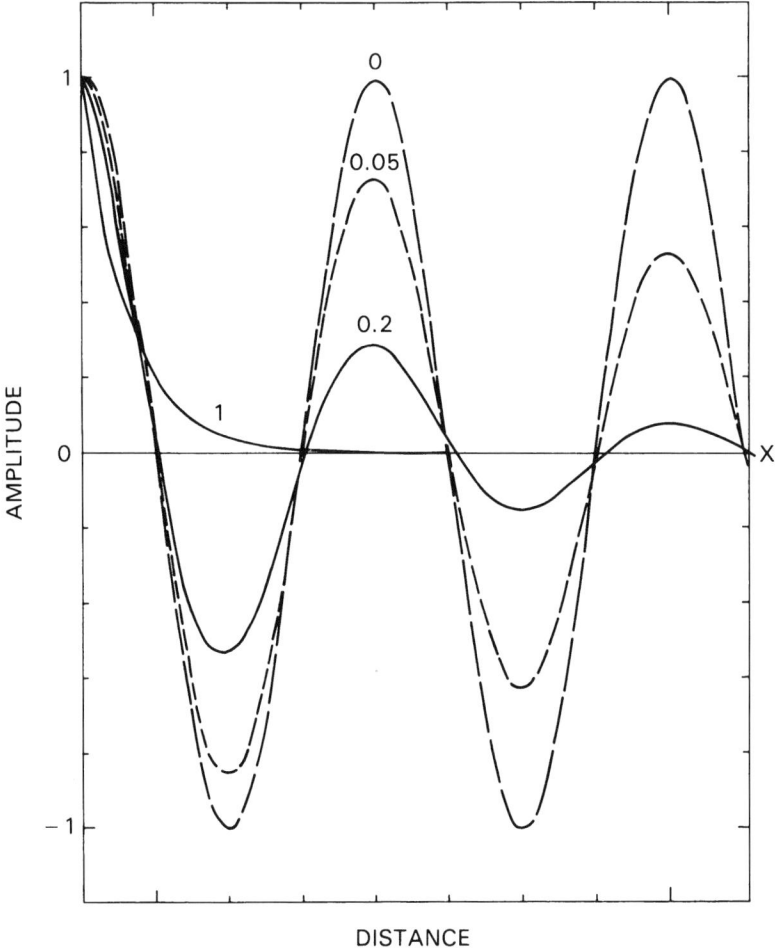

Fig. 1.3.1 Amplitudes of electromagnetic waves propagating in a medium. The parameter refers to the ratio of conduction to displacement current. If this ratio is zero the material is transparent. If this ratio is one or larger, such as in metals, only an exponential decay exists.

but in this case k is complex. We have

$$k = (\varepsilon\mu\omega^2 + i\sigma\mu\omega)^{\frac{1}{2}} = \frac{\omega}{c}(n_r + in_i), \qquad (1.3.24)$$

where n_r is the real and n_i the imaginary part of the refractive index, n. Squaring Eq. (1.3.24) and setting the real and imaginary parts of both sides equal leads to equations for the real part of k,

$$\frac{\omega n_r}{c} = \omega \left(\frac{\varepsilon\mu}{2} \left\{ \left[1 + \left(\frac{\sigma}{\varepsilon\omega} \right)^2 \right]^{\frac{1}{2}} + 1 \right\} \right)^{\frac{1}{2}}, \qquad (1.3.25)$$

1.3 Wave propagation

and for the imaginary part,

$$\frac{\omega n_i}{c} = \omega \left(\frac{\varepsilon\mu}{2}\left\{\left[1+\left(\frac{\sigma}{\varepsilon\omega}\right)^2\right]^{\frac{1}{2}}-1\right\}\right)^{\frac{1}{2}}. \qquad (1.3.26)$$

Therefore, E_y may also be expressed by

$$E_y = e^{\pm i\omega(n_r+in_i)x/c} e^{\pm i\omega t}. \qquad (1.3.27)$$

The term $n = n_r + in_i$ is the complex refractive index, a concept that is used in the discussion of the interaction of radiation with solid matter (Sections 3.7.b and 3.8).

So far we have concerned ourselves with the electric field strength, **E**. Now we return to the magnetic field strength, **H**. Following a similar procedure for **H** as for **E** leads to analogous equations. After multiplication by ε and differentiation with respect to time of Eq. (1.3.2), one obtains

$$\varepsilon\mu\ddot{\mathbf{H}} = -\varepsilon\frac{\partial}{\partial t}(\nabla\times\mathbf{E}). \qquad (1.3.28)$$

If one applies the curl operator to Eq. (1.3.1) one finds

$$\varepsilon\frac{\partial}{\partial t}(\nabla\times\mathbf{E}) + \sigma(\nabla\times\mathbf{E}) = \nabla\times(\nabla\times\mathbf{H}). \qquad (1.3.29)$$

Multiplication of Eq. (1.3.2) by σ and substitution of this as well as Eq. (1.3.29) into Eq. (1.3.28) yields

$$\varepsilon\mu\ddot{\mathbf{H}} + \sigma\mu\dot{\mathbf{H}} = \nabla^2\mathbf{H}, \qquad (1.3.30)$$

which is identical in form with Eq. (1.3.6) for the electrical field strength. The solution for **H** is, therefore, analogous to that for **E**. For $\sigma = 0$, and for the **E** vector in the y-direction only, a component of **H** in the z-direction exists. With the help of Eq. (1.3.22), Eq. (1.3.2) reduces to

$$\mu\frac{\partial H_z}{\partial t} = -\frac{\partial E_y}{\partial x} = -ike^{\pm i(kx-\omega t)}. \qquad (1.3.31)$$

For a periodic function, integration with respect to time is accomplished by dividing by $(-i\omega)$ and, since $k\mu^{-1}\omega^{-1}$ equals $\varepsilon^{\frac{1}{2}}\mu^{-\frac{1}{2}}$,

$$H_z = \frac{k}{\mu\omega}e^{\pm i(kx-\omega t)} = \left(\frac{\varepsilon}{\mu}\right)^{\frac{1}{2}} E_y = mE_y. \tag{1.3.32}$$

The factor m has the dimension of a conductance or, equivalently, of a reciprocal resistance. This resistance is called the wave resistance or, more generally, the optical wave impedance of the medium. For free space the wave resistance is $\sim 377\,\Omega$. For maximum efficiency transmitting and receiving antennas must be matched to that impedance. Similarly, electrical transmission lines must be terminated by their conjugate wave impedances to avoid reflections. In optics an analogous situation exists. No reflection takes place at the interface of two media if their wave impedances are matched, a consideration important for the design of antireflection coatings.

A wave ($\sigma = 0$) traveling in the x-direction, such as described by Eq. (1.3.22), is displayed in Fig. 1.3.2. The electric field strength **E** has a component only in the y-direction and the magnetic vector **H** has one only in the z-direction. The Poynting vector **S** lies along the x-axis. In the time dt the whole pattern moves the distance dx with velocity v in the direction of **S**.

The case shown in Fig. 1.3.2 is for a nonabsorbing medium. To find the relationship between H_z and E_y for an absorbing medium we apply the solution for E_y (Eq. 1.3.27) to the second of Maxwell's equations (Eq. 1.3.2) and find

$$\mu\frac{\partial H_z}{\partial t} = -\frac{\partial E_y}{\partial x} = -i\frac{\omega}{c}(n_r + in_i)E_y, \tag{1.3.33}$$

which leads after integration (division by $-i\omega$) to

$$H_z = \frac{(n_r + in_i)}{\mu c} E_y. \tag{1.3.34}$$

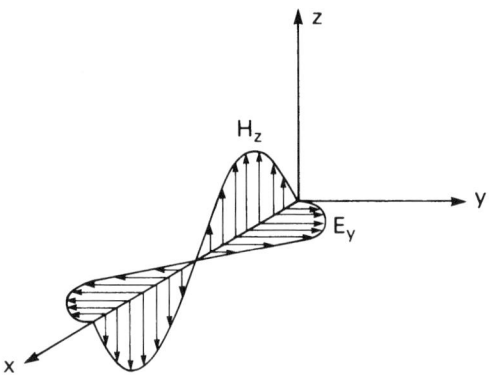

Fig. 1.3.2 Electric (E_y) and magnetic (H_z) vectors in a linearly polarized electromagnetic wave propagating along the x-axis.

Since $(n_r + in_i)$ can be expressed by an amplitude, $(n_r^2 + n_i^2)^{\frac{1}{2}}$, and a phase angle, $\gamma = \text{arctg}\, n_i/n_r$, we obtain

$$H_z = \pm \frac{1}{\mu c}(n_r^2 + n_i^2)^{\frac{1}{2}} e^{i\gamma} E_y. \tag{1.3.35}$$

In a conductive material E_y and H_z are still at right angles to each other and to **S**, but they are phase shifted by an angle γ, and not in phase as shown for a nonabsorbing medium in Fig. 1.3.2.

1.4 Polarization

Now we return to waves in nonabsorbing media. The wave shown in Fig. 1.3.2 is linearly polarized in the y-direction. Traditionally, the direction of the electric field strength, **E**, and the Poynting vector define the plane of polarization. Linearly polarized waves are also possible in the z-direction or at any angle in the y–z plane. The vector **E** may be decomposed into its y- and z-components,

$$\mathbf{E} = \hat{\mathbf{j}} E_y + \hat{\mathbf{k}} E_z. \tag{1.4.1}$$

A linearly polarized wave with an arbitrary plane of polarization may be visualized as the superposition of two waves of the same frequency and phase, one linearly polarized in the y- and the other in the z-direction. But what is the consequence when two waves, E_y and E_z, of the same frequency, both linearly polarized, but with a distinct difference in phase and with different amplitudes, are superimposed? By phase difference we mean differences between the **E** vectors and not between **E** and **H**, which occur only in absorbing media. Since Maxwell's equations are linear, the corresponding vectors, E_y and E_z, of the two waves must be added. The resulting vector sum, **E**, is then the combined field strength. The same applies to the **H** vectors; **E** and **H** are still orthogonal. However, the tip of **E** does not describe a strictly sinusoidal curve in a plane, as shown in Fig. 1.3.2, but rather a curve in space that progresses uniformly along the x-axis; the projection in the y–z plane is not a straight line but an ellipse. We call such a wave elliptically polarized (Fig. 1.4.1). Conversely, an elliptically polarized wave may be decomposed into two linearly polarized waves. If the amplitudes of both superimposed waves are equal, the ellipse becomes a circle and we speak of circular polarization. In that case the end point of the **E** vector travels on a spiral of constant radius around the x-axis. The end point of a circularly or elliptically polarized wave can form a right- or a left-handed spiral. Unfortunately, according to tradition, a right-handed spiral is called a left-handed polarization because in the nineteenth century right- and left-handedness was judged by the observer facing the beam of light. Polarization phenomena play important

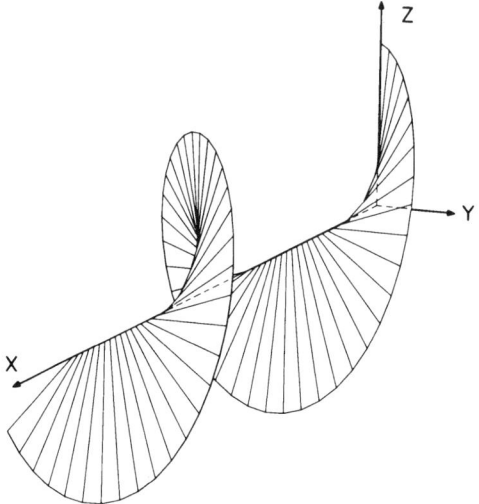

Fig. 1.4.1 Electric vector in an elliptically polarized wave propagating in the x-direction. The magnetic vector is orthogonal to the electric vector.

roles in instrument design, in the theory of reflection and refraction, and in theories of scattering of radiation by particles.

1.5 Boundary conditions

So far we have discussed electromagnetic phenomena in a homogenous medium; now we consider two media and the conditions at their interface. We restrict the discussion to transparent substances. In the media (medium 1 above and medium 0 below the boundary) there exist electric and magnetic fields. In this section the index zero does not refer to free space. At the dividing surface between both domains the fields can be decomposed into two components normal and tangential to the boundary. Consider first the normal component of **B**. To deal with the discontinuity in ε and μ across the dividing surface we consider a small volume that contains a small region of the surface between media 1 and 0 (Fig. 1.5.1). The area of this volume element exposed to medium 1 is δA_1 plus the circumference, s, times $\delta h/2$. The area exposed to medium 0 is δA_0 plus the other half of the circumferential area. Instead of the abrupt change at the boundary we let B_n change gradually from the value $B_n^{(1)}$ at the surface δA_1 to the value $B_n^{(0)}$ at the surface δA_0. Applying Gauss' theorem to this volume yields

$$\int_{\text{Volume}} (\nabla \cdot \mathbf{B})\, dV = \int_{\text{Surface}} \mathbf{B} \cdot d\mathbf{A}. \tag{1.5.1}$$

1.5 Boundary conditions

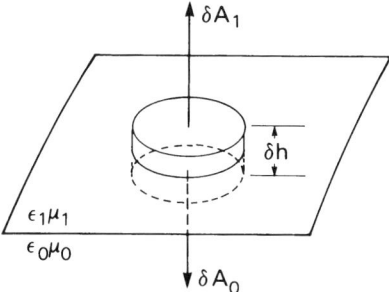

Fig. 1.5.1 Surface element of the boundary of two media of different electromagnetic properties. One half of the volume element is in medium 1, and the other half is in medium 0.

Since $(\nabla \cdot \mathbf{B})$ is zero [Eq. (1.1.4)] the integrals in Eq. (1.5.1) must also be zero. The right side may be expressed by

$$\int \mathbf{B} \cdot d\mathbf{A} = B_n^{(1)} \cdot \delta A_1 - B_n^{(0)} \cdot \delta A_0 + \left(B_t^{(1)} + B_t^{(0)}\right)s\frac{\delta h}{2} = 0. \qquad (1.5.2)$$

Let δh become very small; the contribution from the circumferential area diminishes. Since the areas δA_1 and δA_0 are equal

$$B_n^{(1)} - B_n^{(0)} = 0. \qquad (1.5.3)$$

At the interface the normal components of the induction are identical in both media; B_n is continuous across the boundary.

The behavior of the component of **D** normal to the boundary may be treated similarly, except that the integrals are not necessarily zero. In this case the charge density ρ must be taken into account. In the transition from the volume element to the surface element, the volume density becomes a surface density, ρ_{surf}, given by

$$D_n^{(1)} - D_n^{(0)} = \rho_{\text{surf}}. \qquad (1.5.4)$$

In the presence of a surface charge the normal component of the electric displacement changes abruptly. In the absence of a surface charge, D_n is continuous across the boundary.

To investigate the tangential components of **E** and **H** consider a closed loop (Fig. 1.5.2). The loop consists of the elements δs_1, δs_0, and two short connectors, each of length δh. The surface normal of the loop $d\mathbf{A}$ is in the direction of unit

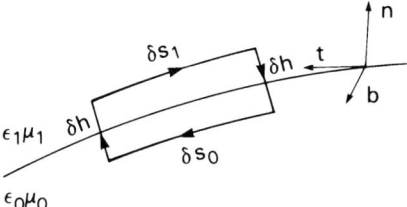

Fig. 1.5.2 Loop at the interface between two media. The vectors **n**, **t**, and **b** indicate the directions normal to the interface surface, tangential to the surface, but in the plane of the loop, and the orthogonal direction, also tangential to the interface, but normal to the loop area.

vector **b**. Applying Stokes' theorem to the loop, one finds

$$\int_{\text{loop area}} (\nabla \times \mathbf{E}) \cdot d\mathbf{A} = \int_{\text{contour}} \mathbf{E} \cdot d\mathbf{s}. \tag{1.5.5}$$

The integration path of the contour integral is along δs_1, δh, δs_0, and δh, as indicated in Fig. 1.5.2. By replacing the contour integral by its elements, the second of Maxwell's equations, Eq. (1.1.2), yields

$$-\int \dot{B}_b \, \delta s \, \delta h = E_t^{(1)} \delta s_1 - E_n \delta h + E_t^{(0)} \delta s_0 + E_n \delta h. \tag{1.5.6}$$

Upon once again letting δh approach zero, the integral over the area of the loop vanishes ($\dot{B}$ is assumed to be finite) and, considering that δs_1 and δs_0 are opposite in sign, we find

$$E_t^{(1)} - E_t^{(0)} = 0. \tag{1.5.7}$$

The tangential component of the electric field strength is continuous across the boundary. Following a similar procedure for the tangential component of **H** one finds:

$$H_t^{(1)} - H_t^{(0)} = j_{\text{surf}}. \tag{1.5.8}$$

The tangential component of the magnetic field strength changes abruptly in the presence of a surface current, but it is continuous in the absence of such a current.

1.6 Reflection, refraction, and the Fresnel equations

With the boundary conditions established, one may examine an electromagnetic wave striking the interface between two media. As before, we assume both media to be nonconductive and located above and below a flat surface, which we choose to be the x–z plane. The dividing surface between both media is assumed to be free of charges and currents, which implies that the normal components of **D** and **B** and the tangential components of **E** and **H** are continuous across the boundary. Medium 1 has the dielectric constants ε_1 and the permeability μ_1; medium 0 has the properties ε_0 and μ_0. We consider a plane wave with Poynting vector **S** incident on the interface; the plane containing **S** and the normal to the interface is called the plane of incidence. Here, we assume this is the x–y plane ($S_z = 0$), and that the electric field vector is perpendicular to this plane; later we consider the case where the electric vector lies in the plane of incidence. The incident wave will be split at the interface into a reflected and a transmitted (refracted) wave. In medium 1 the superposition of the incoming and the reflected wave is

$$E_z(y \geq 0) = B_1\, e^{ik_1(x \sin \phi_1 - y \cos \phi_1)} + C_1\, e^{ik_1(x \sin \phi_1' + y \cos \phi_1')}. \tag{1.6.1}$$

The refracted wave in the lower half-space is

$$E_z(y \leq 0) = B_0\, e^{ik_0(x \sin \phi_0 - y \cos \phi_0)}. \tag{1.6.2}$$

The factor $\exp(i\omega t)$ has been omitted for simplicity, but the amplitudes, B_1, C_1, and B_0 have been written explicitly. At $y = 0$, continuity of the tangential component, E_z, across the boundary requires

$$B_1\, e^{ik_1 x \sin \phi_1} + C_1\, e^{ik_1 x \sin \phi_1'} = B_0\, e^{ik_0 x \sin \phi_0}. \tag{1.6.3}$$

Since this equation must be valid for all values of x, all exponentials must be the same, which leads to two conditions:

$$\phi_1 = \phi_1', \tag{1.6.4}$$

which expresses the law of reflection, and

$$k_1 \sin \phi_1 = k_0 \sin \phi_0 \tag{1.6.5}$$

or

$$\frac{\sin\phi_1}{\sin\phi_0} = \frac{k_0}{k_1} = \left(\frac{\varepsilon_0\mu_0}{\varepsilon_1\mu_1}\right)^{\frac{1}{2}} = \frac{n_0}{n_1} = n_{10}, \quad (1.6.6)$$

which is the law of refraction; n_{10} is the relative refractive index between media 1 and 0. For these conditions, Eq. (1.6.3) reduces to

$$B_1 + C_1 = B_0. \quad (1.6.7)$$

The tangential component of **H** provides another set of equations for the amplitudes B_1, C_1, and B_0. According to Eq. (1.3.32), the amplitude of **H** can be found by multiplying **E** by $\pm m$. The right-hand rule for the vectors **E**, **H**, and **S** determines the choice of sign of m.

$$H_x(y \geq 0) = m_1 \cos\phi_1 \, e^{ik_1 x \sin\phi_1}(-B_1 \, e^{-ik_1 y \cos\phi_1} + C_1 \, e^{ik_1 y \cos\phi_1}) \quad (1.6.8)$$

and

$$H_x(y \leq 0) = -m_0 \cos\phi_0 \, e^{ik_0 x \sin\phi_0} B_0 \, e^{-ik_0 y \cos\phi_0}. \quad (1.6.9)$$

For $y = 0$ the tangential components of **H** must be the same for both media, which leads to

$$m_1 \cos\phi_1(-B_1 + C_1) = -m_0 \cos\phi_0 B_0, \quad (1.6.10)$$

where m_1 and m_0 are the conductances of medium 1 and 0, respectively [see Eq. (1.3.32)]. Combining Eqs. (1.6.10) and (1.6.7) permits elimination of B_0 or C_1. The relative amplitudes of the transmitted ($T_\perp = B_0/B_1$) and reflected ($R_\perp = C_1/B_1$) waves are

$$T_\perp = \frac{2m_1 \cos\phi_1}{m_1 \cos\phi_1 + m_0 \cos\phi_0} \quad (1.6.11)$$

and

$$R_\perp = \frac{m_1 \cos\phi_1 - m_0 \cos\phi_0}{m_1 \cos\phi_1 + m_0 \cos\phi_0}. \quad (1.6.12)$$

1.6 Reflection, refraction, and the Fresnel equations

Now we consider the case of the magnetic vector normal to the plane of incidence; i.e., only H_z exists. $\mathbf{E}$ is orthogonal to $\mathbf{H}$ and, therefore, in the plane of incidence. The polarization of this wave is orthogonal to that of the first case. With similar considerations one finds $B_1 + C_1 = m_{10} B_0$ and $B_1 - C_1 = B_0 \cos\phi_0 / \cos\phi_1$. Solving for $T_\| = B_0/B_1$ and $R_\| = C_1/B_1$ yields in this case

$$T_\| = \frac{2m_1 \cos\phi_1}{m_0 \cos\phi_1 + m_1 \cos\phi_0} \tag{1.6.13}$$

and

$$R_\| = \frac{m_0 \cos\phi_1 - m_1 \cos\phi_0}{m_0 \cos\phi_1 + m_1 \cos\phi_0}. \tag{1.6.14}$$

The transmitted and the reflected fractional amplitudes of the incident radiation polarized perpendicular to the plane of incidence are $T_\perp$ and $R_\perp$, respectively. The components polarized in the plane are $T_\|$ and $R_\|$, respectively. Equations (1.6.11) through (1.6.14) are the Fresnel equations (Fresnel, 1816).

Since the emissivity of a surface is one minus the square of the amplitude ratio, $(R_\perp)^2$ or $(R_\|)^2$, the thermal emissivity depends also on the refractive index and the emission angle. Consider the case of a nonmagnetic homogeneous layer of refractive index $n_0 = n$ bounded by a vacuum, $n_1 = 1$. With the help of Eq. (1.6.6) we can eliminate ϕ_0 from the reflection ratios, Eqs. (1.6.12) and (1.6.14); calling $\phi_1 = \phi$ we obtain for the emissivities

$$\varepsilon_\perp = 1 - (R_\perp)^2 = 1 - \left[\frac{\cos\phi - (n^2 - \sin^2\phi)^{\frac{1}{2}}}{\cos\phi + (n^2 - \sin^2\phi)^{\frac{1}{2}}}\right]^2 \tag{1.6.15}$$

and

$$\varepsilon_\| = 1 - (R_\|)^2 = 1 - \left[\frac{n^2 \cos\phi - (n^2 - \sin^2\phi)^{\frac{1}{2}}}{n^2 \cos\phi + (n^2 - \sin^2\phi)^{\frac{1}{2}}}\right]^2. \tag{1.6.16}$$

The emissivities of substances with refractive indices of 2 or 4, bordered by a vacuum, are shown in Fig. 1.6.1 for both planes of polarization as functions of the emission angle, ϕ. The emitted radiation from a smooth surface is polarized, except for the case of normal incidence. The emission maximum of $\varepsilon_\|$ corresponds to the reflection minimum at the Brewster angle. To find the hemispherical emissivity one has to integrate $\varepsilon_\|$ and $\varepsilon_\perp$ over the whole hemisphere and average the results for both planes of polarization.

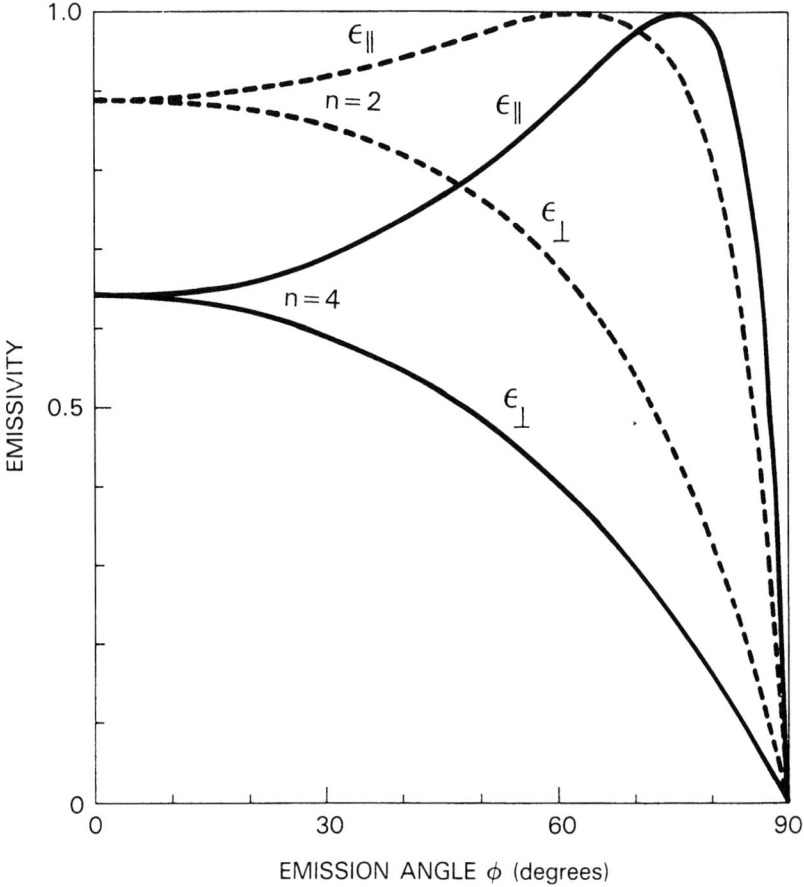

Fig. 1.6.1 Emissivity of a smooth flat surface with an index of refraction, n, as a function of emission angle. Both planes of polarization are shown for n equal to 2 (dashed lines) and 4 (solid lines).

For normal incidence ($\cos\phi_1 = \cos\phi_0 = 1$) and nonmagnetic materials ($m_1/m_0 = n_1/n_0$) the Fresnel equations simplify to

$$T_\perp = T_\| = \frac{2n_1}{n_1 + n_0} \tag{1.6.17}$$

and

$$R_\perp = -R_\| = \frac{n_1 - n_0}{n_1 + n_0}. \tag{1.6.18}$$

If the second medium is metal the same equations are valid; however, n_0 becomes complex [see Eq. (1.3.24)]. For $n_1 = 1$ and $n_0 = n_r + in_i$ the ratio of amplitudes in Eq. (1.6.18), is

$$R = \frac{1 - n_r - in_i}{1 + n_r + in_i}. \tag{1.6.19}$$

Again, n_r is the real part and n_i the imaginary part of the complex refractive index, n. Since the intensity is proportional to the square of the amplitude, the reflectivity, r, is given by the equation

$$r = |R|^2 = \frac{(n_r - 1)^2 + n_i^2}{(n_r + 1)^2 + n_i^2} = 1 - \frac{4n_r}{(n_r + 1)^2 + n_i^2}. \tag{1.6.20}$$

For a good conductor σ is large and both n_r and n_i approach a common limit [see Eqs. (1.3.25) and (1.3.26)],

$$n_{r_{(\sigma \to \infty)}} = n_{i_{(\sigma \to \infty)}} = c \left(\frac{\mu \sigma}{2\omega} \right)^{\frac{1}{2}}, \tag{1.6.21}$$

which implies both n_r and n_i become large. Consequently, $|R|^2$ approaches unity for a good conductor. A gold surface, evaporated onto a well-polished substrate, may have a reflectivity as high as 0.98, which corresponds to $n_r \sim n_i \sim 100$. Since silver and copper have higher conductivities than gold, their far infrared reflectivities are even higher. However, since silver and copper tend to tarnish in the atmosphere and gold is stable, gold is generally preferred as a reflecting surface layer throughout the infrared. In the visible (and near infrared), however, these metals behave differently, as also is evident from their colors. Recent progress in the manufacturing of superconductive materials for operation at temperatures of almost 100 K opens the possibility of constructing totally reflecting mirrors. If superconducting coatings can be found for operation at ambient temperature, a major impact is expected on the design of optical instruments.

1.7 The Planck function

Maxwell's equations describe the propagation of electromagnetic radiation as waves within the framework of classical physics; however, they do not describe emission phenomena. The search for the law that defines the energy distribution of radiation from a small hole in a large isothermal cavity gave rise to quantum theory. The function that describes the frequency distribution of blackbody radiation was the first result of that new theory (Planck, 1900, 1901).

Measurements of the total emission from a small hole in a heated cavity showed thermal radiation to be proportional to the fourth power of the cavity temperature (Stefan, 1879); Boltzmann (1884) derived this power law from thermodynamic considerations. Nine years later, Wien (1893) found that the product of the wavelength at the radiation maximum and the cavity temperature was the same for a wide range of temperatures; he also proposed an exponential radiation law, which was in good agreement with available measurements at short wavelengths (Wien, 1896). Shortly thereafter, Lummer and Pringsheim (1897, 1899) made fairly precise measurements of blackbody emission between 100 °C and 1300 °C. By the end of the nineteenth century an extensive set of experimental evidence was available on the spectral distribution and temperature dependence of blackbody radiation.

At the same time the theoretical understanding of that type of radiation was lagging. Based on concepts of classical physics, a theory was developed by Rayleigh (1900) and Jeans (1905). They started by counting the number of possible modes of standing electromagnetic waves in a cube with opaque and reflecting walls. The walls of the cube and possible small specks of dust inside (to facilitate energy exchange between individual modes) are assumed to be in thermal equilibrium at a temperature, T. If one considers all three dimensions of the cube, and both planes of polarization, the number, $N(f)$, of all possible modes in the range between f and $f + df$, is

$$N(f)\,df = \frac{8\pi V f^2}{c^3} df, \qquad (1.7.1)$$

where V is the volume of the cavity and $f = c\nu$ the frequency in hertz. Although Eq. (1.7.1) has been derived for a cubical volume with reflecting walls, it is equally valid for a cavity of arbitrary shape with partially absorbing walls. To find the energy density, $\rho(f, T)$, inside the cavity one has to multiply the number of standing waves per frequency interval by the energy of each wave and divide by the volume. According to the classical equipartition law, the total (kinetic and potential) energy per degree of freedom is kT, where k is the Boltzmann constant. A standing wave inside the cube can be regarded as a harmonic oscillator of frequency f with one degree of freedom. The energy density in an isothermal cavity is then, according to classical theory,

$$\rho(f, T)\,df = \frac{8\pi f^2 kT}{c^3} df. \qquad (1.7.2)$$

The energy density in this expression increases with the square of frequency, contrary to common experience that shows that blackbodies at a few hundred degrees

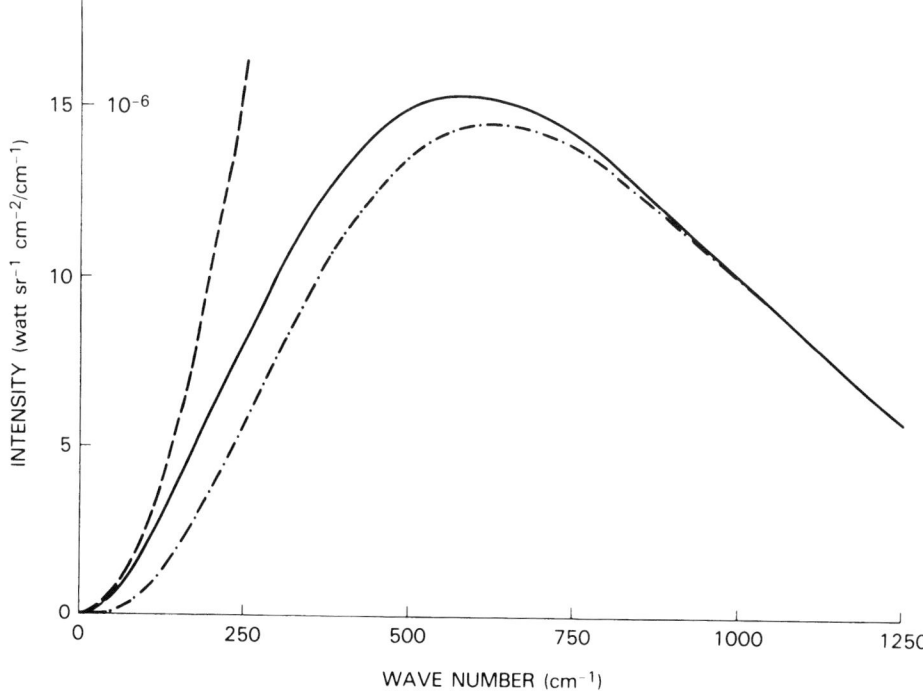

Fig. 1.7.1 Spectral intensity of a blackbody of 300 K as a function of wavenumber. The solid curve is the Planck function, the dashed curve the Rayleigh–Jeans, and the dash-dot curve the Wien approximation.

kelvin do not emit visible light, while they emit strongly in the infrared, at lower frequencies. Furthermore, the integral of the energy density over all frequencies must be a finite value and not infinite as an integration of Eq. (1.7.2) would imply. At the turn of the century it was quite clear that the classical theory of radiation was in conflict with experimental results. As shown in Fig. 1.7.1, neither the classical Rayleigh–Jeans law nor the radiation law of Wien is valid over the whole frequency range, although both seem to be good approximations at opposite ends of the spectrum.

Planck realized that the equipartition law, which assigns equal energy to each standing wave, could not be valid; he also realized that the roll-off in the energy distribution at high frequencies could be obtained with the assumption that the energy of a harmonic oscillator cannot take on any value, as is assumed in the classical equipartition law, but that it is quantized; a harmonic oscillator can absorb and emit energy only in finite steps,

$$\Delta E = nhf, \qquad (1.7.3)$$

where n is an integer and h is now called the Planck constant. Planck (1900, 1901) postulated this rather reluctantly, only after he had exhausted all possible explanations based on classical theory. With the assumption of energy quantization the energy density inside the cavity becomes

$$\rho(f, T)\,df = \frac{8\pi h f^3}{c^3(e^{hf/kT} - 1)}df. \tag{1.7.4}$$

To obtain the spectral distribution of thermal emission emerging at velocity c from a blackbody into a steradian, the energy density must be multiplied by $c/4\pi$, yielding

$$B(f, T)\,df = \frac{2hf^3}{c^2(e^{hf/kT} - 1)}df, \tag{1.7.5}$$

where $B(f, T)$ is the Planck function. It has been found to be in excellent agreement with measurement over wide ranges of temperature and wavenumber. More detailed derivations of the Planck law and that of Rayleigh–Jeans can be found in the published lectures of Planck (1913) or in textbooks on quantum mechanics such as that by Eisberg & Resnick (1974).

By permitting either h or f to approach zero in Eq. (1.7.4) the classical solution of the energy density according to the Rayleigh–Jeans law, Eq. (1.7.2), is obtained. By permitting f in Eq. (1.7.4) to approach very high values the Wien distribution function results and the energy density approaches zero as required by energy conservation. Integration of the Planck function over all frequencies leads to the Stefan–Boltzmann fourth power relationship, and multiplication of the wavelength at the radiation maximum by the blackbody temperature yields a constant, expressing the displacement law of Wien. The Planck formula includes all previously found radiation laws as special cases; moreover, the empirical factors that appeared in these older laws could now be expressed in terms of physical constants containing the Planck constant (h), the velocity of light (c), and the Boltzmann constant (k). Despite this success it was only gradually appreciated that the quantum concept was a major revolution in physics. Today it is fully accepted as a more general framework in which classical physics appears as a special case that is valid only when the small but finite energy steps, given by the value of h, can be considered a continuum.

If one replaces the frequency f by the wavenumber ν ($f = c\nu$) the Planck function takes the form

$$B(\nu, T)\,d\nu = \frac{2hc^2\nu^3}{(e^{hc\nu/kT} - 1)}d\nu. \tag{1.7.6}$$

The Planck function can also be expressed in terms of wavelength ($\lambda \nu = 1$; $\lambda\, d\nu + \nu\, d\lambda = 0$)

$$B(\lambda, T)\, d\lambda = \frac{2hc^2}{\lambda^5 (e^{hc/\lambda kT} - 1)} d\lambda. \quad (1.7.7)$$

The Planck function appears in many aspects of the theory of radiative transfer and the design of infrared instrumentation, as is discussed in forthcoming chapters.

1.8 The Poynting vector, specific intensity, and net flux

The energy flux of a plane, monochromatic wave is represented by the Poynting vector as discussed in Section 1.2. In this section we relate the Poynting vector to other quantities used in the description of the radiative transfer of energy in planetary atmospheres and from surfaces.

Strictly monochromatic radiation propagating in a unique direction (e.g., from a point source) is never realized. A monochromatic wave implies a periodic process of infinite duration. Such waves do not exist, although the signal from a stable, single-mode laser provides a fair approximation. Ordinary incoherent radiation emitted and reflected from real atmospheres and surfaces consists of individual wave packets of finite length and duration; a few meters and $\sim 10^{-8}$ seconds are typical values. Similarly, point sources are replaced by extended sources in practice. Radiation from such sources tends to be incoherent and covers a range of frequencies and directions. Thus, it is more convenient to work with a distribution of plane waves and their associated Poynting vectors.

Consider radiation incident on an element of area da with unit normal vector $\hat{\mathbf{n}}$ as shown in Fig. 1.8.1. The radiation can be regarded as an incoherent superposition of plane waves, each with an associated Poynting vector $\mathbf{S}(\nu, \hat{\mathbf{k}})$ where ν is the wavenumber of the wave and $\hat{\mathbf{k}}$ is a unit vector defining the direction of propagation. Let $N(S, \nu, \hat{\mathbf{k}})\, dS\, d\nu\, d\omega$ be the number of plane waves with Poynting vector

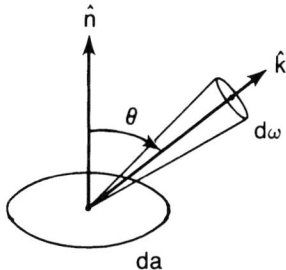

Fig. 1.8.1 Element of solid angle, dω, in direction of unit vector $\hat{\mathbf{k}}$ inclined with respect to surface normal $\hat{\mathbf{n}}$ of area element da.

magnitude between S and $S + \mathrm{d}S$, wavenumber between ν and $\nu + \mathrm{d}\nu$, and within solid angle $\mathrm{d}\omega$ in direction $\hat{\mathbf{k}}$. The net energy in wavenumber interval $\mathrm{d}\nu$ passing through $\mathrm{d}a$ in time $\mathrm{d}t$ from the direction $\hat{\mathbf{k}}$ is then

$$\mathrm{d}E_\nu = \int_0^\infty \mathrm{d}S\, N(S, \nu, \hat{\mathbf{k}}) \mathbf{S}(\nu, \hat{\mathbf{k}}) \cdot \hat{\mathbf{n}}\, \mathrm{d}\omega\, \mathrm{d}\nu\, \mathrm{d}a\, \mathrm{d}t$$

$$= \int_0^\infty \mathrm{d}S\, N(S, \nu, \hat{\mathbf{k}}) S(\nu, \hat{\mathbf{k}}) \cos\theta\, \mathrm{d}\omega\, \mathrm{d}\nu\, \mathrm{d}a\, \mathrm{d}t. \qquad (1.8.1)$$

The second form is obtained by noting from Fig. 1.8.1 that $\mathbf{S}(\nu, \hat{\mathbf{k}}) = S(\nu, \hat{\mathbf{k}})\hat{\mathbf{k}}$ and $\hat{\mathbf{k}} \cdot \hat{\mathbf{n}} = \cos\theta$. This suggests introducing the definition

$$I_\nu(\hat{\mathbf{k}}) = \int_0^\infty \mathrm{d}S\, N(S, \nu, \hat{\mathbf{k}}) S(\nu, \hat{\mathbf{k}}) \qquad (1.8.2)$$

so that Eq. (1.8.1) can be written

$$\mathrm{d}E_\nu = I_\nu \cos\theta\, \mathrm{d}\omega\, \mathrm{d}\nu\, \mathrm{d}a\, \mathrm{d}t. \qquad (1.8.3)$$

Thus I_ν is the rate at which radiant energy confined to a unit solid angle and unit wavenumber interval crosses unit surface area normal to the direction of incidence, and is called the specific intensity. Typical units are W cm^{-2} sr^{-1}/cm^{-1}.

Another important quantity is the monochromatic net flux, πF_ν, which is the rate at which energy per unit wavenumber interval flows across a surface of unit area in all directions,

$$\pi F_\nu = \int_{4\pi} I_\nu \cos\theta\, \mathrm{d}\omega, \qquad (1.8.4)$$

where the integration is performed over all solid angles. In effect, Eq. (1.8.4) gives the difference between the upward and downward fluxes across a horizontal surface of unit area. The net flux plays an important role in determining the magnitudes of radiative heating and cooling rates, as will be discussed in Section 9.1.

2
Radiative transfer

Various physical processes modify a radiation field as it propagates through an atmosphere. The rate at which the atmosphere emits depends on its composition and thermal structure, while its absorption and scattering properties are defined by the prevailing molecular opacity and cloud structure.

Independently of whether the radiation field is generated internally or is imposed externally, the study of how it interacts with the atmosphere is embodied in the theory of radiative transfer. Many authors have dealt with this theory in various contexts. Monographs include those by Kourganoff (1952), Woolley & Stibbs (1953), Goody (1964), and Goody & Yung (1989). A standard text is by Chandrasekhar (1950), which treats the subject as a branch of mathematical physics. The emphasis is on scattered sunlight in planetary atmospheres and on various problems of astrophysical interest.

Our own approach is somewhat different and emphasizes spectra produced by thermal emission from planetary atmospheres, especially as observed from space platforms. In order to demonstrate the connection between the thermal radiation giving rise to these spectra and the physical state of the atmosphere under consideration, it is necessary to examine how the transport of this radiation is effected. Only then is it possible to have a clear understanding of how the structure of an atmosphere leads to its spectral appearance, a topic considered at length in Chapter 4. Once this is accomplished a reversal of the procedure is feasible, and in Chapters 6 through 9 we demonstrate how the observed characteristics of the radiation field imply the underlying physical structure and the state of the interacting atmosphere.

Our aim in this chapter is to develop the mathematical formalism that serves as the foundation for all our analyses involving the radiation field in sufficient depth to be essentially autonomous, though our indebtedness to some of the procedures developed by Chandrasekhar (1950) is obvious. The equation of transfer is derived in Section 2.1, and formal solutions are found in Section 2.2. Very general techniques for solving the transfer equation numerically are developed in Section 2.3. Though powerful, these techniques are often cumbersome, and Section 2.4 discusses the simplifications permitted when multiple scattering is unimportant. Finally, an approximate analytic solution for thick, scattering atmospheres, which can be very useful where physical insight is needed, is developed in Section 2.5.

2.1 The equation of transfer

a. Definitions and geometry

Because radiation tends to be modified when it interacts with matter, it is possible to infer certain physical properties of planetary atmospheres and surfaces by studying their reflected and emitted radiation. Although these modifications are macroscopic in nature (they are manifested over an extended volume), their origins are contained in the processes of absorption, scattering, and emission of radiant energy on a microscopic scale. A quantitative assessment of the relation between these interactions and the resulting radiation field is known as the theory of radiative transfer. It is the purpose of this section to develop the equation central to this theory.

We begin by considering a volume element, dV, containing N_0 particles, either cloud particles or molecules in the vapor phase, located a distance z above the planetary surface. Directions at dV are specified by μ (the cosine of the zenith angle θ) and the azimuthal angle ϕ (Fig. 2.1.1). The emission angle θ is measured positively from zero (the zenith) to π (the nadir); the corresponding range for μ is from $+1$ to -1. The azimuthal angle ϕ is measured through 2π radians in the plane of stratification

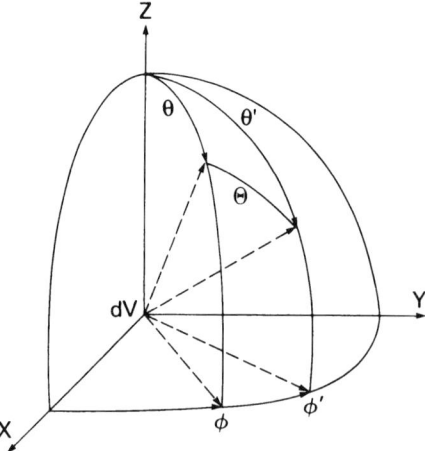

Fig. 2.1.1 Illustration of the relation between the scattering angle Θ and the coordinate angles θ, θ', ϕ, and ϕ'. Radiation is incident on the volume element dV in the direction (μ, ϕ) and scattered by dV through the angle Θ into the direction (μ', ϕ'), where $\mu = \cos\theta$ and $\mu' = \cos\theta'$.

from an arbitrary angle ϕ_0. Directions are specified by the symbol (μ, ϕ). For example, $I(z, \mu, \phi)$ is the intensity of radiation at dV (at a level z) in the direction (μ, ϕ).

The specific intensity, also known as the spectral radiance, has been defined by Eq. (1.8.3). Another parameter of interest is the phase function for single scattering, $p(\cos \Theta)$, which describes the angular distribution of radiation scattered once through the angle Θ. If E_ν represents the fraction of energy per unit time incident on dV in the direction (μ, ϕ) that is either absorbed or scattered in all directions, and dE_ν is that fraction of E_ν scattered into the direction (μ', ϕ') contained in the solid angle $d\omega'$, then $p(\cos \Theta)$ is defined by

$$\frac{dE_\nu(z, \mu', \phi')}{E_\nu(z, \mu, \phi)} = p(\cos \Theta) \frac{d\omega'}{4\pi}, \qquad (2.1.1)$$

where Θ is the angle between the directions of incidence (μ, ϕ) and scattering (μ', ϕ'). The explicit relation between Θ and the variables μ, ϕ, μ', ϕ' is found from spherical trigonometry (see Fig. 2.1.1) to be

$$\cos \Theta = \mu\mu' + (1 - \mu^2)^{\frac{1}{2}}(1 - \mu'^2)^{\frac{1}{2}} \cos(\phi' - \phi). \qquad (2.1.2)$$

The phase function can be written

$$p(\cos \Theta) = p(\mu', \phi'; \mu, \phi), \qquad (2.1.3)$$

where radiation originally in the direction (μ, ϕ) has been scattered into the direction (μ', ϕ'). It should be noted from Eq. (2.1.2) that $p(\mu', \phi'; \mu, \phi)$ is symmetric in the pair of variables μ, ϕ; i.e.,

$$p(\mu, \phi; \mu', \phi') = p(\mu', \phi'; \mu, \phi). \qquad (2.1.4)$$

If the albedo for single scattering, $\tilde{\omega}_0$, is defined to be the ratio of radiant power scattered in all directions to that extinguished (absorbed plus scattered), we have, from Eqs. (2.1.1) and (2.1.3),

$$\tilde{\omega}_0 = \frac{1}{E_\nu(z, \mu, \phi)} \int dE_\nu(z, \mu', \phi') = \int_0^{4\pi} p(\mu', \phi'; \mu, \phi) \frac{d\omega'}{4\pi}. \qquad (2.1.5)$$

b. Microscopic processes

Two points of view are possible in describing radiation–matter interactions on a microscopic scale. In the Lagrangian point of view the movements of individual photons are followed as they interact with matter contained in the volume element dV. 'Monte Carlo' programs are computer versions of the Lagrangian method. However, the observable parameter of interest, the specific intensity, I_ν, is a field of these photons, and we are interested in following the time-averaged variations of this field as it interacts with the matter in dV. For this purpose it is more practical to study local variations of I_ν without regard to the individual history of each photon; this is the Eulerian point of view. In our development we examine both viewpoints and illustrate their equivalence.

Consider the photons of wavenumber ν interacting with dV to be classified according to the interactions they undergo as well as upon the intrinsic characteristics of the photons themselves. We restrict ourselves to one field at a time, which in essence is the same as restricting our attention to one photon of this field at a time. Thus the field from the Lagrangian point of view (in a looser sense of the phrase) is followed.

Since each photon behaves by definition like every other photon in the field, certain criteria must be met. Each photon must be identical with every other photon of this class, and the system of particles with which this class of photons interacts must be composed of exactly the same kind of individual particles (in terms of dimensions, refractive index, etc.) in order that the separate interactions be identical. If more than one interaction per photon takes place, the order, number, and character of these interactions must be the same for each photon. In order to circumvent the inordinate complexity imposed by the last requirement, the volume element, dV, must be restricted to dimensions considerably smaller than the mean free path of an individual photon, so that only single interactions are possible in dV.

Construct a convex closed surface, δS, around dV such that the volume enclosed by δS is large compared with dV but small otherwise (see Figs. 2.1.2 and 2.1.3). Let

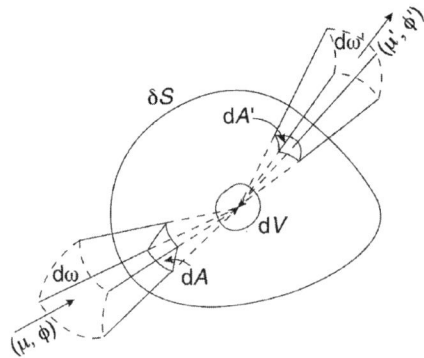

Fig. 2.1.2 Elements of solid angle dω and dω' subtended at the volume element dV by dA and dA', elements of the convex bounding surface δS.

2.1 The equation of transfer

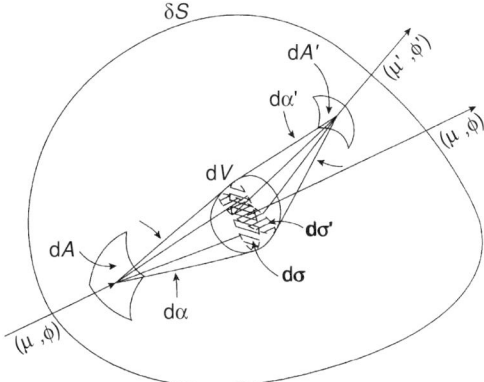

Fig. 2.1.3 Illustration of the elements of solid angle dα and dα' subtended at dA and dA', respectively, and by dσ and dσ' cross sections of dV in the directions (μ, ϕ) and (μ', ϕ').

dA and dA' be elements of δS such that the direction from dA to dV is (μ, ϕ) and the direction from dV to dA' is (μ', ϕ'). Further let dω and dω' be respectively the elements of solid angle containing dA and dA' as seen from dV, and let dα and dα' be the elements of solid angle containing dV as seen respectively from dA and dA'.

Consider that system of photons where each photon is contained in the wavenumber range $(\nu, \nu + d\nu)$ and interacts with a particle of homogeneous composition having an extinction (absorption plus scattering) cross section in the range $(\chi, \chi + d\chi)$. The cross section is generally wavenumber dependent and, if the particle is not spherical, may depend on direction as well. We divide the system of photons under consideration into four classes in the sense defined above, where the classes are distinguishable only through the types of interactions they undergo (see Figs. 2.1.2 and 2.1.3). The classes are:

(1) That class of photons incident on dV in a time dt and in the direction (μ, ϕ) contained in the solid angle dα that is singly scattered into the solid angle dω' in the direction (μ', ϕ') by interactions in dV with particles in the cross section range $(\chi, \chi + d\chi)$. This process can be considered either as a scattering of a certain fraction of the number of incident photons into dω', or as a redirection of a fraction of the incident energy into dω'.
(2) That class of photons incident on dV in a time dt in the direction (μ, ϕ) that is absorbed by the particles under consideration in dV. This process can be considered either as an absorption of a certain fraction of incident photons or as a diminution of the incident energy by some fractional amount.
(3) That class of photons incident on dV in a time dt in the direction (μ', ϕ') that is singly scattered into the solid angle dω in the direction (μ, ϕ). Again, this process can be considered either as a scattering of a certain fraction of incident photons into dω, or a redirection of a fraction of the incident energy into dω.

(4) That class of photons thermally emitted from the particles under consideration in dV into the solid angle dω in the direction (μ, ϕ) in a time dt. This process can be considered either as an emission of many individual photons in the wavenumber range $(\nu, \nu + d\nu)$, or as a source of energy emitted into dω in the direction (μ, ϕ) in a time dt. Nonthermal emission is also possible, though generally is not significant in the investigations considered in this book.

The first two classes of photons are lost from the radiation field in the direction (μ, ϕ) by scattering and absorption. The last two classes consist of photons gained by the radiation field in the direction (μ, ϕ) by scattering and emission. These are not all the losses and gains of the radiation field, however, since only interactions with particles in the cross section range $(\chi, \chi + d\chi)$ have been considered. Integrations over cross section and orientation remain to be performed.

Referring to Figs. 2.1.2 and 2.1.3 and Eq. (1.8.3), it is seen that the amount of energy $\delta E_\nu(z, \mu, \phi)$ in the wavenumber range $(\nu, \nu + d\nu)$ crossing dV in a time dt, which has originated outside δS and has also crossed dA, is

$$\delta E_\nu(z, \mu, \phi) = I_\nu(z, \mu, \phi) \mu_1 \mathrm{d}A \, \mathrm{d}\alpha \, \mathrm{d}\nu \, \mathrm{d}t, \tag{2.1.6}$$

where μ_1 is the cosine of the angle between the direction (μ, ϕ) and the normal to dA. Of all the energy crossing dA in a time dt contained in the solid angle dα, a certain fraction is singly scattered by dV into the solid angle dω'. Analogous to Eqs. (2.1.1) and (2.1.3) this fraction is

$$\frac{\mathrm{d}[\delta E_\nu(z, \mu', \phi')]}{\delta E_\nu(z, \mu, \phi)} = P(\chi) \, p_\chi(\mu', \phi'; \mu, \phi) \frac{\mathrm{d}\omega'}{4\pi}, \tag{2.1.7}$$

where $P(\chi)$ is the probability that any one photon incident on dV is extinguished (either absorbed or scattered) by a particle of cross section χ, and $p_\chi(\mu', \phi'; \mu, \phi)$ is a function of χ and is normalized to $\tilde{\omega}_0(\chi)$ [cf. Eq. (2.1.5)].

Now $P(\chi)$ is just the ratio of the total available effective extinction cross section of all particles in the cross section range $(\chi, \chi + d\chi)$ contained in dV to the geometrical cross section dσ of dV as seen in the direction (μ, ϕ), and this is given by

$$P(\chi) = \frac{N(\chi) \chi \, \mathrm{d}\chi \, \mathrm{d}V}{\mathrm{d}\sigma}, \tag{2.1.8}$$

where $N(\chi)$ is the number of particles per unit volume per unit cross section range centered about χ. Equation (2.1.8) is valid so long as there is no 'shadow' effect; i.e., the probability that any one particle is screened off from any other particle is

2.1 The equation of transfer

negligible. This requires that only single interactions occur in dV, or, put another way, $P(\chi) \ll 1$. Collecting Eqs. (2.1.6) through (2.1.8) yields

$$d[\delta E_\nu(z, \mu', \phi')] = \frac{N(\chi)\chi d\chi dV}{d\sigma} I_\nu(z, \mu, \phi) \mu_1 \, dA \, d\alpha \, d\nu \, dt \, p_\chi(\mu', \phi'; \mu, \phi) \frac{d\omega'}{4\pi}. \quad (2.1.9)$$

In the following we indicate losses and gains of the intensity by the symbols δ_- and δ_+, respectively. Clearly, there must be a loss of intensity $d[\delta_- I_S(z, \mu, \phi)]$ from the direction (μ, ϕ) associated with the loss of energy from this same direction, which in turn corresponds to the energy gain $d[\delta E_\nu(z, \mu', \phi')]$ in the direction (μ', ϕ'), and this, according to Eq. (2.1.6), is

$$d[\delta E_\nu(z, \mu', \phi')] = d[\delta_- I_S(z, \mu, \phi)] \mu_1 dA \, d\alpha \, d\nu \, dt. \quad (2.1.10)$$

Comparing Eqs. (2.1.9) and (2.1.10) we obtain

$$d[\delta_- I_S(z, \mu, \phi)] = \frac{N(\chi)\chi d\chi dV}{d\sigma} I_\nu(z, \mu, \phi) p_\chi(\mu', \phi'; \mu, \phi) \frac{d\omega'}{4\pi}. \quad (2.1.11)$$

This equation is valid for the scattering of incident photons in the wavenumber range $(\nu, \nu + d\nu)$ into the solid angle $d\omega'$ by particles in the cross section range $(\chi, \chi + d\chi)$. In order to obtain the scattering loss into all directions by particles of all sizes, Eq. (2.1.11) must be integrated over all ω' and χ. Implicit in the χ integration is averaging over orientation for particles lacking spherical symmetry. Thus, the total intensity in the wavenumber range $(\nu, \nu + d\nu)$ and in the direction (μ, ϕ), which is lost from this direction by scattering in dV, is

$$\delta_- I_S(z, \mu, \phi) = \frac{N_0 dV}{d\sigma} I_\nu(z, \mu, \phi) \int_0^\infty \int_0^{4\pi} D(\chi) \chi p_\chi(\mu', \phi'; \mu, \phi) \frac{d\omega'}{4\pi} d\chi, \quad (2.1.12)$$

where N_0 is the total number of particles per unit volume and $D(\chi)$ is the normalized distribution function of particle cross sections; i.e., $N(\chi) = N_0 D(\chi)$.

Consider now the second class of photons defined previously. From Eqs. (2.1.6) and (2.1.7) and the related discussion it follows that the energy in the wavenumber range $(\nu, \nu + d\nu)$ crossing dA (in Fig. 2.1.2) in a time dt, which is absorbed by particles in the cross section range $(\chi, \chi + d\chi)$, is

$$d[\delta E_\nu(z, \mu, \phi)] = P_A(\chi) I_\nu(z, \mu, \phi) \mu_1 \, dA \, d\alpha \, d\nu \, dt, \quad (2.1.13)$$

where $P_A(\chi)$, the probability of absorption of a photon in the wavenumber range $(\nu, \nu + d\nu)$, is

$$P_A(\chi) = [1 - \tilde{\omega}_0(\chi)]P(\chi) = \frac{[1 - \tilde{\omega}_0(\chi)]N(\chi)\chi \, d\chi \, dV}{d\sigma}. \quad (2.1.14)$$

By analogy with Eq. (2.1.9), the loss $d[\delta_- I_A(z, \mu, \phi)]$ of the intensity $I_\nu(z, \mu, \phi)$ incident on dV in the direction (μ, ϕ) is given by

$$d[\delta E_\nu(z, \mu, \phi)] = d[\delta_- I_A(z, \mu, \phi)]\mu_1 \, dA \, d\alpha \, d\nu \, dt. \quad (2.1.15)$$

Intercomparing Eqs. (2.1.13) through (2.1.15) yields

$$d[\delta_- I_A(z, \mu, \phi)] = \frac{[1 - \tilde{\omega}_0(\chi)]N(\chi)\chi \, d\chi \, dV}{d\sigma} I(z, \mu, \phi), \quad (2.1.16)$$

and this integrated over all particle cross sections becomes

$$\delta_- I_A(z, \mu, \phi) = \frac{N_0 dV}{d\sigma} I_\nu(z, \mu, \phi) \int_0^\infty [1 - \tilde{\omega}_0(\chi)]D(\chi)\chi \, d\chi. \quad (2.1.17)$$

This equation expresses the total intensity in the wavenumber range $(\nu, \nu + d\nu)$ lost from the direction (μ, ϕ) by absorption in dV.

In order to establish the gain to the radiation field by scattering of the third class of photons into the direction (μ, ϕ), it is necessary only to reverse the initial and final directions of the radiation field. By once again tracing out the consequences of the scattering process it is evident that Eqs. (2.1.6) through (2.1.9) remain valid upon an interchange of the primed and unprimed quantities. Thus, the gain of energy $d[\delta E_\nu(z, \mu, \phi)]$ in a time dt and in the direction (μ, ϕ) contained in the solid angle $d\omega$, which has resulted from a scattering of the energy by particles in the cross section range $(\chi, \chi + d\chi)$ contained in dV, and which was originally in the direction (μ', ϕ') and contained in the solid angle $d\alpha'$ is

$$d[\delta E_\nu(z, \mu, \phi)] = \frac{N(\chi)\chi \, d\chi \, dV}{d\sigma'} I_\nu(z, \mu', \phi')\mu'_1 \, dA' \, d\alpha' \, d\nu \, dt \, p_\chi(\mu, \phi; \mu', \phi') \frac{d\omega}{4\pi}. \quad (2.1.18)$$

Here $d\sigma'$ is the geometrical cross section of dV as seen in the direction (μ', ϕ'), $d\alpha'$ is the solid angle subtended by $d\sigma'$ at dA', and μ'_1 is the cosine of the angle contained between the direction (μ', ϕ') and the normal to dA'.

2.1 The equation of transfer

The corresponding gain in intensity at dV in the direction (μ, ϕ) must be given by

$$d[\delta E_\nu(z, \mu, \phi)] = d[\delta_+ I_S(z, \mu, \phi)] d\sigma \, d\omega \, d\nu \, dt, \qquad (2.1.19)$$

since $d[\delta E_\nu(z, \mu, \phi)]$ is just the energy in a time dt that crosses normal to the surface element $d\sigma$ at dV and is contained in $d\omega$. Comparing Eqs. (2.1.18) and (2.1.19) we obtain

$$d[\delta_+ I_S(z, \mu, \phi)] = \frac{N(\chi)\chi d\chi dV}{d\sigma} I_\nu(z, \mu', \phi') p_\chi(\mu, \phi; \mu', \phi') \frac{\mu'_1 dA' d\alpha'}{4\pi d\sigma'}. \qquad (2.1.20)$$

If the distance between dV and dA' is denoted by r, then, from the geometry,

$$\mu'_1 dA' = r^2 d\omega' \qquad (2.1.21)$$

and

$$d\sigma' = r^2 d\alpha'. \qquad (2.1.22)$$

With the aid of Eqs. (2.1.21) and (2.1.22), Eq. (2.1.20) becomes

$$d[\delta_+ I_S(z, \mu, \phi)] = \frac{N(\chi)\chi d\chi dV}{d\sigma} p_\chi(\mu, \phi; \mu', \phi') I_\nu(z, \mu', \phi') \frac{d\omega'}{4\pi}. \qquad (2.1.23)$$

In order to obtain the total contribution to the intensity in the direction (μ, ϕ) by scattering from dV, Eq. (2.1.23) must be integrated over all solid angles ω' and all particle cross sections χ [see Eq. (2.1.12)];

$$\delta_+ I_S(z, \mu, \phi) = \frac{N_0 dV}{d\sigma} \int_0^\infty \int_0^{4\pi} D(\chi) \chi p_\chi(\mu, \phi; \mu', \phi') I_\nu(z, \mu', \phi') \frac{d\omega'}{4\pi} d\chi. \qquad (2.1.24)$$

The fourth class of photons describes the contribution to the radiation field by thermal emission from dV. In order to calculate this contribution, consider the surface δS in Fig. 2.1.2 to be a perfectly insulating enclosure maintaining the particles in dV at a constant temperature, T. Since the radiation field within the enclosure is in equilibrium with its surroundings and is isotropic, the amount of energy in the wavenumber interval $(\nu, \nu + d\nu)$ that would be emitted by particles in the cross section range $(\chi, \chi + d\chi)$ into the direction (μ, ϕ) contained in the solid angle $d\omega$, upon an instantaneous removal of the enclosure, is given by [cf. Eqs. (2.1.13) and (2.1.14)]

$$d[\delta E_\nu(z, \mu, \phi)] = B_\nu(T)N(\chi)[1 - \tilde{\omega}_0(\chi)]\chi \, d\chi \, dV \, d\omega \, d\nu \, dt, \quad (2.1.25)$$

where $B_\nu(T)$ is the Planck function. The term $[1 - \tilde{\omega}_0(\chi)]\chi$ in Eq. (2.1.25) is an emission cross section, which is identical to the absorption cross section, as required by the first law of thermodynamics.

Corresponding to previous arguments we must also have [see Eq. (2.1.15)]

$$d[\delta E_\nu(z, \mu, \phi)] = d[\delta_+ I_E(z, \mu, \phi)] \, d\sigma \, d\omega \, d\nu \, dt, \quad (2.1.26)$$

where $d[\delta_+ I_E(z, \mu, \phi)]$ refers to the gain in intensity in the direction (μ, ϕ) due to thermal emission from dV. If the total number of particles of all sizes and shapes per unit volume is N_0, and the cross section distribution in dV is given as before by $D(\chi)$, then the total contribution in the direction (μ, ϕ) contained in the solid angle $d\omega$ of radiation thermally emitted from all the particles in dV according to Eq. (2.1.25) is

$$\int_{E_\nu} d[\delta E_\nu(z, \mu, \phi)] = N_0 B_\nu(T) \, dV \, d\omega \, d\nu \, dt \int_0^\infty [1 - \tilde{\omega}_0(\chi)]D(\chi)\chi \, d\chi. \quad (2.1.27)$$

Comparing Eqs. (2.1.26) and (2.1.27) yields

$$\delta_+ I_E(z, \mu, \phi) = \int_{I_E} d[\delta_+ I_E(z, \mu, \phi)]$$

$$= \frac{N_0 dV}{d\sigma} B_\nu(T) \int_0^\infty [1 - \tilde{\omega}_0(\chi)]\chi D(\chi) \, d\chi. \quad (2.1.28)$$

If the enclosure is not replaced, dV will be subjected to the local anisotropic radiation field of arbitrary energy density. What happens now is largely a function of the relative importance of: (1) collisions between molecules, either in the gaseous or condensed state, compared with (2) interactions between these molecules and the radiation field, as a cause of molecular absorptions and emissions. If interactions with the radiation field dominate, the emission will consist essentially of spontaneous emission of photons from excited molecules, and induced emission through perturbations due to the radiation field. The latter type of emission is proportional to the incident intensity and is therefore anisotropic in the same sense as the surrounding radiation field. The molecules are excited through the absorption of incident radiation, and the local temperature depends mainly on the photon density. Thus the radiation emitted from dV cannot be isotropic because of the contribution from induced emission unless the radiation field itself is strictly isotropic.

At the other extreme collisional battering of molecules dominates, and thermal (isotropic) emission far outweighs emission induced by the radiation field. This occurs where the density of molecules is high enough so that the frequency of encounters among molecules is much larger than the frequency of encounters between molecules and incident photons. Such a condition should always occur inside liquid and solid particles. However, if the particles are molecules in the gas phase, greater care is required where N_0 is sufficiently small that induced emission becomes important.

If collisions between molecules dominate, Eq. (2.1.28) is acceptable as it stands upon a removal of the enclosure. The temperature depends only upon the energy available through collisions, and the radiation field does not have to be in equilibrium with the surrounding medium. The limiting case where scattering becomes negligible [$\tilde{\omega}_0(\chi) = 0$] is sometimes referred to as the condition of local thermodynamic equilibrium.

c. The total field

We are now in a position to evaluate the net change in intensity in the direction (μ, ϕ) due to the presence of the volume element dV. Adding all gains and losses from Eqs. (2.1.12), (2.1.17), (2.1.24), and (2.1.28) yields

$$\delta I_\nu(z, \mu, \phi) = -\delta_- I_S(z, \mu, \phi) - \delta_- I_A(z, \mu, \phi) + \delta_+ I_S(z, \mu, \phi) + \delta_+ I_E(z, \mu, \phi)$$

$$= -\frac{N_0 dV}{d\sigma} I_\nu(z, \mu, \phi) \int_0^\infty \int_{\omega'} D(\chi)\, \chi\, p_\chi(\mu', \phi'; \mu, \phi) \frac{d\omega'}{4\pi} d\chi$$

$$- \frac{N_0 dV}{d\sigma} I_\nu(z, \mu, \phi) \int_0^\infty [1 - \tilde{\omega}_0(\chi)] D(\chi)\, \chi\, d\chi$$

$$+ \frac{N_0 dV}{d\sigma} \int_0^\infty \int_{\omega'} D(\chi) \chi p_\chi(\mu, \phi; \mu', \phi') I_\nu(z, \mu', \phi') \frac{d\omega'}{4\pi} d\chi$$

$$+ \frac{N_0 dV}{d\sigma} B_\nu(T) \int_0^\infty [1 - \tilde{\omega}_0(\chi)] D(\chi)\, \chi\, d\chi. \qquad (2.1.29)$$

The terms on the right side of Eq. (2.1.29) are negative or positive depending on whether they are, respectively, losses or gains.

Equation (2.1.29) can be simplified by defining a normalized effective phase function, $p_0(\mu, \phi; \mu', \phi')$, by

$$p_0(\mu, \phi; \mu', \phi') \int_0^\infty \tilde{\omega}_0(\chi)\, D(\chi)\, \chi\, d\chi = \int_0^\infty D(\chi)\, \chi p_\chi(\mu, \phi; \mu', \phi')\, d\chi.$$

$$(2.1.30)$$

Upon multiplying both sides of Eq. (2.1.30) by $d\omega'/4\pi$, integrating over all solid angles, and remembering that $\int_{\omega'} p_\chi(\mu, \phi; \mu', \phi')\,d\omega'/4\pi = \tilde{\omega}_0(\chi)$, we infer

$$\int_0^{4\pi} p_0(\mu, \phi; \mu', \phi')\frac{d\omega'}{4\pi} = 1. \tag{2.1.31}$$

Substituting in Eq. (2.1.29) according to Eq. (2.1.30) and using Eq. (2.1.31) we obtain

$$\frac{\delta I_\nu(z, \mu, \phi)}{-\dfrac{N_0 dV}{d\sigma}\int_0^\infty D(\chi)\chi\,d\chi} = I_\nu(z, \mu, \phi) - \frac{\int_0^\infty \tilde{\omega}_0(\chi)D(\chi)\chi\,d\chi}{\int_0^\infty D(\chi)\chi\,d\chi}$$

$$\times \int_{\omega'} p_0(\mu, \phi; \mu', \phi')\, I_\nu(z, \mu', \phi')\frac{d\omega'}{4\pi}$$

$$- \frac{\int_0^\infty [1 - \tilde{\omega}_0(\chi)]D(\chi)\chi\,d\chi}{\int_0^\infty D(\chi)\chi\,d\chi} B_\nu(T). \tag{2.1.32}$$

Further simplifications of the terms in Eq. (2.1.32) are possible. We define an effective extinction cross section χ_E and an effective single scattering albedo $\tilde{\omega}_0$ of dV by

$$\chi_E = \int_0^\infty D(\chi)\chi\,d\chi \tag{2.1.33}$$

and

$$\tilde{\omega}_0 \int_0^\infty D(\chi)\chi\,d\chi = \int_0^\infty \tilde{\omega}_0(\chi)D(\chi)\chi\,d\chi, \tag{2.1.34}$$

respectively, and a single scattering phase function $p(\mu, \phi; \mu', \phi')$ normalized to $\tilde{\omega}_0$ such that

$$p(\mu, \phi; \mu', \phi') = \tilde{\omega}_0 p_0(\mu, \phi; \mu', \phi'). \tag{2.1.35}$$

If the atmosphere is plane-parallel (infinitely extended in the x- and y-directions and variable in the z-direction only), a volume element cylinder dV of height dz and base area $d\sigma_z$ is given by

$$dV = d\sigma_z dz. \tag{2.1.36}$$

If the linear dimensions of $d\sigma_z$ are made arbitrarily much larger than dz, but still small enough so that $d\sigma_z$ remains an element of surface area, then the sides of the cylinder can be neglected relative to $d\sigma_z$ in determining the effective cross section of dV as seen along a slant-path in the direction (μ, ϕ). The geometric cross section $d\sigma$ of dV is

$$d\sigma = \mu d\sigma_z, \tag{2.1.37}$$

and the denominator of the left side of Eq. (2.1.32) can be reduced to

$$-\frac{N_0 dV}{d\sigma} \int_0^\infty D(\chi) \chi d\chi = -\frac{1}{\mu} N_0 \chi_E dz. \tag{2.1.38}$$

At this point it is convenient to define a normal optical depth τ_ν measured from the top of the atmosphere inward such that

$$d\tau_\nu = -N_0 \chi_E dz. \tag{2.1.39}$$

Thus $d\tau_\nu$ is the optical cross section of dV in the z-direction.

Upon collecting Eqs. (2.1.33) through (2.1.39), introducing the relations $\mu = \cos\theta$ and $d\omega = \sin\theta \, d\theta \, d\phi$ (Fig. 2.1.1), and letting the ratio $\delta I_\nu(z, \mu, \phi)/d\tau_\nu$ approach its limit as $d\tau_\nu \to 0$, Eq. (2.1.32) becomes

$$\mu \frac{dI(\tau_\nu, \mu, \phi)}{d\tau_\nu} = I(\tau_\nu, \mu, \phi)$$
$$- \frac{1}{4\pi} \int_0^{2\pi} \int_{-1}^{+1} p(\mu, \phi; \mu', \phi') I(\tau_\nu, \mu', \phi') d\mu' d\phi'$$
$$- (1 - \tilde{\omega}_0) B_\nu(T). \tag{2.1.40}$$

This is the equation of transfer for an arbitrary, monochromatic field of radiation. In practice this field consists of a diffuse component that originates from thermal emission of the atmosphere and the planetary surface, as well as a component (both diffuse and direct) that originates from the Sun. The former component dominates in the middle and far infrared, whereas the latter component is the sole contributor at visible wavelengths. Both components are important in a transition range in the near infrared.

d. The diffuse field

Upon solving the transfer equation in the context of various physical problems it is evident that the intensity of the diffuse radiation field is the quantity most easily handled in the equation, because the boundary conditions are much simpler to impose than for the case in which the intensity of the total radiation field is the dependent variable. In particular it is convenient to distinguish between the reduced incident radiation field from the Sun, which penetrates to the level τ_ν without suffering any scattering or absorption processes, and the diffuse field, which has arisen as a result of one or more scattering and emission processes.

In order to separate the diffuse field from the directly transmitted radiation we consider a collimated beam of radiation (sunlight is a fair approximation) of flux πF_0 crossing a unit surface area normal to the beam. The magnitude of the flux in the downward direction crossing a unit area contained in a plane at the top of the atmosphere is $\mu_0 \pi F_0$, where μ_0 is the cosine of the zenith angle of the point source, and this is [see Eq. (1.8.4)],

$$\mu_0 \pi F_0 = \int_0^{2\pi} \int_{-1}^{+1} \mu I(0, \mu, \phi) \, d\mu \, d\phi, \quad (2.1.41)$$

where $I(0, \mu, \phi)$ is the downward intensity of radiation in the direction (μ, ϕ) at $\tau_\nu = 0$. Since the only contribution from the point source is in the direction $(-\mu_0, \phi_0)$, the intensity $I(0, \mu, \phi)$ should be of the form

$$I(0, \mu, \phi) = C \delta(\mu + \mu_0) \delta(\phi - \phi_0), \quad (2.1.42)$$

where $\delta(\mu + \mu_0)$ and $\delta(\phi - \phi_0)$ are Dirac δ-functions and C is a normalization constant. Substituting Eq. (2.1.42) into Eq. (2.1.41) readily yields the value $C = \pi F_0$.

The total intensity $I(\tau_\nu, \mu, \phi)$ associated with Eq. (2.1.40) is the sum of the intensity $I_D(\tau_\nu, \mu, \phi)$ arising from the diffuse radiation field and the intensity directly transmitted from the point source to the level τ_ν. By analogy with Eq. (2.1.42) and the related discussion this latter intensity may be expressed by

$$I_T(\tau_\nu, \mu, \phi) = \pi F_0 \delta(\mu + \mu_0) \delta(\phi - \phi_0) h(\tau_\nu), \quad (2.1.43)$$

where $h(\tau_\nu)$ is a function of τ_ν alone yet to be determined. After some reduction,

Eq. (2.1.40) becomes

$$\mu \frac{dI_D(\tau_\nu, \mu, \phi)}{d\tau_\nu} + \pi F_0 \mu \delta(\mu + \mu_0) \delta(\phi - \phi_0) \frac{dh(\tau_\nu)}{d\tau_\nu}$$
$$= I_D(\tau_\nu, \mu, \phi) + \pi F_0 \delta(\mu + \mu_0) \delta(\phi - \phi_0) h(\tau_\nu)$$
$$- \frac{1}{4\pi} \int_0^{2\pi} \int_{-1}^{+1} p(\mu, \phi; \mu', \phi') I_D(\tau_\nu, \mu', \phi') d\mu' d\phi'$$
$$- \frac{F_0}{4} h(\tau_\nu) p(\mu, \phi; -\mu_0, \phi_0) - (1 - \tilde{\omega}_0) B_\nu(T). \quad (2.1.44)$$

We suppose that $dh(\tau_\nu)/d\tau_\nu$ and $h(\tau_\nu)$ are in general non-zero. Then, when $\mu = -\mu_0$ and $\phi = \phi_0$, Eq. (2.1.44) reduces to

$$-\mu_0 \delta(\mu + \mu_0) \delta(\phi - \phi_0) \frac{dh(\tau_\nu)}{d\tau_\nu} = \delta(\mu + \mu_0) \delta(\phi - \phi_0) h(\tau_\nu), \quad (2.1.45)$$

yielding

$$h(\tau_\nu) = C_0 e^{-\tau_\nu/\mu_0}, \quad (2.1.46)$$

where C_0 is the constant of integration. Upon replacing $h(\tau_\nu)$ in Eq. (2.1.43) with its equivalent in Eq. (2.1.46) and letting $\tau_\nu \to 0$, we find $C_0 = 1$. Equation (2.1.46) is equivalent to what is sometimes referred to as Beer's law of exponential attenuation.

Upon dropping the subscripts D and ν, Eq. (2.1.44) for the *diffuse* radiation field becomes

$$\mu \frac{dI(\tau, \mu, \phi)}{d\tau} = I(\tau, \mu, \phi) - \frac{1}{4\pi} \int_0^{2\pi} \int_{-1}^{+1} p(\mu, \phi; \mu', \phi') I(\tau, \mu', \phi') d\mu' d\phi'$$
$$- \frac{F_0}{4} e^{-\tau/\mu_0} p(\mu, \phi; -\mu_0, \phi_0) - (1 - \tilde{\omega}_0) B(\tau), \quad (2.1.47)$$

where $B(T)$ has been replaced with $B(\tau)$ to emphasize that the temperature is a function only of optical depth in plane-parallel atmospheres. It remains understood that the various parameters are in general functions of ν.

Equation (2.1.47) is the basic equation of transfer considered in this book. Solutions to Eq. (2.1.47) are sought in the context of specific problems as they appear in the course of investigation. In the remainder of this chapter we first derive formal solutions, and then examine explicit solutions that are possible, either because certain approximations are invoked or because at some point numerical procedures are introduced.

2.2 Formal solutions

A first-order linear integro-differential equation, such as Eq. (2.1.47), can be solved formally by finding an appropriate integrating factor. Multiplying both sides of Eq. (2.1.47) by $e^{-\tau/\mu}$, transferring the first term on the right side to the left, and gathering terms yields

$$\mu \frac{d[e^{-\tau/\mu} I(\tau, \mu, \phi)]}{d\tau} = -\frac{1}{4\pi} e^{-\tau/\mu} \int_0^{2\pi} \int_{-1}^{+1} p(\mu, \phi; \mu', \phi') I(\tau, \mu', \phi') d\mu' d\phi'$$
$$- \frac{F_0}{4} \exp\left[-\left(\frac{1}{\mu} + \frac{1}{\mu_0}\right)\tau\right] p(\mu, \phi; -\mu_0, \phi_0)$$
$$- (1 - \tilde{\omega}_0) e^{-\tau/\mu} B(\tau). \quad (2.2.1)$$

Integrating between τ and τ_1, where τ_1 is the optical depth at the surface (or lower boundary), and rearranging terms yields the upward intensity at τ (multiplied by the integrating factor $e^{-\tau/\mu}$);

$$I(\tau, \mu, \phi) e^{-\tau/\mu} = I(\tau_1, \mu, \phi) e^{-\tau_1/\mu}$$
$$+ \frac{1}{4\pi \mu} \int_\tau^{\tau_1} e^{-\tau'/\mu} \int_0^{2\pi} \int_{-1}^{+1} p(\mu, \phi; \mu', \phi') I(\tau', \mu', \phi') d\mu' d\phi' d\tau'$$
$$+ \frac{F_0}{4\mu} \int_\tau^{\tau_1} \exp\left[-\left(\frac{1}{\mu} + \frac{1}{\mu_0}\right)\tau'\right] p(\mu, \phi; -\mu_0, \phi_0) d\tau'$$
$$+ \frac{1}{\mu} \int_\tau^{\tau_1} (1 - \tilde{\omega}_0) e^{-\tau'/\mu} B(\tau') d\tau'. \quad (2.2.2)$$

This is only a formal solution, since the unknown intensity is itself contained in this solution. By analogy with our treatment of direct sunlight in Section 2.1, it is useful to separate the intensity arising from the lower boundary from the rest of the radiation field. The direct contribution to $I(\tau', \mu', \phi')$ from the surface is the intensity $I(\tau_1, \mu', \phi')$, attenuated along the path length $(\tau_1 - \tau')/\mu'$ by the factor $\exp[(\tau_1 - \tau')/\mu']$. Thus Eq. (2.2.2) can be written

$I(\tau, \mu, \phi) =$

(1) $I(\tau_1, \mu, \phi) e^{-(\tau_1 - \tau)/\mu}$

(2) $+ \dfrac{1}{4\pi\mu} \int_\tau^{\tau_1} \int_0^{2\pi} \int_0^1 e^{-(\tau' - \tau)/\mu} e^{-(\tau_1 - \tau')/\mu'} p(\mu, \phi; \mu', \phi') I(\tau_1, \mu', \phi') d\mu' d\phi' d\tau'$

(3) $+ \dfrac{F_0}{4\mu} \int_\tau^{\tau_1} e^{-\tau'/\mu_0} e^{-(\tau' - \tau)/\mu} p(\mu, \phi; -\mu_0, \phi_0) d\tau'$

2.2 Formal solutions

$$(4) +\frac{1}{\mu}\int_{\tau}^{\tau_1}(1-\tilde{\omega}_0)\,e^{-(\tau'-\tau)/\mu}B(\tau')\,d\tau'$$

$$(5) +\frac{1}{4\pi\mu}\int_{\tau}^{\tau_1}\int_{0}^{2\pi}\int_{-1}^{+1}e^{-(\tau'-\tau)/\mu}p(\mu,\phi;\mu',\phi')\,I(\tau',\mu',\phi')\,d\mu'\,d\phi'\,d\tau' \quad (2.2.3)$$

where the diffuse field $I(\tau',\mu',\phi')$ is zero in the upward directions along the boundary ($\tau' = \tau_1$).

The various terms for the upward intensity in the direction (μ, ϕ) at a level τ represent:

(1) radiation originating directly from the lower boundary (generally the surface or a compact cloud deck) that is attenuated by the overlying atmosphere between the levels τ and τ_1;
(2) radiation originating from the lower boundary in the direction (μ', ϕ') that is scattered at τ' into the direction (μ, ϕ);
(3) radiation from the Sun that has penetrated to the level τ' before undergoing a scattering process;
(4) radiation that is thermally emitted at the level τ'; and
(5) radiation that has undergone one or more scattering processes before being scattered at τ' into the direction (μ, ϕ).

The downward intensity can be treated in like manner. The integration is now from $\tau = 0$ to τ, and the formal solution to Eq. (2.2.1) becomes [cf. Eq. (2.2.3)]

$$I(\tau, -|\mu|, \phi) =$$

$$(1) \; \frac{1}{4\pi|\mu|}\int_{0}^{\tau}e^{-(\tau-\tau')/|\mu|}\int_{0}^{2\pi}\int_{-1}^{+1}p(-|\mu|,\phi;\mu',\phi')\,I(\tau',\mu',\phi')\,d\mu'\,d\phi'\,d\tau'$$

$$(2) +\frac{F_0}{4|\mu|}\int_{0}^{\tau}e^{-\tau'/\mu_0}e^{-(\tau-\tau')/|\mu|}p(-|\mu|,\phi;-\mu_0,\phi_0)\,d\tau'$$

$$(3) +\frac{1}{|\mu|}\int_{0}^{\tau}(1-\tilde{\omega}_0)\,e^{-(\tau-\tau')/|\mu|}B(\tau')\,d\tau', \quad (2.2.4)$$

where $-|\mu|$ replaces μ as a reminder that downward directions are being considered. The terms for the downward intensity in the direction $(-|\mu|, \phi)$ at a level τ represent:

(1) radiation that has undergone one or more scattering processes before being scattered at τ' into the direction $(-|\mu|, \phi)$;
(2) radiation from the Sun that has penetrated to the level τ' before undergoing a scattering process; and
(3) radiation thermally emitted at τ'.

Downward intensity terms at the upper boundary ($\tau = 0$) analogous to the upward intensity terms at the lower boundary [$\tau = \tau_1$; see Eq. (2.2.4)] do not exist, because the diffuse field is zero in all downward directions at $\tau = 0$.

Reduction of the formal solutions to explicit solutions is complicated, and no known analytic solutions exist in the general case. The major difficulty centers around terms (5) in Eq. (2.2.3) and (1) in (2.2.4), which contain the unknown intensity field and arise as a result of multiple scattering. Three approaches are possible: (1) direct numerical solutions, (2) approximate analytic solutions, and (3) exact analytic solutions for thin layers (where multiple scattering is unimportant), coupled with numerical procedures for developing solutions for thick layers from the starting analytic solutions.

Wherever possible we have chosen the latter two approaches in this book in order to associate the solutions more directly with physical processes. Explicit representation of the parameters involved enables the reader to follow the qualitative way in which changes in the physical state of a system affect the appearance of the outgoing radiation field. Where an accurate quantitative assessment is required we supplement analytic solutions with numerical methods. The remainder of this chapter is directed toward these ends.

2.3 Invariance principles

a. Definitions

A very useful concept for arriving at more explicit solutions to the transfer equation is that of the invariance of the radiation field emerging from a layer of finite thickness to the nature of the incoming radiation field at the layer boundaries. For example, the angular distribution of intensity and the net flux of the outgoing radiation field at the top of the atmosphere do not depend on the explicit nature of the underlying surface. It could be either a solid surface or a vacuum, as long as the upwelling radiation fields at $\tau = \tau_1$ are identical, regardless of the sources of those fields.

We apply this concept first to a single layer, then to two layers, and finally we generalize to many layers with arbitrary thermal and scattering properties. Ultimately, this leads to a procedure for computing the outgoing radiation field from any vertically inhomogeneous atmosphere.

Consider an atmospheric layer that absorbs, emits, and scatters radiation, and that is bounded above and below by the intensity fields $I(\tau_0, -\mu, \phi)$ and $I(\tau_1, \mu, \phi)$, respectively. Define a scattering function $S(\tau_1 - \tau_0; \mu, \phi; \mu', \phi')$ such that the intensity of diffusely reflected radiation in the direction (μ, ϕ) at $\tau = \tau_0$, which arises as a result of the presence of the radiation field $I(\tau_0, -\mu', \phi')$, is given by

$$I_S(\tau_0, \mu, \phi) = \frac{1}{4\pi\mu} \int_0^{2\pi} \int_0^1 S(\tau_1 - \tau_0; \mu, \phi; \mu', \phi') I(\tau_0, -\mu', \phi') \, d\mu' \, d\phi'. \tag{2.3.1}$$

Also define a transmission function $T(\tau_1 - \tau_0; \mu, \phi; \mu', \phi')$ such that the intensity of diffusely transmitted radiation in the direction $(-\mu, \phi)$ at $\tau = \tau_1$, which arises as a result of this same downward radiation field at $\tau = \tau_0$, is given by

$$I_T(\tau_1, -\mu, \phi) = \frac{1}{4\pi\mu} \int_0^{2\pi} \int_0^1 T(\tau_1 - \tau_0; \mu, \phi; \mu', \phi') I(\tau_0, -\mu', \phi') \, d\mu' \, d\phi'. \tag{2.3.2}$$

Further let $\tilde{S}$ and $\tilde{T}$ be the corresponding scattering and transmission functions when the layer is illuminated from below. Thus the intensity of diffusely reflected radiation in the direction $(-\mu, \phi)$ at $\tau = \tau_1$, and the intensity of diffusely transmitted radiation in the direction (μ, ϕ) at $\tau = \tau_0$, are given respectively by

$$I_S(\tau_1, -\mu, \phi) = \frac{1}{4\pi\mu} \int_0^{2\pi} \int_0^1 \tilde{S}(\tau_1 - \tau_0; \mu, \phi; \mu', \phi') I(\tau_1, \mu', \phi') \, d\mu' \, d\phi' \tag{2.3.3}$$

$$I_T(\tau_0, \mu, \phi) = \frac{1}{4\pi\mu} \int_0^{2\pi} \int_0^1 \tilde{T}(\tau_1 - \tau_0; \mu, \phi; \mu', \phi') I(\tau_1, \mu', \phi') \, d\mu' \, d\phi'. \tag{2.3.4}$$

The total outgoing intensities at the top and bottom of the layer are given by

$$I(\tau_0, \mu, \phi) = I_E(\tau_0, \mu, \phi) + I_S(\tau_0, \mu, \phi) + I_T(\tau_0, \mu, \phi) \tag{2.3.5}$$

and

$$I(\tau_1, -\mu, \phi) = I_E(\tau_1, -\mu, \phi) + I_S(\tau_1, -\mu, \phi) + I_T(\tau_1, -\mu, \phi), \tag{2.3.6}$$

where I_E is the intensity of radiation thermally emitted from the layer itself.

b. The stacking of layers

Now consider two stacked layers of thicknesses τ_0 and $\tau_1 - \tau_0$, the scattering properties of which are generally different. Let the radiation fields incident on either side of the composite, $I(0, -\mu, \phi)$ and $I(\tau_1, \mu, \phi)$, be identically zero. By analogy with Eqs. (2.3.1) through (2.3.6), the upward and downward intensities at the common boundary τ_0 are given by

$$I(\tau_0, \mu, \phi) = I_E(\tau_0, \mu, \phi) + \frac{1}{4\pi\mu} \int_0^{2\pi} \int_0^1 S(\tau_1 - \tau_0; \mu, \phi; \mu', \phi')$$
$$\times I(\tau_0, -\mu', \phi') \, d\mu \, d\phi', \qquad (2.3.7)$$

and

$$I(\tau_0, -\mu, \phi) = I_E(\tau_0, -\mu, \phi) + \frac{1}{4\pi\mu} \int_0^{2\pi} \int_0^1 \tilde{S}(\tau_0; \mu, \phi; \mu', \phi')$$
$$\times I(\tau_0, \mu', \phi') \, d\mu \, d\phi'. \qquad (2.3.8)$$

Equations (2.3.7) and (2.3.8) are coupled integral equations defining the intensities $I(\tau_0, \pm\mu, \phi)$, and can be solved numerically provided I_E, S, and $\tilde{S}$ are known functions. By analogy with Eq. (2.2.3), part (1), and Eqs. (2.3.1) through (2.3.6), the outgoing emitted intensities from the two-layer composite at the top ($\tau = 0$) and bottom ($\tau = \tau_1$) are, respectively,

$$I(0, \mu, \phi) = I_E(0, \mu, \phi) + e^{-\tau_0/\mu} I(\tau_0, \mu, \phi)$$
$$+ \frac{1}{4\pi\mu} \int_0^{2\pi} \int_0^1 \tilde{T}(\tau_0; \mu, \phi; \mu', \phi') I(\tau_0, \mu', \phi') \, d\mu' \, d\phi' \qquad (2.3.9)$$

and

$$I(\tau_1, -\mu, \phi) = I_E(\tau_1, -\mu, \phi) + e^{-(\tau_1-\tau_0)/\mu} I(\tau_0, -\mu, \phi)$$
$$+ \frac{1}{4\pi\mu} \int_0^{2\pi} \int_0^1 T(\tau_1 - \tau_0; \mu, \phi; \mu', \phi') I(\tau_0, -\mu', \phi') \, d\mu' \, d\phi'.$$
$$(2.3.10)$$

The first term on the right side of Eq. (2.3.9) is due to thermal emission from the layer defined by the boundaries $\tau = 0$ and $\tau = \tau_0$. The second and third terms are due, respectively, to direct and diffuse transmission through the same layer. Analogous interpretations with respect to the layer defined by the boundaries $\tau = \tau_0$ and $\tau = \tau_1$ are valid for Eq. (2.3.10).

Equations (2.3.9) and (2.3.10) define the outgoing upward and downward radiation fields emitted from a composite inhomogeneous layer. Repeated applications of the procedure admit solutions for an atmosphere composed of any finite number of layers with individually different thermal and scattering properties. It is also necessary, however, to obtain values for I_E for single layers and develop an algorithm for determining S, $\tilde{S}$, T, and $\tilde{T}$ for both single and composite layers.

c. Composite scattering and transmission functions

Consider the two original individually homogeneous layers with thicknesses τ_0 and $\tau_1 - \tau_0$. Let an outside point source of intensity $\pi F_0 \delta(\mu - \mu_0)\delta(\phi - \phi_0)$ irradiate the composite from above, and ignore all thermally emitted radiation. By analogy with Eqs. (2.1.43), (2.1.46), and (2.3.1), the intensity at τ_0 in the direction (μ, ϕ), in the absence of thermal emission, is given by

$$I(\tau_0, \mu, \phi) = \frac{F_0}{4\mu} e^{-\tau_0/\mu_0} S(\tau_1 - \tau_0; \mu, \phi; \mu_0, \phi_0)$$
$$+ \frac{1}{4\pi\mu} \int_0^{2\pi} \int_0^1 S(\tau_1 - \tau_0; \mu, \phi; \mu', \phi') I(\tau_0, -\mu', \phi') d\mu' d\phi', \quad (2.3.11)$$

where the exponential in the first term on the right side is due to attenuation of directly transmitted radiation through the upper layer.

The intensity at τ_0 in the direction $(-\mu, \phi)$ is given by

$$I(\tau_0, -\mu, \phi) = \frac{F_0}{4\mu} T(\tau_0; \mu, \phi; \mu_0, \phi_0)$$
$$+ \frac{1}{4\pi\mu} \int_0^{2\pi} \int_0^1 \tilde{S}(\tau_0; \mu, \phi; \mu', \phi') I(\tau_0, \mu', \phi') d\mu' d\phi', \quad (2.3.12)$$

where the first term on the right side arises from radiation from the point source that is diffusely transmitted through the upper layer, and the second term denotes radiation diffusely reflected by the upper layer into the direction $(-\mu, \phi)$. Eqs. (2.3.11) and (2.3.12) are coupled integral equations that can be solved numerically for $I(\tau_0, \pm\mu, \phi)$.

Finally, in the absence of thermal emission, the outgoing radiation fields at either boundary of the composite are given by

$$I(0, \mu, \phi) = \frac{F_0}{4\mu} S(\tau_0; \mu, \phi; \mu_0, \phi_0) + e^{-\tau_0/\mu_0} I(\tau_0, \mu, \phi)$$
$$+ \frac{1}{4\pi\mu} \int_0^{2\pi} \int_0^1 \tilde{T}(\tau_0; \mu, \phi; \mu', \phi') I(\tau_0, \mu', \phi') d\mu' d\phi' \quad (2.3.13)$$

and

$$I(\tau_1, -\mu, \phi) = \frac{F_0}{4\mu} e^{-\tau_0/\mu_0} T(\tau_1 - \tau_0; \mu, \phi; \mu_0, \phi_0) + e^{-(\tau_1-\tau_0)/\mu} I(\tau_0, -\mu, \phi)$$
$$+ \frac{1}{4\pi\mu} \int_0^{2\pi} \int_0^1 T(\tau_1 - \tau_0; \mu, \phi; \mu', \phi') I(\tau_0, -\mu', \phi') d\mu' d\phi'.$$
(2.3.14)

The terms on the right side of Eq. (2.3.13) refer to: (1) the diffuse reflection from the upper layer of radiation originating with the point source, and (2) the direct and (3) diffuse transmission through the upper layer of the diffuse radiation field originating at the level τ_0 and directed into the upper hemisphere. The terms on the right side of Eq. (2.3.14) from left to right refer to: (1) the diffuse transmission through the lower layer of radiation originating with the point source that has been directly transmitted through the upper layer, and (2) the direct and (3) diffuse transmission through the lower layer of the diffuse radiation field originating at the level τ_0 that is directed into the lower hemisphere.

On the other hand, for an outside point source of intensity $\pi F_0 \delta(\mu - \mu_0) \delta(\phi - \phi_0)$, it is seen by analogy with Eqs. (2.3.1) and (2.3.2) that the diffuse scattering and transmission functions for a two-layer composite of total thickness $\tau = \tau_1$ can be defined by

$$I(0, \mu, \phi) = \frac{F_0}{4\mu} S(\tau_1; \mu, \phi; \mu_0, \phi_0) \quad (2.3.15)$$

and

$$I(\tau_1, -\mu, \phi) = \frac{F_0}{4\mu} T(\tau_1; \mu, \phi; \mu_0, \phi_0). \quad (2.3.16)$$

Thus, according to Eqs. (2.3.11) through (2.3.16), the S and T functions for any two-layer composite can be found if the S, T, $\tilde{S}$, and $\tilde{T}$ functions are known for the individual layers. Clearly the same holds true for the composite $\tilde{S}$ and $\tilde{T}$ functions; the point source is positioned below the composite and the process repeated. In this way the scattering and transmission functions for any combination of individual layers can be determined.

Summarizing the adding process, scattering and transmission functions are obtained from Eqs. (2.3.11) through (2.3.16). Emission intensities are built up from Eqs. (2.3.1) through (2.3.10). The outgoing intensity fields for a plane-parallel atmosphere of any degree of vertical complexity can thus be calculated. All that is required as input are values for I_E, S, $\tilde{S}$, T, and $\tilde{T}$ for the individual layers.

d. Starting solutions

The connection between these functions and the microphysical properties of the layers can be found by returning to the original formal solutions of the transfer equation, Eqs. (2.2.3) and (2.2.4). If the layer is thin enough, the Planck intensity B is essentially constant throughout, and the diffuse field becomes vanishingly small. Hence term (4) in Eq. (2.2.3), and term (3) in Eq. (2.2.4), suffice to describe the emitted radiation fields in the upward and downward directions, respectively. Letting $\tilde{\omega}_0$ and B be independent of τ (valid for a sufficiently thin layer), the solution for Eq. (2.2.3) becomes

$$I(\tau, \mu, \phi) = (1 - \tilde{\omega}_0) B \left[1 - e^{-(\tau_1 - \tau)/|\mu|}\right], \qquad (2.3.17)$$

and the solution to Eq. (2.2.4) is

$$I_E(\tau, -|\mu|, \phi) = (1 - \tilde{\omega}_0) B (1 - e^{-\tau/|\mu|}). \qquad (2.3.18)$$

The solutions for the outgoing radiation fields are obtained by setting $\tau = 0$ in Eq. (2.3.17) and $\tau = \tau_1$ in Eq. (2.3.18). Thus the outgoing intensities emitted from a thin layer are

$$I_E(0, \mu, \phi) = I_E(\tau_1, -|\mu|, \phi) = (1 - \tilde{\omega}_0) B (1 - e^{-\tau_1/|\mu|}). \qquad (2.3.19)$$

These values can be used as starting values for I_E in the repeated applications of Eqs. (2.3.7) through (2.3.10).

On the other hand, the solution to Eq. (2.2.3) for a thin layer in the presence of an outside point source of radiation is given by

$$I_S(\tau, \mu, \phi) = \frac{F_0}{4\mu} \int_\tau^{\tau_1} e^{-\tau'/\mu_0} e^{-(\tau' - \tau)/\mu} p(\mu, \phi; -\mu_0, \phi_0) d\tau', \qquad (2.3.20)$$

which, because p is essentially independent of τ' across this layer, reduces to

$$I_S(\tau, \mu, \phi) = \frac{F_0}{4} p(\mu, \phi; -\mu_0, \phi_0) \frac{\mu_0}{\mu + \mu_0} \left[e^{-\tau/\mu_0} - e^{-\tau_1/\mu_0} e^{-(\tau_1 - \tau)/\mu}\right]. \qquad (2.3.21)$$

Upon letting $\tau = 0$, the solution for the upwelling radiation field is:

$$I_S(0, \mu, \phi) = \frac{F_0}{4} \frac{\mu_0}{\mu + \mu_0} p(\mu, \phi; -\mu_0, \phi_0) \left\{1 - \exp\left[-\left(\frac{1}{\mu} + \frac{1}{\mu_0}\right)\tau_1\right]\right\}. \qquad (2.3.22)$$

By analogy with Eq. (2.3.15), the scattering function for a thin homogeneous layer is

$$S(\tau_1; \mu, \phi; \mu_0, \phi_0) = \frac{\mu\mu_0}{\mu + \mu_0} p(\mu, \phi; -\mu_0, \phi_0) \left\{ 1 - \exp\left[-\left(\frac{1}{\mu} + \frac{1}{\mu_0}\right)\tau_1\right]\right\}. \tag{2.3.23}$$

In like manner, from Eq. (2.2.4), upon letting $\tau = \tau_1$, and by analogy with Eq. (2.3.16), the transmission function for a thin homogeneous layer is found to be

$$T(\tau_1; \mu, \phi; \mu_0, \phi_0) = \frac{|\mu|\mu_0}{|\mu| - \mu_0} p(-|\mu|, \phi; -\mu_0, \phi_0) \left(e^{-\tau_1/|\mu|} - e^{-\tau_1/\mu_0}\right). \tag{2.3.24}$$

Comparable expressions for $\tilde{S}$ and $\tilde{T}$ are readily developed merely by inverting the layer and repeating the process. It is found, by virtue of the layer being homogeneous, that $S = \tilde{S}$ and $T = \tilde{T}$, although some care is required in defining the sign of μ. This completes the discussion of how the macroscopic properties of the outgoing radiation field can be inferred for any atmosphere with vertically inhomogeneous microscopic scattering and emission properties. The surface can be treated as a semi-infinite layer with its own characteristic properties.

2.4 Special cases

a. Nonscattering atmospheres

The general solution of the transfer equation is highly complicated. Most of that complexity arises from the inclusion of scattering processes. Fortunately, at least in the thermal part of the spectrum where absorption and emission by atmospheric gases dominate and solar radiation is negligible, scattering processes can often be neglected and the solution simplifies considerably. If the atmosphere is nonscattering, the solution for the outgoing radiation field at the top of the atmosphere reduces to [see Eq. (2.2.3)]

$$I(0, \mu, \phi) = I(\tau_S, \mu, \phi) e^{-\tau_S/\mu} + \int_0^{\tau_S} e^{-\tau/\mu} B(\tau) \frac{d\tau}{\mu}, \tag{2.4.1}$$

where τ_S is the optical depth at the planetary surface.

It is interesting to compare this result with that obtained by repeated application of Eq. (2.3.9). Imagine an atmosphere divided into many thin layers. The top layer has an optical thickness τ_1 (not to be confused with the total optical thickness of the atmosphere in this context), the top two layers have a composite thickness τ_2,

2.4 Special cases

and so on. According to Eq. (2.3.17) the emitted intensity from the ith layer is

$$I_E(\tau_{i-1}, \mu) = B(\tau_i)\left[1 - e^{-(\tau_i - \tau_{i-1})/\mu}\right], \quad (2.4.2)$$

where the azimuth-independent nature of the solution is indicated by deleting ϕ from the notation. It readily follows by repeated application of Eq. (2.3.9) that the outgoing intensity at the top of an atmosphere composed of n layers is

$$I(0, \mu) = I_E(\tau_n, \mu)e^{-\tau_n/\mu} + \sum_{i=1}^{n} I_E(\tau_{i-1}, \mu)e^{-\tau_{i-1}/\mu}, \quad (2.4.3)$$

or, from Eq. (2.4.2),

$$I(0, \mu) = B(\tau_n)e^{-\tau_n/\mu} + \sum_{i=1}^{n} e^{-\tau_{i-1}/\mu} B(\tau_{i-\frac{1}{2}})\left[1 - e^{-(\tau_i - \tau_{i-1})/\mu}\right], \quad (2.4.4)$$

where $\tau_0 = 0$, $B(\tau_{i-\frac{1}{2}})$ is a mean value between τ_i and τ_{i-1}, τ_n corresponds to τ_S in Eq. (2.4.1), and $B(\tau_n)$ is the Planck intensity of the surface. Eq. (2.4.4) is the finite difference counterpart of Eq. (2.4.1). We use these simpler equations or their equivalent for nonscattering atmospheres in this book, although there are occasions when scattering is important and more extended solutions are necessary.

b. Optically thin atmospheres

Another special case is the restriction of the general solution to objects that do not possess an atmosphere at all or that have only thin atmospheres that are transparent in some spectral regions. Examples of the first group are Mercury, the Moon, and most planetary satellites, although Titan is an important exception. Examples of the second group are Mars and to some degree the Earth; both have more or less transparent atmospheric transmission windows over restricted wavenumber ranges. Sometimes the emissivity of the surface is less than unity because partial reflection takes place. The outgoing intensity at the surface then is

$$I(\tau_{n-1}, \mu, \phi) = B(\tau_{n-1})\left[1 - \frac{1}{4\pi\mu}\int_0^{2\pi}\int_0^1 S(\tau_n; \mu, \phi; \mu', \phi')\,d\mu'\,d\phi'\right]$$

$$+ \frac{F_0}{4\mu}e^{-\tau_{n-1}/\mu}S(\tau_n; \mu, \phi; \mu_0, \phi_0)$$

$$+ \frac{1}{4\pi\mu}\int_0^{2\pi}\int_0^1 S(\tau_n; \mu, \phi; \mu', \phi')\,I(\tau_{n-1}, -\mu', \phi')\,d\mu'\,d\phi', \quad (2.4.5)$$

where S is a scattering function describing reflection from the surface. The first term on the right side is the product of the Planck intensity and the surface emissivity. The form for the emissivity arises from the requirement that the sum of emissivity and reflectivity equals unity. The second term accounts for the reflection of reduced incident sunlight, while the third term refers to backscattering of diffuse atmospheric radiation. For planetary satellites without atmospheres only the first two terms are appropriate, with $\tau_{n-1} = 0$.

2.5 Scattering atmospheres; the two-stream approximation

a. Single scattering phase function

Up to this point, exact analytic expressions for thin layers have been developed, as well as numerical procedures for extending these solutions to thick layers. A straightforward procedure for solving the transfer equation for thick nonscattering atmospheres has been developed. It remains to find a satisfactory approximate solution for thick scattering atmospheres, since exact analytic solutions do not exist.

The two-stream approximation is just such a solution. The continuous radiation field is replaced by two directed beams, one up and one down. Because the solution is analytic, it is helpful to separate the azimuthal and polar angles. We write the phase function [cf. Eq. (2.1.3)] as a finite series of Legendre polynomials:

$$p(\cos \Theta) = \sum_{\lambda=0}^{N} \tilde{\omega}_\lambda P_\lambda(\cos \Theta), \qquad (2.5.1)$$

where the coefficients $\tilde{\omega}_\lambda$ are constants. Integrating Eq. (2.5.1) over all solid angles yields

$$\int_\omega p(\cos \Theta) \frac{d\omega}{4\pi} = \frac{1}{4\pi} \int_0^{2\pi} \int_0^{\pi} \left[\sum_{\lambda=0}^{N} \tilde{\omega}_\lambda P_\lambda(\cos \Theta) \right] \sin \Theta \, d\Theta \, d\beta, \qquad (2.5.2)$$

where Θ is the polar angle, and the azimuthal angle β is the angle of rotation about the axis of symmetry defined by $\Theta = 0$. Using the relation (Appendix 1)

$$\frac{1}{2}(2\lambda + 1) \int_{-1}^{+1} P_m(\alpha) P_\lambda(\alpha) \, d\alpha = \delta_{m,\lambda}, \qquad (2.5.3)$$

we establish the identity

$$\int_\omega p(\cos\Theta)\frac{d\omega}{4\pi} = \tilde{\omega}_0, \qquad (2.5.4)$$

where, from Eq. (2.1.5), $\tilde{\omega}_0$ is the single scattering albedo. Expanding $P(\cos\Theta) = P_\lambda[\mu\mu' + (1-\mu^2)^{\frac{1}{2}}(1-\mu'^2)^{\frac{1}{2}}\cos(\phi'-\phi)]$ in accordance with the addition theorem of spherical harmonics we infer

$$p(\mu,\phi;\mu',\phi') = \sum_{\lambda=0}^{N} \tilde{\omega}_\lambda \bigg[P_\lambda(\mu)P_\lambda(\mu') + 2\sum_{m=1}^{\lambda} \frac{(\lambda-m)!}{(\lambda+m)!}$$

$$\times P_\lambda^m(\mu)P_\lambda^m(\mu')\cos m(\phi'-\phi) \bigg]. \qquad (2.5.5)$$

Inverting the order of summation yields

$$p(\mu,\phi;\mu',\phi') = \sum_{m=0}^{N}(2-\delta_{0,m})\sum_{\lambda=m}^{N} \tilde{\omega}_\lambda^m P_\lambda^m(\mu)P_\lambda^m(\mu')\cos m(\phi'-\phi), \qquad (2.5.6)$$

where

$$\tilde{\omega}_\lambda^m = \tilde{\omega}_\lambda \frac{(\lambda-m)!}{(\lambda+m)!}. \qquad (2.5.7)$$

b. Separation of variables

We try a relation for $I(\tau,\mu,\phi)$ of the form

$$I(\tau,\mu,\phi) = \sum_{m=0}^{N} I^{(m)}(\tau,\mu)\cos m(\phi_0-\phi). \qquad (2.5.8)$$

With the aid of the relation

$$\int_0^{2\pi} \cos k(\phi'-\phi)\cos n(\phi_0-\phi')\,d\phi' = \begin{cases} 0 & (k \neq n) \\ \pi\cos n(\phi_0-\phi) & (k=n \neq 0) \\ 2\pi & (k=n=0) \end{cases} \qquad (2.5.9)$$

we find, upon substituting Eqs. (2.4.1) and (2.5.9) into Eq. (2.1.47) and requiring that Eq. (2.1.47) be valid for all ϕ, that Eq. (2.1.47) separates into the $(N+1)$ independent equations:

$$\mu \frac{dI^{(m)}(\tau,\mu)}{d\tau} = I^{(m)}(\tau,\mu) - \frac{1}{2} \sum_{\lambda=m}^{N} \tilde{\omega}_{\lambda}^{m} P_{\lambda}^{m}(\mu) \int_{-1}^{+1} P_{\lambda}^{m}(\mu') I^{(m)}(\tau,\mu') d\mu'$$

$$- \frac{1}{4} F_0 e^{-\tau/\mu_0} (2 - \delta_{0,m}) \sum_{\lambda=m}^{N} \tilde{\omega}_{\lambda}^{m} P_{\lambda}^{m}(\mu) P_{\lambda}^{m}(-\mu_0)$$

$$- \delta_{0,m}(1 - \tilde{\omega}_0) B(\tau) \quad (m = 0, \ldots, N). \tag{2.5.10}$$

c. Discrete streams

In the two-stream approximation the continuous radiation field is replaced by one traveling in only two directions. Hence the integral in Eq. (2.5.10) is replaced by a sum,

$$\int_{-1}^{+1} P_{\lambda}^{m}(\mu') I^{(m)}(\tau,\mu') d\mu' \sim \sum_{j} a_j P_{\lambda}^{m}(\mu_j) I^{(m)}(\tau,\mu_j), \tag{2.5.11}$$

where $j = \pm 1$. If Gaussian quadrature is used, $a_1 = a_{-1} = 1$ and

$$\mu_1 = -\mu_{-1} = \frac{1}{\sqrt{3}}. \tag{2.5.12}$$

It can be shown in the two-stream approximation that we should restrict N to ≤ 1. Hence, from Eqs. (2.5.6) and (2.5.7),

$$p(\cos\Theta) = \tilde{\omega}_0 + \tilde{\omega}_1 \mu \mu' + \tilde{\omega}_1 (1-\mu^2)^{\frac{1}{2}} (1-\mu'^2)^{\frac{1}{2}} \cos(\phi' - \phi). \tag{2.5.13}$$

In the thermal infrared the radiation field tends to be axially symmetric because no off-axis localized sources (such as the Sun) contribute significantly. Thus, only the azimuth-independent ($m = 0$) transfer equation is required. We have, from Eqs. (2.5.10) through (2.5.13),

$$\mu_i \frac{dI(\tau,\mu_i)}{d\tau} = I(\tau,\mu_i) - \frac{1}{2} \sum_{j} (\tilde{\omega}_0 + \tilde{\omega}_1 \mu_i \mu_j) I(\tau,\mu_j)$$

$$- \frac{1}{4} F_0 e^{-\tau/\mu_0} (\tilde{\omega}_0 - \tilde{\omega}_1 \mu_i \mu_0) - (1 - \tilde{\omega}_0) B(\tau) \quad (j = \pm 1). \tag{2.5.14}$$

2.5 Scattering atmospheres; the two-stream approximation

d. Homogeneous solution

Consider first the homogeneous part of Eq. (2.5.14),

$$\mu_i \frac{dI(\tau, \mu_i)}{d\tau} = I(\tau, \mu_i) - \frac{1}{2} \sum_j (\tilde{\omega}_0 + \tilde{\omega}_1 \mu_i \mu_j) I(\tau, \mu_j). \qquad (2.5.15)$$

The appearance of Eq. (2.5.15) suggests a relation of the form

$$I(\tau, \mu_i) = L g(\mu_i) e^{-k\tau}, \qquad (2.5.16)$$

where L and k are constants. Substituting into Eq. (2.5.15) we find

$$g(\mu_i) = \frac{c_0 + c\mu_i}{1 + k\mu_i}, \qquad (2.5.17)$$

where

$$c_0 = \frac{1}{2} \tilde{\omega}_0 \sum_j g(\mu_j); \quad c = \frac{1}{2} \tilde{\omega}_1 \sum_j \mu_j g(\mu_j) \quad (j = \pm 1). \qquad (2.5.18)$$

Substituting Eq. (2.5.17) into Eq. (2.5.18) yields the pair of equations [cf. Eq. (2.5.12)]

$$c_0 = \tilde{\omega}_0 \left(\frac{c_0 - ck\mu_1^2}{1 - k^2 \mu_1^2} \right); \quad c = \mu_1^2 \tilde{\omega}_1 \left(\frac{c - c_0 k}{1 - k^2 \mu_1^2} \right). \qquad (2.5.19)$$

Reduction of Eqs. (2.5.19) leads to

$$k = \pm \left[3(1 - \tilde{\omega}_0)\left(1 - \frac{1}{3}\tilde{\omega}_1\right) \right]^{\frac{1}{2}}. \qquad (2.5.20)$$

Because Eqs. (2.5.19) are linear and homogeneous, c_0 and c are not both uniquely defined. Choosing

$$c_0 = \tilde{\omega}_0, \qquad (2.5.21)$$

we find

$$c = \mp \frac{\tilde{\omega}_1}{\sqrt{3}} \left(\frac{1 - \tilde{\omega}_0}{1 - \frac{1}{3}\tilde{\omega}_1} \right)^{\frac{1}{2}}, \qquad (2.5.22)$$

depending on the sign of k. According to Eq. (2.5.20) k is double-valued. Hence, from Eqs. (2.5.16) and (2.5.17),

$$I(\tau, \mu_i) = \sum_\alpha L_\alpha \frac{c_0 + c_\alpha \mu_i}{1 + k_\alpha \mu_i} e^{-k_\alpha \tau} \quad (i = \pm 1; \alpha = \pm 1), \qquad (2.5.23)$$

where

$$\left.\begin{array}{l} k_1 = -k_{-1} = \left[3(1 - \tilde{\omega}_0)\left(1 - \tfrac{1}{3}\tilde{\omega}_1\right)\right]^{\tfrac{1}{2}} \\[1ex] c_1 = -c_{-1} = -\dfrac{\tilde{\omega}_1}{\sqrt{3}}\left(\dfrac{1 - \tilde{\omega}_0}{1 - \tfrac{1}{3}\tilde{\omega}_1}\right)^{\tfrac{1}{2}} \end{array}\right\}. \qquad (2.5.24)$$

The arbitrary constants L_α ($\alpha = \pm 1$) are determined by the boundary conditions in the context of specific problems as they arise.

e. Outside point source

Particular integrals associated with the two inhomogeneous terms in Eq. (2.5.14) are also required to complete the solution. Before a specific analytic solution can be found for the integral containing the Planck intensity, it is necessary to specify the explicit form of $B(\tau)$. This is deferred to the appropriate sections where such specific forms are required.

The integral containing F_0 can be dealt with immediately. Writing Eq. (2.5.14) without the Planck intensity term yields

$$\mu_i \frac{dI(\tau, \mu_i)}{d\tau} = I(\tau, \mu_i) - \frac{1}{2} \sum_j (\tilde{\omega}_0 + \tilde{\omega}_1 \mu_i \mu_j) I(\tau, \mu_j)$$

$$- \frac{1}{4} F_0 e^{-\tau/\mu_0}(\tilde{\omega}_0 - \tilde{\omega}_1 \mu_i \mu_0) \quad (j = \pm 1). \qquad (2.5.25)$$

We try a solution of the form

$$I(\tau, \mu_i) = \frac{1}{4} F_0 h(\mu_i) e^{-\tau/\mu_0} \quad (i = \pm 1). \qquad (2.5.26)$$

Substituting Eq. (2.5.26) into Eq. (2.5.25) we find

$$h(\mu_i) = \mu_0 \frac{\gamma_0 + \gamma_1 \mu_i}{\mu_i + \mu_0} \quad (i = \pm 1), \qquad (2.5.27)$$

where the constants γ_0 and γ_1 are given by

$$\gamma_0 = \tilde{\omega}_0 \left[1 + \frac{1}{2} \sum_j h(\mu_j)\right] \qquad (2.5.28)$$

2.5 Scattering atmospheres; the two-stream approximation

and

$$\gamma_1 = \tilde{\omega}_1 \left[-\mu_0 + \frac{1}{2} \sum_j \mu_j \, h(\mu_j) \right] \quad (j = \pm 1). \quad (2.5.29)$$

Substituting Eq. (2.5.27) into Eqs. (2.5.28) and (2.5.29) yields, after some reduction,

$$\gamma_0 \left[(\mu_0^2 - \mu_1^2) - \tilde{\omega}_0 \mu_0^2 \right] + \gamma_1 \tilde{\omega}_0 \mu_0 \mu_1^2 = \tilde{\omega}_0 (\mu_0^2 - \mu_1^2) \quad (2.5.30)$$

and

$$\gamma_0 \, \tilde{\omega}_1 \mu_0 \mu_1^2 + \gamma_1 \left[(\mu_0^2 - \mu_1^2) - \tilde{\omega}_1 \mu_0^2 \mu_1^2 \right] = -\tilde{\omega}_1 \mu_0 (\mu_0^2 - \mu_1^2). \quad (2.5.31)$$

Further reduction yields

$$\gamma_0 = \frac{\tilde{\omega}_0 (\mu_0^2 - \mu_1^2)}{\mu_0^2 (1 - \tilde{\omega}_0)(1 - \mu_1^2 \tilde{\omega}_1) - \mu_1^2} \quad (2.5.32)$$

and

$$\gamma_1 = -\frac{\tilde{\omega}_1 (1 - \tilde{\omega}_0) \mu_0 (\mu_0^2 - \mu_1^2)}{\mu_0^2 (1 - \tilde{\omega}_0)(1 - \mu_1^2 \tilde{\omega}_1) - \mu_1^2}. \quad (2.5.33)$$

Therefore Eq. (2.5.27) becomes

$$h(\mu_i) = \frac{\mu_0^2 - \mu_1^2}{\mu_0^2 (1 - \tilde{\omega}_0)(1 - \mu_1^2 \tilde{\omega}_1) - \mu_1^2} \frac{\mu_0}{\mu_0 + \mu_i} [\tilde{\omega}_0 - \mu_i \mu_0 \tilde{\omega}_1 (1 - \tilde{\omega}_0)]$$

$$(i = \pm 1), \quad (2.5.34)$$

which allows the solution for Eq. (2.5.26).

Summarizing this chapter, we have derived both analytic and numerical procedures for calculating the emerging radiation field provided we can specify the vertical distributions of the temperature as well as the gas and particle compositions. It is also necessary to know the absorption and scattering properties of atmospheric volume elements on a microscopic scale. In the next chapter we discuss these properties before proceeding with the task of computing the intensity of the outgoing radiation field.

3
Interaction of radiation with matter

The discussions of the equation of transfer and the solution of this equation in Chapter 2 rest entirely on concepts of classical physics. Such treatment was possible because we considered a large number of photons interacting with a volume element that, although it was assumed to be small, was still of sufficient size to contain a large number of individual molecules. But with the assumption of many photons acting on many molecules we have only postponed the need to introduce quantum theory. Single photons do interact with individual atoms and molecules. The optical depth, $\tau(\nu)$, depends on the absorption coefficients of the matter present, which must fully reflect quantum mechanical concepts. The role of quantum physics in the derivation of the Planck function has already been discussed in Section 1.7. Both the optical depth and the Planck function appear in the radiative transfer equation (2.1.47).

The interaction of radiation with matter can take many forms. The photoelectric effect, the Compton effect, and pair generation–annihilation are processes that occur at wavelengths shorter than those encountered in the infrared. Infrared photons can excite rotational and vibrational modes of molecules, but they are insufficiently energetic to excite electronic transitions in atoms, which occur mostly in the visible and ultraviolet. Therefore, a discussion of the interaction of infrared radiation with matter in the gaseous phase needs to consider only rotational and vibrational transitions, while in the solid phase lattice vibrations in crystals must be included.

In the following sections on the interaction of radiation with gas molecules we begin with an overview of the physical principles of radiative transitions in molecules in Sections 3.1 and 3.2, proceed to discussions of the properties of diatomic and polyatomic molecules in Sections 3.3 and 3.4, and, finally, examine line strengths in Section 3.5 and line shapes in Section 3.6. Interactions of radiation with solid and liquid surfaces, as well as cloud particles, are the subject of Sections 3.7 and 3.8. For further information on molecular spectroscopy we refer the reader to text books, such as Pauling & Wilson (1935), Herzberg (1939, 1945, 1950), Townes & Schawlow (1955), or Steinfeld (1974). The book by Murcray & Goldman (1981) is an excellent reference showing many laboratory spectra of planetary interest. The reader may also consult the compendium of infrared spectra by Sadtler (1972).

3.1 Absorption and emission in gases

a. The old quantum theory

Gas molecules can alter their states of vibration and rotation by exchanging energy with the radiation field. This exchange occurs in discrete quantities, resulting in modifications to the field at specific frequencies associated with resonances in the molecular structure. As a consequence, molecules absorb and emit radiation in a complex pattern of discrete lines that deviate significantly from a blackbody spectrum. Because each type of molecule has a unique structure and, therefore, unique energies of motion, the pattern of observed lines is characteristic of the matter present.

Remote sensing by means of spectroscopy began early in the nineteenth century when Josef Fraunhofer (1817) observed dark lines in the solar spectrum. Although the visible spectra of many substances were known to be unique for each element (Bunsen & Kirchhoff, 1861, 1863), the origin of the lines was not fully understood until the early twentieth century. Now we know the lines reveal the presence of certain atomic gases, including hydrogen, sodium, calcium, and magnesium, in cooler regions of the solar atmosphere.

Niels Bohr (1913) provided the first successful explanation of the atomic hydrogen spectrum. Hydrogen has one proton and one electron, and exhibits a relatively simple pattern of spectral lines. By postulating that the energy of the atom is quantized, and that the atom absorbs and emits radiation by making transitions among the quantized energy levels, Bohr was able to explain the narrowness of the lines and their regular pattern. These lines coincide with the discrete energy differences, ΔE, among the levels, with the frequency of the emitted or absorbed radiation given by $f = \Delta E/h$, where h is the Planck constant. In analogy with the Solar System, Bohr proposed a model of the hydrogen atom with a proton in the center and an electron traveling around in the field of a central Coulomb force, $-e^2/4\pi\varepsilon_0 r^2$, where e is the charge of the electron, ε_0 the dielectric constant (permittivity) of free space, and r the radius of the electron orbit. By introducing the additional assumption that the angular momentum of the electron, L, can only take on integral multiples of $h/2\pi$,

$$L = mvr = \frac{h}{2\pi}n \quad (n = 1, 2, 3, \ldots), \tag{3.1.1}$$

where m and v are the electron mass and velocity, respectively, Bohr derived the energy levels of atomic hydrogen and calculated the hydrogen spectrum with good accuracy.

In his thesis Louis de Broglie (1924) recognized that the condition that the electron angular momentum must be an integral multiple of $h/2\pi$ implied the electron motion to be governed by a wave, similar to that for a photon. He postulated that a moving particle with momentum p and energy E has wavelike characteristics with wavelength $\lambda = h/p$ and frequency $f = E/h$. The Bohr angular momentum of the electron in a hydrogen atom is then

$$L = pr = \frac{h}{\lambda}r = \frac{h}{2\pi}n, \qquad (3.1.2)$$

which gives

$$2\pi r = n\lambda. \qquad (3.1.3)$$

The Bohr rule that the angular momentum must be an integral multiple of $h/2\pi$ can, therefore, be interpreted as a requirement that the electron orbital circumference must equal an integral number of wavelengths. In other words, after each orbital revolution the wave that describes the electron motion must constructively interfere with itself. This is a strong indication that the motion of material particles has wavelike properties.

The Bohr theory, and its interpretation in terms of waves by de Broglie, was successful in accounting for the spectra of atoms with only one electron, and it worked approximately for alkali elements (Li, Na, Rb, Cs). However, it could not be applied to complex systems, such as molecules, and it did not constitute a dynamical theory of particle motion, nor did it provide an understanding of the interaction of radiation with matter. It did, however, set the stage for the development of modern quantum theory.

b. The Schrödinger equation

Major progress in quantum theory came with the formulation of wave mechanics by Erwin Schrödinger (1926). An equivalent matrix theory of quantum mechanics was simultaneously developed by Werner Heisenberg (1925). Schrödinger accepted de Broglie's postulate that the motion of a particle is governed by a wave; his aim was to find a form of the wave equation applicable to quantum physics. Schrödinger began with the classical expression of the total energy, E, of a particle being the sum of kinetic and potential energy,

$$E = \frac{p^2}{2m} + V, \qquad (3.1.4)$$

3.1 Absorption and emission in gases

where p is the momentum and m the mass of the particle. A de Broglie wave traveling in the x-direction with wavelength $\lambda = h/p$ and frequency $f = E/h$ can be expressed by a wave function

$$\Psi(x, t) = e^{i(2\pi/h)(px - Et)}. \tag{3.1.5}$$

The form of this function suggests the following associations between differential operators and the particle momentum and energy:

$$\frac{\partial \Psi}{\partial x} = i\frac{2\pi}{h} p\Psi \rightarrow p = -i\frac{h}{2\pi}\frac{\partial}{\partial x} \tag{3.1.6}$$

and

$$\frac{\partial \Psi}{\partial t} = -i\frac{2\pi}{h} E\Psi \rightarrow E = i\frac{h}{2\pi}\frac{\partial}{\partial t}. \tag{3.1.7}$$

Substituting these operators in Eq. (3.1.4) produces a wave equation

$$\frac{ih}{2\pi}\frac{\partial \Psi}{\partial t} = -\frac{h^2}{8\pi^2 m}\frac{\partial^2 \Psi}{\partial x^2} + V(x,t)\Psi. \tag{3.1.8}$$

This equation of motion of a particle moving in one dimension subject to a potential V can be generalized to three dimensions:

$$\frac{ih}{2\pi}\frac{\partial \Psi}{\partial t} = -\frac{h^2}{8\pi^2 m}\nabla^2 \Psi + V(x, y, z, t)\Psi. \tag{3.1.9}$$

The Laplace operator, ∇^2, is defined in Appendix 1.

Schrödinger derived Eq. (3.1.9), although in a different way, as the equation of motion of a microscopic particle under the influence of a force with potential V. This equation provides complete dyamical constraints on the particle. The wave function $\Psi(x, y, z, t)$ that satisfies this equation describes the particle motion in the sense, proposed by Max Born (1926, 1927), that the square of its absolute value $|\Psi|^2 = \Psi^*\Psi$ represents a probability distribution. The asterisk denotes the complex conjugate value. The probability of finding the particle within a small volume $\Delta x \Delta y \Delta z$ is

$$\Psi^*(x, y, z, t)\Psi(x, y, z, t)\Delta x \Delta y \Delta z, \tag{3.1.10}$$

provided the wave function is normalized such that

$$\int_{-\infty}^{\infty}\int_{-\infty}^{\infty}\int_{-\infty}^{\infty} \Psi^*\Psi \, dx \, dy \, dz = 1. \quad (3.1.11)$$

This normalization is necessary because the particle must be found somewhere within all space with certainty.

This approach to quantum mechanics constitutes a comprehensive physical theory, eliminating the shortcomings of the older Bohr theory. It applies to all physical systems, including many-electron atoms and polyatomic molecules. Importantly, because the Schrödinger equation is time-dependent, it treats the *rates* of physical processes, including the interaction between matter and radiation. Therefore, the strengths of spectral lines can be predicted by quantum mechanics.

c. Energy levels and radiative transitions

If the potential V is time-independent, wave functions satisfying the Schrödinger equation can be factored into a space function and one periodic in time,

$$\Psi_n(x, y, z, t) = \psi_n(x, y, z) e^{-i(2\pi/h)E_n t}, \quad (3.1.12)$$

where the energy, E_n, is constant. Each ψ_n describes a stationary state. The probability density, $|\psi_n|^2$, has a spatial distribution, but is constant in time. Substituting ψ_n in Eq. (3.1.9) and writing $V = V(x, y, z)$ yields the time-independent Schrödinger equation

$$-\frac{h^2}{8\pi^2 m}\nabla^2 \psi_n + V(x, y, z)\psi_n = E_n \psi_n. \quad (3.1.13)$$

To find a time-dependent solution of the Schrödinger equation we apply perturbation theory. We assume that a transition between two energy levels is caused by a time-dependent influence, which we represent as a small additive potential, $v(x, y, z, t)$. The Schrödinger equation is then

$$\frac{ih}{2\pi}\frac{\partial \Psi}{\partial t} = -\frac{h^2}{8\pi^2 m}\nabla^2 \Psi + V(x, y, z)\Psi + v(x, y, z, t)\Psi. \quad (3.1.14)$$

The time-dependent wave function, Ψ, can be expressed as a linear combination of the Ψ_n in Eq. (3.1.12). The integral

$$\int \psi_n^* v(x, y, z, t) \psi_m \, d\tau = V_{nm}(t); \quad d\tau = dx \, dy \, dz \quad (3.1.15)$$

is a time-dependent perturbation matrix that causes transitions between energy levels E_m and E_n. The ψ_n are the time-independent wavefunctions. The perturbation represents a coupling between the radiation field and the molecule or atom. This matrix element is used later to identify allowed transitions and to find their rates.

To examine the behavior during a transition, consider a molecule that is initially in state ψ_1. What is the probability of finding the molecule in another state ψ_2 at a later time t? In a radiation field the molecule will be subjected to an oscillating electric field,

$$\mathbf{E}(t) = \mathbf{E}_0 \cos 2\pi f t. \tag{3.1.16}$$

$\mathbf{E}$ and $\mathbf{E}_0$ are vectors in x, y, and z. The interaction energy is given by the scalar product of this field and the electric dipole moment, $\mathbf{M}$, of the molecule,

$$v(t) = \mathbf{M} \cdot \mathbf{E}_0 \cos 2\pi f t. \tag{3.1.17}$$

The dipole moment is defined as

$$\mathbf{M} = \sum_i q_i \mathbf{r}_i, \tag{3.1.18}$$

where q_i and $\mathbf{r}_i$ are the charges and position vectors of all nuclei and electrons in the molecule. The matrix element, Eq. (3.1.15), also called the transition moment, is then

$$V_{21}(t) = \int \psi_2^* v(t) \psi_1 \, d\tau = \mathbf{E}_0 \left(\int \psi_2^* \mathbf{M} \psi_1 \, d\tau \right) \cos 2\pi f t$$
$$= \mathbf{E}_0 \cdot \mathbf{R}_{21} \cos 2\pi f t. \tag{3.1.19}$$

The quantity $\mathbf{R}_{21}$, thus defined, is the electric dipole moment matrix. The probability of finding the molecule in state ψ_2 with energy E_2 after time t is given by (see Steinfeld, 1974),

$$P_{21}(t) = \frac{4\pi^2}{h^2} I |R_{21}|^2 \frac{\sin^2\left(\frac{1}{2} W t\right)}{W^2}, \tag{3.1.20}$$

where $I = |E_0|^2$ and

$$W = 2\pi \left[f - \frac{1}{h} |E_1 - E_2| \right]. \tag{3.1.21}$$

The probability only reaches large values when W^2 is near zero, that is, when $f=|E_1-E_2|/h$. This is just the emission frequency given by Bohr's original theory. Usually the observed transition rate $T_{21} = P_{21}/t$ is an average over frequency,

$$T_{21} = \frac{1}{t}\int_{-\infty}^{+\infty} P_{21}(f,t)\,df = \frac{8\pi^3}{h^2}I|R_{21}|^2; \qquad (3.1.22)$$

T_{21} times t is the probability of finding the molecule in ψ_2 after time t when the radiation energy is distributed uniformly over a range of frequencies. The transition rate T_{21} is, therefore, proportional to the intensity of the radiation field and the square of the dipole moment matrix. The lifetime of the transition, or time per transition, is the reciprocal of T_{21}.

Two types of transition have rates described by Eq. (3.1.22), depending on the intensity and the square of the dipole moment matrix. One is absorption, where the molecule goes from a lower to an upper state with removal of a photon, and the other is stimulated emission, where the molecule changes from an upper to a lower state under influence of the radiation field with the generation of a photon. A third type of transition, spontaneous emission, is not described by Eq. (3.1.22). In this process, a molecule in an upper state decays to a lower state and emits a photon. The rate of spontaneous emission is independent of the radiation field. The relationships among the three types of transition are treated in the discussion of line strength in Section 3.5.

Electric dipole transitions, that is, transitions resulting from the interaction of the radiation field with the molecular electric dipole moment, are the most common transitions encountered in planetary spectra. There are other, generally weaker, types of transitions, however. The molecule might have an electric quadrupole moment or magnetic dipole moment, both of which interact with the radiation field. In particular, electric quadrupole transitions are important in diatomic molecules of identical atoms, since these have no dipole moment. Magnetic dipole transitions are rare in planetary spectra. Weak electric dipoles can also be induced in molecules during collisions with other molecules, and these are treated in Subsection 3.3.d.

3.2 Vibration and rotation of molecules

A molecule can be visualized as an aggregate of atoms bound together by a balance of mutually attractive and repulsive forces. Individual atoms vibrate with respect to one another while the molecule as a whole rotates about any spatial axis. Both types of motion occur simultaneously, and transitions between pairs of vibration–rotation states create the characteristic patterns of infrared spectra.

Sharing of valence electrons, that is, electrons in the outer shell among their orbitals, binds atoms into a molecular structure. Electronic binding forces create a three-dimensional potential energy distribution in which the atoms move. Although the electronic structure of a molecule may be in one of many states, thermal excitation encountered by the molecule in a planetary atmosphere almost always leaves the molecule in the ground electronic state. All motions of a given molecule occur, therefore, within the same ground state potential distribution. Because electrons travel much faster than nuclei, the assumption is usually made (called the adiabatic approximation) that the electronic energy depends only on the positions of the nuclei and not on their own velocities. The potential energy distribution is then a function of the internuclear distances alone. The potential energy has a minimum at locations where the attractive binding forces balance repulsive internuclear forces. A typical potential energy curve for a diatomic molecule (hydrogen chloride, HCl) is shown in Fig. 3.2.1. Without vibration the atoms are at the minimum and the molecule is in its equilibrium configuration with internuclear separation r_e.

For small displacements from equilibrium a molecule can be regarded as a group of atomic or nucleonic masses linked by springs; the atoms behave as a set of coupled harmonic oscillators. Each atom is in a part of the potential that is approximately parabolic and nearly obeys Hooke's law. The potential energy, V, is then

$$V = \tfrac{1}{2}k(q - q_e)^2, \tag{3.2.1}$$

where k is the force constant, q is a molecular coordinate such as an interatomic separation or an angle between atomic bonds, and q_e is the equilibrium value of the coordinate. The overall vibration of the molecule is a linear combination of several fundamental, or 'normal' modes of vibration, each having a well-defined vibration frequency. These frequencies can be estimated from a knowledge of the potential

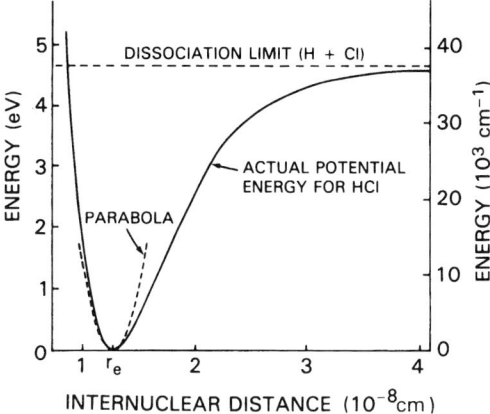

Fig. 3.2.1 Potential energy for hydrogen chloride (HCl). The curve is approximately parabolic near minimum and deviates from this harmonic dependence as the nuclear separation is increased or decreased.

curve shown in Fig. 3.2.1. Although this figure is for a diatomic molecule, the forces encountered in a polyatomic configuration are similar in shape and magnitude. From the change of slope of the curve with increasing displacement the force constant k is found to be of the order 10^{15} electron-volts per square meter, or about 10^{28} atomic mass units per square second. The frequency, $\omega = 2\pi f$ (in radians per second), of a simple harmonic oscillator with this force constant is

$$\omega_v = (k/m)^{\frac{1}{2}}, \quad (3.2.2)$$

where m is an effective mass for the oscillator, and is similar in magnitude to the atomic masses. Vibration wavenumbers of molecular modes normally fall in the range 50–5000 cm^{-1}.

As for rotation, if the molecule is regarded as a rigid aggregate of atoms rotating as a unit, its angular momentum is

$$L = I\omega_r, \quad (3.2.3)$$

where I is the moment of inertia about the rotation axis and ω_r is the rotation frequency. The Bohr quantum theory (Eq. 3.1.1) requires the angular momentum to be an integral multiple of $h/2\pi$. Therefore, the rotation frequency can take on values

$$\omega_r = n\frac{h}{2\pi I} \quad (n = 1, 2, 3, \ldots). \quad (3.2.4)$$

The moment of inertia, I, is of the order of the molecular mass multiplied by the square of the molecular radius, or about 10^{-46} kg m^2. At temperatures commonly found in planetary atmospheres molecules only reach the lower levels of n. Typical rotation wavenumbers range from 1 to 300 cm^{-1}, which is generally below the range of vibration wavenumbers. Rotational levels form a low-energy series beginning at zero, and also create a series within each vibrational energy level. The infrared spectrum of a molecule appears as a set of bands corresponding to changes in vibration states, with a fine structure of lines within each band due to changes in rotation states. Vibration and rotation of molecules are first discussed in Section 3.3 for diatomic and then in Section 3.4 for polyatomic molecules.

3.3 Diatomic molecules

The simplest type of molecule is the diatomic configuration, where two atoms, either of the same element (H_2, N_2) or of different elements (HCl, CO), join together. The

3.3 Diatomic molecules

derivation of spectral characteristics of diatomic molecules is an elementary, but nevertheless important, application of the Schrödinger equation because diatomic species are abundant in planetary atmospheres. Indeed, the atmospheres of the Earth, the giant planets, and Titan have diatomic molecules as their major constituents (H_2, N_2, O_2).

a. Vibration

In a diatomic molecule all vibratory motion takes place along the line joining both atoms. The vibration can be understood as oscillations of the two nuclei in one dimension, as shown in Fig. 3.3.1. Choosing this dimension to be the x-axis and designating the masses of the nuclei m_1 and m_2, the time-independent Schrödinger Eq. (3.1.13) for the motion of the nuclei is

$$-\frac{h^2}{8\pi^2}\left(\frac{1}{m_1}\frac{\partial^2}{\partial x_1^2} + \frac{1}{m_2}\frac{\partial^2}{\partial x_2^2}\right)\psi + V(x_2 - x_1)\psi = E\psi. \qquad (3.3.1)$$

The potential energy V is a function of only one parameter, $r = x_2 - x_1$, the distance between the nuclei. Near the equilibrium distance, r_e, when the molecule is at the bottom of the potential well (Fig. 3.2.1), the potential is approximately parabolic, and the vibration approximates that of a harmonic oscillator, just as if the nuclei were connected by a linear spring. If we write $q = r - r_e$ for the displacement from equilibrium, Eq. (3.3.1) becomes

$$\frac{h^2}{8\pi^2\mu}\frac{\partial^2\psi}{\partial q^2} + \left(E - \tfrac{1}{2}kq^2\right)\psi = 0, \qquad (3.3.2)$$

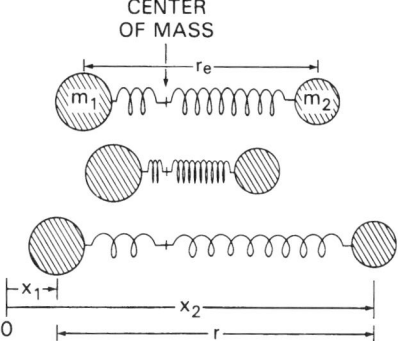

Fig. 3.3.1 Oscillation of a diatomic molecule. The vibration can be described by a single parameter, r, the nuclear separation. The separation r_e is the equilibrium distance, where the net force is zero.

where μ is the reduced mass,

$$\mu = \frac{m_1 m_2}{m_1 + m_2}. \tag{3.3.3}$$

The problem of two masses coupled by a springlike force has thus been transformed into the equivalent problem of a single mass, μ, moving on the same potential curve with its position coordinate equal to the displacement of the two atoms from their equilibrium distance.

Equation (3.3.2) can be solved for the vibration energy levels accessible to any diatomic molecule undergoing simple harmonic motion. The solution proceeds by trying wavefunctions of the form

$$\psi(q) = A e^{-\alpha q^2/2} H\left(\alpha^{\frac{1}{2}} q\right) \tag{3.3.4}$$

where

$$\alpha = \frac{2\pi}{h}(\mu k)^{\frac{1}{2}} \tag{3.3.5}$$

and A is a constant. Substituting Eq. (3.3.4) into Eq. (3.3.2) produces an equation for H, which turns out to be the Hermite differential equation, whose solution may be found in textbooks (e.g., Courant & Hilbert, 1931). The equation is satisfied by a set of functions $H_v(\alpha^{\frac{1}{2}} q)$, called Hermite polynomials ($v = 0, 1, 2, \ldots$), each of which gives a corresponding solution $\psi_v(q)$ to Eq. (3.3.2). The coefficient A is chosen to normalize ψ_v so that $\int |\psi_v|^2 \, dq = 1$. The energies $E(v)$, which are eigenvalues associated with the ψ_vs, have the discrete values

$$E(v) = \frac{h}{2\pi}\left(\frac{k}{\mu}\right)^{\frac{1}{2}}\left(v + \tfrac{1}{2}\right) = \frac{h}{2\pi}\omega\left(v + \tfrac{1}{2}\right). \tag{3.3.6}$$

The coefficient $\omega = 2\pi f$ corresponds to the classical oscillation frequency of the vibration in radians per second. The vibrational quantum number, v, can only be zero or a positive integer. The energy levels form a series,

$$\frac{1}{4\pi}h\omega, \frac{3}{4\pi}h\omega, \frac{5}{4\pi}h\omega, \ldots, \tag{3.3.7}$$

located at half-integer units of $h\omega/2\pi$. The zero-point energy at $v = 0$ is not zero but $h\omega/4\pi$. The energy levels of a simple harmonic oscillator are shown in Fig. 3.3.2. Table 3.3.1 lists the vibration frequencies of several diatomic molecules.

Table 3.3.1 *Vibration frequencies and rotation constants of several diatomic molecules*

Molecule	$\nu(\text{cm}^{-1})$	$B(\text{cm}^{-1})$
H_2	4395	60.8
N_2	2360	2.01
O_2	1580	1.45
CO	2170	1.93
SiO	1242	0.73
SO	1124	0.71
HCl	2990	10.6

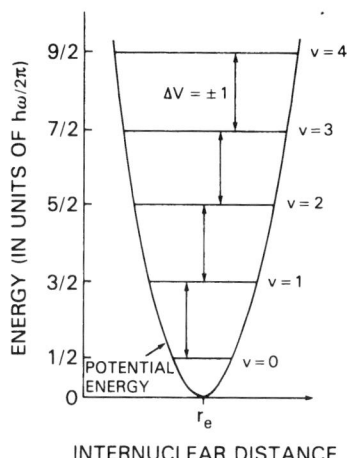

Fig. 3.3.2 Energy levels of a simple harmonic oscillator. The potential energy for harmonic motion is parabolic. The internuclear distance, r_e, is the equilibrium separation.

A molecule in one state can make a transition to another state by emitting or absorbing a photon. For a purely harmonic oscillator this can only occur in units of $\hbar\omega/2\pi$. This selection rule follows from the way the interaction energy couples two states of the molecule. The interaction energy is $\mathbf{E} \cdot \mathbf{M}$, where $\mathbf{E}$ is the electric field vector of the radiation and $\mathbf{M}$ is the electric dipole moment, $\mathbf{M} = \Sigma e_n \cdot \mathbf{X}_n$. The sum is over all charged particles in the system, each with charge e_n and coordinate vector $\mathbf{X}_n$. If the molecular axis is chosen to be the x-axis, then $M_y = M_z = 0$ and, to a good approximation, $M_x = M_0 + M_1 q$. This is an approximate expression for the dipole moment consisting of a constant, M_0, and a linear term, $M_1 q$, which represents the change in dipole moment with nuclear displacement q. The probability of a radiative transition between $\psi_{v'}$ and $\psi_{v''}$ is proportional to the electric dipole moment matrix [compare with Eq. (3.1.19)]. Only the x-components are nonzero in our simplified case, so the matrix elements are (setting $M = M_x$)

$$R_{v'v''} = \int \psi_{v'}^* M_x \psi_{v''} \, dq = M_0 \int \psi_{v'}^* \psi_{v''} \, dq + M_1 \int q \psi_{v'}^* \psi_{v''} \, dq. \quad (3.3.8)$$

A transition is allowed if $R_{v'v''} \neq 0$. The orthogonality of the terms $\psi_{v'}$ and $\psi_{v''}$ makes the first integral nonzero only if $v' = v''$; therefore, the first term with M_0 does not couple different states. However, the properties of the Hermite polynomials make the second term nonzero only if $v' = v'' \pm 1$. This selection rule, $\Delta v = \pm 1$, holds strictly for a harmonic oscillator when $M_x = M_0 + M_1 q$.

A diatomic molecule consisting of dissimilar atoms has a permanent dipole moment M_0 and a dipole moment derivative M_1. In general, the more dissimilar the electronegativities (abilities to attract a shared electron pair) of the atoms, the larger M_0 and M_1, and the stronger the observed transitions tend to be; for instance, HCl absorbs more radiation per molecule than does CO. Moreover, a homonuclear diatomic molecule has a completely symmetric structure and no dipole moment, that is, M_0 and M_1 are zero. Therefore, H_2, N_2, and O_2 do not normally exhibit infrared spectra due to electric dipole transitions. However, if the molecular number density is sufficiently high, they can undergo transitions resulting from collisionally induced dipoles. This is important in the H_2 atmospheres of the giant planets, as discussed in Section 3.3.d. Homonuclear diatomic molecules can also undergo much weaker (by 10^{-7}–10^{-9}) electric quadrupole transitions. The derivation of the vibrational selection rule for these transitions follows analogously to that for the electric dipole case. Substituting the quadrupole moment in Eq. (3.3.8) leads to the same result: $\Delta v = \pm 1$.

A simulated vibrational spectrum of carbon monoxide is shown in Fig. 3.3.3. The strong fundamental transition at $v_0 = 2143$ cm^{-1} corresponds to a change in vibrational quantum number from 1 to 0. If carbon monoxide were a perfect harmonic oscillator, all transitions of the type $v = 2$–1, 3–2, etc. would fall exactly at that position, rather than being offset slightly as they appear in measured spectra. In addition, the overtone transitions at $2v_0$ and $3v_0$ corresponding to $\Delta v = \pm 2$ and ± 3, would not be allowed by the harmonic selection rule.

Because the transition at v_0 dominates the spectrum, the harmonic oscillator seems to be a valid approximation for diatomic molecules, but it is clear from Fig. 3.3.3 that anharmonic effects will have to be included if we wish to reproduce observed spectra in detail. In all diatomic molecules the deviation of their potential energy curve from a strictly parabolic shape causes the departure from harmonicity. The curve for HCl in Fig. 3.2.1 is only approximately parabolic near minimum and deviates more so at large nuclear displacements. Indeed, a parabola cannot describe the actual curve; as the nuclear separation is increased indefinitely the molecule

3.3 Diatomic molecules

splits into two isolated atoms and the potential energy approaches a dissociation limit, whereas the parabola goes to infinity.

When the potential energy function, V, is generalized, the vibration energies given in Eq. (3.3.6) become

$$E(v) = h\omega/2\pi \left[\left(v + \tfrac{1}{2}\right) - a\left(v + \tfrac{1}{2}\right)^2 + b\left(v + \tfrac{1}{2}\right)^3 + \cdots \right], \quad (3.3.9)$$

where successive terms in the series diminish rapidly in magnitude. The coefficient a is written with a minus sign because the second-order term generally subtracts from the energy in real molecules. Only a few terms are needed in Eq. (3.3.9) to describe observed spectra to within experimental accuracies. The coefficients in Eq. (3.3.9) are commonly found by measuring the precise frequencies of spectral lines and then deriving the energy levels.

The energy levels given by Eq. (3.3.9) are not equally spaced, as in the harmonic case. They are separated by intervals slightly less than $h\omega/2\pi$, which become smaller as v increases. In the CO spectrum shown in Fig. 3.3.3, the weaker lines near the fundamental at $v_0 = 2143$ cm^{-1} form a series toward lower wavenumbers.

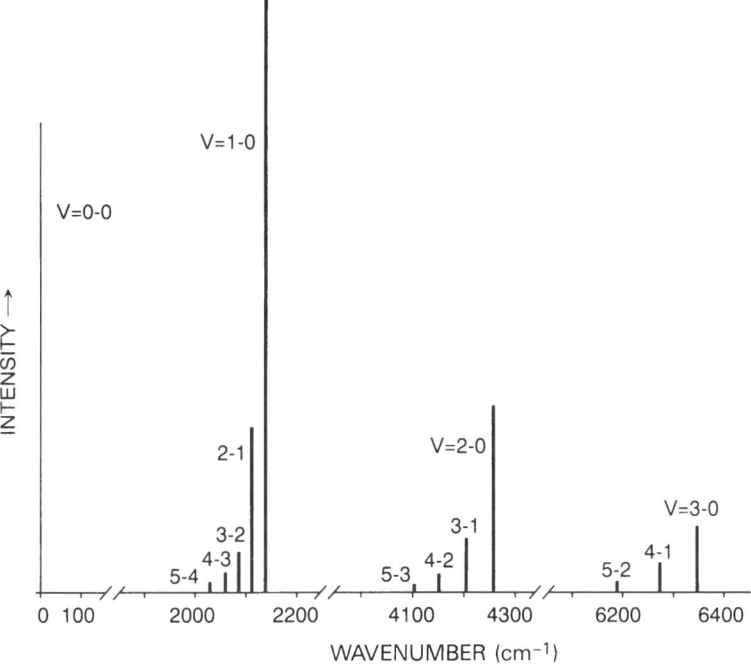

Fig. 3.3.3 Simulation of the vibrational spectrum of carbon monoxide (CO). Only the positions of the vibration frequencies are shown (without rotational structure). The wavenumber scale has been segmented to more clearly show the effects of anharmonicity. Purely rotational transitions with $\Delta v = 0$ all occur at zero wavenumber, while $\Delta v = 1, 2, 3, \ldots$ are split by anharmonic motion. Intensities as shown are not intended to be accurate.

The transitions $v = 1\text{–}0, 2\text{–}1, 3\text{–}2, \ldots$ correspond to progressively smaller changes in energy. This also occurs in the overtones ($v = 2\text{–}0, 3\text{–}1, 4\text{–}2, \ldots$ and $v = 3\text{–}0, 4\text{–}1, 5\text{–}2, \ldots$).

The overtones, which were strictly forbidden in the harmonic oscillator, are allowed in the anharmonic case because small transition probabilities now exist between any two states. The selection rule on Δv becomes relaxed to include

$$\Delta v = \pm 1, \pm 2, \pm 3, \ldots. \tag{3.3.10}$$

As can be seen from the relative transition intensities in Fig. 3.3.3, the fundamental $\Delta v = \pm 1$ is far more intense than the overtones, indicating that the vibration of the molecule is close to harmonic.

b. Rotation

Rotation of a diatomic molecule can be treated approximately by regarding the molecule as two masses m_1 and m_2 at a fixed separation r_e. In this model the molecule rotates about an axis perpendicular to the line joining the nuclei. If r_1 and r_2 are the distances of m_1 and m_2 from the center of mass, and ω is the rotation frequency, the classical expression for the energy is

$$E = \frac{L^2}{2I}, \tag{3.3.11}$$

where $L = I\omega$ is the angular momentum and $I = m_1 r_1^2 + m_2 r_2^2$ is the moment of inertia about the rotation axis. The moment of inertia can also be expressed in terms of $r_e = r_1 + r_2$ and the reduced mass, μ,

$$I = \frac{m_1 m_2}{m_1 + m_2} r_e^2 = \mu r_e^2. \tag{3.3.12}$$

Although in classical theory E and ω can assume any value, in quantum mechanics the rigid rotator can only exist in discrete energy states. To find these we use the time-independent Schrödinger equation (3.1.13), which we now write for the nuclei and without a potential energy term (the masses are assumed to have a fixed separation),

$$\frac{h^2}{8\pi^2}\left(\frac{1}{m_1}\nabla_1^2 + \frac{1}{m_2}\nabla_2^2\right)\psi + E\psi = 0. \tag{3.3.13}$$

Similar to the treatment of the harmonic oscillator, this equation of motion can be transformed into one for a mass μ moving at radius $r_e = (x^2 + y^2 + z^2)^{\frac{1}{2}}$ about the

3.3 Diatomic molecules

origin. The orientation of the molecule can be specified in polar coordinates θ and ϕ. In these coordinates the Schrödinger equation is

$$\frac{1}{\sin\theta}\frac{\partial}{\partial\theta}\left(\sin\theta\frac{\partial\psi}{\partial\theta}\right) + \frac{1}{\sin^2\theta}\frac{\partial^2\psi}{\partial\phi^2} + \frac{8\pi^2\mu r_e^2}{h^2}E\psi = 0. \quad (3.3.14)$$

The solutions to this equation are

$$\psi_{JM} = \left[\frac{(2J+1)(J-|M|)!}{4\pi(J+|M|)!}\right]^{\frac{1}{2}} e^{im\phi} P_J^{|M|}(\cos\theta), \quad (3.3.15)$$

where M and J are integers obeying $|M| \leq J$; J is always positive or zero. The $P_J^{|M|}$ are associated Legendre polynomials. The corresponding energies of the rotating molecule are

$$E = \frac{h^2}{8\pi^2\mu r_e^2} J(J+1) \quad (J = 0, 1, 2, \ldots). \quad (3.3.16)$$

Comparison with Eq. (3.3.11) shows that the classical angular momentum L increases approximately as J. J can be identified as the quantum number for the angular momentum while M can be regarded as the projection of J on the polar axis. Thus, the angular momentum is

$$L = \frac{h}{2\pi}[J(J+1)]^{\frac{1}{2}} \sim \frac{h}{2\pi}J, \quad (3.3.17)$$

which is similar to the axiom of the Bohr quantum theory, Eq. (3.1.1). It differs from the Bohr version in that $[J(J+1)]^{\frac{1}{2}}$ is not an integer, and J and L can be zero. The classical rotation frequency ω is related to the quantum number J by

$$\omega = \frac{L}{\mu r_e^2} = \frac{h[J(J+1)]^{\frac{1}{2}}}{2\pi\mu r_e^2}, \quad (3.3.18)$$

demonstrating a nearly linear dependence of ω on J. In units of cm^{-1} the energy expression, Eq. (3.3.16), is

$$F(J) = \frac{E}{hc} = BJ(J+1), \quad (3.3.19)$$

where B is called the rotational constant,

$$B = \frac{h}{8\pi^2 c \mu r_e^2} = \frac{h}{8\pi^2 c I}. \tag{3.3.20}$$

The selection rule for a transition between one state J' and another J'' can be derived from the wavefunctions given by Eq. (3.3.15). Again the interaction energy between two states is proportional to the dipole moment **M**, which is constant for a rigid molecule. The selection rule on J is found from the electric dipole moment matrix [see Eq. (3.1.19)],

$$\mathbf{R}_{J'M'J''M''} = \int \psi^*_{J'M'} \mathbf{M} \psi_{J''M''} \, d\tau. \tag{3.3.21}$$

Relationships among the associated Legendre functions contained in the ψ_{JM} terms [see Eq. (3.3.15)] cause the integral in Eq. (3.3.21) to be nonzero only if $J' = J'' \pm 1$. This selection rule, $\Delta J = \pm 1$, means that the angular momentum can only change in units of $h/2\pi$.

The selection rule for electric quadrupole transitions, the strongest allowed for homonuclear diatomic molecules, can be derived using a similar, though more complex, approach. The result is $\Delta J = \pm 2$; the odd- and even-J levels do not mix.

Applying the selection rule $\Delta J = 1$ for purely rotational transitions to the rotational energy [Eq. (3.3.19)], the allowed transition wavenumbers are

$$\nu = F(J') - F(J'') = F(J+1) - F(J) = 2B(J+1), \tag{3.3.22}$$

where J refers to the lower state. This series begins with a transition at $2B$ and continues at equal intervals; it is the spectrum of a rigid rotator. Figure 3.3.4 shows the pattern formed by the purely rotational transitions in carbon monoxide. The spectrum appears as nearly equally spaced lines increasing in intensity to a maximum and then decreasing gradually to zero as J becomes large. The distribution of intensities in rotational spectra are discussed in Section 3.5.

The line spacing, $2B$ for a particular diatomic molecule, can be estimated from Eq. (3.3.18) by substituting the atomic masses and separations. In general, heavier molecules will have smaller rotational B constants and smaller line spacings. Table 3.3.1 lists the rotational constants for several diatomic molecules.

Although the rigid rotator model predicts a line structure that agrees reasonably well with observed spectra, a detailed comparison shows that the resemblance is not perfect. In Fig. 3.3.4 the lines appear equally spaced, but in reality decrease slightly in spacing as J increases. This is not surprising, since the molecule is not perfectly rigid, but stretches slightly in response to centrifugal forces. The rotational energy

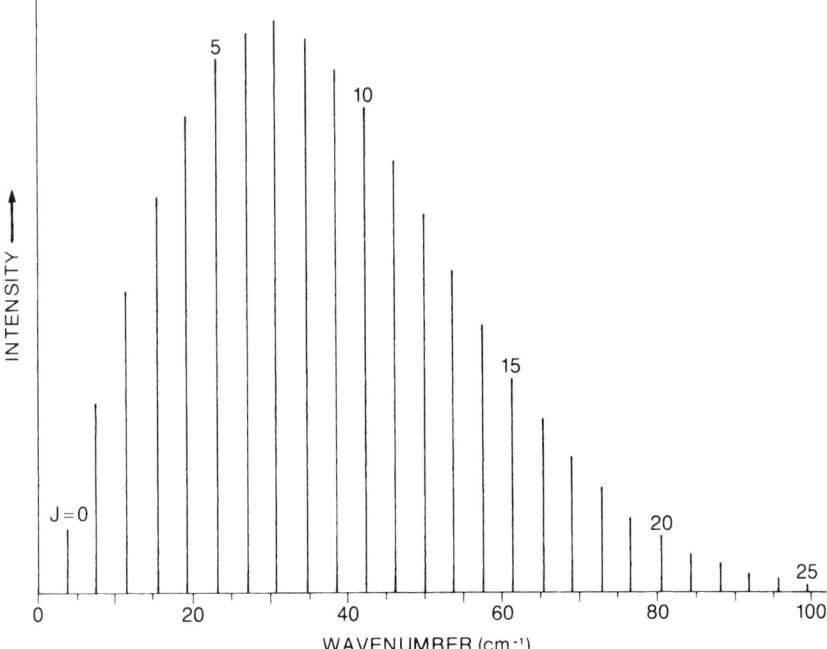

Fig. 3.3.4 Simulation of the purely rotational spectrum of carbon monoxide. The relative line intensities are approximately correct for a temperature of 300 K. The rotational quantum number J is for the lower state of the transition.

levels corrected for centrifugal effects are given by

$$F(J) = \frac{E}{hc} = BJ(J+1) - DJ^2(J+1)^2. \tag{3.3.23}$$

The coefficient D is called the centrifugal distortion constant. The negative sign preceding D is conventionally written explicitly to yield positive values of D. The effect of D is to lower each energy level slightly as compared to that of a rigid rotator. A single correction term is usually sufficient to describe rotational structure observed in moderate resolution spectra, but at high resolution a higher order term in $J^4(J+1)^4$ is often needed.

c. Vibration–rotation interaction

So far we have treated the vibration and rotation of a diatomic molecule as two separate motions. Vibration and rotation occur simultaneously, of course, and transitions between energy levels can involve changes in both vibration and rotation quantum numbers, v and J. The vibration–rotation combination gives rise to observed spectra with not only the expected vibration transitions, but also rotational fine structure around each of the vibrational wavenumbers. Calculation of vibration–rotation structure is the final step in predicting the detailed positions of lines in observed spectra.

As a first approximation the rotational energy simply adds to the vibrational energy. With anharmonicity and centrifugal distortion, the energy in wavenumbers is

$$F(v, J) = \frac{E}{hc} = v_0\left(v + \tfrac{1}{2}\right) - v_0 a\left(v + \tfrac{1}{2}\right)^2 + BJ(J+1) - DJ^2(J+1)^2, \tag{3.3.24}$$

where $v_0 = \omega/2\pi c$ is the classical vibration wavenumber. To be completely valid, however, the energy expression for combined vibration and rotation also must include interaction between both types of motion. To see why, consider the consequences to the rotational constant B as the vibrational state is changed. As discussed earlier, the frequency of vibration is typically much higher than that of rotation, so that the B value results from an average molecular configuration and moment of inertia, as the dimensions change during vibration. B is therefore a function of vibrational state, and will be progressively smaller as the vibrational quantum number v increases. $B(v)$ can be approximated as a linear function of $(v + \tfrac{1}{2})$,

$$B(v) = B_e - \alpha\left(v + \tfrac{1}{2}\right). \tag{3.3.25}$$

B_e is the equilibrium value of the rotational constant corresponding to the equilibrium internuclear separation r_e. The coefficient α is small compared to B_e because the displacement during vibration is generally a small fraction of r_e. The negative sign is conventionally used to make α positive; $B(v)$ is never equal to B_e, even in the ground vibrational state $v = 0$.

A similar dependence on vibrational state applies to the centrifugal distortion constant D. As the molecule rotates the 'stretching' distortion of the internuclear separation is characterized by D, and this effect will change from one vibrational state to another. This vibrational dependence can again be approximated as a linear function of $(v + \tfrac{1}{2})$

$$D(v) = D_e + \beta\left(v + \tfrac{1}{2}\right). \tag{3.3.26}$$

Again, β is much smaller than the equilibrium distortion constant D_e.

By introducing $B(v)$ and $D(v)$ into Eq. (3.3.24) we obtain the vibration–rotation energy, including interactions between vibration and rotation

$$\begin{aligned}F(v, J) = &\ v_0\left(v + \tfrac{1}{2}\right) - v_0 a\left(v + \tfrac{1}{2}\right)^2 \\&+ B_e J(J+1) - \alpha\left(v + \tfrac{1}{2}\right)J(J+1) \\&- D_e J^2(J+1)^2 - \beta\left(v + \tfrac{1}{2}\right)J^2(J+1)^2.\end{aligned} \tag{3.3.27}$$

This expression combines the overall dependence of the energy levels on both vibration and rotation. The pattern of energy levels can be thought of as a ladder of widely spaced vibration energies, identified by v, each split into a more closely spaced ladder of rotation energies, identified by J.

As can be seen from the last two terms in Eq. (3.3.24), the rotational energy level structure in each vibrational state resembles that for the vibrationless nonrigid rotator in Eq. (3.3.23). The rotational ladder in each vibration state begins (for $J = 0$) at the energy given by the first two, purely vibrational terms in Eq. (3.3.24). The rotational levels within each vibrational level follow a pattern similar to that described by Eq. (3.3.23), except that the rotational coefficients are now assigned their vibrationally dependent values $B(v)$ and $D(v)$. This overall vibration–rotation dependence is embodied in Eq. (3.3.27).

A transition between two vibration–rotation energy levels will be accompanied by emission or absorption of a photon at wavenumber

$$v = \frac{E'}{hc} - \frac{E''}{hc} = F(v', J') - F(v'', J''), \qquad (3.3.28)$$

where $F(v, J)$ is given by Eq. (3.3.27). The selection rules $\Delta J = \pm 1$ and $\Delta v = 0, \pm 1, \pm 2, \ldots$ apply in the case of combined vibration–rotation, where $\Delta v = 0$ is for pure-rotation transitions. Each transition appears in the spectrum as a band, or set of lines corresponding to the many rotational transitions that may accompany a vibrational transition. The separations between rotational lines are generally much smaller than the separations between vibration bands. Each band is centered at its vibrational wavenumber $v_i \sim v_0(v' - v'')$, and is composed of two series of rotational lines. These are described approximately by

$$v = v_i + 2B(J + 1) \quad \text{for } \Delta J = +1 \qquad (3.3.29)$$

and

$$v = v_i - 2B(J + 1) \quad \text{for } \Delta J = -1. \qquad (3.3.30)$$

The series produced by $\Delta J = +1$ transitions extends to higher wavenumbers and is referred to as the R-branch. The $\Delta J = -1$ transitions form a series toward lower wavenumbers and is called the P-branch. Lines in the P and R branches are separated by approximately $2B$. Figure 3.3.5 shows the $v = 1–0$ spectrum of carbon monoxide. The variation in strength among the rotational lines will be discussed in Section 3.5.

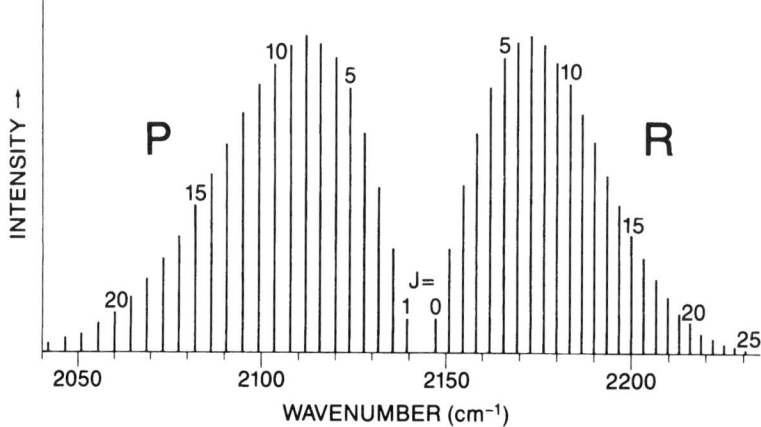

Fig. 3.3.5 Simulation of the $v = 1$–0 band of carbon monoxide. The band center is at 2143 cm^{-1}, and the P- and R-branch rotational series extend toward lower and higher wavenumbers. The rotational quantum number shown (J) is for the lower state of the transition. The line intensities are approximately correct for 300 K.

d. Collision-induced transitions

Homonuclear diatomic molecules (H_2, N_2, O_2) have no permanent dipole moment, and, therefore, no electric dipole transitions. However, at high pressure and long path length, electric dipole absorption is observed. This absorption results from a short-time collisional interaction between molecules. Radiative transitions among rotational, vibrational, and translational states of colliding pairs of molecules can take place, even though they are not allowed in the isolated molecules. This process is especially important in the atmospheres of the giant planets, where collision-induced absorption by molecular hydrogen dominates the far infrared spectrum (Trafton, 1966). This spectrum has been studied in the laboratory by Birnbaum (1978) and Bachet et al. (1983), and has been characterized by Birnbaum & Cohen (1976).

During a collision a transient dipole moment arises as the electron distribution is distorted by long-range forces or overlapping charge densities. The magnitude of the dipole moment is extremely small: 10^{-3}–10^{-2} in units of ea_0, the product of the charge of the electron and the Bohr radius of the hydrogen atom. The shape and intensity of observed spectra are determined by the induced dipole moment $\mu(\mathbf{R}, \mathbf{r}_1, \mathbf{r}_2)$, and the interaction potential $V(\mathbf{R}, \mathbf{r}_1, \mathbf{r}_2)$ where $\mathbf{R}$ is the intermolecular separation and $\mathbf{r}_i$ are the vibrational coordinates of molecule i. Both μ and V depend on the orientations of the molecules. Collision-induced transitions arise from free pairs and from bound pairs (dimers). The induced dipole moment can arise in collisions between two molecules (H_2–H_2, H_2–N_2, H_2–CH_4), a molecule and an atom (H_2–He), or between two dissimilar atoms (He–Ar).

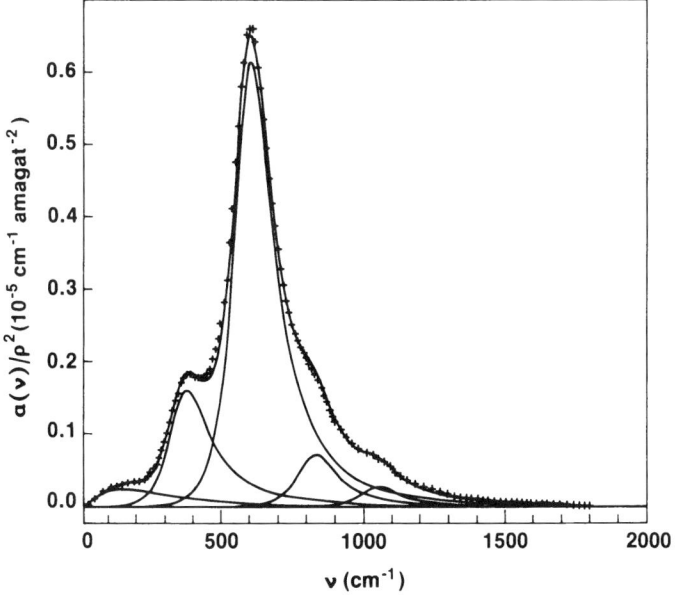

Fig. 3.3.6 The collision-induced spectrum of H_2 at a temperature of 195 K. Experimental data are shown with + marks. The ordinate is in units of absorption strength per density squared. Computed curves are shown for the translational spectrum, and for the pure-rotation $J = 0 \to 2$, $1 \to 3$, $2 \to 4$, and $3 \to 5$ lines (after Bachet et al., 1983).

Collision-induced absorption from free pairs of molecules appear as broad lines or bands located at the wavenumbers of the pure-rotation or vibration–rotation transitions in the participating individual molecules. Figure 3.3.6 shows the spectrum for H_2–H_2 collisions (Bachet et al., 1983) [see also Courtin (1988)]. In the far infrared (below 200 cm^{-1}) a weak translational band is also present. In H_2 the prominent features in planetary atmospheres occur at the pure-rotation $J = 0 \to 2$ and $1 \to 3$ transitions located at 354 and 587 cm^{-1}. The widths of collision-induced features are extremely large, about 100 cm^{-1} or more, because the time during the collision in which the partners are interacting is very short ($\sim 10^{-12}$ seconds or less). The width of a spectral line is related to the reciprocal of the collision duration.

The details of interaction between two molecules or an atom and molecule are complex, and depend on the minimum separation of the partners during a collision. If the separation is small enough to allow temporarily some van der Waals binding of the partners, a longer-lived molecular complex, a dimer, is formed. In contrast to the free pair case, the partners in a dimer arrange in a quasi-stable geometry and the complex behaves similarly to a large molecule. The interaction time increases giving rise to much narrower (~ 1 cm^{-1}) lines. These may appear near the centers of the free pair induced-dipole lines. Such lines in

H$_2$–H$_2$ have been observed on Jupiter and Saturn (McKellar, 1984; Frommhold *et al.*, 1984).

The absorption in the collision-induced features, both from free pairs and dimers, grows in proportion to both the number of molecules per volume element and the number of collision partners. Therefore, collision-induced absorption depends on the product of the densities of the partners. When the partners are the same (H$_2$–H$_2$, N$_2$–N$_2$), the absorption depends on the square of the density. The absorption strength also increases at lower gas temperatures, since this corresponds to higher densities at a given pressure, and because the molecules are thermally distributed over fewer energy levels. The theory and spectroscopy of collision-induced absorption have been reviewed by Welsh (1972) and by Birnbaum (1985).

3.4 Polyatomic molecules

Molecules with three or more atoms exhibit spectra of greater complexity than those with only two, but much of the overall structure of polyatomic spectra can be understood by generalizing the basic principles developed for the diatomic case. For example, the polyatomic molecule can be viewed in first approximation as a set of masses linked by springs undergoing simple harmonic oscillations while its rotation can be approximated by that of a rigid aggregate of atoms, just as with the diatomic molecules. Moreover, anharmonic effects and centrifugal distortion arise in polyatomic molecules for similar reasons as in diatomic molecules. Following the approach of Section 3.3 we begin with the classical picture of vibration and rotation and then introduce quantum mechanics to derive the true appearance of molecular spectra.

a. Vibration

Whereas the diatomic molecule has only one fundamental oscillation frequency, the polyatomic molecule generally vibrates in several modes, each with a different frequency. The number of possible vibration modes depends on the number of nuclei. Each of the N nuclei can move in the x-, y-, and z-directions, giving a total of $3N$ degrees of freedom. Six of those degrees, however, correspond to the motion of the molecule as a whole. These are, for instance, the three coordinates of its center of mass and the three angles defining its orientation in space. Therefore, the total number of degrees of freedom available for vibration is $3N - 6$. For a linear molecule only two angles are needed to specify its orientation, so the number of vibrational degrees of freedom is $3N - 5$ in that case. This rule also applies to diatomic molecules, which are linear by definition, and have only $3 \times 2 - 5 = 1$ degree of freedom.

3.4 Polyatomic molecules

As an example consider ammonia (NH_3), which is a nonlinear molecule consisting of four atoms. The relative positions of the three hydrogen nuclei are specified by the three distances between H–H pairs, and the position of the nitrogen nucleus is given by the three N–H distances. This permits a total of $3 \times 4 - 6 = 6$ vibrational degrees of freedom. An example of a simple linear molecule is hydrogen cyanide (HCN) with three atoms. Here the relative positions of the three nuclei during vibration are given by the H–C and C–N separations, plus the projections of the H–C–N angle on two fixed orthogonal planes intersecting on the H–C axis. The result is four vibrational degrees of freedom; as discussed below, two of these have the same frequency.

The number of degrees of freedom equals the number of normal modes of vibration. The normal modes, also called fundamental modes, are a set of harmonic motions, each independent of the others and each having a distinct frequency. It is possible for two or more of the frequencies to be identical, and the corresponding modes are said to be degenerate. However, the total number of modes in the individual degenerate states are counted separately and still total $3N - 6$ for nonlinear and $3N - 5$ for linear molecules. A set of coordinates can be defined, each of which gives the displacement in one of the normal modes of vibration. The normal coordinates can be expressed as combinations of the x-, y-, and z-coordinates of the individual nuclei.

Consider now carbon dioxide (CO_2), a linear triatomic molecule. Figure 3.4.1 shows the $3 \times 3 - 5 = 4$ normal modes of vibration. The directions and relative amplitudes of displacements in the four modes are indicated by arrows. In the first mode the carbon nucleus is stationary while the two oxygen nuclei oscillate symmetrically. Because this symmetric motion does not generate a dipole moment, infrared transitions to this state are forbidden. In the second and third modes the molecule bends in two, orthogonal planes. The second and third modes are, therefore, degenerate. The last mode has the carbon nucleus moving alternately toward one and then the other of the two oxygen nuclei. The normal modes are conventionally labeled ν_1, ν_2, and ν_3, respectively, with ν_2 referring to the combination of the two degenerate bending modes. The water molecule (H_2O) provides an example of a nonlinear triatomic molecule. Figure 3.4.2 shows the $3 \times 3 - 6 = 3$ normal modes of vibration. No analog exists here to the degenerate modes in CO_2. Again, the modes are conventionally labeled ν_1, ν_2, and ν_3.

The important characteristic of a normal mode is the harmonic motion of each nucleus; the overall motion of the molecule in each mode is then also harmonic. The classical picture of the molecule is, therefore, a set of independent harmonic oscillators. Each oscillator has its own frequency, ω_i, which, by analogy with Eq. (3.2.2), can be related to an effective mass, m_i, and spring constant, k_i, by $\omega_i = (k_i/m_i)^{\frac{1}{2}}$. The effective mass is related to the nuclear masses, and the spring constant arises from the strengths of the interatomic bonds. The time dependence of the normal

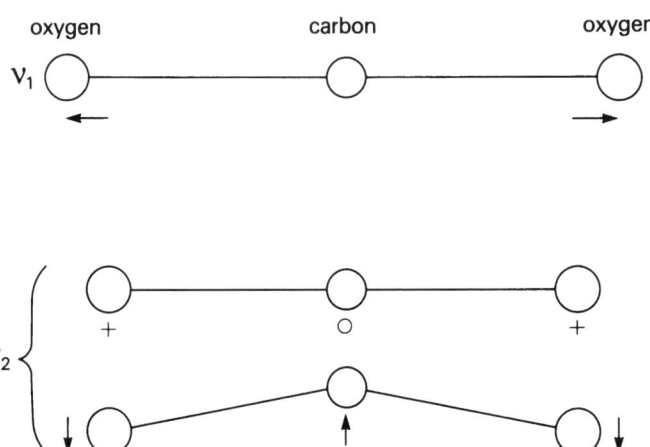

Fig. 3.4.1 The normal modes of vibration of carbon dioxide (CO_2). This molecule has two stretching modes, v_1 and v_3, and two degenerate bending modes, v_2. The upper component shown for v_2 has motion into (+) and out of (○) the plane of the paper.

coordinate q_i for a particular mode can be written

$$q_i = q_i^0 \cos(\omega_i t + \delta_i), \quad (3.4.1)$$

where q_i^0 is the maximum displacement from the equilibrium position and δ_i is a phase.

Normal molecular modes of vibration are found classically by treating the equations of motions of all nuclei as a set of linear differential equations. When expressed in normal coordinates the equations of motion are decoupled, and each can be written in terms of only one coordinate. However, this approach is limited to strictly harmonic cases. In a real molecule these oscillators are somewhat anharmonic. Then the normal oscillators are coupled, giving rise to oscillations with combinations of the fundamental vibration frequencies.

In degenerate modes only the displacements and phases of the motions differ. Therefore, when two or more of these modes are excited, the nuclei move together in a simple normal mode. For this reason degenerate modes are conventionally referred to as a single mode and given a single designation, such as with v_2 in CO_2 in the above example. Because the relative phases among the component modes

3.4 Polyatomic molecules

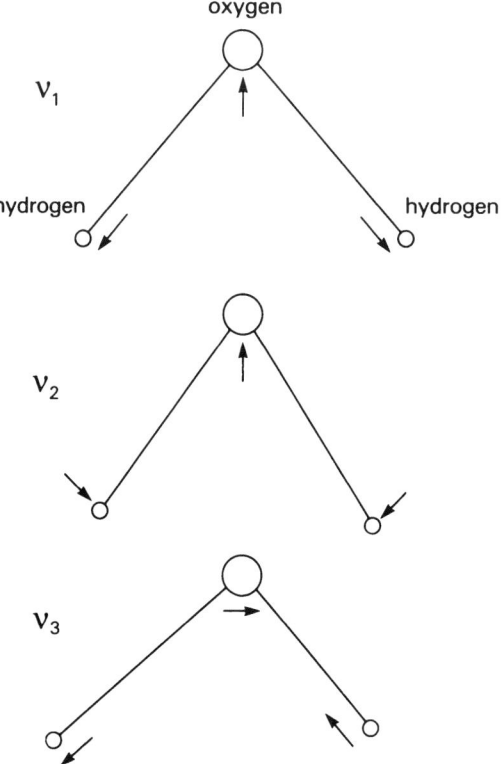

Fig. 3.4.2 The normal modes of vibration of water vapor (H_2O). Since H_2O is nonlinear there are only three modes, and none is degenerate.

can vary continuously, a combined degenerate mode, such as in CO_2, describes an infinite variety of motions. A special case exists when the two component modes are in phase and the combined motion is a bending in a plane other than the two shown in Fig. 3.4.1. In another interesting case the two component modes differ in phase by 90°. The motion is then an apparent rotation of the bent molecule, with each nucleus describing a circle or ellipse. In this type of motion vibration actually produces angular momentum and must be given an angular momentum quantum number. Nonlinear molecules and molecules with more than three nuclei can also have vibrational angular momentum as the result of phased degenerate vibrations.

Observed vibrational spectra can only be described by introducing quantum mechanics. To do this, the normal modes are regarded as a set of harmonic oscillators, each obeying a wave equation of the form

$$\frac{h}{8\pi^2 \mu_i} \frac{d^2 \psi_i}{dq_i^2} + \left(E_i - \tfrac{1}{2} k_i q_i^2\right) \psi_i = 0. \qquad (3.4.2)$$

Here μ_i and k_i are the effective reduced mass and spring constants for the modes.

84   *Interaction of radiation with matter*

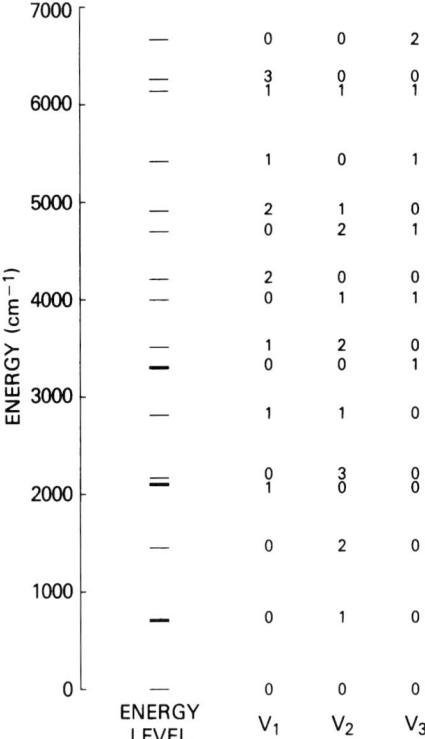

Fig. 3.4.3 Energy levels of hydrogen cyanide (HCN). The vibrational quantum numbers for the three normal modes are v_1, v_2, and v_3. Only levels below 7000 cm^{-1} with three or fewer vibrational quanta are shown. Heavy lines mark the fundamental energy levels. Energies are shown as differences from the lowest level $v_1 = v_2 = v_3 = 0$.

This equation is similar to Eq. (3.3.2) for the diatomic molecule. The quantum mechanical energy levels of each mode are given by eigenvalues of Eq. (3.4.2),

$$E_i = \frac{h}{2\pi}\omega_i\left(v_i + \tfrac{1}{2}\right) \quad (v_i = 0, 1, 2, \ldots). \tag{3.4.3}$$

Here ω_i is the oscillation frequency of mode i and v_i is the vibrational quantum number. The total vibrational energy of the molecule is the sum of the energies of the individual normal modes,

$$E(v_1, v_2, \ldots) = \frac{h}{2\pi}\sum_i \omega_i\left(v_i + \tfrac{1}{2}\right). \tag{3.4.4}$$

Even a simple triatomic molecule, such as HCN, with only three normal modes (two degrees of freedom are degenerate) can take on a complex variety of energies as shown in Fig. 3.4.3. Not only can each normal mode be excited into a series of

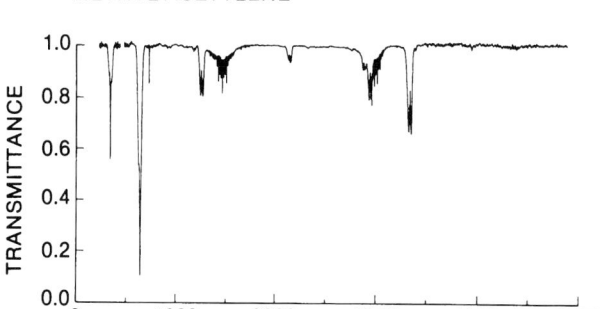

Fig. 3.4.4 Spectrum of methyl acetylene (C_3H_4) showing the structure of vibrational bands. The sample was placed in a 3 cm long cell at 47 mbar. The spectral resolution is 2 cm^{-1}. Detailed shapes of the vibrational bands are due to unresolved rotational lines.

levels when the other two normal modes are in the $v_i = 0$ level, but all three modes can assume $v_i \neq 0$ simultaneously. The lowest level, when all $v_i = 0$, is the zero-point vibrational energy, similar to the diatomic case. The zero-point energy is generally ignored because only energy differences are of interest.

The vibrational selection rule for the harmonic oscillator, $\Delta v_i = \pm 1$, applies to polyatomic molecules just as it did to diatomic molecules. Vibrational energy can, therefore, change in units of $h\omega_i/2\pi$. Transitions in which one of the three normal modes of energy changes by $\Delta v_i = +1$ (for example: $v_1 = 0 \to 1$, $v_2 = v_3 = 0$; or $v_1 = 1$, $v_2 = 3$, $v_3 = 2 \to 3$) result from absorption of a photon having one of three fundamental frequencies of the molecule. In the actual case, anharmonicities also allow transitions with $\Delta v_i = \pm 2, \pm 3, \ldots$ so that, for example, weak absorption also occurs at $2\omega_i$, $3\omega_i$, etc. and at $\omega_i + \omega_j$, $2\omega_i + \omega_j$, etc. These weaker vibrational transitions often play major roles in planetary spectroscopy.

For larger polyatomic molecules the vibrational motions can become very complicated, with dozens of normal modes contributing to the observed spectra. However, the basic principles outlined here apply, at least as a first order approximation, to the motion of molecules as large as methyl acetylene (C_3H_4), shown in Fig. 3.4.4, and propane (C_3H_8), shown in Fig. 3.4.5. For some large molecules with simple structures, such as ring-shaped benzene (C_6H_6), the observed spectra are often quite simple (see, for example, Sadtler, 1972).

Small changes in the atomic masses in a molecule can cause great changes in the appearance of spectra. These are the effects of isotopic substitution, such as deuterium for hydrogen or ^{13}carbon for ^{12}carbon. Vibrations that involve a motion of the substituted atom will change frequency, and changes in molecular symmetry may alter the appearance of spectra drastically or cause the emergence of new lines

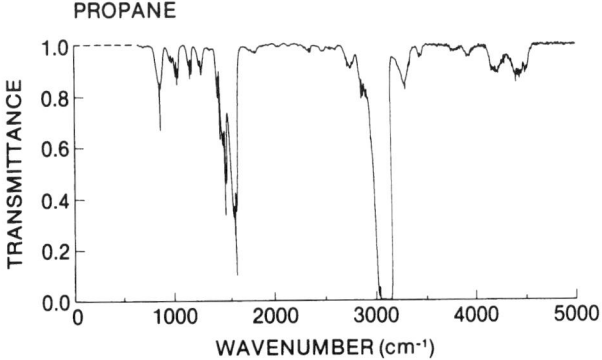

Fig. 3.4.5 Spectrum of propane (C_3H_8) showing the structure of vibrational bands. The sample was placed in a 3 cm long cell at 400 mbar. The spectral resolution is 2 cm^{-1}. Detailed shapes of the vibrational bands are due to unresolved rotational lines.

where none were originally. Changes in spectra among the isotopic variants of a molecule provide a powerful tool for identifying isotopic constituents.

b. Rotation

As stated earlier, in considering rotation of polyatomic molecules we can follow the approximation used for diatomic molecules, that the rotation can be treated as independent of vibration. Much of the structure of rotational spectra of polyatomic molecules can be understood by using as a model an aggregate of nuclear masses connected rigidly at their equilibrium positions. In contrast to diatomic molecules where only one axis of rotation is required, in polyatomic molecules we must consider rotation about any axis.

The moment of inertia of the rigid molecule about axis α is

$$I_\alpha = \sum_i m_i r_i^2, \qquad (3.4.5)$$

where m_i is the mass and r_i the distance of nucleus i from the axis. If the axis α is an arbitrary axis through the center of mass of the molecule and the direction of α is varied, we find three mutually orthogonal directions in which the moment of inertia, I, has a local extremum. These principal axes form a natural internal coordinate system for the molecule. The moments about these axes are the principal moments of inertia.

If the molecule has an axis of symmetry, it will always be a principal axis. An axis of symmetry is one about which the molecule can be rotated by some rational fraction of 360° (180°, 120°, 90°, etc.) and thereby returned to its original configuration. H_2O has such an axis for 180° rotations and NH_3 has one for 120°. Similarly, a reflection plane of symmetry, one in which the molecule can be reflected and thereby returned to itself, will always contain two principal axes.

A molecule for which the three principal moments of inertia are different is called an asymmetric rotor or an asymmetric top. Examples of asymmetric rotors are water (H_2O), ozone (O_3), and propane (C_3H_8). If two of the principal moments are identical, the molecule is a symmetric rotor. Examples of symmetric rotors are ammonia (NH_3), deuterated methane (CH_3D), ethane (C_2H_6), and methyl acetylene (C_3H_4). These molecules each have a threefold (120°) axis of symmetry, and the other two axes have equal moments of inertia. A molecule with all moments equal is called a spherical rotor; examples are methane (CH_4) and germane (GeH_4). Finally, there is the case of linear molecules such as carbon dioxide (CO_2), acetylene (C_2H_2), and cyanogen (C_2N_2). Linear molecules have one zero moment of inertia, corresponding to the axis through the nuclei, while the other two moments are equal.

The classical rotational energy is given by

$$E = \tfrac{1}{2} I_a \omega_a^2 + \tfrac{1}{2} I_b \omega_b^2 + \tfrac{1}{2} I_c \omega_c^2. \tag{3.4.6}$$

By convention $I_a \leq I_b \leq I_c$. To obtain the quantum mechanical formulation replace the energy and angular momenta by their corresponding operators. For our purposes the important quantum mechanical properties are: first, the square of the total angular momentum, $L^2 = L_a^2 + L_b^2 + L_c^2$, has the values

$$L^2 = \frac{h^2}{4\pi^2} J(J+1) \tag{3.4.7}$$

and, second, the angular momentum along a symmetry axis has the values

$$L_\alpha = \frac{h}{2\pi} K. \tag{3.4.8}$$

The rotational quantum numbers J and K can have integral values $0, 1, 2, \ldots$ with both $\pm K$ allowed. The quantum numbers obey the constraint $|K| \leq J$.

The rotational energy levels of a linear molecule are given by setting $I_a = 0$ and $I_b = I_c = I$ in Eq. (3.4.6). The total angular momentum is then $L^2 = L_b^2 + L_c^2$ and has the quantum mechanical values given by Eq. (3.4.7). The rotational energy is, therefore,

$$\frac{E(J)}{hc} = BJ(J+1), \tag{3.4.9}$$

where

$$B = \frac{h}{8\pi^2 cI}. \qquad (3.4.10)$$

The same expressions have been obtained for diatomic molecules [see Eqs. (3.3.19) and (3.3.20)]. The molecule rotates around any axis perpendicular to the line through the nuclei and passing through the center of mass. As with a diatomic molecule, the nonrigid nature of the molecule requires a small correction term due to centrifugal distortion, $DJ^2(J+1)^2$, to be subtracted from the right side of Eq. (3.4.9).

The rotational energy of a spherical top molecule is found from Eq. (3.4.6) by setting $I_a = I_b = I_c = I$. By using Eq. (3.4.7) for the total angular momentum, the energy levels are found to be identical to Eq. (3.4.9). A spherical molecule can rotate about any axis that passes through the center of mass. The correction due to centrifugal distortion is again $-DJ^2(J+1)^2$. Some of the degeneracy of spherical tops is split, which gives rise in moderate resolution spectra to multiple spectral lines for each J value.

The form of the energy level expression for symmetric top molecules is different from that of linear and spherical molecules. It follows again from the classical energy of a rigid rotor, Eq. (3.4.6). However, for the symmetric top $I_a = I_A$ and $I_b = I_c = I_B$ (prolate case) or $I_c = I_C$ and $I_a = I_b = I_B$ (oblate case). The angular momenta can be written in terms of the total angular momentum $L^2 = L_a^2 + L_b^2 + L_c^2$ and the component of angular momentum in the axial direction L_a^2 or L_c^2. Thus, for the prolate case

$$E = \frac{L_a^2}{2I_A} + \frac{L_b^2 + L_c^2}{2I_B} = \frac{L^2}{2I_B} + \left(\frac{1}{2I_A} - \frac{1}{2I_B}\right)L_a^2, \qquad (3.4.11)$$

and if L^2 and L_a^2 are replaced with quantum mechanical operators, their values are given by Eqs. (3.4.7) and (3.4.8). The energy levels are, therefore,

$$\frac{E(J,K)}{hc} = BJ(J+1) + (A-B)K^2, \qquad (3.4.12)$$

where

$$B = \frac{h}{8\pi^2 cI_B} \quad \text{and} \quad A = \frac{h}{8\pi^2 cI_A}. \qquad (3.4.13)$$

The oblate forms of Eqs. (3.4.11) through (3.4.13) are obtained by substituting C

for A and c for a. For the prolate symmetric top the second term in Eq. (3.4.12) adds energy in the K series. For the oblate case the second term subtracts energy in the K series. Because both $\pm K$ are allowed, the levels are twofold degenerate. Methyl acetylene (C_3H_4) is an example of a prolate top and ammonia (NH_3) is an oblate top. When the nonrigid effects of centrifugal distortion are included, extra terms must be added to the energy of the form

$$-D_J J^2(J+1)^2 - D_{JK} J(J+1)K^2 - D_K K^4. \qquad (3.4.14)$$

The contribution of these terms is usually small compared with the rigid-rotor energy in Eq. (3.4.12), but produces observable effects in moderately resolved spectra.

For an asymmetric rotor, $I_a \neq I_b \neq I_c$ in Eq. (3.4.7), and, therefore, no simplification of the energy expression is possible. Although the total angular momentum is constant and has the quantum mechanical values of Eq. (3.4.7), no principal axis of the molecule exists along which the projection of angular momentum is constant. This complicates the quantitative treatment of the energy levels. A qualitative understanding can be arrived at, however, by considering the limits in which asymmetric molecules approach axial symmetry as the moments of inertia are varied. Since, by convention, the magnitudes of the moments of inertia have the relationship $I_a < I_b < I_c$, the molecule is approximately a symmetric top when $I_b \sim I_a$ or $I_b \sim I_c$. The former case is the oblate and the latter the prolate limit. In these limits the energy level expression for symmetric rotors Eq. (3.4.12) holds approximately. For the oblate limit it is

$$\frac{E(J,K)}{hc} = BJ(J+1) + (C-B)K^2, \qquad (3.4.15)$$

and for the prolate limit

$$\frac{E(J,K)}{hc} = BJ(J+1) + (A-B)K^2. \qquad (3.4.16)$$

Here, A, B, and C are related to the moments of inertia in the manner of Eq. (3.4.13). As I_b is varied between I_a and I_c the $J+1$ twofold degenerate rotational levels of the symmetric top for each J ($K = 0, \pm 1, \pm 2, \ldots, \pm J$) split into $2J+1$ levels; each level with $K \neq 0$ splits into two levels. Quantitative treatments of the general asymmetric rotor have been presented by Wang (1929) and Ray (1932) and reviewed by Herzberg (1945). We refer the reader to these references for more detailed discussions of asymmetric rotors.

c. Vibration–rotation transitions

The motion of a molecule involves vibration and rotation simultaneously. As in the diatomic case, the overall energy of a polyatomic molecule is approximately the sum of the vibration and rotation energies (expressed in wavenumbers $F = E/hc$)

$$F(v, J, K) = F_{\text{vib}}(v) + F_{\text{rot}}(J, K) \tag{3.4.17}$$

where F_{vib} and F_{rot} are those given in Subsections 3.4.a and 3.4.b. The vibration energy F_{vib} is a sum over all normal modes. The rotation energy F_{rot} may be a function of J only, as in the cases of linear and spherical top molecules, or both J and K, as in symmetric top molecules. In asymmetric top molecules K is not a good quantum number, but can still be used near the oblate and prolate limits. If higher order terms are included in Eq. (3.4.17) the energy cannot be divided into separate vibration and rotation parts. In particular, the values of the rotational constants B, D, etc. depend slightly on vibrational state.

The vibrational term F_{vib} in Eq. (3.4.17) is to first order just the summation

$$F_{\text{vib}}(v) = \sum_i v_i \left(v_i + \frac{g_i}{2} \right) \tag{3.4.18}$$

where the v_i are the normal vibration wavenumbers and the g_i are the mode degeneracies ($g = 1$ for nondegenerate, $g = 2$ for doubly degenerate, etc.). This expression serves approximately for all types of polyatomic molecules.

The rotational term F_{rot} in Eq. (3.4.17) is different for each type of molecule as discussed in Subsection 3.4.b above. The forms of F_{rot} for linear, spherical, symmetric, and asymmetric molecules can be constructed using Eqs. (3.4.9), (3.4.12), (3.4.15), and (3.4.16), with the addition in each case of the appropriate centrifugal distortion terms. The rotation constants A, B, C and centrifugal distortion constant D are functions of the vibrational state. In practice, higher order interactions among motions within the molecule cause perturbations of the rotational energy, which changes the simple form of these equations. The perturbations introduce additional terms in the energy expressions, which may produce relatively large effects.

A transition between two vibration–rotation levels in a polyatomic molecule causes emission or absorption of a photon at wavenumber

$$\nu = F(v', J', K') - F(v'', J'', K''), \tag{3.4.19}$$

where $F(v, J, K)$ is given by Eq. (3.4.17). Again, as in diatomic molecules, the change in vibrational energy is usually much greater than the change in rotational

3.4 Polyatomic molecules

energy. This gives rise to a pattern of bands, each centered near the wavenumber of a vibrational transition, with the rotational transitions forming a series of lines spreading out from the band center. The selection rules for polyatomic molecules are $\Delta v_i = 0, \pm 1, \pm 2, \ldots$ and $\Delta J = 0, \pm 1$ (with $\Delta K = 0, \pm 1$ for symmetric rotors). The ΔJ selection rule permits not only R- and P-branches for $\Delta J = \pm 1$, as there were for diatomic molecules, but also series with $\Delta J = 0$, called Q-branches, clustered near the center of each band.

Bands of linear and spherical molecules have this general appearance, with a central Q-branch, and P- and R-branch wings. At high resolution each line in the P-, Q-, and R-branches of a spherical top is separated into several lines by higher order interactions. Spectra of acetylene (C_2H_2) and methane (CH_4) are shown in Figs. 3.4.6 and 3.4.7.

Symmetric rotors have two types of bands, corresponding to $\Delta K = 0$ and $\Delta K = \pm 1$. If each set of P-, Q-, and R-branches for a single value of K is called a sub-band, then all sub-bands in a $\Delta K = 0$ type band are centered at the same location. These

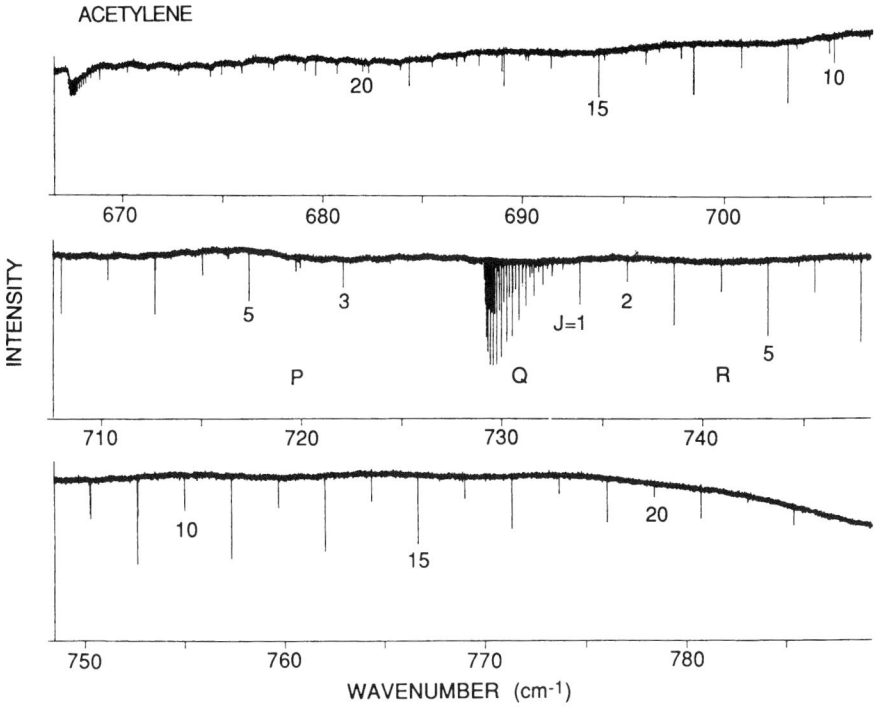

Fig. 3.4.6 The ν_5 vibration–rotation band of acetylene (C_2H_2) recorded at high spectral resolution (0.003 cm^{-1}). The P-, Q-, and R-branches are indicated, and the lower-state J value is shown for lines in the P- and R-branches. The sample was in a 1 cm cell at 0.162 mbar and 296 K. Structure in the continuum level is due to variations in the response of the spectrometer. Lines of CO_2 in the optical path appear in the 666–690 cm^{-1} region.

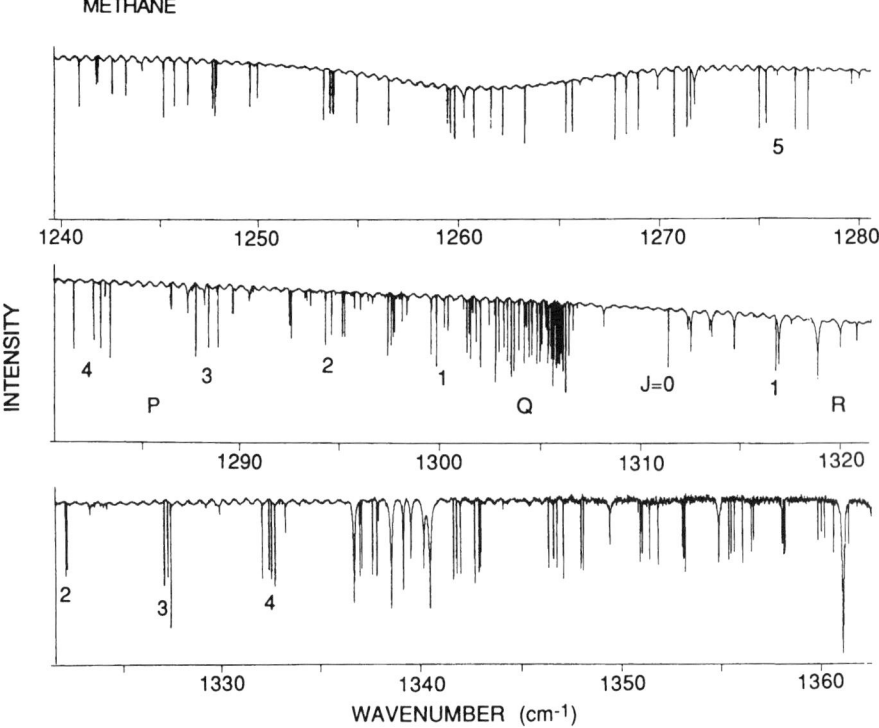

Fig. 3.4.7 The ν_4 vibration–rotation band of methane (CH$_4$) recorded at high spectral resolution (0.003 cm^{-1}). The P-, Q-, and R-branches are indicated, and the lower-state J value is shown for lines in the P- and R-branches. The sample was in a 2.4 m cell at 0.079 mbar and 296 K. Structure in the continuum level is due to variations in the response of the spectrometer. Broadened lines are due to H$_2$O in the optical path.

are called parallel bands. In the $\Delta K = \pm 1$ type bands, called perpendicular bands, the sub-band centers are spread out in a series under the influence of Coriolis forces. Parallel and perpendicular bands have distinctly different appearances, with parallel bands exhibiting a single prominent Q-branch composed of all the overlapping sub-band Q-branches, and perpendicular bands appearing as a regular series of Q-branches extending over tens of wavenumbers. In both cases the P- and R-branches are also present. A perpendicular band of ethane (C$_2$H$_6$) is shown in Fig. 3.4.8.

The spectral bands of asymmetric rotors are classified in three types labeled A, B, and C. These designations correspond to differences in symmetry selection rules arising from the orientation of the transition moment with respect to the molecular moments of inertia. Type A bands can appear similar to either parallel or perpendicular bands depending on the moments of inertia (that is, parallel for the prolate and perpendicular for the oblate limit). Type C bands are analogous to

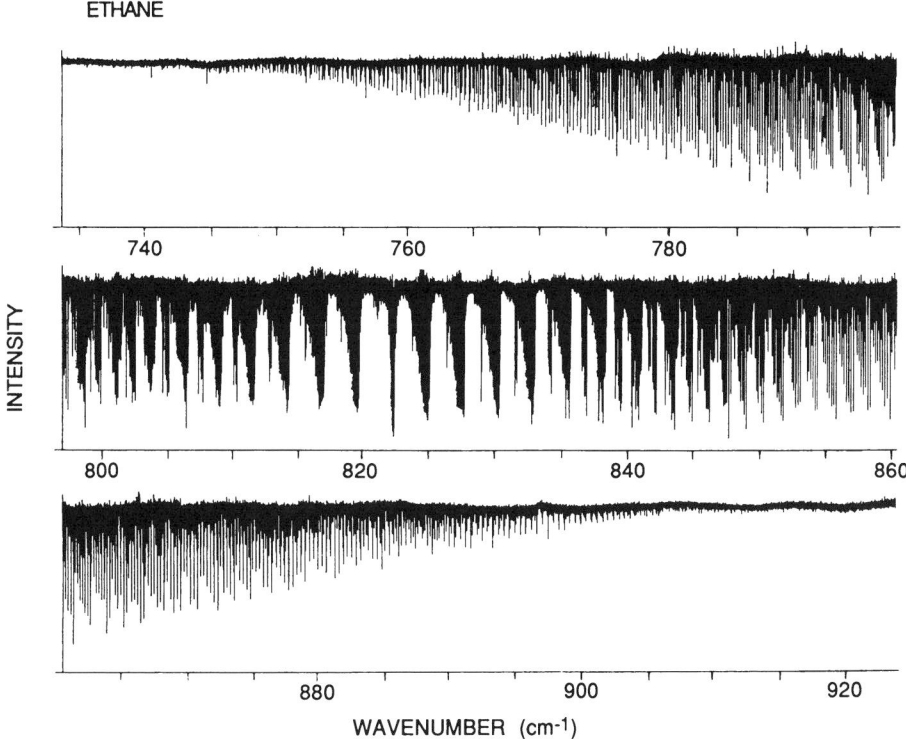

Fig. 3.4.8 The ν_9 vibration–rotation band of ethane (C_2H_6) recorded at high spectral resolution (0.003 cm^{-1}). This is a perpendicular band, having a series of regularly-spaced Q-branches. The band is centered at 822 cm^{-1}, and the Q-branch series is visible in the 790–850 cm^{-1} region. P- and R-branch lines are most clearly visible in the wings of the band. The sample was in a 1.5 m cell at 2.05 mbar and 296 K.

type A bands. Type B bands appear similar to perpendicular bands, with a series of Q-branches spread across the band. A type C band of propane (C_3H_8) is shown in Fig. 3.4.9.

3.5 Line strength

Both line positions and line strengths must be understood in order to interpret the structure of observed spectra and derive abundances and temperatures of contributing molecules. Following Planck (1901) and Einstein (1906b), we treat radiation as composed of photons with energy $E = hc\nu$, where h is the Planck constant and $c\nu$ is the frequency of the radiation in hertz. The intensity is proportional to the arrival rate of photons. If these photons originate from molecules in a small volume, the intensity is proportional to the number of molecules in the optical path undergoing transitions at that frequency.

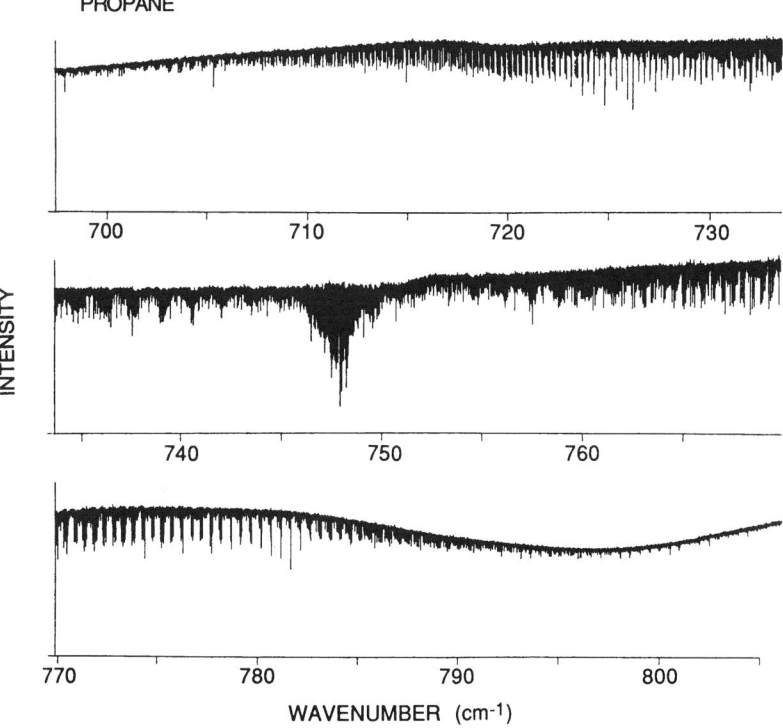

Fig. 3.4.9 The ν_{21} vibration–rotation band of propane (C_3H_8) recorded at high spectral resolution (0.003 cm^{-1}). This is a type C band with a parallel-like structure, having apparent P-, Q-, and R-branches. The sample was in a 1.5 m cell at 3.95 mbar and 296 K. Structure in the continuum level is due to variations in the response of the spectrometer.

The electric dipole transition rate between two energy levels of a molecule, E_n and E_m, depends on the probability per second that a molecule in E_n will make the transition to E_m, and on the number of molecules in the initial state. The strengths of emission or absorption lines between an upper level, E_n, and a lower level, E_m, are given by

$$I_{\text{em}} = N_n A_{nm} hc\nu = N_n \frac{64\pi^4 c \nu^4}{3} |R_{nm}|^2 \tag{3.5.1}$$

$$I_{\text{abs}} = I_0 N_m B_{mn} h\nu = I_0 N_m \frac{8\pi^3 \nu}{3hc} |R_{nm}|^2. \tag{3.5.2}$$

A_{nm} and B_{mn} are the Einstein coefficients for spontaneous emission and for absorption and stimulated emission, respectively. The Einstein coefficients are defined here in terms of the square of the electric dipole moment matrix, $|R_{nm}|^2$

[see Eq. (3.1.19)], and the wavenumber ν; N_n and N_m are the populations of the states. Strictly speaking, Eq. (3.5.2) gives the intensity absorbed per unit length in a thin layer, so that $I_{\text{abs}} \ll I_0$. The absorption in a thick layer may be obtained by integrating through that layer, as discussed in Subsection 2.4.a.

The number of molecules N_i in the initial state depends on the total number of molecules, the distribution of energy levels, the degeneracy of individual levels (number of levels of identical energy), and the temperature of the gas. In equilibrium the initial state population is

$$N_i = \frac{N d_i}{Q(T)} e^{-E_i/kT}, \qquad (3.5.3)$$

where N is the total number of molecules, d_i the degeneracy, E_i the energy of the level, T the absolute temperature, and k the Boltzmann constant. The exponential $\exp(-E_i/kT)$ is the Boltzmann factor. $Q(T)$ is the partition function,

$$Q(T) = \sum_{n=0}^{\infty} d_n e^{-E_n/kT}, \qquad (3.5.4)$$

which is the sum of Boltzmann factors weighted by their degeneracies.

We illustrate the main characteristics of line strength by considering diatomic molecules. The vibrational states (without rotation) are all nondegenerate, so $d_n = 1$. Ignoring anharmonicities, vibrational energies are given by Eq. (3.3.6),

$$E(v) = \frac{h}{2\pi} \omega \left(v + \tfrac{1}{2} \right) \quad (v = 0, 1, 2, \ldots). \qquad (3.5.5)$$

If we eliminate a factor $\exp(-h\omega/4\pi kT)$ from both numerator and $Q(T)$ in Eqs. (3.5.3) and (3.5.4) the number of molecules in state v of a diatomic molecule is then

$$N_v = \frac{N}{Q_{\text{vib}}(T)} e^{-(h/2\pi)(\omega v/kT)}; \qquad (3.5.6)$$

where the partition function is now written

$$Q_{\text{vib}}(T) = \sum_{v=0}^{\infty} e^{-(h/2\pi)(\omega v/kT)} = \left[1 - e^{-(h/2\pi)(\omega/kT)} \right]^{-1}. \qquad (3.5.7)$$

Equation (3.5.6) has several implications. At high temperatures $Q_{\text{vib}}(T)$ is larger than unity and the Boltzmann factor decreases gradually, so that several levels are populated. At low temperatures $Q_{\text{vib}}(T)$ is ~ 1 and the fraction of molecules in $v = 0$ approaches unity; very few molecules are in $v = 1$ and higher levels. For example, in carbon monoxide (CO) at 300 K the fractions of molecules in $v = 1$ and $v = 2$ are about 10^{-5} and 10^{-9}, respectively. At the much higher temperature of the Sun (~ 5000 K), fractions in these states are 0.29 and 0.18, respectively.

The expression for N_v can be generalized by including the rotational energy and the rotational partition function in Eq. (3.5.6). The rotational energies of a diatomic molecule, given for the rigid rotator by Eq. (3.3.19), are

$$E(J) = hcBJ(J+1); \quad J = 0, 1, 2, \ldots. \tag{3.5.8}$$

The degeneracy of level J is $d = 2J + 1$. Since we are discussing electric dipole transitions, the diatomic molecule must be heteronuclear. The population in the initial state is then

$$N_{v,J} = \frac{N(2J+1)}{Q_{\text{vib}}(T)Q_{\text{rot}}(T)} e^{-h[(\omega v/2\pi)+cBJ(J+1)]/kT}. \tag{3.5.9}$$

The rotational partition function can be approximated by an integral

$$Q_{\text{rot}}(T) \simeq \int_0^\infty (2J+1) e^{-hcBJ(J+1)/kT} \, dJ = \frac{kT}{hcB}. \tag{3.5.10}$$

The level populations, therefore, have the dependence

$$N_{v,J} \simeq \left[\frac{1}{Q_{\text{vib}}(T)} e^{-(h/2\pi)(\omega v/kT)} \right] N \frac{hcB}{kT} (2J+1) e^{-hcBJ(J+1)/kT}. \tag{3.5.11}$$

The factor in brackets gives the fractional population in a given vibrational level, and the factor outside the brackets gives the distribution over rotational levels within that vibration state. $Q_{\text{vib}}(T)$ can be set to unity at planetary temperatures.

The degeneracy factor $(2J + 1)$ in Eq. (3.5.11) has a strong influence on the rotational population distribution within a vibrational level. At a particular temperature the degeneracy factor causes the population to be proportional to hcB/kT at $J = 0$, and to increase for small J. As J increases further the effect of the exponential form of the Boltzmann factor becomes more and more dominant, causing the population to approach zero for high J values. Figure 3.5.1 shows this distribution of rotational

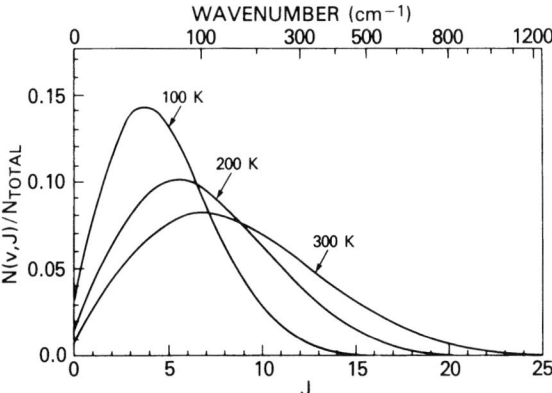

Fig. 3.5.1 Fractional populations $N(v, J)/N_{\text{TOTAL}}$ of the $v = 0$ state of rotational levels in carbon monoxide at 100, 200, and 300 K. The lower scale is the rotational quantum number J. The upper scale is the energy in cm^{-1} for the rotational levels.

populations for the ground vibrational ($v = 0$) state of CO at 100, 200, and 300 K. The peak value of J is given by

$$J_{\max} \simeq \left(\frac{kT}{2hcB}\right)^{\frac{1}{2}}. \tag{3.5.12}$$

Therefore, by observing the rotational line structure an estimate of the temperature in the line forming region may be obtained. The temperature so derived is called the rotational temperature. The strengths of emission and absorption lines are obtained by substituting the initial state population [Eq. (3.5.11)], in Eqs. (3.5.1) and (3.5.2):

$$I_{\text{em}} = \frac{F_{\text{em}}}{Q_{\text{rot}}(T)} v^4 (2J+1) \, e^{-hcBJ(J+1)/kT} \tag{3.5.13}$$

$$I_{\text{abs}} = I_0 \frac{F_{\text{abs}}}{Q_{\text{rot}}(T)} v(2J+1) \, e^{-hcBJ(J+1)/kT}. \tag{3.5.14}$$

Here F_{em} and F_{abs} are factors containing the dipole moment matrix for the vibrational transition, and the population of the initial vibrational level. These factors depend only weakly on J and can be considered approximately constant for all rotational lines within each vibrational transition. The J quantum number in Eqs. (3.5.13) and (3.5.14) refers to the initial rotational level.

The overall strength of a vibrational transition is determined by the transition matrix and the level population; the strength is distributed among the rotational lines within the vibrational transition, producing the intensity pattern in the vibrational band. The selection rules on J for a diatomic molecule together with the degeneracy and Boltzmann factors in Eqs. (3.5.13) and (3.5.14) produce the observed shape of the band.

When the diatomic molecule makes an electric dipole transition from one vibrational state, v, to another state, v', the rotational quantum number J can change by $\Delta J = \pm 1$. If $v = v'$, that is, if the transition is purely rotational, a series of lines is formed that begins at $2B$ wavenumbers with a separation between lines of approximately $2B$. In the case of absorption all transitions are $\Delta J = +1$, while for emission $\Delta J = -1$ applies. The pattern is shown for CO with $v = 0$ in Fig. 3.3.4. The strengths of the lines increase as the degeneracy increases up to $J_{\max}$, given by Eq. (3.5.12), and then decrease as the Boltzmann population decreases. If $v \neq v'$, where the vibrational state changes along with the rotational state, the change in rotational quantum number can be either $\Delta J = +1$ or -1 in both absorption and emission. This gives rise to a series of lines centered at $\nu = F(v', 0) - F(v, 0)$. In emission, the lower wavenumber series corresponds to $\Delta J = +1$ from the upper to a lower state and is called the P-branch. The higher wavenumber $\Delta J = -1$ series is called the R-branch. The pattern is shown for CO $v' = 1$ and $v = 0$ in Fig. 3.3.5. Again, lines are separated by about $2B$, and J increases away from the band center. The line strengths increase until $J_{\max}$ [see Eq. (3.5.12)] is reached and then decrease with the Boltzmann factor.

For polyatomic molecules, line strengths in the vibration–rotation spectra follow the general behavior of those in diatomic molecules. The position of lines and bands of polyatomic molecules are determined by the energy levels and selection rules described in Section 3.4. The theory predicting line intensities is similar to that described above for diatomic molecules, in that the intensity is proportional to the population in the initial state and the square of the electric dipole moment matrix. The level population is again a product of level degeneracy and Boltzmann factor. The P-, Q-, and R-branches each have line strengths that increase with rotational quantum number up to a maximum, and then decrease and gradually approach zero. The appearance of a polyatomic spectrum is more complex than that of a diatomic molecule, because a polyatomic molecule has more internal degrees of freedom and, therefore, has more modes of vibration. Also additional rotational quantum numbers exist with associated selection rules. Moreover, symmetries in molecules cause some vibrational bands to be absent in the infrared, or some rotational lines to be missing. Also, mixing among rotation–vibration states that lie close together in energy can greatly modify line intensities and frequencies for transitions involving those levels. The various types of polyatomic molecules – linear, spherical top, symmetric top, and asymmetric top – each have their characteristic band structures. Detailed discussions of line strengths in polyatomic molecules can be found in Allen & Cross (1963), for example.

3.6 Line shape

Thus far, we have discussed the processes that determine the spectral positions of lines and their strengths. Another important characteristic is line shape, which must be understood if one wishes to interpret measured planetary spectra. The line shape determines the wavenumber dependence of a spectral feature in the vicinity of the line position. Even if perfectly resolved by a spectrometer, a line is spread over a finite wavenumber range and does not appear as an infinitely narrow feature. Several mechanisms are responsible for that; some are of a fundamental quantum mechanical nature (natural line broadening) and some are functions of the environment of the emitting or absorbing molecules (collision and Doppler broadening). In this section we discuss the most important of these mechanisms. A more complete treatment of line shapes is given by Goody & Yung (1989).

Natural line broadening is usually much smaller than the broadening seen in planetary spectra, but it is of fundamental importance as the ultimate limit of the narrowness of a spectral line. Natural line broadening arises from the finite coherence length of the wave train associated with a transition. Fourier analysis of radiation composed of a wave train of finite time, Δt, shows a frequency spread, Δf, that is related to this time by

$$\Delta f \Delta t \sim 1. \tag{3.6.1}$$

The line width Δf is just the range of radiation frequencies that have a high probability of interacting with the molecule. The time Δt can be understood as the spontaneous emission lifetime of the transition discussed in Section 3.5. The spontaneous decay has an exponential dependence, and therefore the line shape due to natural line broadening is Lorentzian. The natural line width,

$$w_n = \tfrac{1}{2}\Delta \nu \sim \frac{1}{c\Delta t}, \tag{3.6.2}$$

thus varies according to the decay time of the interaction. For a molecular transition near 1000 cm^{-1} the natural line width can be of the order 10^{-7} cm^{-1}. This is much narrower than the other line broadening processes that we discuss. Natural line broadening in the infrared is only observed under special laboratory conditions. From the strong spectral dependence of spontaneous emission, as shown by the Einstein coefficient [see Eq. (3.5.1)], the natural line width is much larger in the visible or ultraviolet spectral region.

From a practical point of view collision broadening is a more important mechanism. If one molecule in the process of a transition collides with another molecule, the phase continuity of the transition is interrupted, thereby reducing Δt in

Eq. (3.6.1) from the natural lifetime of the transition to the mean time between collisions, leading in this case to a larger line width, $\Delta \nu = w_c$. Under atmospheric conditions of interest, the time between collisions is usually much shorter than the natural transition lifetime. Collision broadening in the infrared can thus be orders of magnitudes larger than natural line broadening. The line profile generated by collisions is

$$I(\nu) = K \frac{w_c}{(\nu - \nu_0)^2 + w_c^2/4} \quad (3.6.3)$$

where K is the spectrally integrated line intensity. In Eq. (3.6.3) $I(\nu)$ is the line intensity at wavenumber ν, w_c is the full-width of the line at half-maximum, and ν_0 is the line center. The collision shape, also called Lorentzian line shape, is shown along with other line shapes in Fig. 3.6.1. The collision line width increases with density as the mean time between collisions decreases. Therefore, at a given temperature, w_c is proportional to pressure, P, which leads to the definition of a

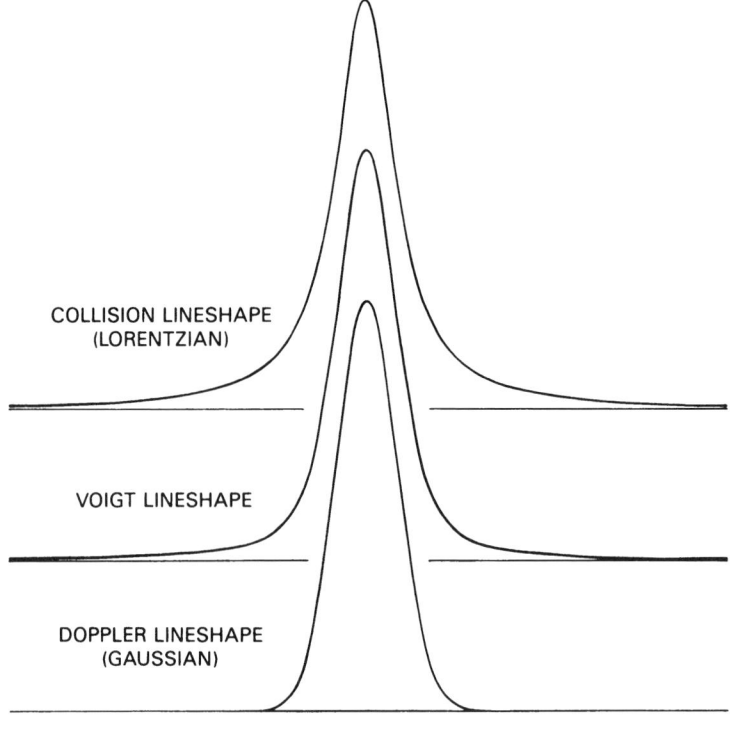

Fig. 3.6.1 Collision, Voigt, and Doppler line shapes. All three are shown with the same maximum amplitude, and with the same width at half maximum amplitude. The Voigt line shape is one of a continuum of profiles between Gaussian and Lorentzian, depending on the degree of collision broadening.

collision broadening coefficient

$$\gamma = w_c/P. \tag{3.6.4}$$

The collision broadening coefficient varies among molecules and collision partners, but is usually close to 3 gigahertz (0.1 cm^{-1}) at one bar. Under typical conditions found in planetary tropospheres and stratospheres pressure broadening usually dominates other forms of line broadening.

Natural and collision broadening are examples of homogeneous broadening, that is, a process where each molecule contributes to all frequencies under the distribution function. An important example of an inhomogeneous broadening effect is Doppler broadening. In that case the overall line shape consists of contributions from many molecules, or many segments of trajectories between collisions of one molecule, some moving towards and some away from the observer.

The frequency at which a particular transition interacts with the radiation is thereby shifted slightly by

$$\frac{\Delta f}{f_0} = -\frac{\Delta v}{c}, \tag{3.6.5}$$

where f_0 is the rest frequency of the transition, Δv is the component of the velocity along the line of sight, and c is again the speed of light. The velocity components along any line of sight have a Maxwellian distribution, which results in the observed line having a Gaussian shape,

$$I(\nu) = K \frac{2(\ln 2)^{\frac{1}{2}}}{\pi w_D} e^{-4\ln 2(\nu-\nu_0)^2/w_D^2}, \tag{3.6.6}$$

where the symbols have the same meanings as in Eq. (3.6.3). This is called the Doppler line shape. The width, w_D, is determined by the spread in velocities in the Maxwellian distribution and these depend on the temperature T of the gas and the molecular weight, M. The Doppler line width, or full width at half-maximum intensity of the Gaussian profile, is given by (see Townes & Shawlow, 1955),

$$w_D = \frac{\nu}{c}\left(8 \ln 2 k N_0 \frac{T}{M}\right)^{\frac{1}{2}} = 7.16 \times 10^{-7} \left(\frac{T}{M}\right)^{\frac{1}{2}} \nu. \tag{3.6.7}$$

Here k is the Boltzmann constant and N_0 is the Avogadro number. Whereas collision broadening is independent of the location of the line in the spectrum, Doppler broadening is proportional to wavenumber. For a particular molecule, at a low wavenumber the Doppler width can be narrower than the collision width, while at higher wavenumbers the opposite may be true. The Doppler width for methane at a temperature of 200 K is 3.3×10^{-3} cm^{-1} at 1300 cm^{-1} and 7.6×10^{-3} cm^{-1} at 3300 cm^{-1}. An example of a Doppler broadened line is also shown in Fig. 3.6.1.

In most cases in which line profiles are completely resolved by the infrared spectrometer both collision and Doppler broadening contribute to the line shape. The function that describes the composite line profile is a convolution of a Gaussian and a Lorentzian function,

$$I(\nu) = \int_{\nu'} G(\nu) L(\nu - \nu') \, d\nu'. \qquad (3.6.8)$$

The width of the Gaussian function, $G(\nu)$, is the Doppler width, w_D, and that of the Lorentzian function, $L(\nu - \nu')$, is the collision width, w_c. This generalized line shape, introduced by Voigt (1912), has no analytical expression, but it can easily be computed numerically using the convolution represented in Eq. (3.6.8). Its shape is shown in Fig. 3.6.1. An empirical expression relating w_D and w_c to the total Voigt width, w_V, has been introduced by Whiting (1968),

$$w_V = \frac{w_c}{2} + \left(\frac{w_c^2}{4} + w_D^2 \right)^{\frac{1}{2}}. \qquad (3.6.9)$$

The Voigt line shape closely describes line profiles measured under most laboratory and atmospheric conditions.

Other processes also alter the observed line shape in planetary spectra. For example, a line that forms at different altitudes will have a profile with a broad base due to higher pressure at lower altitudes, and a narrow center corresponding to the lower pressure at higher altitudes. Similarly, scattering due to cloud particles in a real planetary atmosphere can affect the line shape. Examples of scattering effects are discussed in Chapter 4. Another example is Doppler broadening of a line due to planetary rotation. This phenomenon can be observed at high spectral resolution when the field of view of the instrument covers an area on the planet with a range of velocities.

Finally, we should mention the effect of far-wing absorption, an important but poorly understood aspect of line shape. Collision-induced opacity is observed in spectral regions well-separated by as much as 10 to 100 cm^{-1} from the line center. This very weak absorption is not described by the wings of a simple Lorentz line shape discussed above; it shows an exponentially decreasing dependence on the separation from the line center (see, e.g., Birnbaum, 1979). The anomalous far-wing absorption is due to inadequacies in the hard-sphere collision model used in the derivation of the Lorentz line shape, which assumes instantaneous interaction. Far-wing absorption usually produces measurable effects only in spectral regions that are relatively free of absorption lines, so-called windows, but where many strong lines are not far away. Examples are the atmospheric window between 800 and 1250 cm^{-1} in the Earth atmosphere, where numerous strong water vapor and carbon dioxide lines contribute to a continuum opacity, and the 5 μm window at Jupiter, where many nearby lines of ammonia and methane have a noticeable effect.

3.7 Solid and liquid surfaces

a. Solid and liquid phases

The study of the interaction of electromagnetic radiation with solid or liquid matter requires some understanding of these phases. In the discussion of the interaction of radiation with gases it is generally sufficient to consider the energy levels of an individual molecule of a particular gas. Collision-induced phenomena, where at least two gas molecules are involved in a transition, provide an important exception to this rule (Subsection 3.3.d). However, in most cases the interaction of radiation with a gas can be adequately understood by considering quantum processes involving only one molecule. Such is not the case in interactions of radiation with solids or liquids.

In a solid body atoms are often arranged in a well-ordered crystal structure. The motion of each atom in the crystal lattice is not isolated, but is coupled to that of all other atoms in the structure, though the influence of each atom is felt most strongly in its immediate neighborhood. The entire crystal is a set of coupled oscillators. In addition to the previously discussed vibration of the molecule as a whole, the atoms or groups of atoms oscillate about an equilibrium position in the crystal structure, giving rise to quantized lattice modes. The coupling within the crystal also perturbs the individual molecular vibration modes. Coupling within the crystal and the absence of rotation cause a loss of the sharp line features so characteristic of the gas phase.

The coupling of the individual oscillators complicates the theoretical treatment of solid matter. Further complexity arises in the treatment of crystal boundaries and of other disturbances of the structure. Lattice periodicity may be interrupted by imperfections or impurities. However, where the crystal periodicity is well-preserved, certain simplifications are permitted in the theoretical treatment. Unfortunately, on real planetary or satellite surfaces matter is rarely found in the form of large crystals. In surface rocks the structure usually consists of small, sometimes minute crystals. Very small dust particles of micrometer size are frequently formed by meteoritic impact and by erosion resulting from thermal stresses, wind, or water.

Good examples of dusty surfaces are found on Mars and the Moon. The theoretical treatment of amorphous substances and liquids is even more difficult. There the periodic lattice structure does not exist at all, but the interaction among molecules is nevertheless strong.

Solids can be classified according to the dominant binding forces that hold them together. Molecules are attracted to each other by these forces, but intermolecular, short-range repulsion prevents collapse of matter beyond a certain packing density. In this densely packed state the energy per unit volume has a minimum. Additional energy must be supplied if further compression is to be accomplished or if molecules are to be separated. A stable configuration results with a well-defined energy minimum that is not too different in concept from the situation found in gas molecules (see Fig. 3.2.1).

The weakest bond is that of van der Waals forces in which molecules are attracted by fluctuations in dipole moments. Ices of hydrogen and the noble gases (except helium) are examples of van der Waals binding. Since the forces are relatively weak (inverse seventh power in the separation distance) these ices are mechanically relatively fragile, with their strength increasing as temperature drops far below the triple point. In the fluctuating dipoles that create the van der Waals attraction electrons stay attached to the individual molecules; therefore, ices are relatively poor conductors of heat or electricity. Another example of a weak bond is the hydrogen bond, which holds together the molecules of water ice. Tough weakly bound, water ice can form mountain ridges of considerable height at low temperatures and in a low-gravity environment. Steep cliffs nearly 15 km high on Miranda (Fig. 5.4.12), are believed to be of water ice.

In ionic crystals the main binding force is electrostatic attraction between ions of opposite polarity. A good example is sodium chloride, which forms a cubic crystal of alternating positive sodium and negative chlorine ions. Since electric forces are stronger than van der Waals forces, ionic crystals are generally more rigid and have higher melting temperatures than ices. With their electrons tightly bound to the anions they are electric insulators and poor heat conductors. Once melted or in aqueous solutions, however, ionic materials may conduct electricity well.

Covalent solids are formed by sharing valence electrons among atoms in a crystal. The resulting forces are quite analogous to the binding forces in covalent gas molecules. The covalent bond is very strong, which accounts for the high mechanical strength and high melting point of such crystals. Most such crystals are insulators and poor thermal conductors. Important examples of covalent solids are quartz and rock-forming silicates in general. Some covalent substances, such as germanium or silicon, can become semiconductive, particularly if the regular crystal structure is interrupted by impurities. In other cases of covalent binding, outer-shell electrons move freely in the lattice of positive ions. Such substances are called metals, and are good conductors of electric current and heat. The high electric conductivity

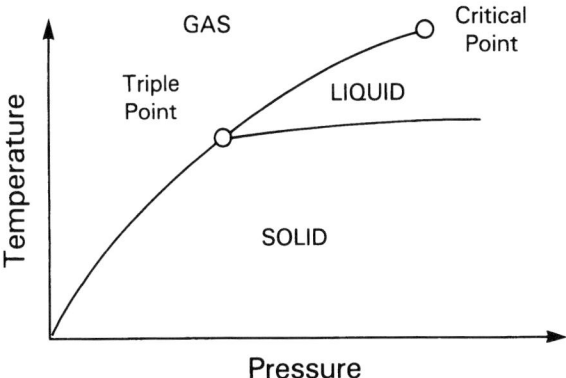

Fig. 3.7.1 Typical phase diagram. Below the triple point matter exists only as a solid or a gas. Between the triple point and the critical point a liquid phase may also exist. Above the critical point matter is in a fluid phase.

(large σ; see Chapter 1) also makes metals good reflectors of electromagnetic radiation. Metals are essential to the construction of instruments, but are very rarely found in pure form on planetary surfaces. Metallic hydrogen exists in the interiors of Jupiter and Saturn, and iron, nickel, and other metals are expected to exist in all planetary cores.

All solid matter can change phase if heated to sufficiently high temperatures at low enough pressures. At high temperature, the energy of molecular thermal agitation may exceed the crystal binding energy, so that molecules may leave the crystal formation. A schematic phase diagram is shown in Fig. 3.7.1. Below the triple point pressure, changes occur between the solid and the gaseous phases only. Above the triple point pressure a solid may melt first, and with further heating may reach a second boundary between the liquid and gaseous phases. However, above the critical pressure liquid and gas are not separated by a phase transition; in this condition the material is in a fluid state. For example, the critical point of hydrogen is at 33 K and 13 bar. Therefore, it is more precise to speak of a fluid envelope for the bulk of the giant planets, rather than a gaseous one. On the other hand, the critical temperature and pressure of nitrogen are at 126 K and 33.9 bar, so that the atmospheres of Earth and Titan are properly called gaseous. At very high pressures many substances undergo further phase transitions. Water ice, for example, has at least seven phases. The phase changes of hydrogen–helium mixtures, so important for the study of the interior of the giant planets, are discussed by Stevenson (1982).

b. Complex refractive indices

To gain an understanding of the interaction of radiation with solids and liquids we use the complex index of refraction, n, introduced in Section 1.3. The real part of the index is defined as the ratio of the speed of light in a vacuum to that in the material, or $n_r = c/v$ [see Eq. (1.3.17)]. Its value for a vacuum is unity, and for other media it is larger than one, except for the case of anomalous dispersion discussed later. According to Maxwell's theory, n is related to the dielectric constant ε and magnetic permeability μ of the material by $n^2 = \varepsilon_{rel}\mu_{rel}$. As pointed out before, μ_{rel} is close to unity for all substances of interest in remote sensing.

Equations (1.3.24) to (1.3.26) express the real and imaginary parts of the complex index of refraction in terms of the electric conductivity σ, the dielectric constant ε, and the magnetic permeability μ of the medium. The real part of the index is called the propagation constant and the imaginary part the absorption constant. Neither n_r nor n_i are true constants, but are functions of wavenumber. The electric field of radiation traveling in the x-direction is [see Eq. (1.3.27)]

$$\mathbf{E} = \mathbf{E}_0 \, e^{(i\omega/c)[(n_r + in_i)x - ct]} \qquad (3.7.1)$$

$$= \mathbf{E}_0 \, e^{-(\omega/c)n_i x} e^{(i\omega/c)(n_r x - ct)}. \qquad (3.7.2)$$

The second exponential in Eq. (3.7.2) is the usual wave function; the first exponential represents an attenuation of the wave amplitude as it travels through the medium. Since exponentially increasing amplitudes are not allowed on physical grounds, only positive values of n_i are accepted for positive values of x (although a negative n_i is sometimes used to represent gain in a medium, such as a laser). For the case of $n_i = 0$ in Eq. (3.7.2) the wave function corresponds to propagation in a transparent medium, whereas the case of $n_i > 0$ corresponds to attenuation in an absorbing medium. The coefficient $\omega n_i/c$ in Eq. (3.7.2) is called the extinction coefficient.

Since the index of refraction is known to be a function of wavenumber, the dielectric constant, ε, also depends on wavenumber. This follows from the atomic and molecular composition of matter. The material consists of molecules or atoms in which the charges are bound, and therefore acts as a collection of oscillators. Consider the case of charges in the molecules and atoms subjected to polarized radiation traveling in the x-direction with electric field $\mathbf{E} = \mathbf{E}_0 \, e^{i\omega t}$. Each molecule experiences a force due to the radiation field of the form $\mathbf{F} = q\mathbf{E}_0 \cos \omega t$, where q is the electric charge. This force causes small displacements, $\mathbf{r}$, of the charges, and thereby gives rise to a polarization of the material. The total polarization, $\mathbf{P}$, is the volume sum of all individual dipole moments $\mathbf{p}_i = q_i \mathbf{r}_i$ generated by the field,

$$\mathbf{P} = \sum_{i=1}^{N} \mathbf{p}_i = Nq\mathbf{r}, \qquad (3.7.3)$$

3.7 Solid and liquid surfaces

where in the right-most part the summation has been eliminated by assuming that the individual qs and $\mathbf{r}$s are the same. Solution of the equations of motion for the oscillating charges, including damping (due to collisions, radiative losses, etc.), yields

$$\mathbf{r} = \frac{q}{m} \frac{\mathbf{E}}{\omega_0^2 - \omega^2 - i\omega d}, \tag{3.7.4}$$

where m is the mass of the displaced particle and d is called the damping parameter. The resonance frequency, ω_0, of the oscillating charge depends on the physical properties of the material. According to Eqs. (3.7.3) and (3.7.4) the polarization in the material is related to the electric field,

$$\mathbf{P} = \frac{Nq^2}{m} \frac{\mathbf{E}}{\omega_0^2 - \omega^2 - i\omega d}. \tag{3.7.5}$$

Moreover, the polarization and electric field are related through the dielectric constant by $\mathbf{P} = (\varepsilon - \varepsilon_0)\mathbf{E}$, so that Eq. (3.7.5) also gives the dielectric constant in terms of frequency.

Maxwell's equations for a polarized medium lead to the wave equation

$$\nabla^2 \mathbf{E} = \frac{1}{c^2} \left[\frac{\partial^2 \mathbf{E}}{\partial t^2} + \frac{1}{\varepsilon_0} \frac{\partial^2 \mathbf{P}}{\partial t^2} \right]. \tag{3.7.6}$$

For a plane wave travelling in the x-direction, and with the help of Eq. (3.7.5), Eq. (3.7.6) reduces to

$$\frac{\partial^2 \mathbf{E}}{\partial x^2} = \frac{1}{c^2} \left(1 + \frac{Nq^2}{m\varepsilon_0} \frac{1}{\omega_0^2 - \omega^2 - i\omega d} \right) \frac{\partial^2 \mathbf{E}}{\partial t^2}. \tag{3.7.7}$$

The solution to this equation is given by Eq. (3.7.1) or (3.7.2). Recognizing that the coefficient of the time-derivative term is n^2/c^2, we arrive at the expression for the complex index of refraction,

$$n = n_\mathrm{r} + i n_\mathrm{i} = \left[1 + \frac{Nq^2}{m\varepsilon_0} \left(\frac{1}{\omega_0^2 - \omega^2 - i\omega d} \right) \right]^{\frac{1}{2}}. \tag{3.7.8}$$

Since the damping term, $-i\omega d$, results in a complex quantity inside the parentheses, the square root is also complex, and yields formulas for n_r and n_i comparable to Eqs. (1.3.25) and (1.3.26).

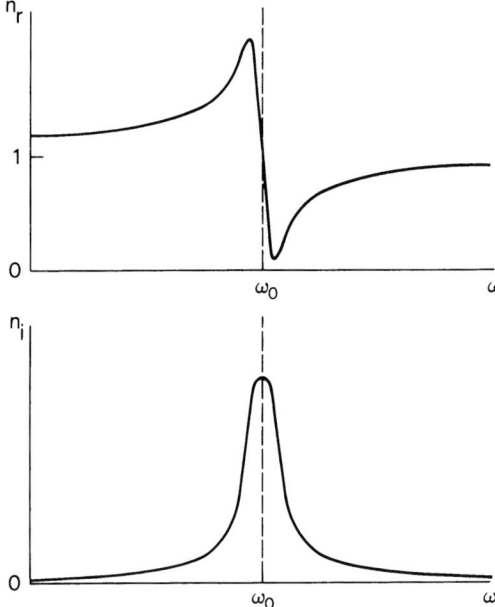

Fig. 3.7.2 Behavior of the real, n_r, and imaginary, n_i, parts of the refractive index with frequency. The real part generally increases with frequency, but goes through a rapid decrease in the vicinity of ω_0. The imaginary part is maximum at ω_0.

Figure 3.7.2 shows the typical dependence of n_r and n_i on radiation frequency. The imaginary part of the index of refraction is symmetrical and has a maximum at the resonance frequency, ω_0. The real part of n goes through a narrow, sharp inflection close to ω_0. An increase of n_r with ω is the usual case and corresponds to higher wavenumbers being refracted more than lower wavenumbers. However, in the region near ω_0 where n_r decreases with ω, the opposite is true and the material is said to exhibit anomalous dispersion. Anomalous dispersion can be observed if the absorption is not too high at ω_0, and can be interpreted as due to phase velocities exceeding the speed of light in vacuum. Figure 3.7.3 shows measured curves of n_r and n_i for potassium bromide obtained by Bell (1971).

The limit of n for ω approaching zero yields the static dielectric constant of the material. As can be seen from Eq. (3.7.8), the static constant is real ($i\omega d = 0$), and has the form

$$n_0 = \left(1 + \frac{Nq^2}{m\varepsilon_0 \omega_0^2}\right)^{\frac{1}{2}}. \tag{3.7.9}$$

3.7 Solid and liquid surfaces

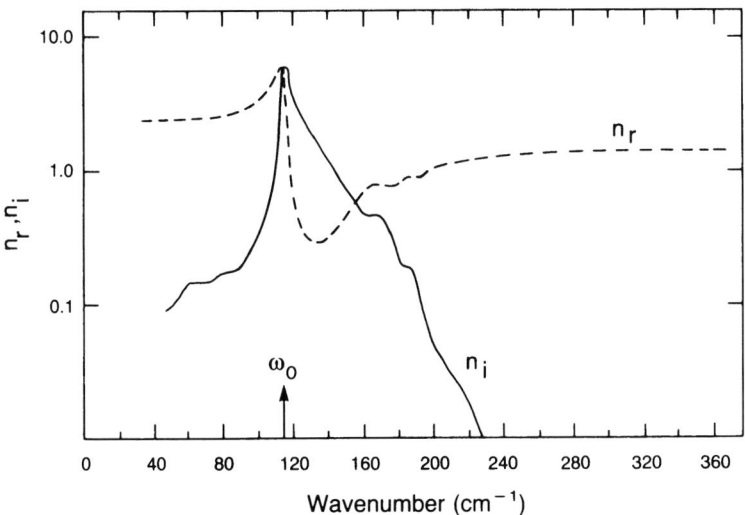

Fig. 3.7.3 The real, n_r, and imaginary, n_i, parts of the refractive index of potassium bromide. Anomalous dispersion exists in n_r near ω_0, where n_i is near maximum (Bell, 1971).

A material typically has many resonance frequencies, ω_j, associated with the bound charges. To account for these frequencies the theory must be generalized by modifying Eq. (3.7.8) to include a summation over these frequencies. Thus

$$n = \left(1 + \frac{Nq^2}{m\varepsilon_0} \sum_j \frac{s_j}{\omega_j^2 - \omega^2 - i\omega d_j}\right)^{\frac{1}{2}}. \qquad (3.7.10)$$

The quantities s_j are the oscillator strengths, and each is the fraction of the harmonic oscillators which have frequency ω_j. The damping parameters, d_j, are also different for each oscillator frequency.

If the damping parameters are small, the index of refraction is nearly real, i.e.,

$$n \sim n_r \sim \left(1 + \frac{Nq^2}{m\varepsilon_0} \sum_j \frac{s_j}{\omega_j^2 - \omega^2}\right)^{\frac{1}{2}}. \qquad (3.7.11)$$

For many transparent materials the measured index of refraction can be fitted to a curve of this form quite well. The dependence of n_r on frequency is called the dispersion relation and is often written in terms of wavelength λ_j,

$$n_r^2 - 1 = \sum_j A_j \frac{\lambda^2}{\lambda^2 - \lambda_j^2}. \qquad (3.7.12)$$

Tables of A_j and λ_j are given by Wolfe & Zissis (1978), for example.

If either the propagation constant n_r or the absorption constant n_i is known over a wide wavenumber range, the other parameter can be calculated numerically using relations between real and imaginary parts of a function developed by Kramers and Kronig (see for example, Landau & Lifshitz, 1960). These relations follow from the causal restriction that the group velocity of radiation cannot exceed the speed of light in vacuum. The Kramers–Kronig relations provide a powerful method for deriving the complex index of refraction from spectrometric measurements. An alternative technique, illustrated in Fig. 3.7.3, uses amplitude Fourier transform spectroscopy to produce both n_r and n_i simultaneously (Bell, 1971; Bell, 1972).

For the case of conducting materials (metals) Eq. (3.7.8) can be applied by noting that unbound charges correspond to $\omega_0 = 0$. The square of the complex index of refraction is then

$$n^2 = \varepsilon_{\text{rel}} = 1 - \frac{\omega_p^2}{\omega^2 + i\omega/\tau}, \quad (3.7.13)$$

where

$$\omega_p = \left(\frac{Nq^2}{m\varepsilon_0}\right)^{\frac{1}{2}} \quad (3.7.14)$$

is the plasma frequency, and the damping parameter d has been replaced by the reciprocal of the relaxation time, τ, in the medium. The frequency dependence of the coefficients n_r and n_i in metals is completely determined by the plasma frequency and the relaxation time. We will not pursue the optical theory of conductors, since virtually all solid and liquid materials encountered in planetary environments are dielectrics. Further discussions of the complex index of refraction appear in texts such as Born & Wolf (1959) and Strong (1958).

3.8 Cloud and aerosol particles

a. Asymptotic scattering functions

Cloud particles are assemblages of molecules, large enough to treat from classical theory, yet small enough to require considerations of size and shape. If a given particle is much smaller than the wavelength of incident plane-polarized electromagnetic radiation, the particle can be treated as though it were in an applied homogeneous electric field, $\mathbf{E}_0 \, e^{-i\omega t}$, at any instant in time. The resulting induced dipole moment is given by

$$\mathbf{p} \, e^{-i\omega t} = \alpha \mathbf{E}_0 \, e^{-i\omega t}, \tag{3.8.1}$$

where α is the polarizability of the particle. If the particle is spherical and homogeneous, α is a scalar.

The oscillating dipole radiates in all directions, and the resulting emission is called Rayleigh scattering. If $\mathbf{p}$ is perpendicular to the plane of scattering, the amplitude $|a|$ of scattered radiation is independent of the scattering angle Θ. If $\mathbf{p}$ is contained in the plane of scattering, $|a|$ is proportional to the absolute value of the projection of $\mathbf{p}$ onto the direction of scattering; i.e., $|a| \propto |\cos \Theta|$. The scattered flux is proportional to $\sum |a|^2$, averaged over all polarization states.

For incident unpolarized radiation, a complete analysis leads to

$$F = 8 \left(\frac{\pi^2 \alpha}{\lambda^2 r} \right)^2 (1 + \cos^2 \Theta) F_0, \tag{3.8.2}$$

where πF_0 and πF are respectively the fluxes of incident and scattered radiation. The quantities r and Θ refer respectively to the distance and direction from the scattering center. In particular $(1 + \cos^2 \Theta)$ is the unnormalized phase function for single scattering. This factor does not depend on wavelength as long as the particle radius $a \ll \lambda$, although the familiar inverse fourth power wavelength dependence of scattered radiation enters the complete equation.

At the other extreme the particle is large compared with the wavelength ($a \gg \lambda$) and the concepts employed in geometric optics and Fraunhofer diffraction theory can be used to advantage. Let a beam of parallel radiation be incident on any surface element of the spherical particle, and require the width D of the beam to be much larger than λ and much smaller than a; i.e., $\lambda \ll D \ll a$. In geometric optics such a beam is called a ray.

If incidence is grazing, Fraunhofer diffraction will occur around the edge of the particle. Because $D \gg \lambda$, the diffracted radiation will be concentrated in a very narrow cone, the axis of which is along the direction of incidence. On the other hand, if the point of incidence is not near the edge of the particle, diffraction is unimportant and only refraction and reflection in accordance with Snell's laws [Eqs. (1.6.4) and (1.6.6)] with certain modifications need be considered.

Figure 3.8.1 illustrates the geometry. A particle of radius a is irradiated by a ray in the direction $\Theta = 0$. Upon contact with the surface of the particle, part of the incident radiation is reflected into a new direction $\Theta = \Theta_0$, designated in

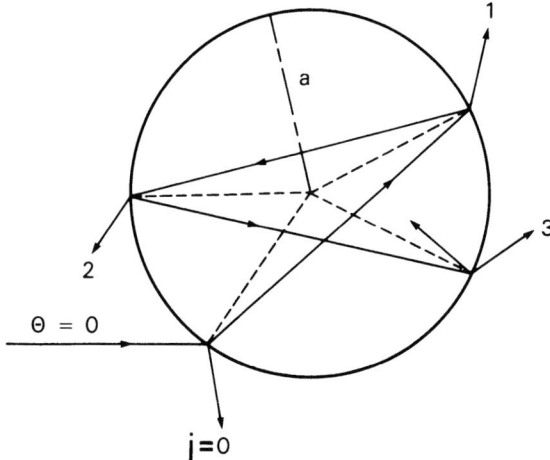

Fig. 3.8.1 Reflected and refracted components of a ray incident on a homogeneous particle of radius $a \gg \lambda$ in the direction $\Theta = 0$. The components of the scattered radiation field are labeled $j = 0, 1, 2, 3, \ldots$ in the order of scattering.

the figure by the component $j = 0$. According to Snell's law of reflection the angles of incidence and reflection (relative to the surface normal at the point of contact) are equal for this component. The remainder of the ray is directed according to Snell's law for refraction into another direction $\Theta = \Theta'_0$, and travels through the particle with some loss due to absorption until it again contacts the particle's surface. Again, part of the ray is refracted into the direction $\Theta = \Theta_1$ (denoted by $j = 1$ in the figure) and part is internally reflected into the direction $\Theta = \Theta'_1$. This process is repeated until all the radiation has been absorbed by or lost from the particle. The components $j = 0, 1, 2, \ldots$ comprise the 'scattered' radiation.

In practice a few corrections to this procedure are required. The simple theory fails at a focus or focal line, and phase shifts occur throughout the process. If these phase shifts do not average out (i.e., if the coherence of all rays is not completely destroyed), interference between outgoing rays will disallow a simple addition of outgoing fluxes. In order to minimize this complication it is necessary to introduce the further requirement that $a(n_r - 1) \gg \lambda$, where n_r is the real part of the refractive index.

Cloud particles are basically dielectrics and thus tend to be rather poor reflectors. Hence, even in the absence of strong absorption, only the lower values of j will contribute significantly to the scattered radiation field. From the figure it would appear that the component $j = 1$ plus Fraunhofer diffraction contribute almost all the radiation at small scattering angles (forward directions), and that the components $j = 0, 2$ contribute most of the scattered radiation at other angles. It would also seem that the component $j = 1$ dominates all other components in the amount of radiation contributed to the scattered field, and that the 'spread' of the forward peak of scattered radiation is governed primarily by the refractive index; i.e., the larger the real part of the refractive index, the larger the spread will be. Since Fraunhofer diffraction contributes at least half the scattered radiation field, there should be a very intense 'spike' to the angular scattering pattern around $\Theta = 0$ in addition to the forward scattering lobe contributed by the component $j = 1$.

b. Rigorous scattering theory; general solution

Particles of intermediate size are too large to treat as being entirely contained in a homogeneous periodic electric field, and too small to neglect the radius of curvature of a surface element, where such an element is also large enough to contain an optical ray. In addition, Fraunhofer diffraction cannot be separated from refraction and reflection because of mutual interference effects, and phase shifts inside the particle also contribute substantially to interference. A rigorous solution using Maxwell's equations is required, subject to the appropriate boundary conditions. Such a solution for spherical particles was obtained by Mie (1908). An excellent detailed account of the Mie theory is given in Born & Wolf (1959). Below we sketch out the basic principles involved, expanding somewhat on the procedure followed by van de Hulst (1957).

The problem is cast in spherical coordinates in order to ensure the surface of the particle coincides with one of the coordinate surfaces, making it much easier to impose the boundary conditions correctly. Solutions to the scalar wave equation are then used to derive two vector fields, linear combinations of which satisfy Maxwell's equations. Particular integrals are found by requiring continuity of the field components across the particle surface. The scattered field at large distances from the particle is then evaluated, leading to explicit expressions for particle cross sections and the single scattering phase function.

We assume the particle is imbedded in a nonconducting medium, so that the applied electromagnetic field of circular frequency ω is given by

$$\mathbf{E} = \mathbf{E}_0 \, e^{-i\omega t} \tag{3.8.3}$$

and

$$\mathbf{H} = \mathbf{H}_0 \, e^{-i\omega t}, \tag{3.8.4}$$

where $\mathbf{E}_0$ and $\mathbf{H}_0$ are independent of time. From Eqs. (1.3.1) and (1.3.2) it follows that

$$\nabla \times \mathbf{H}_0 = -i\omega\left(\varepsilon + i\frac{\sigma}{\omega}\right)\mathbf{E}_0 \qquad (3.8.5)$$

and

$$\nabla \times \mathbf{E}_0 = i\omega\mu\mathbf{H}_0. \qquad (3.8.6)$$

Because the particles under consideration are spherical, solutions to Eqs. (3.8.5) and (3.8.6) are sought in spherical coordinates (r, θ, ϕ) with the origin at the particle center. We have

$$\left.\begin{array}{l} x = r\sin\theta\cos\phi \\ y = r\sin\theta\sin\phi \\ z = r\cos\theta \end{array}\right\}. \qquad (3.8.7)$$

The components of any vector $\mathbf{A}$ are related in the two systems by

$$\left.\begin{array}{l} A_r = A_x\sin\theta\cos\phi + A_y\sin\theta\sin\phi + A_z\cos\theta \\ A_\theta = A_x\cos\theta\cos\phi + A_y\cos\theta\sin\phi - A_z\sin\theta \\ A_\phi = -A_x\sin\phi + A_y\cos\phi \end{array}\right\}. \qquad (3.8.8)$$

Let ψ be a solution to the scalar wave equation

$$\frac{1}{r}\frac{\partial^2(r\psi)}{\partial r^2} + \frac{1}{r^2\sin\theta}\frac{\partial}{\partial\theta}\left(\sin\theta\frac{\partial\psi}{\partial\theta}\right) + \frac{1}{r^2\sin^2\theta}\frac{\partial^2\psi}{\partial\phi^2} + k^2\psi = 0, \qquad (3.8.9)$$

where the propagation constant k is given by Eq. (1.3.24). Equation (3.8.9) is separable and has elementary single-valued solutions of the form

$$\psi_{mn} \sim \left\{\begin{array}{l} \cos m\phi \\ \sin m\phi \end{array}\right\} P_n^m(\cos\theta)z_n(kr) \quad (n \geq m \geq 0), \qquad (3.8.10)$$

where the P_n^m are associated Legendre polynomials and the z_n, given by

$$\frac{d^2(rz_n)}{dr^2} + \left[k^2 - \frac{n(n+1)}{r^2}\right]rz_n = 0, \qquad (3.8.11)$$

3.8 Cloud and aerosol particles

are generalized spherical functions derived from the ordinary cylindrical (Bessel, Neumann, and Hankel) functions $Z_{n+\frac{1}{2}}$ by the relation

$$z_n(\rho) = \left(\frac{\pi}{2\rho}\right)^{\frac{1}{2}} Z_{n+\frac{1}{2}}(\rho). \tag{3.8.12}$$

Now consider the vector

$$\mathbf{A}(\psi) = \nabla \times (\mathbf{r}\psi). \tag{3.8.13}$$

The curl of any vector

$$\mathbf{A} = \hat{\mathbf{r}} A_r + \hat{\boldsymbol{\theta}} A_\theta + \hat{\boldsymbol{\phi}} A_\phi \tag{3.8.14}$$

in spherical coordinates is given by (see Appendix 1)

$$\nabla \times \mathbf{A} = \frac{\hat{\mathbf{r}}}{r^2 \sin\theta} \left[\frac{\partial}{\partial \theta}(r A_\phi \sin\theta) - \frac{\partial}{\partial \phi}(r A_\theta) \right]$$
$$+ \frac{\hat{\boldsymbol{\theta}}}{r \sin\theta} \left[\frac{\partial}{\partial \phi}(A_r) - \frac{\partial}{\partial r}(r A_\phi \sin\theta) \right]$$
$$+ \frac{\hat{\boldsymbol{\phi}}}{r} \left[\frac{\partial}{\partial r}(r A_\theta) - \frac{\partial}{\partial \theta}(A_r) \right], \tag{3.8.15}$$

where $\hat{\mathbf{r}}$, $\hat{\boldsymbol{\theta}}$, and $\hat{\boldsymbol{\phi}}$ are the basis vectors. Hence, from Eqs. (3.8.13) and (3.8.15),

$$\mathbf{A}(\psi) = \hat{\boldsymbol{\theta}} \frac{1}{\sin\theta} \frac{\partial \psi}{\partial \phi} - \hat{\boldsymbol{\phi}} \frac{\partial \psi}{\partial \theta} \tag{3.8.16}$$

and

$$\nabla \times \mathbf{A}(\psi) = -\frac{\hat{\mathbf{r}}}{r \sin\theta} \left[\frac{\partial}{\partial \theta}\left(\sin\theta \frac{\partial \psi}{\partial \theta}\right) + \frac{\partial}{\partial \phi}\left(\frac{1}{\sin\theta} \frac{\partial \psi}{\partial \phi}\right) \right]$$
$$+ \frac{\hat{\boldsymbol{\theta}}}{r} \frac{\partial}{\partial r}\left(r \frac{\partial \psi}{\partial \theta}\right) + \frac{\hat{\boldsymbol{\phi}}}{r \sin\theta} \frac{\partial}{\partial r}\left(r \frac{\partial \psi}{\partial \phi}\right), \tag{3.8.17}$$

or, from Eq. (3.8.9),

$$\nabla \times \mathbf{A}(\psi) = \hat{\mathbf{r}} \left[\frac{\partial^2}{\partial r^2}(r\psi) + k^2 r\psi \right]$$
$$+ \frac{\hat{\boldsymbol{\theta}}}{r} \frac{\partial}{\partial r}\left(r \frac{\partial \psi}{\partial \theta}\right) + \frac{\hat{\boldsymbol{\phi}}}{r \sin\theta} \frac{\partial}{\partial r}\left(r \frac{\partial \psi}{\partial \phi}\right). \tag{3.8.18}$$

Applying the curl operator to both sides of Eq. (3.8.18) yields, after some reduction,

$$\nabla \times [\nabla \times \mathbf{A}(\psi)] = \frac{\hat{\boldsymbol{\theta}}}{\sin\theta} k^2 \frac{\partial \psi}{\partial \phi} - \hat{\boldsymbol{\phi}} k^2 \frac{\partial \psi}{\partial \theta}. \tag{3.8.19}$$

For any vector $\mathbf{q}$, $\nabla \cdot (\nabla \times \mathbf{q}) \equiv 0$. Thus [see Eq. (3.8.13)]

$$\nabla \cdot \mathbf{A}(\psi) \equiv 0. \tag{3.8.20}$$

Hence, from Eqs. (1.3.5), (3.8.16), and (3.8.19),

$$\nabla^2 \mathbf{A}(\psi) + k^2 \mathbf{A}(\psi) = 0, \tag{3.8.21}$$

i.e., $\mathbf{A}(\psi)$ is a solution of the vector wave equation.

In like manner the vector

$$k\mathbf{C}(\psi) = \nabla \times \mathbf{A}(\psi) \tag{3.8.22}$$

can also be shown to be a solution of Eq. (3.8.21), leading to

$$k\mathbf{A}(\psi) = \nabla \times \mathbf{C}(\psi), \tag{3.8.23}$$

as can be verified directly by applying the curl operator to both sides of Eq. (3.8.22) and comparing with Eq. (3.8.21). In making the comparison, we use Eqs. (1.3.5), (3.8.20), and

$$\nabla \cdot \mathbf{C}(\psi) \equiv 0, \tag{3.8.24}$$

which follows immediately from Eq. (3.8.22).

Thus, if $\psi = u$ and $\psi = v$ are two independent solutions of the scalar wave equation (3.8.9), the associated vector fields are $\mathbf{A}(u)$, $\mathbf{C}(u)$, $\mathbf{A}(v)$, and $\mathbf{C}(v)$. Direct substitution using Eqs. (1.3.24), (3.8.5), (3.8.6), (3.8.22), and (3.8.23) demonstrates that

$$\mathbf{E}_0 = -\omega\mu[\mathbf{A}(v) + \mathrm{i}\mathbf{C}(u)] \tag{3.8.25}$$

and

$$\mathbf{H}_0 = k[-\mathbf{A}(u) + \mathrm{i}\mathbf{C}(v)] \tag{3.8.26}$$

are solutions to Maxwell's equations. The full vector components are [see Eqs. (3.8.16), (3.8.18), and (3.8.22)]

$$\left. \begin{array}{l} A_r(\psi) = 0; \; kC_r(\psi) = \dfrac{\partial^2(r\psi)}{\partial r^2} + k^2 r\psi \\[6pt] A_\theta(\psi) = \dfrac{1}{r\sin\theta}\dfrac{\partial(r\psi)}{\partial\phi}; \quad kC_\theta(\psi) = \dfrac{1}{r}\dfrac{\partial^2(r\psi)}{\partial r \partial\theta} \\[6pt] A_\phi(\psi) = -\dfrac{1}{r}\dfrac{\partial(r\psi)}{\partial\theta}; \quad kC_\phi(\psi) = \dfrac{1}{r\sin\theta}\dfrac{\partial^2(r\psi)}{\partial r \partial\phi} \end{array} \right\}. \qquad (3.8.27)$$

This demonstrates that the components of $\mathbf{E}_0$ and $\mathbf{H}_0$ can be expressed in terms of the scalars u and v [see Eq. (3.8.10)] and their first and second derivatives.

c. Particular solutions and boundary conditions

Now let the origin of the coordinate system be the particle center, and the direction of propagation of the incident radiation be along the z- (or $\theta = 0$) axis in the positive direction. Further, let the electric vibration of the incident wave be in the x–z plane ($\phi = 0°$), and k_1 and k_2 be the propagation constants outside and inside the particle of radius a, respectively.

It is useful to think of the total field ($\mathbf{E}_0$, $\mathbf{H}_0$) as being composed of three partial fields: an incident and a scattered field outside the particle, as well as a field within the particle. Solutions for these fields can be expressed as expansions in u and v with undetermined coefficients, each term representing a particular integral. The coefficients can then be determined from the boundary conditions, which are that the four tangential components of the total field $E_{0\theta}$, $E_{0\phi}$, $H_{0\theta}$, and $H_{0\phi}$ remain continuous across the spherical surface $r = a$ even though the propagation constant k and magnetic permeability μ are discontinuous. The conditions that the radial components E_{0r} and H_{0r} are also continuous across the surface then follow automatically from Maxwell's equations.

We recall from Eqs. (3.8.3) and (3.8.4) and the related discussion that $\mathbf{E}_0$ and $\mathbf{H}_0$ of the incident field are independent of time, and define the amplitude of the electric vector to be normalized to unity. Because of the coordinate system chosen, the Poynting vector is along the positive z-direction and $\mathbf{E}_0$ is in the x–y plane. Hence, from Eq. (3.8.3), the electric field for the incident wave is [see Eq. (1.3.22)]

$$\mathbf{E} = \hat{\mathbf{i}} e^{ik_1 z - i\omega t}. \qquad (3.8.28)$$

With the aid of Eqs. (3.8.7) and (3.8.8), the time-independent component of $\mathbf{E}$ in the radial direction is

$$E_{0r} = e^{ik_1 r \cos\theta} \sin\theta \cos\phi. \tag{3.8.29}$$

Differentiation of the identity

$$e^{ik_1 r \cos\theta} = \sum_{n=0}^{\infty} i^n (2n+1) P_n(\cos\theta) j_n(k_1 r) \tag{3.8.30}$$

with respect to θ leads to

$$E_{0r} = -\cos\phi \frac{1}{k_1 r} \sum_{n=1}^{\infty} i^{n+1}(2n+1) P_n^1(\cos\theta) j_n(k_1 r), \tag{3.8.31}$$

where we use the relations

$$\frac{\partial P_n(\cos\theta)}{\partial \theta} = -P_n^1(\cos\theta); \qquad P_0^1(\cos\theta) = 0. \tag{3.8.32}$$

On the other hand, from Eqs. (3.8.25) and (3.8.27) we also find

$$E_{0r} = -\frac{i\omega\mu_1}{k_1}\left[\frac{\partial^2(ru)}{\partial r^2} + k_1^2(ru)\right]. \tag{3.8.33}$$

Let

$$u = \sum_{n=1}^{\infty} \alpha_n u_{1n}, \tag{3.8.34}$$

where the α_n are undetermined coefficients. From the forms of Eqs. (3.8.10), (3.8.11), (3.8.31), and (3.8.33) we find

$$E_{0r} = -\frac{i\omega\mu_1}{k_1} \sum_{n=1}^{\infty} \frac{n(n+1)}{r} \alpha_n u_{1n}, \tag{3.8.35}$$

where

$$u_{1n} = \cos\phi \, P_n^1(\cos\theta) j_n(k_1 r) \tag{3.8.36}$$

and

$$\frac{\partial^2(ru_{1n})}{\partial r^2} + k_1^2(ru_{1n}) = \frac{n(n+1)}{r}u_{1n}. \tag{3.8.37}$$

Solving for α_n yields

$$\alpha_n = \frac{i^n}{\omega\mu_1}\frac{2n+1}{n(n+1)}, \tag{3.8.38}$$

leading to

$$u = \frac{1}{\omega\mu_1}\cos\phi\sum_{n=1}^{\infty}i^n\frac{2n+1}{n(n+1)}P_n^1(\cos\theta)j_n(k_1r). \tag{3.8.39}$$

Returning to Eqs. (3.8.6) and (3.8.28), and using Eqs. (3.8.7) and (3.8.8), it follows that the radial component of the magnetic field for the incident wave is

$$H_{0r} = \frac{k_1}{\omega\mu_1}e^{ik_1r\cos\theta}\sin\theta\sin\phi. \tag{3.8.40}$$

An analysis of H_{0r} completely analogous to that just undertaken for E_{0r} reveals that

$$v = -\frac{1}{\omega\mu_1}\sin\phi\sum_{n=1}^{\infty}i^n\frac{2n+1}{n(n+1)}P_n^1(\cos\theta)j_n(k_1r). \tag{3.8.41}$$

Equations (3.8.39) and (3.8.41) together suffice to describe the complete incident field.

From the form of the solutions for the incident wave, and considerations of conditions that the radial components E_{0r} and H_{0r} of the scattered wave must obey, it appears that the scattered field can be constructed from the functions

$$u = -\frac{1}{\omega\mu_1}\cos\phi\sum_{n=1}^{\infty}a_n i^n\frac{2n+1}{n(n+1)}P_n^1(\cos\theta)h_n^{(1)}(k_1r) \tag{3.8.42}$$

and

$$v = \frac{1}{\omega\mu_1}\sin\phi\sum_{n=1}^{\infty}b_n i^n\frac{2n+1}{n(n+1)}P_n^1(\cos\theta)h_n^{(1)}(k_1r). \tag{3.8.43}$$

The spherical Hankel function of the first kind has been chosen because of its asymptotic behavior as $r \to \infty$, which is

$$h_n^{(1)}(x) = \frac{(-i)^{n+1}}{x} e^{ix}, \qquad (3.8.44)$$

and this, when multiplied by $e^{-i\omega t}$, represents an outgoing spherical wave as required. The constants a_n and b_n are coefficients to be determined by the boundary conditions.

The field inside the particle can be constructed from

$$u = \frac{1}{\omega\mu_2} \cos\phi \sum_{n=1}^{\infty} c_n i^n \frac{2n+1}{n(n+1)} P_n^1(\cos\theta) j_n(k_2 r) \qquad (3.8.45)$$

and

$$v = -\frac{1}{\omega\mu_2} \sin\phi \sum_{n=1}^{\infty} d_n i^n \frac{2n+1}{n(n+1)} P_n^1(\cos\theta) j_n(k_2 r). \qquad (3.8.46)$$

The choice of the spherical Bessel function j_n is based on the requirements that k_2 and the fields be finite at the origin. The undetermined coefficients c_n and d_n can be expressed in terms of a_n and b_n.

The conditions at $r = 0$ and $r = \infty$ have just been imposed, which is equivalent to imposing general conditions on E_{0r} and H_{0r} at these locations. The remaining four conditions needed for a complete solution are the requirements that $E_{0\theta}$, $E_{0\phi}$, $H_{0\theta}$, and $H_{0\phi}$ be continuous across the boundary $r = a$. From Eqs. (3.8.25), (3.8.26), and (3.8.27), these components are found to be

$$\left. \begin{aligned} E_{0\theta} &= -\frac{\omega}{\sin\theta} \frac{\partial}{\partial\phi}[\mu v] - \frac{i\omega}{r} \frac{\partial}{\partial\theta}\left[\frac{\mu}{k} \frac{\partial(ru)}{\partial r}\right] \\ E_{0\phi} &= \omega\frac{\partial}{\partial\theta}[\mu v] - \frac{i\omega}{r\sin\theta} \frac{\partial}{\partial\phi}\left[\frac{\mu}{k} \frac{\partial(ru)}{\partial r}\right] \\ H_{0\theta} &= -\frac{1}{\sin\theta} \frac{\partial}{\partial\phi}[ku] + \frac{i}{r} \frac{\partial}{\partial\theta}\left[\frac{\partial(rv)}{\partial r}\right] \\ H_{0\phi} &= \frac{\partial}{\partial\theta}[ku] + \frac{i}{r\sin\theta} \frac{\partial}{\partial\phi}\left[\frac{\partial(rv)}{\partial r}\right] \end{aligned} \right\}. \qquad (3.8.47)$$

The quantities in brackets must each be continuous across the boundary $r = a$. The form of the equations for u and v then guarantees that the derivatives of the brackets

3.8 Cloud and aerosol particles

with respect to the angular coordinates (θ, ϕ) are also continuous at $r = a$, which, according to Eq. (3.8.47), is sufficient to satisfy the boundary conditions.

Let:

$$\left.\begin{array}{ll} x = k_1 a; & y = k_2 a \\[6pt] \psi_n(z) = z j_n(z); & \psi'_n(z) = \dfrac{\partial [z j_n(z)]}{\partial z} \\[10pt] \zeta_n(z) = z h_n^{(1)}(z); & \zeta'_n(z) = \dfrac{\partial [z h_n^1(z)]}{\partial z} \end{array}\right\}. \qquad (3.8.48)$$

In order for the tangential components of the fields inside and outside the particle to match at the surface, it is found from Eqs. (3.8.39), (3.8.41), (3.8.42), (3.8.43), (3.8.45), and (3.8.46) that continuity of the bracketed quantities in Eq. (3.8.47) requires

$$\left.\begin{array}{ll} [\mu v]: & \dfrac{1}{k_1}\psi_n(x) - b_n \dfrac{1}{k_1}\zeta_n(x) = d_n \dfrac{1}{k_2}\psi_n(y) \\[10pt] \left[\dfrac{\mu}{k}\dfrac{\partial (ru)}{\partial r}\right]: & \dfrac{1}{k_1}\psi'_n(x) - a_n \dfrac{1}{k_1}\zeta'_n(x) = c_n \dfrac{1}{k_2}\psi'_n(y) \\[10pt] [ku]: & \dfrac{1}{\mu_1}\psi_n(x) - a_n \dfrac{1}{\mu_1}\zeta_n(x) = c_n \dfrac{1}{\mu_2}\psi_n(y) \\[10pt] \left[\dfrac{\partial (rv)}{\partial r}\right]: & \dfrac{1}{\mu_1}\psi'_n(x) - b_n \dfrac{1}{\mu_1}\zeta'_n(x) = d_n \dfrac{1}{\mu_2}\psi'_n(y) \end{array}\right\}. \qquad (3.8.49)$$

Eliminating c_n and d_n leads to

$$a_n = \frac{\mu_1 k_2 \psi_n(y)\psi'_n(x) - \mu_2 k_1 \psi_n(x)\psi'_n(y)}{\mu_1 k_2 \psi_n(y)\zeta'_n(x) - \mu_2 k_1 \zeta_n(x)\psi'_n(y)} \qquad (3.8.50)$$

and

$$b_n = \frac{\mu_1 k_2 \psi_n(x)\psi'_n(y) - \mu_2 k_1 \psi_n(y)\psi'_n(x)}{\mu_1 k_2 \zeta_n(x)\psi'_n(y) - \mu_2 k_1 \psi_n(y)\zeta'_n(x)}, \qquad (3.8.51)$$

completing the solution for the scattered field.

122   *Interaction of radiation with matter*

d. The far field; phase function and efficiency factors

The scattered field as $r \to \infty$ is the one that is measured. From Eqs. (3.8.42), (3.8.43), (3.8.44), (3.8.47), and the relations

$$\left.\begin{aligned}\pi_n(\cos\theta) &= \frac{1}{\sin\theta}P_n^1(\cos\theta)\\ \tau_n(\cos\theta) &= \frac{d}{d\theta}P_n^1(\cos\theta)\end{aligned}\right\}, \qquad (3.8.52)$$

we find that, as $r \to \infty$,

$$\left.\begin{aligned}E_{0\theta} &= \frac{\omega\mu_1}{k_1}H_{0\phi} = \frac{i}{k_1 r}e^{ik_1 r}\cos\phi\, S_2(\theta)\\ -E_{0\phi} &= \frac{\omega\mu_1}{k_1}H_{0\theta} = \frac{i}{k_1 r}e^{ik_1 r}\sin\phi\, S_1(\theta)\end{aligned}\right\}, \qquad (3.8.53)$$

where

$$\left.\begin{aligned}S_1(\theta) &= \sum_{n=1}^{\infty}\frac{2n+1}{n(n+1)}[a_n\pi_n(\cos\theta) + b_n\tau_n(\cos\theta)]\\ S_2(\theta) &= \sum_{n=1}^{\infty}\frac{2n+1}{n(n+1)}[a_n\tau_n(\cos\theta) + b_n\pi_n(\cos\theta)]\end{aligned}\right\}. \qquad (3.8.54)$$

The radial components (E_{0r}, H_{0r}) are of order $(1/r^2)$ as $r \to \infty$, and thus do not contribute; that is, the scattered wave becomes transverse at large r. In order to obtain the single-scattering phase function we are interested only in relative fluxes, and can set the flux equal to the square of the real amplitude of the electric vector. The two polarization components of the flux are

$$\left.\begin{aligned}\pi F_1 &= |E_{0\phi}|^2 = \frac{1}{k_1^2 r^2}\sin^2\phi\, |S_1(\theta)|^2\\ \pi F_2 &= |E_{0\theta}|^2 = \frac{1}{k_1^2 r^2}\cos^2\phi\, |S_2(\theta)|^2\end{aligned}\right\}. \qquad (3.8.55)$$

Averaging over all states of polarization and adding yields (since $\langle\cos^2\phi\rangle = \langle\sin^2\phi\rangle = \frac{1}{2}$)

$$\pi F = \langle\pi F_1\rangle + \langle\pi F_2\rangle = \frac{1}{2k_1^2 r^2}[|S_1(\theta)|^2 + |S_2(\theta)|^2] \qquad (3.8.56)$$

3.8 Cloud and aerosol particles

for the total scattered flux in the direction θ. The single scattering phase function then becomes

$$cp(\cos\theta) = |S_1(\theta)|^2 + |S_2(\theta)|^2, \qquad (3.8.57)$$

where c is a normalization constant.

The value for c is determined as follows. Let πF_0 be the incident flux and πF be the scattered flux at a distance $r \gg a$ from the particle. According to Eq. (3.8.28), the magnitude of the incident flux is

$$\pi F_0 = |E_{0z}|^2 = 1. \qquad (3.8.58)$$

The fraction of πF_0 scattered by a particle of effective scattering cross section χ_s is $\chi_s \pi F_0$. This flux is scattered in all directions, and the fraction crossing a differential area dA normal to the direction of propagation along r is $\pi F r^2\, d\omega$, where $d\omega = \sin\theta\, d\theta\, d\phi$ is the element of solid angle subtended at the particle by dA. Hence, by the law of the conservation of energy,

$$\chi_s F_0 = \int_\omega r^2 F\, d\omega, \qquad (3.8.59)$$

where the integration is over all solid angles.

We define the efficiency factors for extinction, scattering, and absorption by, respectively,

$$\left. \begin{array}{l} Q_E = \chi_E \pi a^2 \\ Q_S = \chi_S \pi a^2 \\ Q_A = \chi_A \pi a^2 \end{array} \right\}, \qquad (3.8.60)$$

where the χ_j ($j = $ E, S, A) are the corresponding cross sections. By definition

$$Q_E = Q_S + Q_A, \qquad (3.8.61)$$

since the fraction of flux extinguished by the particle is just the sum of the fractions absorbed and scattered.

It follows from Eqs. (3.8.56), (3.8.58), (3.8.59), and (3.8.60) that

$$Q_S = \frac{1}{k_1^2 a^2} \int_0^\pi [|S_1(\theta)|^2 + |S_2(\theta)|^2] \sin\theta\, d\theta. \qquad (3.8.62)$$

Though somewhat detailed, it can be shown from Eqs. (3.8.52) and (3.8.54) that Eq. (3.8.62) reduces to

$$Q_S = \frac{2}{k_1^2 a^2} \sum_{n=1}^{\infty} (2n+1)\left(|a_n|^2 + |b_n|^2\right). \quad (3.8.63)$$

Integration over doubly infinite series are involved, although orthogonality relations between π_n and τ_n result in most of the integrations over products being equal to zero. Integration of Eq. (3.8.57), with the aid of Eqs. (2.1.5), (3.8.62), and the relation $\tilde{\omega}_0 = Q_S/Q_E$, then yields

$$c = \tfrac{1}{2} k_1^2 a^2 Q_E. \quad (3.8.64)$$

The last quantity to be determined is Q_E [Q_A then follows immediately from Eq. (3.8.61)]. Because the incident field consists of a beam of radiation parallel to the z-axis [see Eq. (3.8.28)] the flux, in the absence of the particle, is a constant independent of z. If the particle is then inserted into the beam, the fractional amount by which the flux is decreased at large z is equal to χ_E, the cross section for extinction.

However, the incident and scattered fields along the z-axis ($\theta = 0$) are not completely incoherent. In addition to absorption and Fresnel reflection and refraction, interference effects are important, and it is necessary to add the field amplitudes before computing the flux loss due to extinction by the particle. According to Eqs. (3.8.8), the component $E_{0\theta}^i$ of the incident wave, given by Eq. (3.8.28), is (for $\theta = 0$)

$$E_{0\theta}^i = e^{ik_1 z} \cos\phi, \quad (3.8.65)$$

while that for the scattered wave $E_{0\theta}^s$ is given approximately by [see Eq. (3.8.53)]

$$E_{0\theta}^s = \frac{i}{k_1 z} e^{ik_1 r} \cos\phi\, S(0), \quad (3.8.66)$$

where

$$S(0) = S_1(0) = S_2(0) = \frac{1}{2} \sum_{n=1}^{\infty} (2n+1)(a_n + b_n). \quad (3.8.67)$$

Equation (3.8.66) is very accurate for small, nonzero values of θ, as long as $\cos\theta \sim 1$. However, even though $z \sim r$ for slightly off-axis waves, important interference effects require that the distinction be maintained in the exponentials. Consider a plane normal to the z-axis and intersecting it at large z. Let the intersection be the origin of an (x, y) coordinate system in the plane. Restrict the limits of both x and y to $\pm D$, where D is small compared with z but large compared with $z \tan\theta_0$, where θ_0 is the scattering angle within which interference between the

3.8 Cloud and aerosol particles

incident and scattered waves is basically contained. Then, to a sufficient degree of approximation,

$$r \sim z\left(1 + \frac{x^2 + y^2}{2z^2}\right) \qquad (3.8.68)$$

over the range of x and y of interest.

From Eqs. (3.8.65) through (3.8.68), the total θ-component of the time-independent electric field at large z, with $\theta \leq \theta_0$, is

$$E_{0\theta} = E_{0\theta}^i + E_{0\theta}^s = \cos\phi\, e^{ik_1 z}\left[1 + \frac{i}{k_1 z} S(0) e^{ik_1(x^2+y^2)/2z}\right]. \qquad (3.8.69)$$

In like manner it is readily demonstrated that the ϕ-component of this field is

$$E_{0\phi} = E_{0\phi}^i + E_{0\phi}^s = -\sin\phi\, e^{ik_1 z}\left[1 + \frac{i}{k_1 z} S(0) e^{ik_1(x^2+y^2)/2z}\right]. \qquad (3.8.70)$$

Both the real and imaginary parts of the complex quantity containing $S(0)$ in the brackets are small compared with unity. Hence, from Eqs. (3.8.55) and (3.8.56), the flux at large z is very nearly

$$\pi F = 1 + \frac{2}{k_1 z}\mathrm{Re}\left[iS(0)e^{ik_1(x^2+y^2)/2z}\right]. \qquad (3.8.71)$$

From the previous discussion it is apparent that

$$\chi_E = \int_{-D}^{D}\int_{-D}^{D} (\pi F_0 - \pi F)\, dx\, dy, \qquad (3.8.72)$$

where the integrations are extended over large enough ranges to include all interference effects. Substituting Eqs. (3.8.58) and (3.8.71) into Eq. (3.8.72) yields

$$\chi_E = -\frac{2}{k_1 z}\mathrm{Re}\left[iS(0)\int_{-D}^{D}\int_{-D}^{D} e^{ik_1(x^2+y^2)/2z}\, dx\, dy\right]. \qquad (3.8.73)$$

Letting $D \to \infty$ is now legitimate because, by postulate, almost all interference phenomena have been accounted for within the limits $\pm D$, and extending these limits without bound adds almost nothing to this integral. The value of each integral becomes

$$\int_{-\infty}^{\infty} e^{ik_1 x^2/2z} \, dx = (i+1)\left(\frac{\pi z}{k_1}\right)^{\frac{1}{2}}. \tag{3.8.74}$$

Substituting into Eq. (3.8.73) according to Eqs. (3.8.60) and (3.8.74), we finally obtain

$$Q_E = \frac{4}{k_1^2 a^2} \text{Re}[S(0)]. \tag{3.8.75}$$

Expressions have been derived for Q_S, Q_E, and $p(\cos \Theta)$. These are the main single-scattering parameters needed, along with the molecular absorption coefficients considered earlier in this chapter, to put the theory of radiative transfer developed in Chapter 2 on a quantitative basis. In reviewing the various formulas developed, it is clear that the quantity ka, given by

$$ka = \frac{2\pi a}{\lambda} n, \tag{3.8.76}$$

where $n = n_r + in_i$ is the complex refractive index [cf. Eq. (1.3.24)], plays a central role in scattering theory. The factor $2\pi a/\lambda$ is sometimes called the size parameter.

Figure 3.8.2 illustrates two phase functions, one for water drops large compared with the wavelength of incident radiation, and one for smaller aerosol particles. The one for large water drops shows several interesting features: (1) the strong, narrow Fraunhofer diffraction 'aureole' near $\Theta = 0°$, (2) the broader forward scattering lobe covering the range $5° < \Theta < 80°$, (3) the rainbow, centered about $\Theta = 140°$, and (4) the back scattering 'glory' near $\Theta = 180°$.

The $j = 1$ component in Fig. 3.8.1 is basically responsible for the forward scattering lobe outside the aureole, while the $j = 2$ component gives rise to the rainbow. In the latter case there is a minimum deviation of the once internally reflected ray near $\Theta = 140°$, and this varies with changing refractive index. Much of the radiation comprising the $j = 2$ component passes close to this angle of minimum deviation, leading to an enhanced intensity around this angle. The changing refractive index with wavelength gives rise to the colors observed.

The aureole and glory are both due primarily to interference, although there are basic differences. The aureole depends on the interference of waves diffracted around the particle, whereas the glory is constructed from the interference of waves undergoing refraction. Edifying discussions of the physical principles can be found in van de Hulst (1957).

Figures 3.8.3 and 3.8.4 are plots of the efficiency factors Q_E and Q_A as functions of the size parameter $2\pi a/\lambda$ for two different refractive indices. The difference between Q_E and Q_A is Q_S. The latter figure represents a much more absorbing medium than the former.

3.8 Cloud and aerosol particles

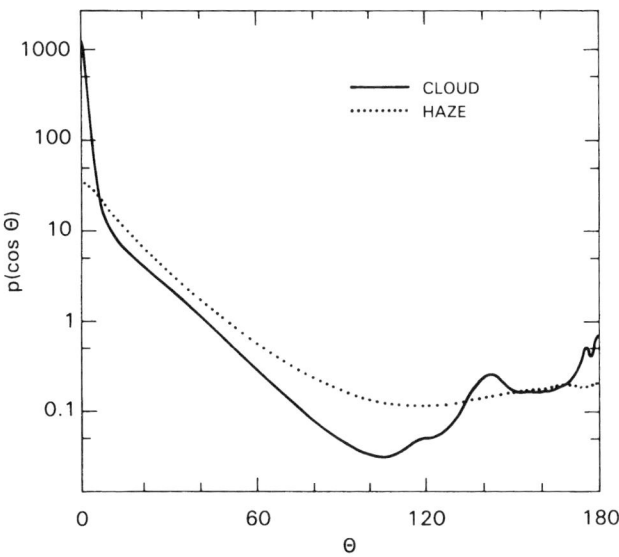

Fig. 3.8.2 Single scattering functions for a water cloud and an aerosol haze, each associated with a real refractive index $n = 1.33$. The cloud and haze particle size distributions peak at particle radii $a \sim 4.0$ and $a \sim 0.05$ μm, respectively. Computations were made for a wavelength $\lambda = 0.8189$ μm (after Hansen, 1969).

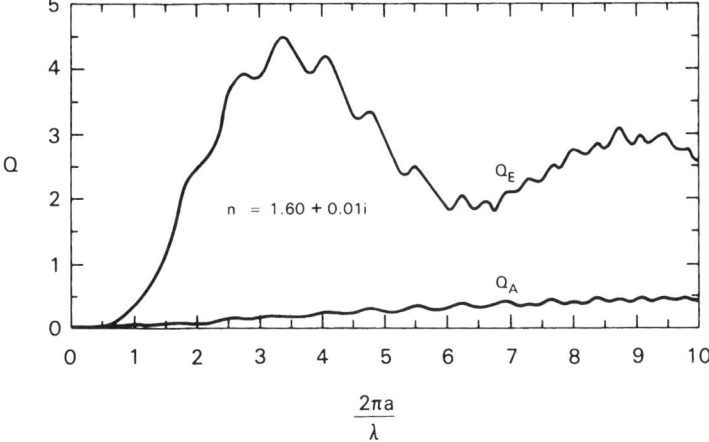

Fig. 3.8.3 Efficiency factors for extinction, Q_E, and absorption, Q_A, for weakly absorbing particles.

Several interesting features are evident in Fig. 3.8.3. The limiting value for Q_E for $a \gg \lambda$ is $Q_E = 2$. This is because the radiation diffracted around the particle is included in the computation for Q_E, and this comprises half the extinguished radiation for particles large compared with the wavelength of incident radiation. A second feature is that several large-scale maxima and minima occur in the extinction curve, decreasing in amplitude with increasing a/λ. The maxima are due to constructive interference between diffracted and transmitted radiation, while the minima are due to destructive interference. Comparable periodic fluctuations do not occur in Fig. 3.8.4, because the transmitted radiation is strongly attenuated in the more absorbing medium, reducing interference to a negligible effect.

Another interesting feature of the Q-functions in Fig. 3.8.3 is a small-amplitude 'ripple' superimposed on each of the main curves. This ripple appears to be the result of interference between the Fraunhofer diffraction peak and a surface wave that takes occasional short-cuts through the particle just under the outer boundary. Because this phenomenon depends on unattenuated transmission to maximize the amplitude of the ripple, the strongly absorbing medium in Fig. 3.8.4 fails to show a corresponding effect.

For particle compositions of interest to us there appears to be at least one maximum in Q_E. It is a general rule that the first maximum is defined by

$$\frac{\pi a}{\lambda}|n-1| \sim 1+\varepsilon, \quad (3.8.77)$$

where $\varepsilon \ll 1$. For particle radii considerably smaller than the value contained in Eq. (3.8.77), extinction cross sections decrease dramatically, and much simpler expressions than required by the Mie theory can often be used to describe the extinction characteristics with adequate precision.

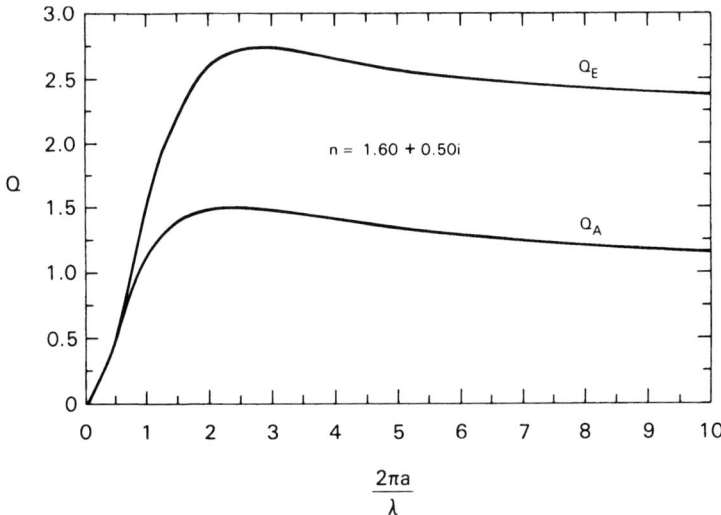

Fig. 3.8.4 Efficiency factors for extinction, Q_E, and absorption, Q_A, for strongly absorbing particles.

4
The emerging radiation field

The last chapter dealt with the interaction of radiation with matter, mostly in the gaseous, but also in the liquid and solid phases. Absorption coefficients of infrared active gases, emission and scattering properties of surfaces, and single scattering albedos and phase functions of aerosols were considered. Applications of these concepts, along with the principles of radiative transfer discussed in Chapter 2, enable us to calculate the emerging radiation field of a planet or satellite.

In this chapter we examine how the physical state of an emitting medium gives rise to the general spectral characteristics of the outgoing radiation field. We begin in Section 4.1 by considering the behavior of a single spectral line in an isothermal atmosphere, both with and without scattering. In Section 4.2 we introduce nonscattering atmospheric models having a more complicated thermal structure, but still consider only a single line. Finally, in Section 4.3 we conclude our investigation of nonscattering models using realistic molecular parameters. Our aim in this chapter is to illustrate the principles behind the analysis of remotely sensed data, especially with regard to how scattering, atmospheric abundances, and thermal structures affect the appearance of the observed spectrum. Later, in Chapter 6, we apply these principles to the descriptions of real planetary spectra.

4.1 Models with one isothermal layer

a. Without scattering

The first model considered consists of a nonscattering gas layer at constant pressure and temperature adjacent to a solid surface of unit emissivity. This model is illustrated in Fig. 4.1.1. We consider the absorbing gas to have only one spectral line. Such a line is not perfectly monochromatic, but, as discussed in Section 3.6, is broadened by various effects. In this section we assume that the gas pressure is sufficiently high and the temperature sufficiently low so that collisional broadening dominates and the line has a Lorentz shape. Other line shapes would serve equally well. For a Lorentz line the normal optical thickness of the gas layer is given by [see Eq. (3.6.3)]

$$\tau(\nu) = \frac{N \bar{\chi} \alpha \Delta z}{\pi[(\nu - \nu_0)^2 + \alpha^2]}, \qquad (4.1.1)$$

where N is the molecular number density, $\bar{\chi}$ the mean cross section per molecule multiplied by the effective line width, Δz the layer thickness, and α the line width appropriate for the assumed pressure and temperature. At the line center ($\nu = \nu_0$) the optical path is maximum.

According to our postulate, scattering does not occur in the layer. If we assume a surface emissivity of unity, the outgoing monochromatic radiation field is given by Eq. (2.4.4), with $n = 1$. Thus $\tau_0 = 0$ and

$$I_\nu(0, \nu) = B_\nu[\tau_{1/2}(\nu)]\left[1 - e^{-\tau_1(\nu)/\mu}\right] + B_\nu(T_S) e^{-\tau_1(\nu)/\mu}, \qquad (4.1.2)$$

where T_S is the surface temperature at τ_1. The first term corresponds to I_E in Fig. 4.1.1, and the second to I_D. Upon substituting $B_\nu(T_A)$ for $B_\nu[\tau_{1/2}(\nu)]$, where T_A is the gas temperature, Eq. (4.1.2) can be written

$$I_\nu(0, \nu) = B_\nu(T_A) - [B_\nu(T_A) - B_\nu(T_S)] e^{-\tau_1(\nu)/\mu}. \qquad (4.1.3)$$

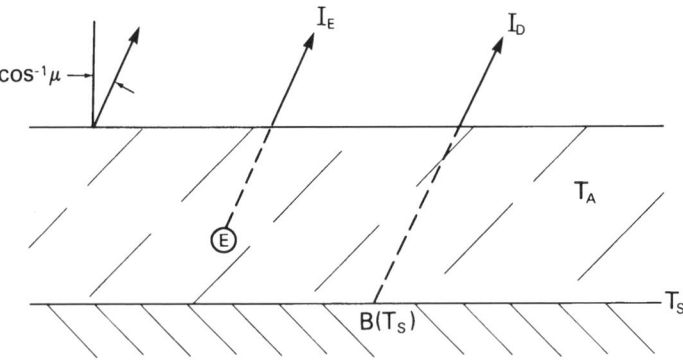

Fig. 4.1.1 Simple nonscattering model. Radiation of intensity I_E is emitted by an isothermal layer of temperature T_A in the direction μ. An underlying surface of unit emissivity and temperature T_S also radiates with its characteristic blackbody intensity $B(T_S)$. After partial absorption a reduced fraction I_D is transmitted through the layer in the same direction. The sum of I_E and I_D is the total outgoing intensity.

4.1 Models with one isothermal layer

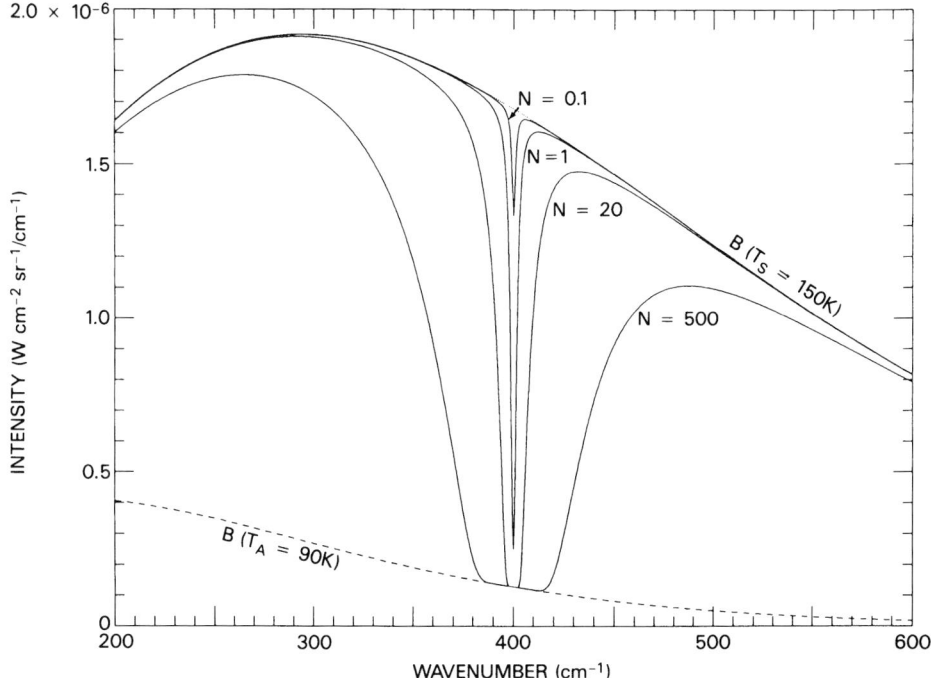

Fig. 4.1.2 Lorentz lines in absorption in a homogeneous isothermal atmosphere, generated according to the model illustrated in Fig. 4.1.1. Normal viewing ($\mu = 1$) is assumed, with surface and atmospheric temperatures of $T_S = 150$ K and $T_A = 90$ K, respectively. The line width α in Eq. (4.1.1) is 1 cm^{-1}, and the line center is at $\nu_0 = 400$ cm^{-1}. The product $\bar{\chi}\Delta z$ is 8 cm^2 per particle, and the particle number density is assigned the four values $N = 0.1$, 1, 20, and 500 particles per cm^3. Planck functions for $T = 150$ K and 90 K serve as limiting boundaries for the lines.

As $\tau_1(\nu)/\mu$ increases, $I_\nu(0, \mu)$ increases if the temperature difference $T_A - T_S$ is positive, or decreases if the difference is negative. Thus spectral lines are seen either in emission or absorption depending on whether $T_A > T_S$ or $T_A < T_S$. If $T_A = T_S$ the line disappears; only the blackbody continuum $B_\nu(T_A)$ is seen.

These principles are illustrated in Figs. 4.1.2 and 4.1.3, which show how absorption or emission lines are formed when T_A is, respectively, less than or greater than T_S. The figures show different line strengths depending on the number density of absorbing molecules. In order to illustrate how the line wings blend into the Planck continuum we have adopted line widths much larger than those associated with real molecules, although exceptions can occur under special circumstances, such as when lines are formed by collision-induced absorption.

If the optical path length at the line center, $\tau_1(\nu_0)/\mu$, is sufficiently small, the line is relatively weak and the area under the curve (defined relative to the continuum)

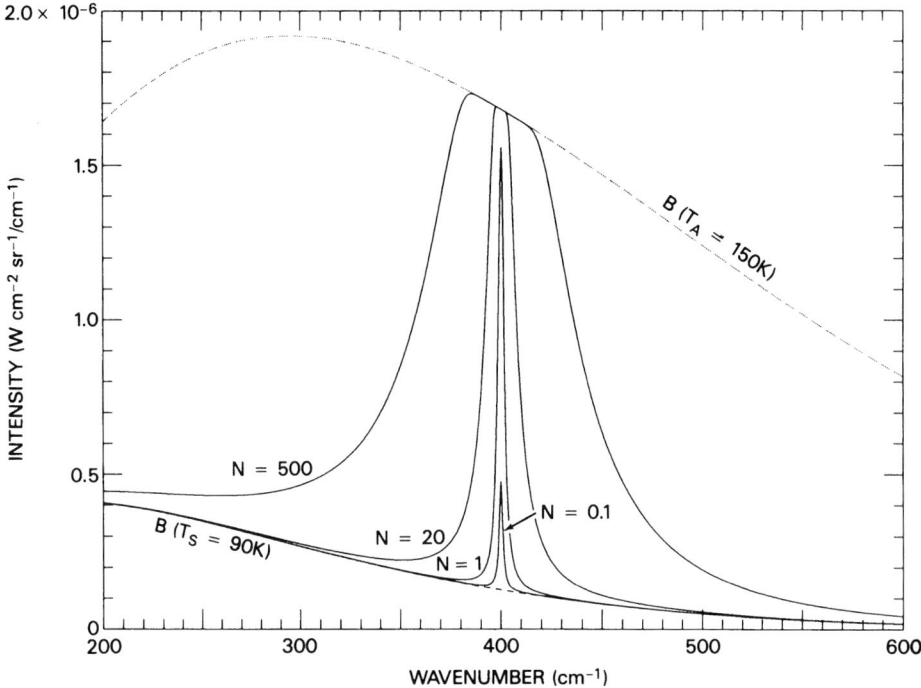

Fig. 4.1.3 As Figure 4.1.2, except the lines are in emission because the surface and atmospheric temperatures are reversed, with $T_S = 90$ K and $T_A = 150$ K, respectively.

is proportional to the column density. Physically this occurs because very few molecules along the line-of-sight are obstructed by other molecules, and almost all contribute to the opacity. As $\tau_1(\nu)/\mu$ becomes large, most of the molecules are masked by others, especially near the line center, where the cross section per molecule is largest. The line approaches saturation and the area under the curve becomes proportional to some fractional power of the column density. The core of the line flattens out because the layer becomes opaque there, and the temperature contrast between the layer and the surface no longer affects the intensity near the line center (the surface cannot be seen). Only in the wings of the line, where the cross section per molecule is still small enough to permit the surface to contribute, does a variation of intensity with wavenumber occur.

b. With scattering

Certain modifications of the line shape and strength occur when scattering is introduced. Figure 4.1.4 illustrates the physical process, and should be contrasted with the nonscattering model shown in Fig. 4.1.1. A useful analytic representation follows from the two-stream approximation developed in Chapter 2. Considering

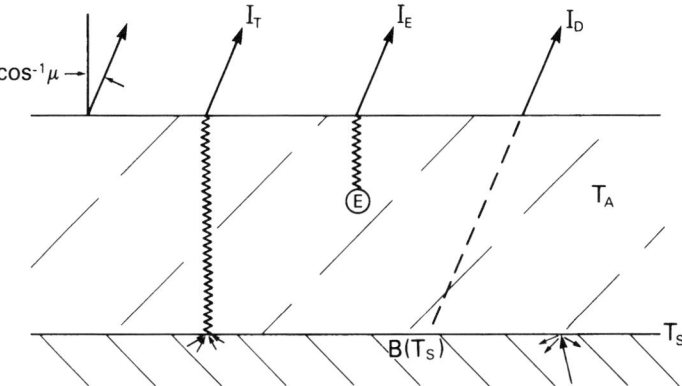

Fig. 4.1.4 Simple scattering model. Radiation of intensity I_E is emitted in the direction μ by an isothermal layer of temperature T_A. Because of multiple scattering, many photons emitted by the layer have originated within the layer from other directions. An underlying surface of unit emissivity and temperature T_S also emits in all directions. Some of the radiation is backscattered by the layer and is reabsorbed by the non-reflecting surface. A component I_T is diffusely transmitted through the layer by multiple scattering and another, I_D, is directly transmitted. The total outgoing intensity in the direction μ is the sum of I_E, I_T, and I_D.

only internal sources, Eq. (2.5.14) becomes

$$\mu_i \frac{dI(\tau,\mu_i)}{d\tau} = I(\tau,\mu_i) - \tfrac{1}{2}\sum_j (\tilde{\omega}_0 + \tilde{\omega}_1 \mu_i \mu_j) I(\tau,\mu_j) - (1-\tilde{\omega}_0) B(\tau)$$

$$(i, j = \pm 1). \quad (4.1.4)$$

In an isothermal layer the Planck function, B, is a constant. In this case it can be shown by direct substitution that

$$I(\tau,\mu_i) = B \quad (4.1.5)$$

is a particular integral of Eq. (4.1.4). Hence, upon adding this integral to the general solution [Eq. (2.5.23)], the complete solution becomes [note also Eqs. (2.5.21), (2.5.22), and (2.5.24)]

$$I(\tau,\mu_i) = \sum_\alpha L_\alpha \frac{\tilde{\omega}_0 + c_\alpha \mu_i}{1 + k_\alpha \mu_i} e^{-k_\alpha \tau} + B \quad (\alpha, i = \pm 1). \quad (4.1.6)$$

The integration constants L_α can be determined from the boundary conditions

$$I(0, -\mu_1) = 0; \quad I(\tau_1, \mu_1) = cB, \quad (4.1.7)$$

where [see Eq. (4.1.3) and the related discussion]

$$c = B(T_S)/B(T_A). \qquad (4.1.8)$$

Thus

$$\left. \begin{array}{l} L_1 \dfrac{\tilde{\omega}_0 - c_1 \mu_1}{1 - k_1 \mu_1} + L_{-1} \dfrac{\tilde{\omega}_0 + c_1 \mu_1}{1 + k_1 \mu_1} = -B \\[1em] L_1 \dfrac{\tilde{\omega}_0 + c_1 \mu_1}{1 + k_1 \mu_1} e^{-k_1 \tau_1} + L_{-1} \dfrac{\tilde{\omega}_0 - c_1 \mu_1}{1 - k_1 \mu_1} e^{k_1 \tau_1} = (c-1)B \end{array} \right\}. \qquad (4.1.9)$$

Define

$$f = \left(\frac{1 - \tilde{\omega}_0}{1 - \tfrac{1}{3}\tilde{\omega}_1} \right)^{\frac{1}{2}}. \qquad (4.1.10)$$

With the aid of Eqs. (2.5.21), (2.5.22), and (2.5.24) we find that

$$\frac{\tilde{\omega}_0 - c_1 \mu_1}{1 - k_1 \mu_1} = 1 + f \qquad (4.1.11)$$

and

$$\frac{\tilde{\omega}_0 + c_1 \mu_1}{1 + k_1 \mu_1} = 1 - f. \qquad (4.1.12)$$

Reduction of Eqs. (4.1.9) then yields

$$L_1 = -B \frac{(c-1)(1-f)e^{-k_1 \tau_1} + (1+f)}{(1+f)^2 - (1-f)^2 e^{-2k_1 \tau_1}} \qquad (4.1.13)$$

and

$$L_{-1} = -B e^{-k_1 \tau_1} \frac{(c-1)(1+f) + (1-f)e^{-k_1 \tau_1}}{(1+f)^2 - (1-f)^2 e^{-2k_1 \tau_1}}. \qquad (4.1.14)$$

Hence, from Eq. (4.1.6), the outgoing intensity becomes

$$I(0, \mu_1) = 2fB \frac{(1+f) + 2(c-1)e^{-k_1 \tau_1} + (1-f)e^{-2k_1 \tau_1}}{(1+f)^2 - (1-f)^2 e^{-2k_1 \tau_1}}. \qquad (4.1.15)$$

4.1 Models with one isothermal layer

The physical meanings of the parameters c, B, $\tilde{\omega}_0$, and τ_1, either explicitly or implicitly contained in Eq. (4.1.15) [see Eqs. (2.5.24) and (4.1.8)], are clear, but $\tilde{\omega}_1$ requires a little more discussion. From Eq. (2.5.1) and (2.5.3) we find

$$\int_\omega \cos\Theta\, p(\cos\Theta)\frac{d\omega}{4\pi} = \tfrac{1}{3}\tilde{\omega}_1. \tag{4.1.16}$$

With the aid of Eq. (2.5.4) we can define an asymmetry factor

$$\langle \cos\Theta \rangle = \frac{\int_\omega \cos\Theta\, p(\cos\Theta)\dfrac{d\omega}{4\pi}}{\int_\omega p(\cos\Theta)\dfrac{d\omega}{4\pi}} = \frac{\tilde{\omega}_1}{3\tilde{\omega}_0}, \tag{4.1.17}$$

which expresses the degree to which radiation is singly scattered into the forward (plus) or backward (minus) directions. The more directionally extreme the scattering, the greater $|\langle \cos\Theta \rangle|$ is. The full range is $-1 \leq \langle \cos\Theta \rangle \leq 1$, although if only two terms are included in $p(\cos\Theta)$ [Eq. (2.5.1)], the phase function itself will be negative over certain ranges of Θ if $|\cos\Theta| > \tfrac{1}{3}$.

Rewriting Eq. (4.1.17) yields

$$\tfrac{1}{3}\tilde{\omega}_1 = \tilde{\omega}_0 \langle \cos\Theta \rangle. \tag{4.1.18}$$

For a given single scattering albedo $\tilde{\omega}_0$, the greater the forward scattering, the larger $\tilde{\omega}_1$ will be. Either $\tilde{\omega}_1$ or $\langle \cos\Theta \rangle$ can be used to describe the degree of forward (or backward) scattering, though we favor $\langle \cos\Theta \rangle$.

After this digression we return to Eq. (4.1.15). Four limiting cases are of particular interest for explaining the principles of line formation in scattering atmospheres: the cases $\tau_1 = 0$ and $\tau_1 = \infty$, and the cases $\tilde{\omega}_0 = 0$ and $\tilde{\omega}_0 = 1$.

Case 1: If the atmospheric layer disappears ($\tau_1 = 0$) we have

$$\lim_{\tau_1 \to 0} I(0, \mu_1) = cB, \tag{4.1.19}$$

i.e., the atmosphere is transparent and the upwelling intensity is just the Planck intensity of the surface.

Case 2: If the atmosphere becomes very deep ($\tau_1 \to \infty$) Eq. (4.1.15) reduces to

$$\lim_{\tau_1 \to \infty} I(0, \mu_1) = \frac{2f}{1+f} B, \tag{4.1.20}$$

where

$$\varepsilon = \frac{2f}{1+f} = 2\left[1 + \left(\frac{1 - \tilde{\omega}_0 \langle \cos \Theta \rangle}{1 - \tilde{\omega}_0}\right)^{\frac{1}{2}}\right]^{-1} \quad (4.1.21)$$

is the atmospheric emissivity.

Case 3: If the particles are completely absorbing ($\tilde{\omega}_0 = 0$), Eq. (4.1.15), with the aid of Eqs. (2.5.24) and (4.1.10), reduces to [note from Eq. (4.1.18) that $\tilde{\omega}_1 \to 0$ as $\tilde{\omega}_0 \to 0$]

$$I(0, \mu_1) = B[1 + (c - 1)e^{-\sqrt{3}\tau_1}]. \quad (4.1.22)$$

This is identical with Eq. (4.1.3), with $\mu = \mu_1 = 1/\sqrt{3}$.

Case 4: If the particles are fully reflecting ($\tilde{\omega}_0 = 1$), both f and k are zero and Eq. (4.1.15) becomes indeterminate. If we let $\tilde{\omega}_0$ approach unity in such a way that, for any $\tau_1, k_1\tau_1 \ll 1$, we can approximate $\exp(-nk_1\tau_1)$ with $(1 - nk_1\tau_1)$. After some reduction Eq. (4.1.15) becomes

$$\lim_{\tilde{\omega}_0 \to 1} I(0, \mu_1) = \lim_{\tilde{\omega}_0 \to 1} \left[B \frac{c - (c - f)k_1\tau_1}{2 + (1 - f)^2 \frac{k_1\tau_1}{f}}\right], \quad (4.1.23)$$

or, because

$$\frac{k_1}{f} = \sqrt{3}(1 - \tilde{\omega}_0 \langle \cos \Theta \rangle) \quad (4.1.24)$$

(which is true independently of the value of $\tilde{\omega}_0$), we finally obtain

$$\lim_{\tilde{\omega}_0 \to 1} I(0, \mu_1) = \frac{cB}{1 + \frac{\sqrt{3}}{2}(1 - \langle \cos \Theta \rangle)\tau_1}. \quad (4.1.25)$$

The curves in the upper parts of Figs. 4.1.5 and 4.1.6 illustrate the formation of absorption lines for moderate to small optical path lengths. If τ_1 is small, Eqs. (4.1.22) and (4.1.25) respectively reduce to

$$\lim_{\tilde{\omega}_0 \to 0} I(0, \mu_1) \sim cB\left[1 - \left(1 - \frac{1}{c}\right)\sqrt{3}\tau_1\right] \quad (4.1.26)$$

and

$$\lim_{\tilde{\omega}_0 \to 1} I(0, \mu_1) \sim cB\left[1 - \frac{\sqrt{3}}{2}(1 - \langle \cos \Theta \rangle)\tau_1\right]. \quad (4.1.27)$$

Hence, for a given (small) value of τ_1, a nonscattering medium will give rise to a

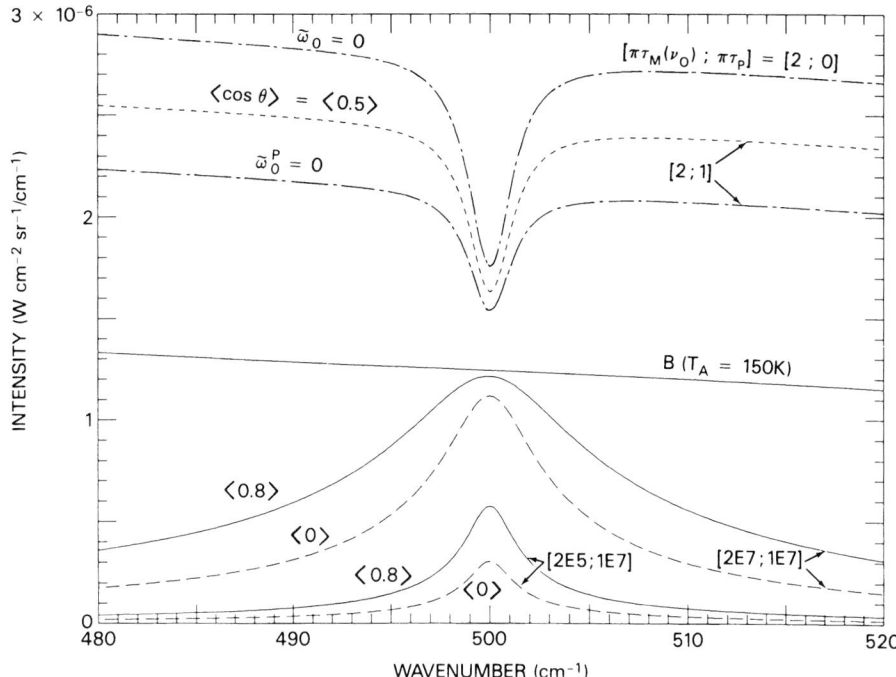

Fig. 4.1.5 Lorentz lines generated in a partially scattering, homogeneous, isothermal medium, illustrated by the model in Fig. 4.1.4. The direction cosine of viewing is $\mu = 1/\sqrt{3}$. Surface and atmospheric temperatures are $T_S = 180$ K and $T_A = 150$ K, respectively. Two components contribute to the total optical thickness of the atmosphere: (1) $\tau_M(\nu)$, due to a purely absorbing molecular medium, given by Eq. (4.1.1), and (2) a wavenumber-independent component, τ_p, due to a uniformly mixed medium of scattering aerosol particles. The line width of the molecular component is $\alpha = 1$ cm^{-1}, and the product $N\bar{\chi}\Delta z$ varies between 2 and 2×10^7 cm^{-1}; the line center is at $\nu_0 = 500$ cm^{-1}. The two components of normal optical thickness at line center are indicated by $[\pi\tau_M(\nu_0), \pi\tau_p]$; the factor π is introduced to eliminate it from the right side of Eq. (4.1.1). Single scattering albedos of the aerosol particles are $\tilde{\omega}_0^p = 0$ for the dot-dash curves and $\tilde{\omega}_0^p = 1$ otherwise. Volume element single scattering albedos (gas plus aerosol) vary with wavenumber for the dashed and solid curves, being smallest at line center. Models for which the asymmetry factor $\langle \cos\Theta \rangle = 0, 0.5$, and 0.8 are illustrated by the long dashed, short dashed, and solid curves, respectively. The slanted line separating emission from absorption features is the Planck intensity for a 150 K blackbody. (2 E 5 = 2×10^5.)

greater or lesser intensity than a scattering medium depending on whether $(1 - 1/c)$ is respectively less than or greater than $\frac{1}{2}(1 - \langle \cos\Theta \rangle)$.

The far wings of the lines in the upper parts of Figs. 4.1.5 and 4.1.6 best demonstrate this point. Here τ_1 is smallest and $\tilde{\omega}_0$ approaches unity in the scattering model. In both figures $c \sim 2$, and in Fig. 4.1.5 $\langle \cos\Theta \rangle = 0.5$. Hence the wings of the line formed by the scattering aerosol are about midway in intensity between those formed by the absorbing aerosol and those formed in the absence

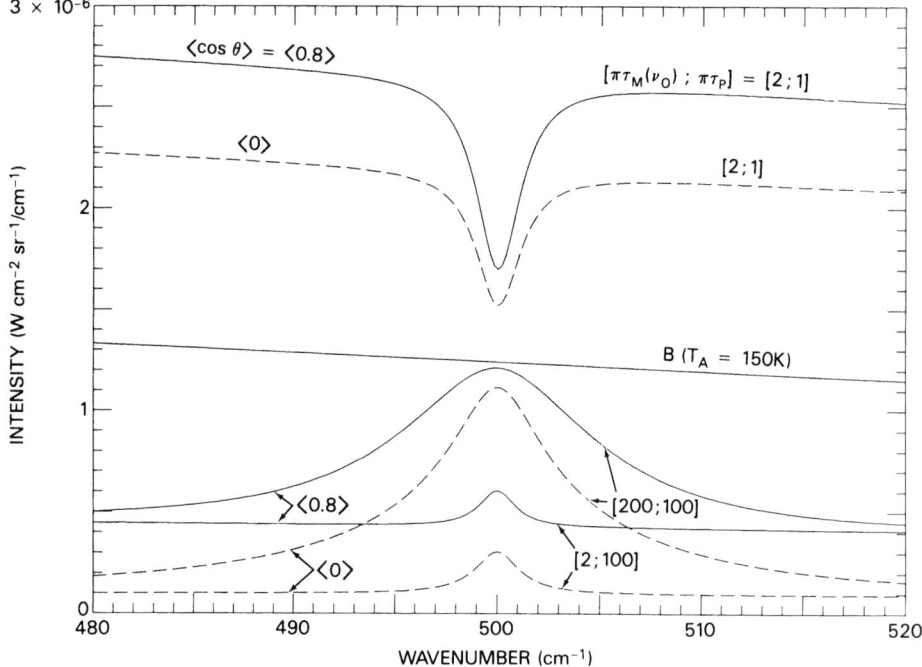

Fig. 4.1.6 As Fig. 4.1.5, except for the absence of models for which (1) $\tilde{\omega}_0^p = 0$ and (2) $\langle \cos \Theta \rangle = 0.5$. The normal optical thicknesses $\tau_M(\nu_0)$ and τ_p in the lower part of the figure are a factor 10^{-5} of those in Fig. 4.1.5. As before, models for which the asymmetry factors are $\langle \cos \Theta \rangle = 0$ and 0.8 are illustrated by the long dashed and solid curves, respectively.

of any aerosol, where $\tau_1 \sim 0$. In this case more radiation is diffusely transmitted through the medium with $\tilde{\omega}_0 \sim 1$ than is emitted by the medium with $\tilde{\omega}_0 = 0$. The corresponding intensity components are I_T in Fig. 4.1.4 and I_E in Fig. 4.1.1, respectively.

On the other hand these same components become comparable when $\langle \cos \Theta \rangle = 0$. This can be seen by comparing Eqs. (4.1.26) and (4.1.27) as well as the relevant curves in Figs. 4.1.5 and 4.1.6. The physical reason is illustrated in Fig. 4.1.4; as $\langle \cos \Theta \rangle$ becomes smaller, more radiation emitted by the surface is scattered by the layer back to the surface, and not so much is diffusely transmitted upward through the medium by multiple scattering in the forward direction.

The lower curves in Fig. 4.1.5 show the consequences of τ_1 becoming extremely large [$\pi \tau_M(\nu_0) = 2 \times 10^5$ and 2×10^7, and $\pi \tau_P = 10^7$]. Because the atmosphere is effectively semi-infinite and isothermal, no thermal contrast is possible, and the line shapes are due solely to variations in multiple scattering. As indicated by Eqs. (4.1.20) and (4.1.21), the emissivity is everywhere less than unity, leading to intensities less than the Planck intensity. At the line centers $\tilde{\omega}_0$ is smallest because

4.1 Models with one isothermal layer

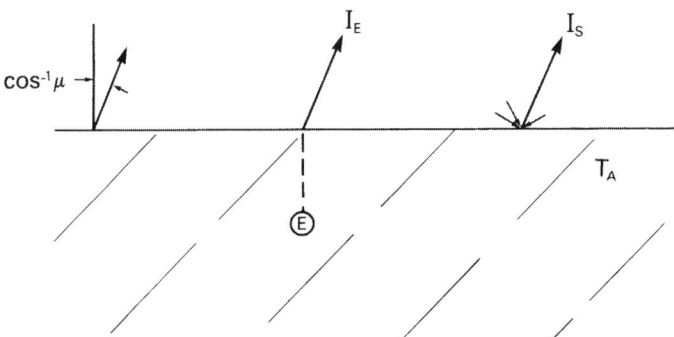

Fig. 4.1.7 Scattering model in thermodynamic equilibrium. An opaque slab is placed over an isothermal semi-infinite partially scattering medium. Both the slab and the medium are held at the same temperature T_A. Radiation from the slab is incident on the medium in all downward directions, and a component of intensity I_S is diffusely reflected by the medium into the direction μ. The intensity of radiation thermally emitted by the medium into the direction μ is I_E. Because the space between the slab and medium is equivalent to a blackbody cavity (see Section 1.7), the sum of I_S and I_E is the Planck intensity $B(T_A)$.

gaseous absorption is strongest at these wavenumbers, leading to maximum emissivities. As the distance from the line center increases, absorption decreases, leading to larger values of $\tilde{\omega}_0$ and correspondingly smaller values for ε. This results in a line shape that resembles emission from a warmer medium above a cooler surface, even though no temperature contrast actually exists.

The strength of the line is also affected by the shape of the phase function for single scattering. From Eq. (4.1.21) we see that ε increases as $\langle\cos\Theta\rangle$ increases. This effect is illustrated in the lower part of Fig. 4.1.5. The physical reason can be understood with the aid of Fig. 4.1.7. Radiation from an overlying slab is incident on a partially scattering semi-infinite medium. If single scattering is predominantly in the forward direction ($\langle\cos\Theta\rangle \sim 1$), any given photon tends to penetrate rather deeply into the medium, even after several scattering events, before being turned around by scattering into a backward direction. Conversely, if single scattering is isotropic ($\langle\cos\Theta\rangle = 0$), a given photon is just as likely as not to be backscattered in any given single scattering interaction. The average path length in the medium is larger in the former case than in the latter, leading to a greater likelihood of absorption; i.e., I_S will decrease as $\langle\cos\Theta\rangle$ increases.

If the slab is now removed, only

$$I_E = \varepsilon B(T_A) = B(T_A) - I_S \qquad (4.1.28)$$

will contribute to the outgoing intensity in Fig. 4.1.7, and I_E must increase as $\langle \cos \Theta \rangle$ increases. This accounts for the differences with $\langle \cos \Theta \rangle$ in the lower part of Fig. 4.1.5.

As an interesting aside, the local reflectivity (or albedo) is given by

$$a = 1 - \varepsilon = \frac{I_S}{B(T_A)}. \quad (4.1.29)$$

From Eqs. (4.1.21) and (4.1.29) it is easily deduced that

$$\frac{1-a}{1+a} = \left(\frac{1 - \tilde{\omega}_0}{1 - \tilde{\omega}_0 \langle \cos \Theta \rangle} \right)^{\frac{1}{2}}, \quad (4.1.30)$$

a relationship connecting the local albedo with the single scattering properties of a homogeneous semi-infinite medium in the two-stream approximation.

Thick, finite media, while exhibiting many characteristics similar to those of semi-infinite media, do show some differences. Comparisons between the lower portions of Figs 4.1.5 and 4.1.6 illustrate these differences. Although the cores of the lines in the cases shown are almost identical, the wings of the lines in finite media manifest relatively larger intensities than those in semi-infinite media. This is because the single scattering albedo is very high far from the line center, and diffuse transmission through the medium is effective. Equation (4.1.25) is the quantitative expression of this statement in the limiting case $\tilde{\omega}_0 = 1$, and demonstrates how forward scattering enhances diffusion through the medium, a fact also illustrated by the lower portion of Fig. 4.1.6. Remarkably thick atmospheres can transmit some radiation if $\tilde{\omega}_0$ is close to unity because almost no absorption per scattered event takes place, and on the average many such events will occur before any given photon is absorbed. Forward scattering is more effective in transmitting radiation than isotropic scattering because it is more directed, and the total path length a multiply scattered photon traverses through the medium is shorter. In the limit, as $\langle \cos \Theta \rangle \to 1$, radiation remains undeviated and is effectively unscattered.

4.2 Models with a vertical temperature structure

A more thermally complex system involves an atmosphere that is vertically variable in temperature and in hydrostatic equilibrium under the force of gravity. The vertical pressure gradient in such an atmosphere is given by

$$\frac{dP}{dz} = -gmN, \quad (4.2.1)$$

where P, N, m, and g are, respectively, the pressure, molecular number density, mean molecular weight, and acceleration due to gravity at the level z. Assuming that the perfect gas law is sufficiently accurate, we also have

$$P = NkT, \qquad (4.2.2)$$

where k is the Boltzmann constant and T the temperature. Upon differentiating Eq. (4.2.2) and substituting into Eq. (4.2.1), we obtain

$$\frac{dN}{dz} = -N\left(\frac{mg}{kT} + \frac{1}{T}\frac{dT}{dz}\right). \qquad (4.2.3)$$

a. Single lapse rate

The simplest example of a nonisothermal atmosphere is one with a single lapse rate,

$$\Gamma = -\frac{dT}{dz} = \text{const.} \qquad (4.2.4)$$

This is also physically realistic for planetary atmospheres over moderate ranges of altitude. Substituting into Eq. (4.2.3), replacing z with T as the independent variable, and integrating, yields

$$N = N_r \left(\frac{T}{T_r}\right)^{-(1-mg/k\Gamma)}, \qquad (4.2.5)$$

where the subscript r refers to any arbitrary reference level. An integration of Eq. (4.2.4) provides the relation between temperature and altitude,

$$T = T_r - \Gamma(z - z_r). \qquad (4.2.6)$$

To calculate the outgoing intensity, we apply the multilayer model of Subsection 2.4.a [see Eq. (2.4.4)]. The relation between optical and geometric thicknesses of the individual layers is given by [cf. Eq. (4.1.1)]

$$\tau_i - \tau_{i-1} = \frac{N_i \bar{\chi} \alpha_i (z_i - z_{i-1})}{\pi\left[(\nu - \nu_0)^2 + \alpha_i^2\right]}. \qquad (4.2.7)$$

Each layer is assumed to be individually isothermal; hence, from the kinetic theory

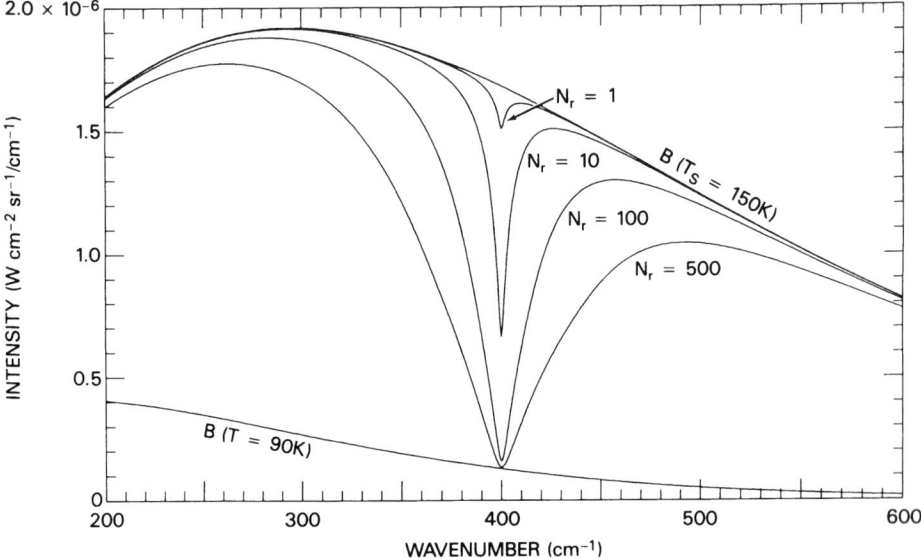

Fig. 4.2.1 Lorentz lines in absorption in an atmosphere with a constant, positive lapse rate. Normal viewing ($\mu = 1$) is assumed. The temperature at the surface and base of the atmosphere is $T = 150$ K, while the temperature at the top of the atmosphere is $T = 90$ K. The line width α is 1 cm^{-1}, and the line center is at 400 cm^{-1}. The product $\bar{\chi} \Delta z$ is 8 cm^2 per particle, and the particle number density at the base (reference level) is assigned the four values $N_r = 1$, 10, 100, and 500 particles per cm^3. Planck functions for $T = 150$ K and 90 K serve as limiting boundaries for the lines.

of gases we have

$$\frac{\alpha_i}{\alpha_r} = \frac{N_i}{N_r} \left(\frac{T_i}{T_r} \right)^{\frac{1}{2}} \tag{4.2.8}$$

where N_i is an appropriate mean number density and T_i is the temperature for the ith layer. The relations between N_i, T_i, and z_i follow from Eqs. (4.2.5) and (4.2.6).

Figures 4.2.1 and 4.2.2 illustrate the effect of a constant lapse rate on the spectrum. A positive lapse rate (negative temperature gradient) is considered in the first figure, and a negative one in the second. The lines are seen, respectively, in absorption and emission and are qualitatively similar to those in Figs. 4.1.2 and 4.1.3. This is largely because the mean atmospheric temperature is less than the surface temperature in the first case, and greater in the second.

There are some differences, however. In particular, lines formed in an atmosphere with a temperature gradient do not exhibit the symmetry in absorption and emission shown in homogeneous isothermal media. According to Eqs. (4.2.5)

4.2 Models with a vertical temperature structure

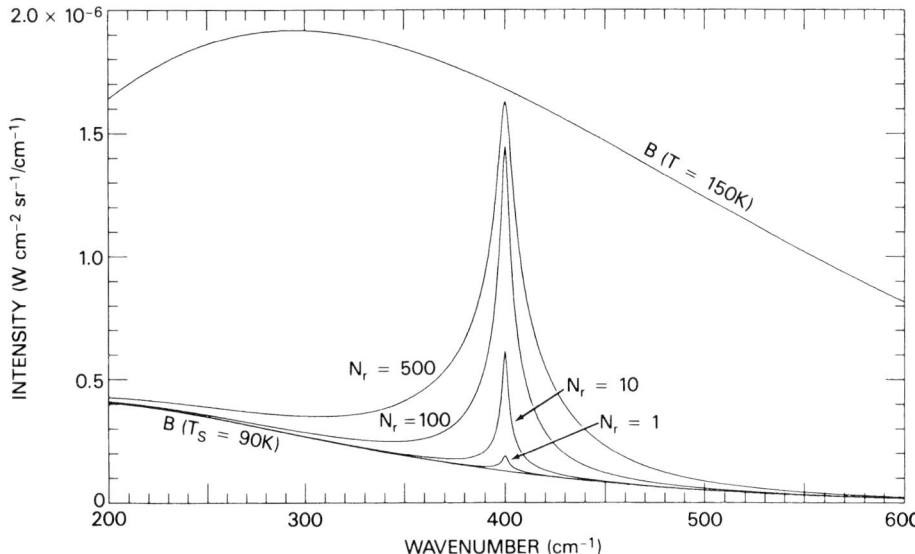

Fig. 4.2.2 Lorentz lines in emission in an atmosphere with a constant, negative lapse rate. The temperature at the surface and base of the atmosphere is 90 K, while the temperature at the top of the atmosphere is 150 K. The remaining parameters are the same as in Fig. 4.2.1.

and (4.2.6) the variation of number density with altitude is more extreme with a negative lapse rate. Hence, if the base of the atmosphere is the reference level, the integrated column density will be less. The lines, seen in emission, will appear weaker than corresponding absorption lines formed in an atmosphere with a positive lapse rate, as long as the molecular number densities at the bases are equal.

A second difference is that lines associated with a temperature gradient tend not to saturate as readily as those associated with an isothermal atmosphere. In the latter case saturation occurs at wavenumbers where the total atmosphere becomes opaque. In the former case, however, the temperature contrast between different levels within the atmosphere is sufficient to cause a spectral variation with wavenumber even though the surface cannot contribute.

The reason can be clarified with the concepts of weighting and contribution functions. If emission from the surface does not depend on azimuth, Eq. (2.4.1) can be written

$$I(z = z_0, \mu) = I(z = 0, \mu) \exp\left(-\frac{1}{\mu} \int_0^{z_0} N\chi \, dz\right) + \int_z^{z_0} B(z')W(z', \mu) \, dz',$$

(4.2.9)

where z_0 is the top of the atmosphere for all practical purposes, and the weighting function $W(z, \mu)$ is given by

$$W(z, \mu) = \frac{\partial}{\partial z}\left[\exp\left(-\frac{1}{\mu} \int_z^{z_0} N\chi \, dz' \right) \right]. \tag{4.2.10}$$

This function is the gradient of the transmittance through the atmosphere in the direction μ. The weighting function is a maximum where the transmittance is changing most rapidly. Hence, from Eq. (4.2.9), if the variation of $B(z)$ with z is small compared with that of $W(z, \mu)$, the level at which $W(z, \mu)$ is maximum is also the level that contributes most to the outgoing intensity $I(z_0, \mu)$. More often, however, the variation $B(z)$ with z is not small, and the maximum of the contribution function,

$$C(z, \mu) = B(z)W(z, \mu), \tag{4.2.11}$$

is more appropriate for defining the level of maximum contribution to $I(z_0, \mu)$. This level will vary with wavenumber because χ and $B(z)$ do. Consequently, the effective level from which the atmosphere emits depends on wavenumber, and, because the various levels are at different temperatures, the outgoing intensity will also vary with wavenumber, even though the atmosphere may be sufficiently opaque to obscure the contribution from the surface. The net effect is to defer saturation in opaque lines arising from an atmosphere with a temperature gradient. These lines appear to be somewhat broader in the wings and less flat near the core than lines generated in an isothermal atmosphere. This is illustrated in Fig. 4.2.3.

b. Multiple lapse rates

The correlation between contribution functions and the observed spectrum becomes most evident when the atmosphere undergoes temperature reversals. Consider the temperature profile shown in Fig. 4.2.4. Three distinct lapse rates define the thermal structure; two positive lapse rates bound a negative one in the middle. Discontinuities in temperature gradient are avoided by rounding the profile at the two extremes.

Outgoing intensity spectra of models using this temperature profile are shown in Fig. 4.2.5. The models are defined by Eqs. (4.2.5) through (4.2.8). The different panels in the figure are associated with different values of the base number density N_r. As N_r grows larger the opacity at a given wavenumber increases,

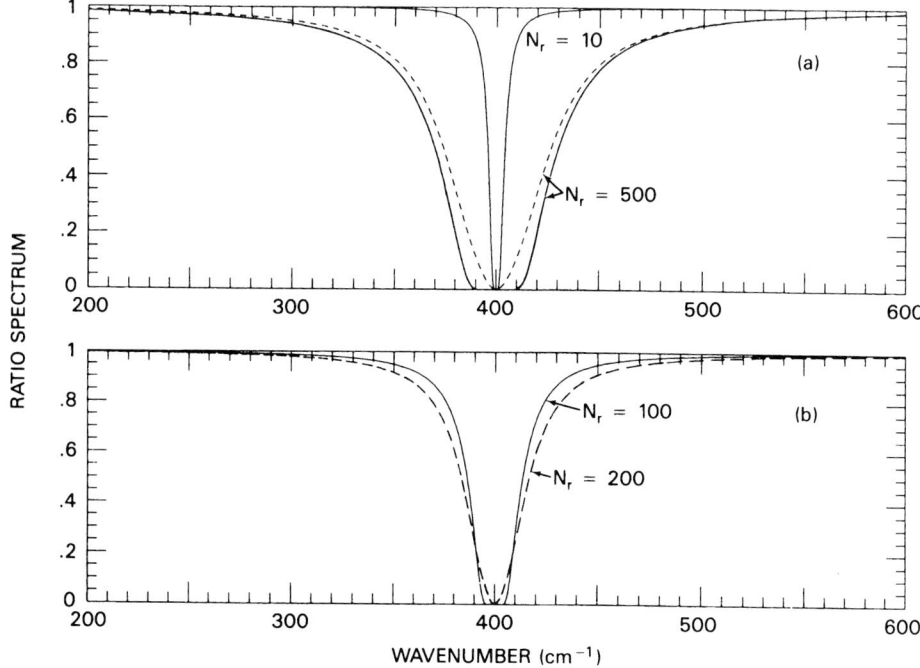

Fig. 4.2.3 Comparisons of Lorentz line ratio spectra for isothermal atmospheres (solid curves) and constant lapse rate atmospheres (dashed curves). All lines are normalized to the continuum. The temperature at the top ($z = 50$ km) of all atmospheres is $T_A = 90$ K; the temperature remains constant throughout the isothermal atmospheres, and increases linearly with decreasing altitude to $T_A = 150$ K at $z = 0$ for the constant lapse rate atmospheres. The surface temperature is $T_S = 150$ K for all models, and the reference level is at $z = 10$ km, where $N = N_r$. The number density is $N = 0$ below 10 km, and is calculated in accordance with Eqs. (4.2.5)–(4.2.6) between 10 and 50 km.

causing the effective emission level z_{eff} at that wavenumber to rise to higher altitudes.

The topmost panel in Fig. 4.2.5 shows a spectrum for which z_{eff} extends over the full altitude range, 0–90 km. This is demonstrated in Fig. 4.2.6, in which four contribution functions associated with four critical wavenumbers are displayed. Three of the wavenumbers ($\nu = 367$, 393, and 400 cm^{-1}) correspond to minima or maxima in the spectrum, while the fourth ($\nu = 200$ cm^{-1}) is located in the far wing.

The atmosphere is very transparent at 200 cm^{-1} and almost opaque at the line center at 400 cm^{-1}. The corresponding contribution functions are relatively narrow, and their peaks are located at the surface and at the top of the atmosphere, respectively. As a result the calculated intensities of the spectrum at these two wavenumbers are almost equal to the Planck intensities associated with the temperature ($T = 150$ K) at the bottom and top of the atmosphere.

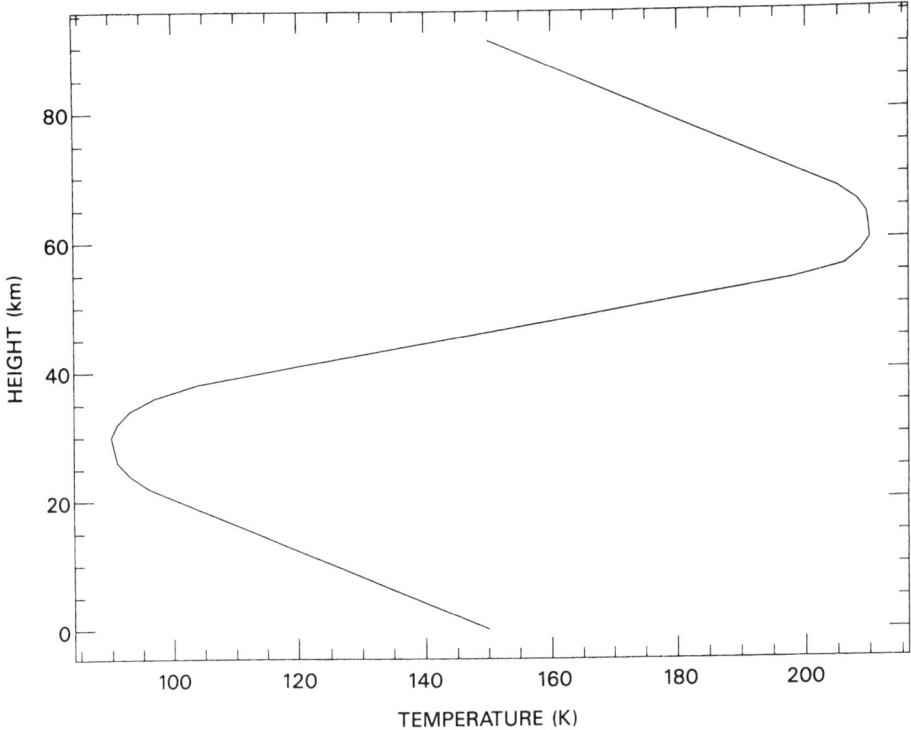

Fig. 4.2.4 Temperature profile used to calculate spectra in Fig. 4.2.5.

The contribution function for $\nu = 393$ cm^{-1} is associated with a peak spectral intensity at this wavenumber, and also has a maximum at $z = 60$ km, the altitude at which the temperature profile is maximum. However, the function itself is rather broad, and fairly large contributions to the outgoing intensity at 393 cm^{-1} arise from a moderate range of altitudes centered about 60 km, over which the temperature is less than maximum. Consequently, the spectral intensity at 393 cm^{-1} is only slightly greater than the Planck intensity for $T = 180$ K, as indicated in Fig. 4.2.5a, rather than that for $T = 210$ K, as implied by Fig. 4.2.4.

A minimum in the spectrum occurs at $\nu = 367$ cm^{-1}, implying the associated weighting function is maximum near $z = 30$ km, where the temperature (and hence the Planck intensity) has a minimum. Thus the weighting function and Planck intensity tend to counteract each other, and their product results in the broad, double-peaked contribution function shown in Fig. 4.2.6. In this case the concept of an effective emission level has little meaning, since there exists a broad altitude range over which individual levels contribute about equally to the outgoing intensity. This phenomenon is characteristic of temperature minima

4.2 Models with a vertical temperature structure

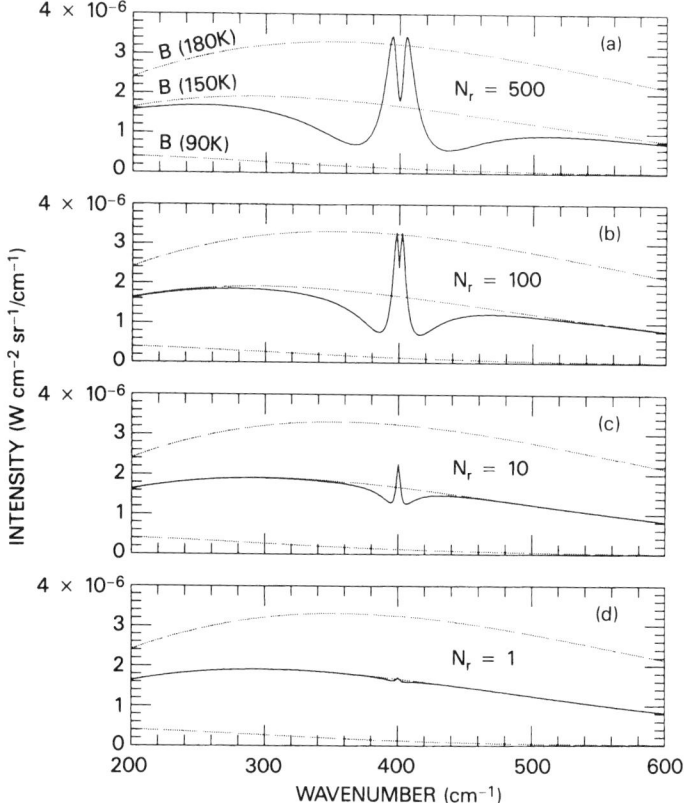

Fig. 4.2.5 Emission spectra of atmospheric models that incorporate the temperature profile in Fig. 4.2.4. The models are defined by Eqs. (4.2.5) through (4.2.8). The reference level, where $N = N_r$, is at $z = 0$; the top in all cases is at $z = 90$ km.

in general, and can lead to difficulty in attempts to infer the detailed thermal structure of real atmospheres from observed spectra, the subject of Section 8.2.

In summary, the spectral intensity at a given wavenumber can, in certain spectral regions, be closely associated with the Planck intensity of the atmosphere at a given effective emission level z_{eff}. In other spectral regions, especially near spectral minima, the association is not as close. To the extent that z_{eff} is meaningful, the emission properties of this level are governed by the optical properties and cross sections of the particles and molecules present at this level, as well as the temperature profile. The sharper the contribution function associated with z_{eff}, the better defined this level is.

If comparisons are being made across a fairly broad spectral range, the Planck intensity will vary with wavenumber at a given z_{eff}. In this case it is frequently useful to plot the spectrum in units of brightness temperature, T_B. This is not a physical

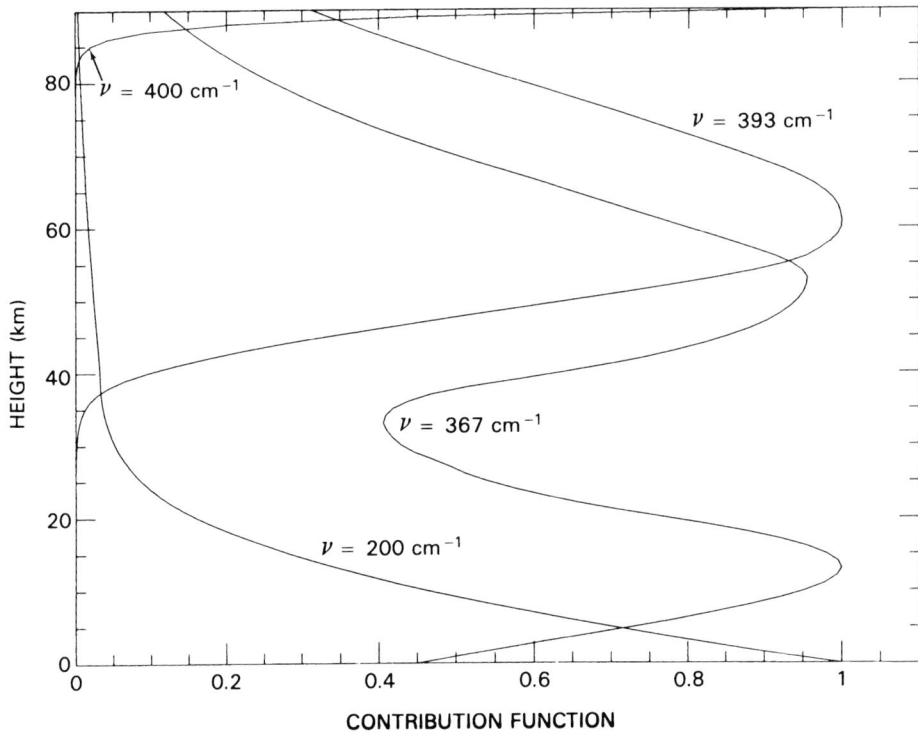

Fig. 4.2.6 Contribution functions at selected wavenumbers, corresponding to the atmospheric model with $N_r = 500$ particles cm^{-3}. The associated emission spectrum is shown in panel (a) of Fig. 4.2.5.

temperature, but rather is the temperature of a blackbody that would radiate with the outgoing intensity at a given wavenumber. Thus T_B is a function of wavenumber, and, provided the thermal structure is known, gives one an easy way to estimate the variation of z_{eff} with wavenumber, or, what amounts to the same thing, to estimate the approximate level at which each point in the spectrum is formed. We use this representation in the last section.

4.3 Model with realistic molecular parameters

So far we have considered only the effect of the atmospheric thermal structure on a single line. However, in the thermal infrared, molecular bands are dominantly responsible for the gaseous opacity, and it is useful to see how they affect the appearance of spectra. We illustrate this with the 667 cm^{-1} band of carbon dioxide (CO_2) and the temperature profile shown in Fig. 4.3.1. This profile qualitatively

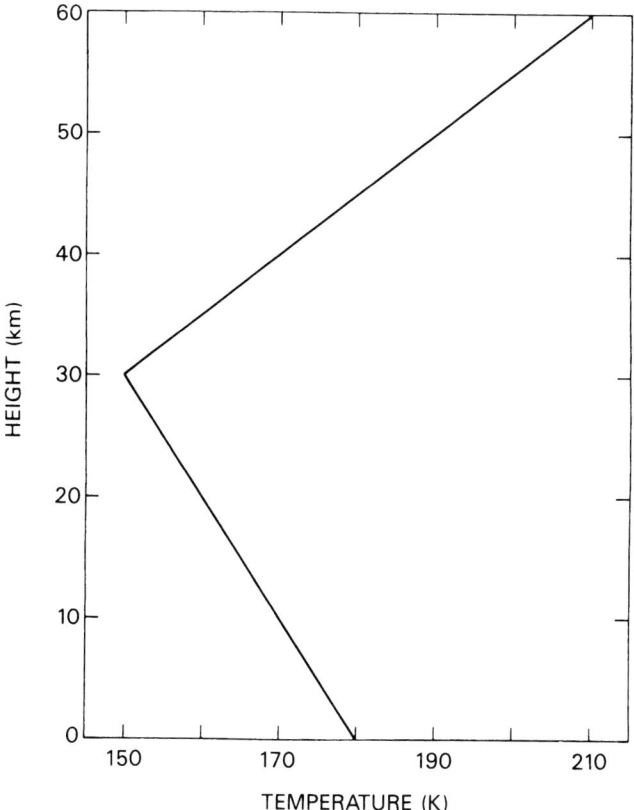

Fig. 4.3.1 Temperature profile used to calculate the CO_2 emission spectrum in Fig. 4.3.2.

resembles those for Earth and Titan (see Chapters 6 and 8). We assume a nitrogen (N_2) atmosphere (mean molecular weight of 28 amu) and add a CO_2 mole fraction of 1×10^{-6}. The spectrum shown in Fig. 4.3.2 is computed at a resolution of 0.01 cm^{-1} and degraded to 1.0 cm^{-1}. Part of the spectrum is shown in the lower panel at a resolution of 0.1 cm^{-1}.

The P- and R-branches of the 667 cm^{-1} CO_2 band display considerable structure, principally in absorption in the ranges 625–664 cm^{-1} and 671–705 cm^{-1}, although individual lines appear in emission. This indicates that the corresponding contribution functions peak in the lower atmosphere where the temperature gradient is negative, except in the line centers where the contribution functions have shifted to a position above the temperature minimum. In the central Q-branch the spectrum is strongly inverted and appears in emission. Here the atmosphere is opaque enough to shift the contribution functions well above the temperature minimum, where the temperature gradient becomes positive.

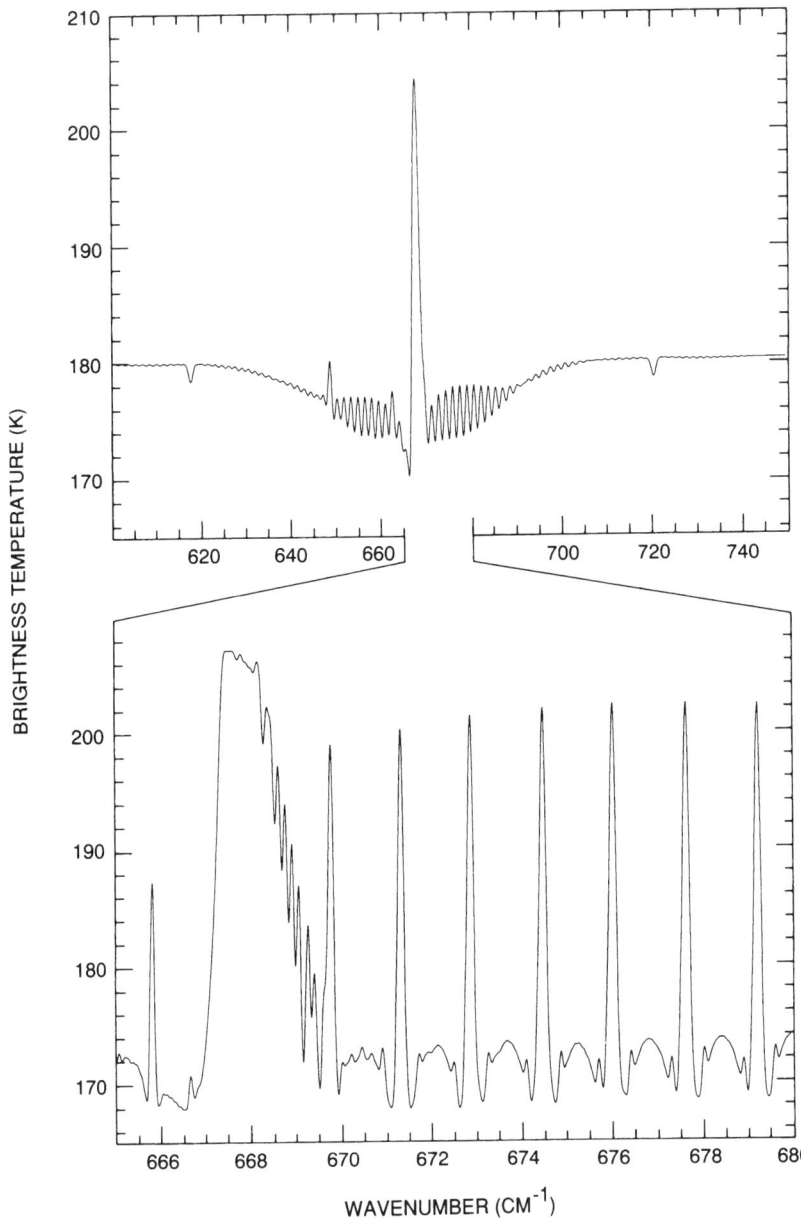

Fig. 4.3.2 Upper panel shows a spectrum calculated for an N_2–CO_2 atmosphere with the temperature profile illustrated in Fig. 4.3.1 and a spectral resolution of 1.0 cm^{-1}. The lower panel is part of the same spectrum but displayed between 666 and 680 cm^{-1} at a resolution of 0.1 cm^{-1}. The CO_2 mole fraction is $q = 1 \times 10^{-6}$ between $z = 0$ and $z = 60$ km, and $q = 0$ above these altitudes.

4.3 Model with realistic molecular parameters

From this example it is clear that some combination of thermal structure and emitting gas abundance can be inferred from observed spectra. As we have seen, a qualitative picture can often be obtained simply by inspecting a display of the spectral data. A more quantitative assessment requires solutions of the equation of transfer. First, however, it is necessary to examine how instrumental effects modify the appearance of planetary spectra. This will be discussed in the next chapter.

5
Instruments to measure the radiation field

In Chapter 4 we constructed examples of planetary spectra by applying solutions of the radiative transfer equation to model atmospheres of assumed composition and temperature structure. Before we can compare the results of such calculations with measured spectra we have to understand the modifications the emerging radiation field experiences in the recording process performed by radiometric instruments. A full comprehension of the detailed functioning of instruments is also necessary for the planning and the design of remote sensing investigations. Therefore, in Chapter 5 we discuss the principles of infrared instrumentation. We concentrate on instruments for space use, but the physical principles are equally applicable to ground-based astronomical sensors. On several occasions we refer to such Earth-based devices. It is neither possible nor useful to mention all infrared instruments ever flown in space or ever used for planetary work with ground-based telescopes. Instead, we analyze the physical concepts of different design approaches. To illustrate these concepts we occasionally show diagrams of specific instruments as well as samples of results obtained with them.

Radiometric devices have certain common characteristics. For example, most radiometric instruments contain optical elements to channel planetary radiation onto a detector. Telescopes are often essential parts of these designs. Following a brief introduction in Section 5.1, the subject of telescopes is discussed in Section 5.2. In the process of imaging a planetary surface element onto the detector, fundamental limits in spatial resolution are encountered. These limits, set by diffraction, are discussed in Section 5.3. Sometimes, radiometric instruments have components that chop the incoming radiation, or allow line by line scanning to form an infrared image, or permit image motion compensation to reduce smear caused by the relative motion of the field of view against the planet during the exposure. Different system configurations have evolved to permit execution of these functions; they are the subject of Section 5.4.

The spectral range of a remote sensing instrument must be restricted to at least one, but sometimes to many, well-defined intervals. Calculations are carried out most conveniently for monochromatic radiation, and nearly monochromatic measurements are often desirable. But instrumental and other realities set practical limits, and for some investigations the optimum spectral interval may be relatively wide. To provide structure to the discussion of the numerous methods used to restrict the spectral range of remote sensing instruments, we divide this subject into four sections: spectral separators based on intrinsic material properties, spectral filters based on interference in thin films, spectral analysers based on diffraction, and Fourier transform spectrometers. Section 5.5 on intrinsic material properties includes a discussion of prism spectrometers and gas cell instruments. In Section 5.6 on interference phenomena we present an analysis of narrow- and wide-band optical filters designed to operate at fixed frequencies as well as tunable filters. Interference filters and Fabry–Perot interferometers are part of this discussion. Grating spectrometers are the subject of Section 5.7. Fourier transform spectrometers, that is, instruments for which the output signal needs to be Fourier transformed to obtain a spectrum, are grouped in Section 5.8. Of course, they are also based on interference phenomena. Michelson interferometers, on space platforms and from the ground, have made significant contributions to our knowledge of planetary atmospheres and surfaces. However, with cryogenically cooled detectors one reaches a point where background noise becomes a limiting factor and where Fourier transform spectrometers lose the important multiplex advantage. The post-dispersion technique, also discussed in Section 5.8, overcomes this limitation. The Martin–Puplett and the lamellar grating interferometers are included in this section because of their significance to present and future space exploration. Today, Fourier transform spectrometers reach spectral resolutions on the order of 10^{-2} cm^{-1}. Heterodyne spectroscopy can reach a spectral resolution as small as 10^{-5} cm^{-1}, although most planetary measurements have been performed with somewhat higher resolution. This powerful technique is discussed in Section 5.9.

All radiometric devices must convert infrared energy into electrical signals. The fundamental properties of infrared converters, commonly called detectors, are analyzed in Section 5.10. In Section 5.11 the operating principles, noise limitations, and several temperature to voltage conversion mechanisms of thermal detectors are treated. Properties and noise characteristics of quantum detectors are the subject of Section 5.12. In many cases radiometric instruments must be calibrated in intensity and wavenumber. For best results calibration techniques are part of the instrument design. Several calibration methods are treated and their merits discussed in Section 5.13. Finally, Section 5.14 deals with considerations encountered in the selection of specific instrument designs; scientific objectives are related to certain instrument parameters.

Chapter 5 is neither a listing of available instruments nor is it an instruction manual for the construction of such devices. Rather, this chapter is intended to provide the necessary background to make an intelligent choice of a particular instrumental technique in the pursuit of a specific scientific goal and to be able to assess instrumental effects in recorded data.

5.1 Introduction to infrared radiometry

Instruments designed to measure infrared radiation are traditionally called radiometers, photometers, or, if they record the polarization, photopolarimeters; if they measure the intensity as a continuous function of wavenumber, or wavelength, the term spectrometer is applicable; if they generate a two-dimensional display of the radiation field, they are called imaging systems or cameras. However, there is no fundamental difference between radiometers, spectrometers, or cameras, and most of what has to be discussed in this chapter applies to all. For example, all radiometric instruments must have detectors, that is, elements that absorb infrared radiation and convert it to another form of energy, which can then be sensed by electronic means. In general, radiometric instruments have other elements as well, such as lenses or mirrors, but only a detector is needed to construct a simple radiometer. The black and white detectors originally designed for the Vanguard program and later flown in modified form on Explorer 7 by Verner Suomi of the University of Wisconsin, for example, consisted only of an absorbing and emitting surface the size and shape of a table tennis ball, with a temperature sensor inside, mounted on the tip of an antenna rod (Weinstein & Suomi, 1961). The spherical detector attempts to reach equilibrium between its own thermal emission and the radiative fluxes from the Sun, the Earth, and the spacecraft. Taking advantage of the spacecraft passing periodically into the Earth's shadow, the individual contributions to the energy balance of the detectors can be separated and the fluxes emitted and reflected by the Earth and its atmosphere determined. The simplicity of the device made it attractive in the early days of space meteorology. Such detectors and several variations of the same idea have been flown repeatedly in Earth's orbit (Weinstein & Suomi, 1961; Hanel, 1961a; Astheimer et al., 1961; Nordberg et al., 1962; Bandeen et al., 1964). However, the precision and the spatial and spectral resolution inherent in such devices are insufficient for more sophisticated investigations. Additional complexity must be tolerated to satisfy scientific requirements. Therefore, most radiometric instruments include optical components to focus radiation from a planetary area onto one or several detectors, elements that limit or otherwise identify the spectral range, and circuitry to amplify and record

5.2 Optical elements 155

the signal. In addition, choppers, shutters, scanning mirrors, image motion compensators, calibration sources, and other components may also be part of a fully functional remote sensing device. Each task – imaging, spectral separation, and detection – can be implemented by a multitude of techniques. Besides the scientific requirements, physical size, weight, power consumption, cryogenic demands, data rates, and other often subtle requirements set further boundaries to the design. The organization of this chapter follows the functional elements of radiometric systems.

5.2 Optical elements

The purpose of optical elements is to channel radiation as efficiently as possible from the observed area on the planetary object onto the detector. The simplest optical elements capable of doing so are lenses and curved reflectors (Fig. 5.2.1). Mirrors

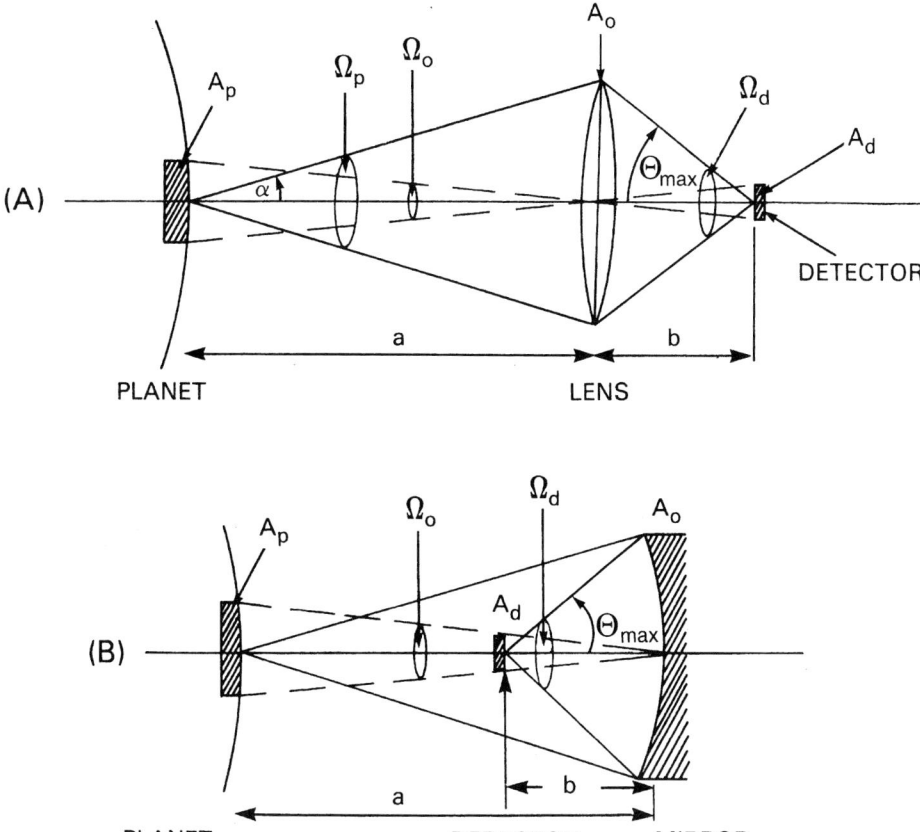

Fig. 5.2.1 Simple optical systems to image a planetary area onto a detector; (A) with a refractive and (B) with a reflective element.

fold back individual rays, in contrast to lenses, which do not; lenses lead, therefore, to more straightforward layouts. Since it is then easier to follow individual rays we use lenses often in the optical schematics designed to demonstrate the principles. However, for every optical layout with lenses, as shown in panel A of the figure, there exists an equivalent layout with mirrors, as illustrated in panel B. Sometimes one configuration is easier to implement than the other.

In Fig. 5.2.1 the lens images the scene of area A_p onto the detector of area A_d. The size of the detector and its distance to the lens determine the field of view given by the solid angle, Ω_0. The detector is illuminated by a much larger solid angle, Ω_d, determined by the maximum half angle, θ_{max}, which in turn is given by the size of the lens and the distance to the detector. For a circular lens

$$\Omega_d = \int_0^{\theta_{max}} \int_0^{2\pi} \sin\theta \, d\phi \, d\theta = 2\pi(1 - \cos\theta_{max}). \tag{5.2.1}$$

The azimuth angle is ϕ. The relationship between Ω and θ_{max} is shown in Fig. 5.2.2. The same figure indicates the f-number of the lens; f-number = focal length ÷ diameter. For small values of θ_{max}, the cosine term may be approximated by $(1 - \theta^2/2)$, which is equivalent to the paraxial approximation, $\sin\theta \sim \theta$, and

$$\Omega \sim \pi \theta_{max}^2. \tag{5.2.2}$$

This approximation, also shown in Fig. 5.2.2, is good for values of θ_{max} up to 40° (80° full cone angle) or an f-number of 0.6. Another often used and convenient approximation, area of aperture divided by the square of the distance between aperture and detector, yields

$$\Omega \sim \pi \tan^2 \theta_{max}; \tag{5.2.3}$$

it is good to about 20° half cone angle, or an f-number of 1.4.

The product of the cross section (area) of the radiation times the solid angle is the étendue, or simply the A-Omega. Unfortunately, no specific name exists for this quantity in the English language. Sometimes the term throughput is suggested, but this seems more appropriate to the quantity $A\Omega\eta$, η being the optical efficiency of the system.

As can be seen from Fig. 5.2.1 and approximation Eq. (5.2.3), the solid angle of the instrument is

$$\Omega_0 = \frac{A_p}{a^2} = \frac{A_d}{b^2}. \tag{5.2.4}$$

5.2 Optical elements

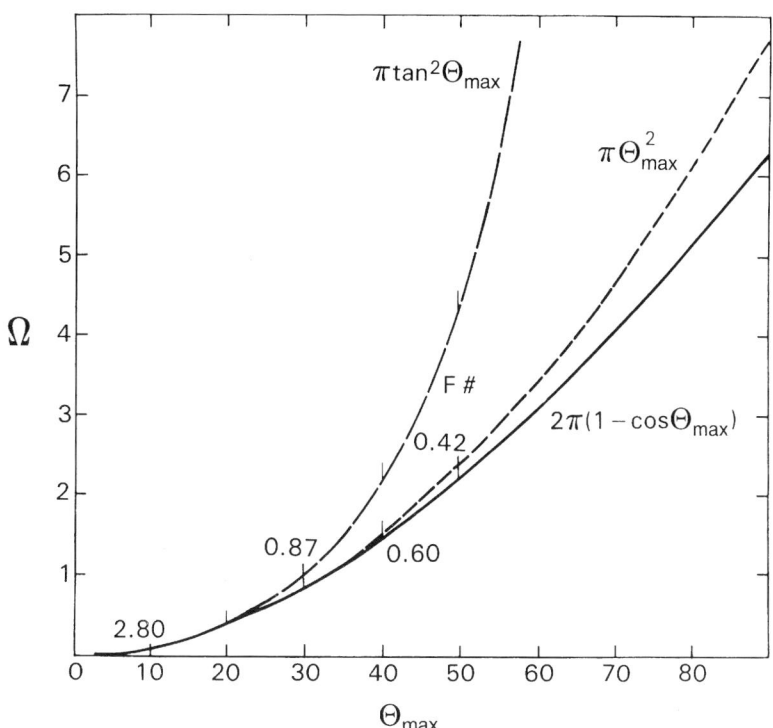

Fig. 5.2.2 Solid angle versus maximum cone angle. The solid curve is the exact formula, the dashed curves are approximations valid for small cone angles. The *f*-numbers of the optical elements are also shown (F#).

If we let the radius of the lens of area A_0 be R_0 we obtain:

$$a \tan \alpha = R_0 = b \tan \theta. \tag{5.2.5}$$

Squaring this equation and multiplying by π yields

$$a^2 \pi \tan^2 \alpha = \pi R_0^2 = b^2 \pi \tan^2 \theta. \tag{5.2.6}$$

With approximation (5.2.3) one obtains

$$a^2 \Omega_p = A_0 = b^2 \Omega_d. \tag{5.2.7}$$

Elimination of a^2 and b^2 with the help of Eq. (5.2.4) yields

$$A_p \Omega_p = A_0 \Omega_0 = A_d \Omega_d. \tag{5.2.8}$$

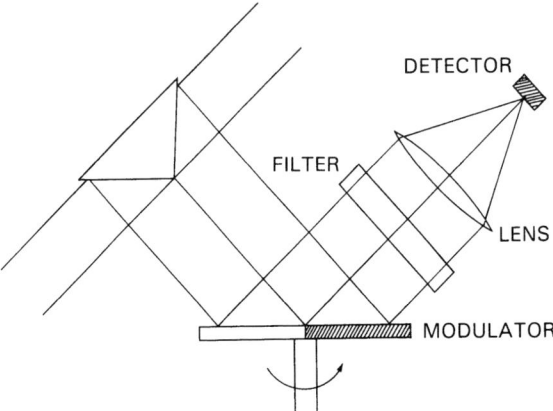

Fig. 5.2.3 Schematic of the infrared radiometer on TIROS 2. Half of the rotating modulator is reflective, the other half is absorbing.

Thus the product $A\Omega$ is invariant in an optical system. Actually the term $n^2 A\Omega$ is invariant, n being the refractive index, but as long as we measure A and Ω in the same medium (air or vacuum, for example) it is sufficient to consider $A\Omega$ to be invariant. An optical system may be compared to an electrical transformer, where the product current times voltage (power) is also invariant. The telescope and the transformer are not without losses; optical and electrical efficiencies are less than unity.

In reality, the planet–lens distance is much larger than the lens–detector distance; for all practical purposes the planet is at infinity and the detector must be placed at the focal point of the lens. Indeed such simple radiometers have been flown in space, for example, on TIROS 2 (Fig. 5.2.3). As the half-reflective, half-black modulator rotates, half of the radiation arriving at the detector originates at the black modulator while the other half alternates between both input beams. The optical element of this instrument consists of a lens, which focuses the planetary radiation (or that from space) onto the detector. A cluster of five such devices, each one with its own spectral filter, formed this five channel radiometer. For each channel the entrance aperture is the lens or, if you consider only one of the entrance beams, one half of the lens. Modulator and prismatically shaped mirror are oversized. The field stop, that is, the aperture that determines the field of view, is the detector. Such an optical layout has the advantage of simplicity, but it has shortcomings as well. The response of infrared detectors is often not uniform across the sensitive area. Some parts of the detector surface may be several times as sensitive as others. The area of the planet to be imaged onto this nonuniform detector is in all likelihood also not uniform. The detector signal is then proportional to the two-dimensional convolution of both distribution patterns. If the field of view and, therefore, also the detector, is small and possibly scanning across the planetary surface, the effect may

5.2 Optical elements

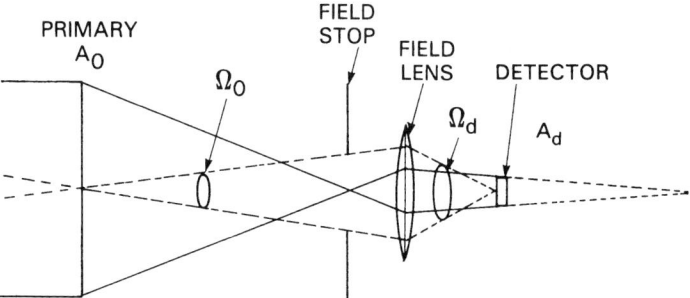

Fig. 5.2.4 Optical layout demonstrating the use of a field lens. The field lens images the primary mirror onto the detector.

not be too detrimental. The one- or two-dimensional image, which can be formed from one or several adjacent scan lines, may show little of the convolution problem. If, just to consider an extreme example, the planet is smaller than the field of view, a nonuniform response may lead to severe errors.

Addition of a field lens overcomes the problem (Fig. 5.2.4). The entrance aperture is still the telescope primary, but the detector is no longer the field stop; the stop is an opening in an otherwise opaque disk located at the focal plane. The field stop can now be circular or of any other shape; it is not restricted to the usually square shape of detectors. At or immediately behind the field stop is the field lens with the focal length chosen to image the primary onto the detector. A bundle of rays that originates from a particular point on the planet within the field of view and that passes through the entrance aperture, forms an image of the point in the plane of the field stop, but it floods all parts of the detector, as bundles from other points within the field also do. In spite of a nonuniform detector sensitivity, the response function of this radiometer across its field of view is theoretically ideal, which means it is constant within the field and zero outside. Of course, this is strictly true under otherwise ideal conditions. In reality, aberrations in the optical components, diffraction at the edges, scattering on optical imperfections, and alignment errors cause deviations from ideal conditions. The addition of a field lens, or field mirror, reduces the overall efficiency by a few percent, depending on the implementation of the design; however, it may be a small sacrifice if a flat field of view response is of concern.

Two variations of the field lens concept are mentioned: the immersion lens and the Winston cone. Both are designed to reduce the required detector area and, thereby, improve the signal-to-noise ratio. The immersion lens (Fig. 5.2.5) increases the speed of a condensing optic beyond that of a conventional field lens by coupling the detector directly to a material of a high refractive index. The immersion lens is the analog to the immersion objective of high-power microscopes. Instead of a transparent oil used with microscope objectives a soft arsenic modified selenium film couples the detector to the plane lens surface. The selenium film is transparent from 1 to 20 μm. No air gap is permitted between lens and detector. The reduction in detector area, in comparison with that of a conventional field lens, is proportional to n^2 of the immersion lens material, a consequence of the invariance of $n^2 A\Omega$. Since detector noise is often proportional to the square root of the area, the improvement in signal-to-noise is proportional to n; therefore, a material with a high refractive index, such as germanium ($n = 4$), is desirable. It is necessary to apply an antireflection coating to the hemispherical surface. Immersion lenses have been used mainly in conjunction with thermistor bolometers (De Waard & Wormser, 1959).

A Winston cone is a non-image-forming concentrator (Fig. 5.2.6). Winston cones are used widely in the far infrared, partly because of their high efficiency, and partly because refractive materials to form field lenses or immersed detectors are not conveniently available in that spectral domain. The cone scrambles the image at the focal plane and, thereby, fulfills the objective of the field lens concept, which is to prevent a direct mapping of the observed scene onto the detector surface. Since cones are reflective elements, they are highly efficient and without wavelength limitations, except for diffraction effects. In addition to the rays shown in Fig. 5.2.6, skew rays must be considered. Design theory and many useful diagrams can be found in the book by Welford & Winston (1978).

In Fig. 5.2.4 the primary optical element defines the entrance aperture. This may not always be the case, as shown in Fig. 5.2.7, where the aperture stop is located after the field stop. If this system is arranged on an optical bench with components

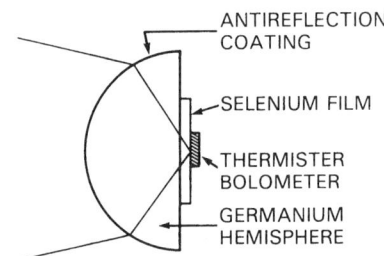

Fig. 5.2.5 Thermistor bolometer attached to a hemispherical immersion lens.

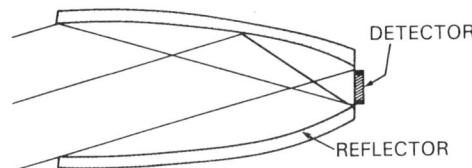

Fig. 5.2.6 Winston cone, a non-image-forming concentrator.

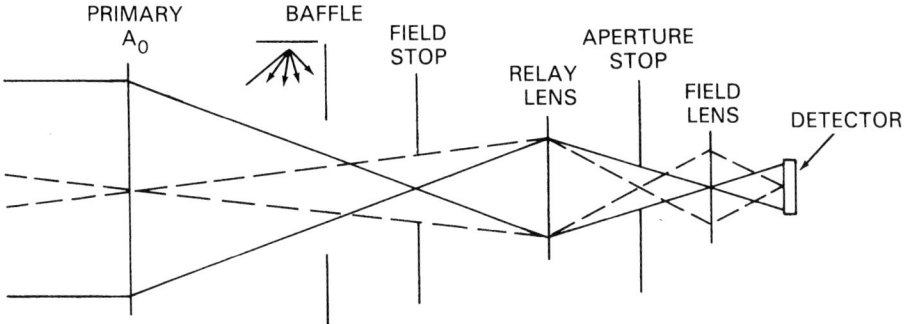

Fig. 5.2.7 Optical layout with aperture stop behind field stop.

transparent to visible radiation and allowed to face a distant scene, a sharp, inverted image of this scene can be observed by holding a piece of white paper at the field stop. If one moves this piece of paper behind the relay lens to the aperture stop, the image of the primary appears at that position. The physical size of the aperture stop must be slightly smaller than this image, or else the primary will be the limiting aperture. A design with the primary not being the aperture stop prevents radiation scattered off the edge of the primary from reaching the detector.

The field stop limits the solid angle of the instrument by eliminating all rays with angles larger than that of the principal rays (dashed lines in Fig. 5.2.7). The principal rays are rays that pass through the center of the aperture stop, have the largest angle with the optical axis, and still reach the detector. The aperture stop is the particular aperture that determines the entrance pupil, A_0, of the instrument. In Fig. 5.2.7 the entrance pupil is the image of the aperture stop at the primary. The entrance pupil need not necessarily be at the position of the primary; it could be located at a distance in front of the primary. The optical elements that do not serve as stops should be sufficiently large to admit all legitimate rays, even those that are slightly diffracted, or else they may become partial stops for extreme rays. If this occurs, the system is said to have vignetting. In an arrangement such as the one shown in Fig. 5.2.7, it may also be advisable to introduce other baffles to reduce light scattered by a tubular housing, for example.

Many instruments designed to analyze spectral information, such as Fabry–Perot and Michelson interferometers, operate best in a collimated beam. Telescopes that produce such a beam are well-known for visual observations; the Galilean and the astronomical telescope are examples. The first uses a convex objective lens and a concave eyepiece, while the latter a similar objective and an eyepiece of convex curvature. In either case, the focal points of both lenses must coincide. Implementations of both systems with mirrors and lenses are shown in Fig. 5.2.8. Afocal systems have the object as well as the image at infinity. The example shown

162   *Instruments to measure the radiation field*

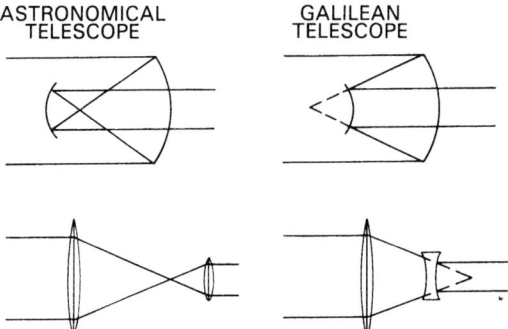

Fig. 5.2.8 Afocal systems. Reflective and refractive versions of the astronomical and Galilean telescopes.

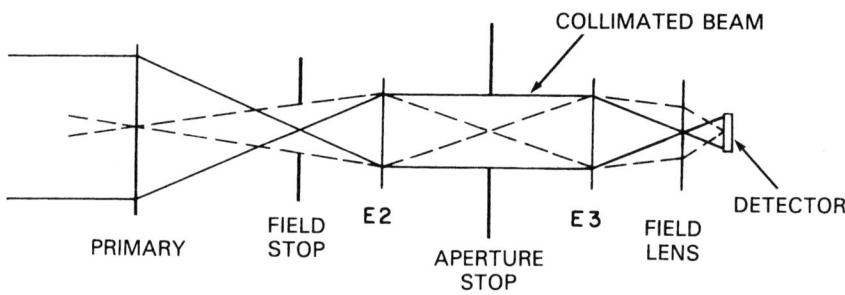

Fig. 5.2.9 Optical layout with a collimated beam at the aperture stop.

in Fig. 5.2.9, based on the astronomical telescope, has a well-defined field stop. The aperture stop has been placed between elements 2 and 3 in the middle of the collimated beam. The aperture stop may coincide with the position of the stationary mirror of a Michelson interferometer or with the etalon (mounted plates) of a Fabry–Perot interferometer. For operation with a single detector, a field lens may be used at or near the focus of the third element. If a multi-element detector array is desired in conjunction with a spectral analyzer to provide simultaneous imaging capability, the array must be placed at the focal point of element 3. The main field stop limits the frame size and reduces out-of-field stray light, but each individual detector element functions as its own field stop.

For systems with large primary elements and long focal lengths, as are often required when good spatial resolution is to be achieved from a great distance, the overall length of a single element telescope is prohibitive. The Cassegrain design combines a long focal length with a short physical dimension. It consists of a concave primary mirror of paraboloidal shape and a convex secondary of

Table 5.2.1 *Parameters (in cm) of Voyager infrared telescope*

Focal length of primary	35.56
Diameter of primary	50.8
Focal length of secondary	−5.751
Diameter of secondary	7.257
Distance primary–secondary	30.48
Diameter of field stop	1.33
Virtual image of primary behind secondary	7.088
Diameter of virtual image	8.861

hyperboloidal configuration. Many large, ground-based astronomical telescopes can be operated at a Cassegrain focus. A schematic cross section of the Voyager infrared spectrometer (IRIS) and dimensions of the Cassegrain telescope are shown in Fig. 5.2.10 and in Table 5.2.1, respectively. The negative focal length of the secondary implies a convex mirror surface. The effective focal length of the IRIS telescope is 305 cm and the f-number is 6. The telescope was designed with a field limiting aperture of 1.33 cm in the focal plane and a 0.25° full field of view. The Composite Infrared Spectrometer (CIRS) on Cassini has a similar telescope.

If in the design of the reflective version of the astronomical telescope (Fig. 5.2.8) the secondary mirror is moved farther from the primary focus, the telescope focal point moves from infinity to a finite position and one obtains a Gregorian telescope. The secondary mirror is of a concave ellipsoidal shape and reimages the real focal point of the primary into that of the telescope. This design leads to a larger primary–secondary separation and, therefore, to a larger overall length than that of the Cassegrain configuration. The only advantage of the Gregorian telescope in its off-axis version is improved rejection of stray light. The Gregorian telescope has seen little use, except where stray light is of concern. Such concerns arise in the measurement of feeble radiation from the limb of a planet (the outer edge of the apparent disk) in the presence of a large and strongly emitting planetary disk or in precise measurements of the cosmic background in the presence of strong sources, such as the Sun, the Earth, the Moon, and the planets. The advantage of an off-axis Gregorian telescope is the accessibility of the prime focus, which permits additional and effective baffling. Residual stray light is caused by the presence of dust particles and quality flaws on the primary mirror surface. If necessary, additional baffles ahead of the primary may be installed. The off-axis configuration leads to a lower *f*-number of the primary mirror, which is often cut from a larger, on-axis paraboloid. An example of an off-axis Gregorian design is the telescope

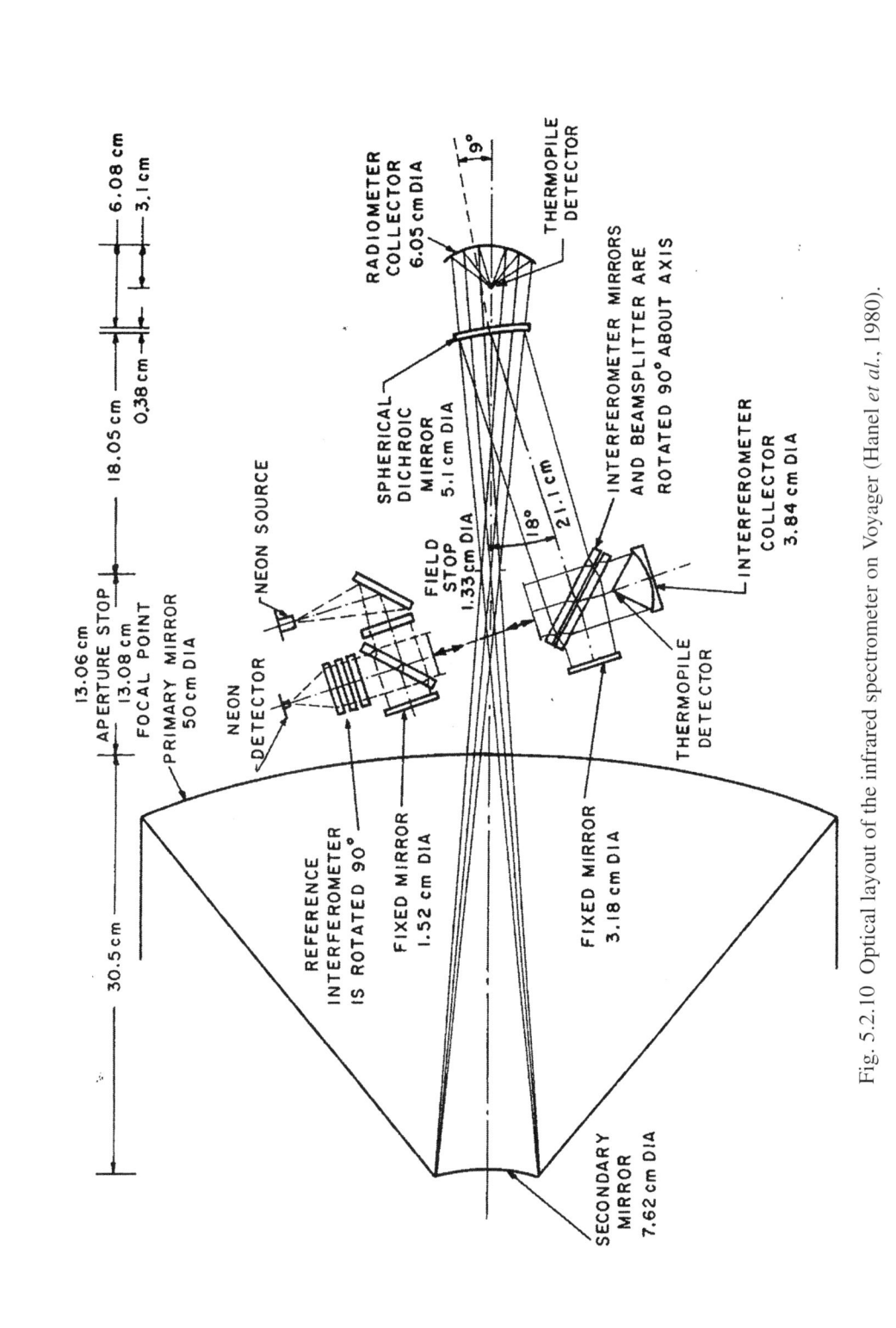

Fig. 5.2.10 Optical layout of the infrared spectrometer on *Voyager* (Hanel *et al.*, 1980).

5.2 Optical elements

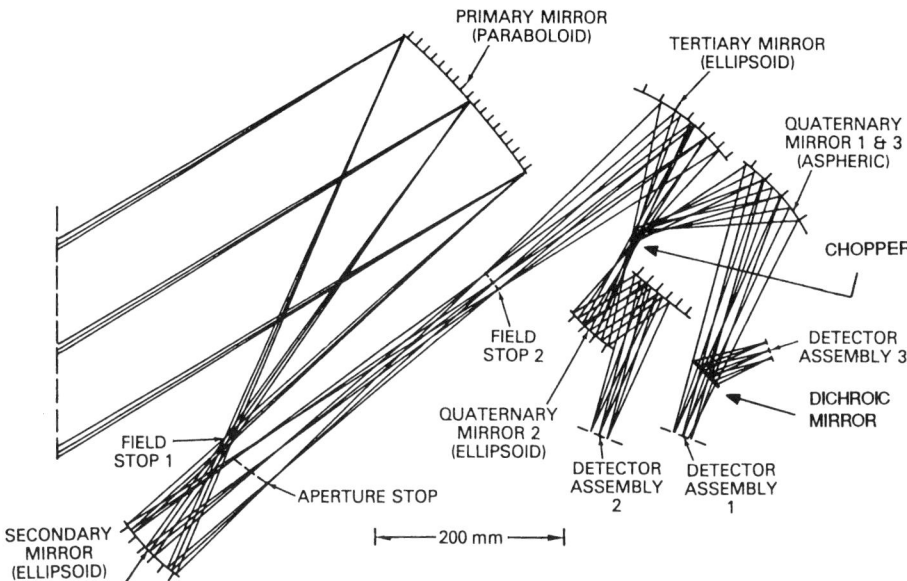

Fig. 5.2.11 Schematics of an off-axis Gregorian telescope, similar to one used on the Cosmic Background Explorer (COBE Project, NASA).

of the infrared photometer flown on the Diffuse Infrared Background Experiment (DIRBE) on the Cosmic Background Explorer (COBE) (Fig. 5.2.11).

The decision as to whether to use mirrors or lenses for the telescope depends on several factors. A first principle in the design of optical systems requires symmetry if at all possible. Symmetrical optical systems are usually much easier to fabricate, to align, and they have large alignment error allocations. The latter is especially important when the operating temperature significantly differs from the alignment temperature of the system. In such a case it is desirable to fabricate the telescope mirrors and the support structure of the same material (aluminum or beryllium, for example). A secondary principle is to avoid obscurations in the beam. The effect of obscurations on the optical efficiency, the generation of side lobes, and the problem of warm structural elements in the field of view of a detector at cryogenic temperatures have to be considered. Both mirror and lens systems can be designed to comply with the first principle; however, symmetrical mirror systems must have obscurations and do not comply with the second principle. In some critical cases even a lens system must be followed by a small obscuration – for example, to block a ghost image formed by multiple reflections between lens surfaces.

Selection of a specific telescope design is, therefore, not straightforward. Large aperture lens systems can become very heavy. The refractive index of materials is usually quite high in the infrared and requires each surface to have multilayer antireflection coatings. In addition, the refractive index of infrared materials is temperature sensitive and the material itself is often not transparent in the visible, which complicates both system alignment at ambient temperatures and control of system aberrations at operational temperatures. Correction of chromatic aberration to acceptable limits near material absorption edges is frequently not possible. On the other hand lenses do offer the possibility of designing a controlled contamination environment because of their obvious sealing characteristics. Restrictions in volume and weight, combined with étendue and pixel resolution requirements, frequently favor a reflecting telescope.

After the conceptual layout of a telescope is complete, the design process must include a study of aberrations, which may involve iterative ray tracing. The reader interested in this subject should consult books on optical design such as Conrady (1957), Strong (1958), Born & Wolf (1959), or *The Infrared Handbook* by Wolfe & Zissis (1978).

5.3 Diffraction limit

The angular (spatial) resolution of a telescope, or of any optical system, is ultimately limited by diffraction. As an example, we calculate these limitations for an instrument with a Cassegrain telescope; however, the theory is directly applicable to other configurations as well.

The entrance aperture of the telescope is represented by a prime mirror of radius a, with a central obscuration of radius $a' = \varepsilon a$. A small field stop is visualized as a small detector at the focal point. The optical axis of the telescope is the x-axis. To study diffraction we let a plane wave arrive at an angle α. The plane wave is expressed by

$$E = e^{i\omega t} e^{-ik(x\cos\alpha + y\sin\alpha)}. \tag{5.3.1}$$

Introducing polar coordinates ($x = r\cos\phi$, $y = r\sin\phi$), and omitting the time factor, one obtains for the amplitude distribution at the entrance aperture:

$$E(x = 0) = e^{-ikr\sin\phi\sin\alpha}. \tag{5.3.2}$$

We are dealing with plane waves and a detector at the focal point; we are, therefore, in the domain of Fraunhofer diffraction. The more general diffraction theory, applicable to a source and a detector at arbitrary positions to both sides of an aperture, is called Fresnel diffraction theory. Only Fraunhofer diffraction is important for planetary telescopes since all objects are effectively at infinity and detectors or field stops are in the focal plane of optical systems.

5.3 Diffraction limit

The resulting signal at the detector may be obtained by integration over the aperture area. Indicating this average by $\bar{E}$, one obtains

$$\bar{E}(\varepsilon, \alpha, a) = \int_{\varepsilon a}^{a} \int_{0}^{2\pi} e^{-ikr \sin\phi \sin\alpha} r \, d\phi \, dr. \quad (5.3.3)$$

Integration over ϕ is accomplished with the help of the Sommerfeld integral representation of Bessel functions

$$\int_{0}^{2\pi} e^{-iz \sin\phi} d\phi = 2\pi J_0(z); \quad z = kr \sin\alpha, \quad (5.3.4)$$

where $J_0(z)$ is the zero-order Bessel function of argument z. Consequently,

$$\bar{E} = 2\pi \int_{\varepsilon a}^{a} J_0(kr \sin\alpha) r \, dr. \quad (5.3.5)$$

With the integral formulas for Bessel functions (Jahnke & Emde, 1933, p. 143) one obtains

$$\bar{E}(\varepsilon, \alpha, a) = a^2 \left[\frac{2J_1(ka \sin\alpha)}{ka \sin\alpha} - \varepsilon^2 \frac{2J_1(k\varepsilon a \sin\alpha)}{k\varepsilon a \sin\alpha} \right]. \quad (5.3.6)$$

The limit of $2J_1(x)/x$ for $x = 0$ is 1. We normalize by dividing by $\bar{E}(\varepsilon, \alpha = 0, a)$ and square the result to obtain the normalized intensity:

$$\bar{I}(\varepsilon, \alpha, a) = \frac{1}{(1-\varepsilon^2)^2} \left[\frac{2J_1(ka \sin\alpha)}{ka \sin\alpha} - \varepsilon^2 \frac{2J_1(k\varepsilon a \sin\alpha)}{k\varepsilon a \sin\alpha} \right]^2. \quad (5.3.7)$$

This function is shown in Fig. 5.3.1 for several values of ε as a function of $ka \sin\alpha$. Since ka equals $\pi D/\lambda$, ka is π times the diameter, D, of the aperture measured in wavelengths. Therefore, the 50 cm telescope of Voyager has at 50 μm the same diffraction pattern as a 0.5 cm lens has with the same relative central obscuration in the visible (0.5 μm).

For $\varepsilon = 0$ one obtains the diffraction pattern of a lens or an off-axis mirror without central obscuration. The first null of $J_1(x)/x$ is at $x_1 = 3.832$. The central part of the diffraction pattern – that is, the portion inside the first null – is said to be inside the Airy disk given by

$$ka \sin\alpha = 3.832. \quad (5.3.8)$$

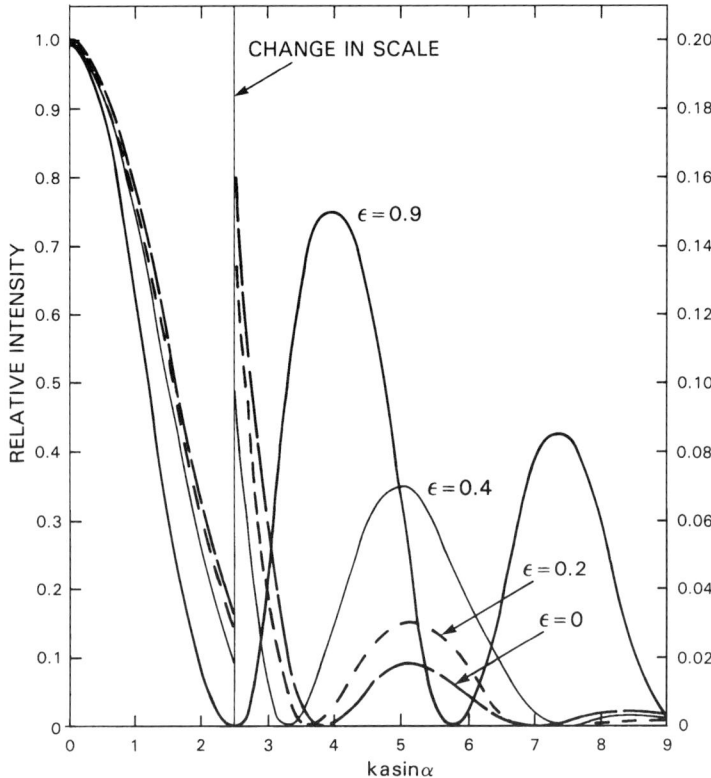

Fig. 5.3.1 Relative intensity of a point source impinging on a telescope with central obscuration ε as a function of $ka\sin\alpha$ (see text).

Since $ka = \pi D/\lambda$ and $\sin\alpha \sim \alpha$ for small angles,

$$\alpha = 1.22\lambda/D, \tag{5.3.9}$$

which is the well-known equation for diffraction limited angular resolution. Because of the A-Omega invariance,

$$\frac{\pi D^2}{4}\pi\alpha^2 = A_d\Omega_d, \tag{5.3.10}$$

the diffraction limited étendue may be expressed by

$$3.7\lambda^2 = A_d\Omega_d. \tag{5.3.11}$$

The half cone angle, $\theta_{\max}$, (Fig. 5.2.2) will seldom exceed 66°; then Ω_d is 3.7 and the dimensions of the detector equal the wavelength. With more moderate angles, say 35° and $\Omega_d = 1.15$, the detector is 1.8 times the wavelength in the

5.3 Diffraction limit

diffraction limited case. Reduction of the detector size to less than the size given by the diffraction limit will only reduce the intercepted energy, but will not improve the angular resolution.

As shown in Fig. 5.3.1, the effect of a central obscuration is not only to reduce the position of the first null of the intensity function, but also to increase the energy contained in the side lobes. Each side lobe or diffraction ring contains a certain fraction of the total energy. By integrating the intensity function with respect to $ka \sin \alpha$ from the center to the first null one obtains the energy contained in the Airy disk, or the first ring. Integration from the first to the second null yields the energy of the second ring and so forth. This integration, as a function of ε, was carried out by Goldberg & McCulloch (1969) (Fig. 5.3.2). Even for $\varepsilon = 0$, the Airy disk contains only 0.84 of the total energy and the second ring about 0.08. For values of ε between 0.3 and 0.4, the first two rings contain most of the energy, ~ 0.89, and little is gained by including the third and fourth ring because their contributions are small. To recover almost 0.90 of the total energy the angular size of the field of view

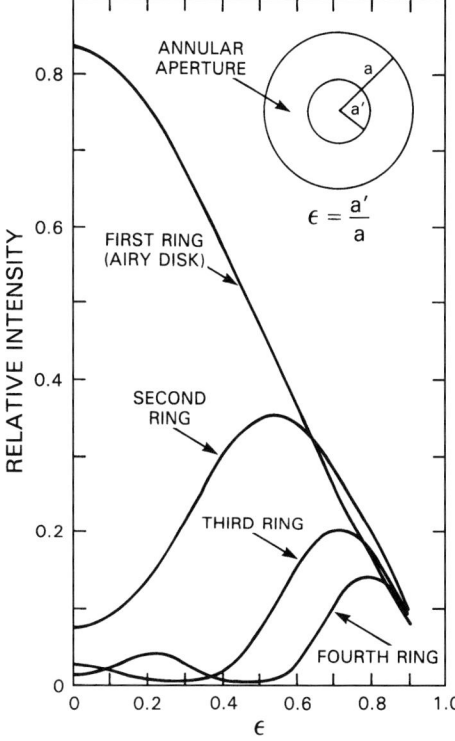

Fig. 5.3.2 Energy distribution of a point source among several diffraction rings; as in Fig. 5.3.1, ε is the ratio of the radii of the central obscuration to the full aperture (Goldberg & McCulloch, 1969).

must be chosen to reach to the second null; for $\varepsilon = 0.4$ of the intensity function.

$$ka \sin \alpha \sim 7.5 \qquad (5.3.12)$$

and

$$\alpha = 2.39 \lambda / D, \qquad (5.3.13)$$

or approximately twice the value of $1.22\lambda/D$. This diffraction criterion is more realistic for telescopes with a sizeable central obscuration and if the total energy of a point source is to be measured with a diffraction limited Cassegrain system.

For a telescope of 50 cm diameter and for a wavelength of 50 μm (as in the case of the Voyager instrument), α, as defined by Eq. (5.3.13), is 2.30×10^{-4}, which corresponds to a half cone angle of $0.014°$. This angle is small in comparison with the $0.125°$ half cone angle of the instrument. The Voyager instrument (IRIS) is almost an order of magnitude from being diffraction limited.

In the above derivations we neglected the effect of the structure necessary to hold the secondary mirror in place. The impact of diffraction effects by these structures has been examined in detail by Harvey & Ftaclas (1988). If the primary is imaged onto the detector, as is often done to obtain a uniform field of view, the sensitivity pattern across the detector element influences the effective diffraction pattern of the system. The integrations of Eq. (5.3.3) should then include the detector sensitivity pattern as a weighting function under the integral. If the sensitivity pattern has pronounced peaks or voids, the effect can be substantial for a diffraction limited system.

5.4 Chopping, scanning, and image motion compensation

In this section we discuss basic radiometer configurations. Unchopped radiometers, also called d.c. radiometers, and instruments where the optical beam is interrupted at a relatively high rate, called a.c. radiometers, have been used widely. Often these devices are employed in a line by line scanning mode with the intent to generate a two-dimensional image as well as to produce data suitable for numerical analysis. Image motion compensation may be required when integration of the signal is necessary over a time interval where smear, caused by the motion of the spacecraft relative to that of the object, cannot be neglected.

a. D.C. radiometers

The basic radiometer configuration shown in Fig. 5.2.1 functions adequately in the visible and near infrared where radiation from the object under study is likely to dominate radiation emitted from the instrument. However, this must not be true in the middle and far infrared. At room temperature and for wavelengths longer than a few micrometers, thermal emission from optical filters, lenses, and mirrors may exceed planetary radiation by a large amount. Several methods may be applied to distinguish between desired radiation from the object under investigation and undesired emission from optical components.

The most straightforward method is to cool the instrument to a temperature low enough to render thermal emission from components insignificant. Although cooling of the whole instrument also reduces detector noise, it is often not convenient, sometimes not even possible to implement. For planetary observations within a narrow spectral band, where the signal even from a relatively warm object may be quite feeble, it would be necessary to cool the instrument, or at least the detector environment and the narrow-band filter, to cryogenic temperatures. This is not possible without severe penalties in weight or power. For a mission to the outer planets, which could easily take a decade to reach the objects of interest, it would be impractical to install a large tank of liquid helium, or to provide enough power to operate a cryogenic refrigerator. Other examples where cooling is impossible are ground-based telescopes, which cannot be operated below ambient air temperature without risking condensation.

An alternative way to separate radiation from the object under study from that of the instrument is to hold the instrument temperature constant and occasionally intersperse measurements of deep space. For most purposes deep space is a nonemitting sink. The weak emission of the cosmic background (~ 2.7 K), of stars, and of galaxies is negligible compared with that of objects in our Solar System. To cancel instrument emission one subtracts deep space readings from those of the object of interest. Deep space observations must occur often enough so that the effect of a residual drift in the instrument temperature, and in atmospheric conditions in case of ground-based observations, can be kept small in comparison with the planetary radiance to be measured. Deep space observations must not occur too often, however, so that data collection can proceed undisturbed in between space observations. The optimum duration of planetary and space measurements depends on the planetary intensity, the instrument temperature, and on the thermal time constants of the components involved.

While scanning across a planetary disk, deep space observations come naturally just before and after the field of view crosses the limb. If the field of view is permitted to complete a full $360°$ rotation at uniform angular speed, sufficient time exists to observe a blackbody of known temperature mounted inside the instrument. Assuming the radiometer response to be linear, the planetary measurements can be scaled to the readings from space and the blackbody, permitting absolute calibration (see Section 5.13).

To demonstrate the principles involved we discuss the conceptual design of a scanning radiometer for a spacecraft in a low, circular orbit. The field of view and the scan rate of that instrument shall be chosen to yield a contiguous pattern near the subspacecraft point. Closer to the horizon the widths of individual scans projected onto the curved planetary surface become wider and adjacent scan patterns overlap. In certain designs the field of view may rotate around the center of the view direction, adding further distortion to picture elements away from the nadir. For a low orbit the spin of the spacecraft or, more likely, the rotation of a mirror in the instrument generates the line by line sweep. The orbital motion provides the progression from one line to the next. A planet-oriented, well-stabilized spacecraft in a circular orbit is best suited for such a mapping radiometer. For a geosynchronous orbit, where the spacecraft remains over nearly the same equatorial point, the line scan can be provided by spinning the spacecraft about an axis parallel to that of the planet. The advancement from one line to the next can be accomplished by stepping a mirror each revolution or by tilting the whole instrument in small increments with respect to the spacecraft spin axis. All these techniques have actually been used.

To an instrument at an altitude, h, above the surface, the planetary disk appears at a half angle, θ,

$$\theta = \arcsin \frac{R}{R+h}, \qquad (5.4.1)$$

where R is the planetary radius ($\sim$6371 km for the Earth). The full angle, horizon to horizon, is then $\sim$125° for a 800 km orbit. The subspacecraft velocity, v_{SS}, is calculated from celestial mechanics,

$$v_{SS} = \sqrt{\frac{GM}{R+h}} \cdot \frac{R}{R+h}, \qquad (5.4.2)$$

where G is the gravitational constant ($G = 6.67 \times 10^{-11}$ m^3 kg^{-1} s^{-2}) and M is the mass of the Earth ($M = 5.98 \times 10^{24}$ kg). The first term with the square root is the orbital velocity, while the second term is the ratio of the projection of the orbital velocity onto the Earth's surface to the orbital velocity itself. For an 800 km orbit the subspacecraft point moves with a speed of 6.63 km s^{-1}; for a lower orbit, say 500 km, the speed is slightly higher, 7.07 km s^{-1}.

Suppose one desires a spatial resolution of 14 km at the subspacecraft point. An instrument with a 1° square field of view in an 800 km orbit has a 'footprint' of just that size (Fig. 5.4.1). It takes the spacecraft 2.1 s to advance the subspacecraft point the width of the field of view, 14 km. The scan mirror must complete a full revolution in 2.1 s to provide a contiguous scan pattern. This implies a mirror rotation rate of 0.47 revolutions s^{-1}. Knowing the field of view and the scan rate, one may derive the electric bandwidth required to transmit the information, provided one also knows the scanning function.

5.4 Chopping, scanning, and image motion compensation 173

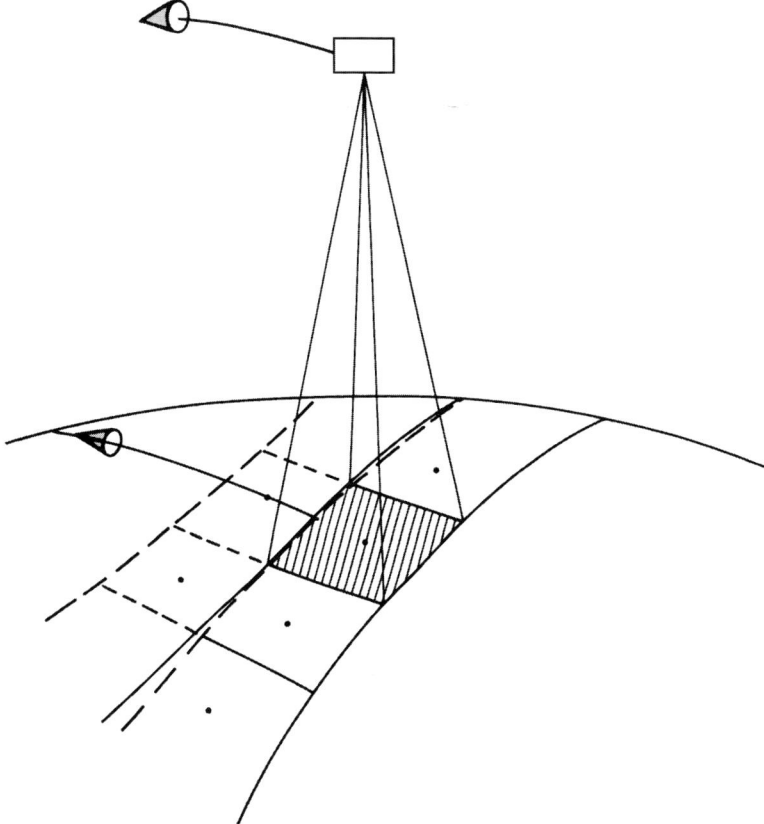

Fig. 5.4.1 Projection of the rectangular field of view of a radiometer scanning perpendicular to the subsatellite path. The scan rate is chosen to produce contiguous coverage on the planetary surface at the subsatellite point.

To derive the scanning function consider an arbitrary intensity distribution in the scan direction represented by a sum of sinusoidal components of an amplitude, I_n, and a spatial wavelength, Λ_n (Fig. 5.4.2). Integration over a square field of view yields the scanning function, $\sin x_n/x_n$, where $x_n = \pi L/\Lambda_n$; L is the length of the footprint in the scan direction (14 km) and Λ_n is the nth Fourier component of the intensity along the same direction. The function $\sin x/x$ has the first null at $x = \pi$. If the field is small in comparison with Λ, degradation of the amplitude does not occur; if the field equals Λ, the signal reduces to zero. The optical scanning process shows complete analogy to magnetic tape recording. The field of view in the scan direction corresponds to the gap width of the playback head and the scan speed to the tape speed.

A circular field of view requires averaging over a circular area of the patterns with wavelength Λ_n. A scanning function $2J_1(x)/x$ is obtained, where $J_1(x)$ is

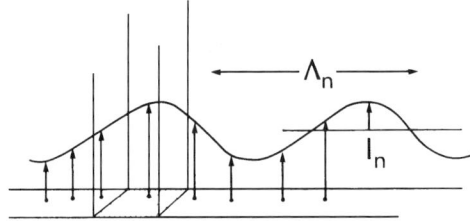

Fig. 5.4.2 Schematic to illustrate variation in intensity along the scanning direction. Only one modulation of periodicity Λ_n is shown.

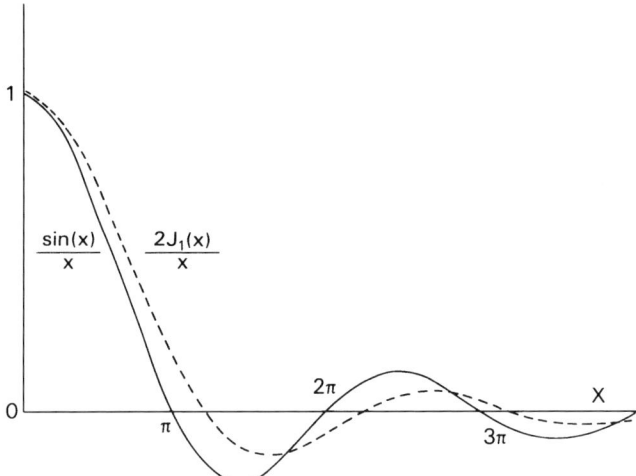

Fig. 5.4.3 Scanning function for a rectangular (solid line) and a circular (dashed line) field of view.

the first-order Bessel function of the argument $x = \pi D/\Lambda$, D being the diameter of the footprint. In the discussion of diffraction in Section 5.3, integration over a circular aperture illuminated by a plane wave resulted in the same function [Eq. (5.3.7)]. Both scanning functions, for square and circular fields of view, are shown in Fig. 5.4.3. The highest signal frequency that should be recorded is a frequency just below the first null of the scanning function. For simplicity, we take the null frequency as the limit of the base band. Higher frequencies, corresponding to side lobes, are usually not of interest and should be suppressed. If the detector output is sampled at discrete intervals, as is normally the case, a sampling rate of at least double the first null frequency is necessary to record the base band. Rejection of frequencies above the first null is important to avoid folding of high frequency interference and noise into the base band, called aliasing.

A square field of 1° has 360 elements in a full circle. With the scan mirror rotating 0.47 revolutions s^{-1} this corresponds to 171 periods s^{-1}, or 171 Hz. The

5.4 Chopping, scanning, and image motion compensation

sampling rate should, therefore, be at least 342 samples s^{-1}. The lowest frequency to be reproduced is d.c. or zero hertz. Therefore, this type of radiometer is called a d.c. radiometer. However, detecting and amplifying d.c. levels is inconvenient. As a compromise, one may limit the low frequency end of the passband at 0.05 Hz, that is, well below that given by the rotation rate of ~ 0.5 Hz in this example. Once each revolution, while viewing deep space, the signal is set to zero electronically, a process called clamping. The accumulative error due to the lack of true d.c. response can thereby be kept small. The permissible error due to insufficient low frequency response must be evaluated for a particular application. Besides the amplitude response the phase characteristic of the system is important. The overall phase characteristic, including that of the detector, amplifier, and electronic filter, must be linear over the whole passband. Then the overall shape of the signal is preserved; otherwise distortion will occur and errors will be introduced. The problem is analogous to that encountered in the reproduction of sound, except that the human ear may be more forgiving than the stringent requirements of an infrared photometer.

A number of scanning radiometers, based on this principle, have flown in low Earth orbits for meteorological purposes. Typical instruments are the two-channel radiometer of ITOS (1970) (Fig. 5.4.4) and the Very High Resolution Radiometer (VHRR) flown first on NOAA 2 in 1972. A more complete listing and several schematics of such instruments can be found in Houghton *et al.* (1984).

An early example of such a scanning d.c. radiometer on a geostationary spacecraft is the Spin Scan Cloud Camera of the Applications Technology Satellite (ATS 1). A color image of the Earth was produced by ATS 3. The spacecraft rotation provides the scanning motion and a stepping motor, tilting the whole instrument by one field of view once every spacecraft rotation, provides the advancement from one scan line to the next. The ATS 1 instrument has a 12.7 cm diameter primary mirror and a 0.1 mrad (0.0057°) field of view. From geosynchronous altitude ($\sim 35\,870$ km) the footprint is 3.6 km wide. With the spacecraft spinning at 100 rpm it takes 20 minutes to generate a 2000 line image of the Earth (Fig. 5.4.5). The electronic bandwidth covers the range from 0.1 Hz to 100 kHz. Low noise of the photomultiplier (0.475–0.630 μm) permits operation at such a low frequency limit. The advantage of geosynchronous observations is apparent from Fig. 5.4.5. A substantial fraction of a hemisphere can be observed every 20 minutes, supporting the analysis of mesoscale weather systems. Three geostationary satellites placed 120° apart can monitor the weather over the whole globe in nearly real time, if a similar camera working in the infrared traces clouds at night. Such a camera was the Visible Infrared Spin Scan Radiometer (VISSR) on the Geostationary Operational Environmental Satellite (GOES) 4, which achieved a 7 km resolution in the thermal infrared.

The most spectacular instruments of this type have been the Thematic Mapper on the Landsat spacecraft and the radiometer on the French satellite SPOT. In the

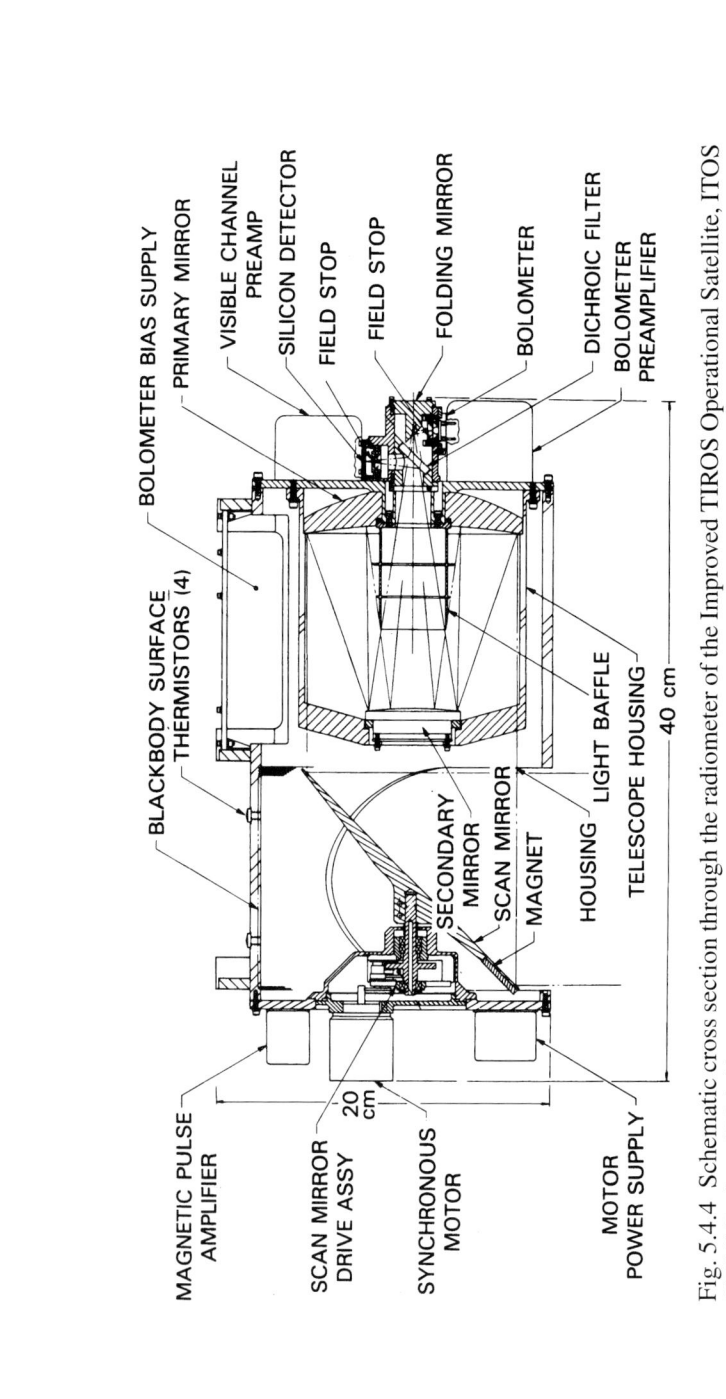

Fig. 5.4.4 Schematic cross section through the radiometer of the Improved TIROS Operational Satellite, ITOS (ITOS Project, NOAA).

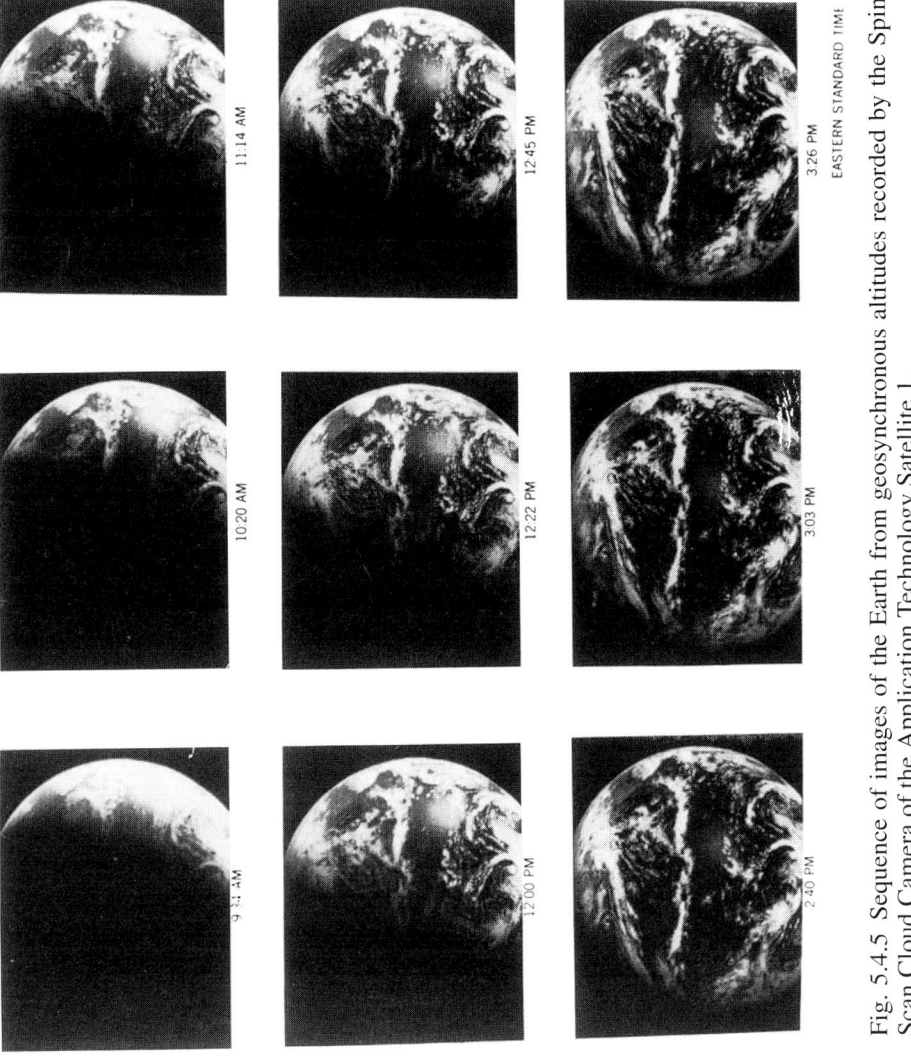

Fig. 5.4.5 Sequence of images of the Earth from geosynchronous altitudes recorded by the Spin Scan Cloud Camera of the Application Technology Satellite 1.

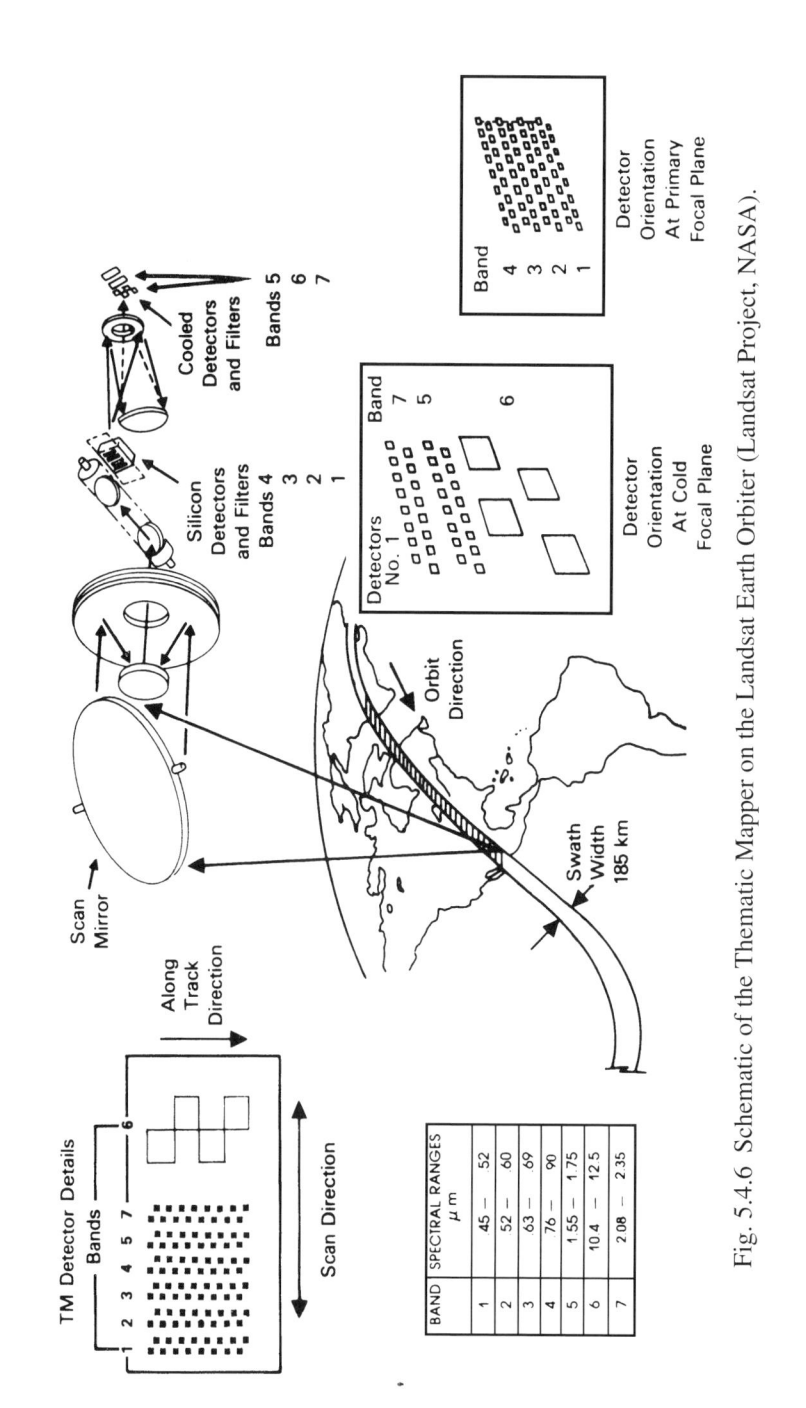

Fig. 5.4.6 Schematic of the Thematic Mapper on the Landsat Earth Orbiter (Landsat Project, NASA).

Thematic Mapper the circular scanning motion through 360° has been replaced by the more efficient motion through a relatively short angular segment. Space and calibration sources are viewed only occasionally. Instead of a single detector for each channel, an array of 16 detectors is used for each visible and near infrared interval, sweeping out 16 scan lines simultaneously. In the middle infrared four detectors form an array as shown schematically in Fig. 5.4.6. By this process a spatial resolution of 30 m has been obtained in the visible and near infrared, and 120 m in the thermal infrared. SPOT obtained an even better resolution of about 15 m. A set of images obtained by the Thematic Mapper is shown in Fig. 5.4.7. Additional information and other images appear in *NASA Conference Publications* (Barker, 1985*a*, *b*) and in the September 1985 issue of *Photogrammetric Engineering & Remote Sensing*.

In the decades that have passed between the first space-borne scanners mapping clouds at night and the Thematic Mapper are many generations of radiometers, often launched for weather observations, but more recently also for geological, agricultural, hydrological, and land use purposes.

b. Chopped or a.c. radiometers

In the far infrared, where signals usually are weak and where detectors may show $1/f$ noise, operation at frequencies as low as a fraction of a hertz is often not possible; for an explanation of $1/f$ noise see page 264. To raise the low frequency cut-off of such a radiometer is also undesirable. Inadequate reproduction of low frequencies affects mostly the overall signal level (the absolute calibration), while inadequate response at high frequencies affects primarily spatial resolution and sharpness of contours in the images.

Another type of radiometer is based on the chopping principle. Comparison between radiation from the object of interest and that of deep space, or another reference source, is accomplished at a rate much faster than the highest signal frequency dictated by the scanning function. The chopper, often implemented by a set of mirrors on a rotating shaft or on the tongues of a tuning fork, is best positioned as the first optical element of the radiometer. For precise measurements it is necessary to reflect the straight beam, that is, the beam that is unaffected by the chopper blades, by another mirror of identical properties and at the same temperature as the chopper. Then the reflected as well as the transmitted beam pass the same number of reflections and the elements emit equal amounts of thermal radiation. Only radiation to be measured is then modulated by the chopper. The constant flux from elements of the instrument is not modulated or is canceled and, therefore, contributes nothing to the alternating signal sensed by the detector. The amplitude of the detected a.c. signal is proportional to the difference in radiance

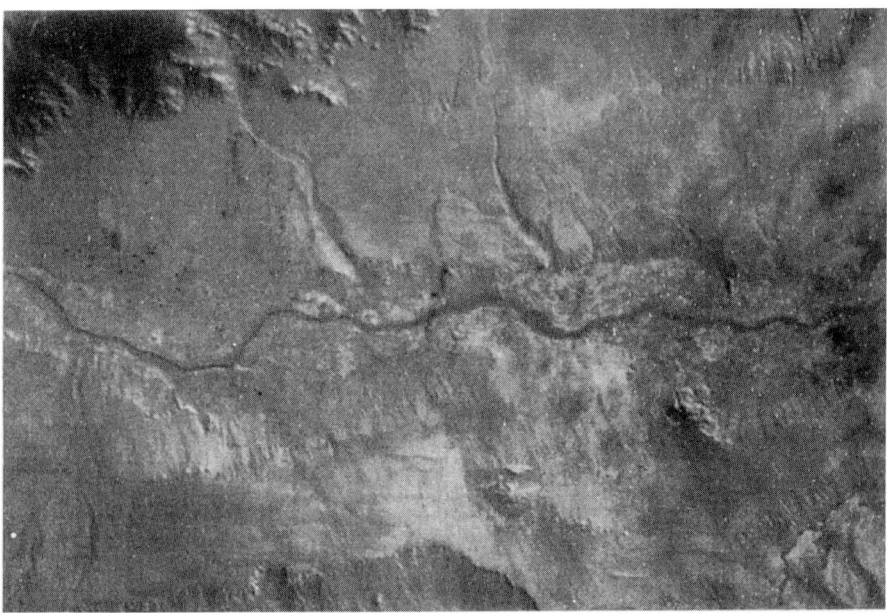

Fig. 5.4.7 Images of an area near Quemado, New Mexico. The upper panel shows channel 6 (10.4–12.5 μm) and the lower panel channel 7 (2.08–2.35 μm) of the Thematic Mapper. The Sun is to the right. Local time is about 0930.

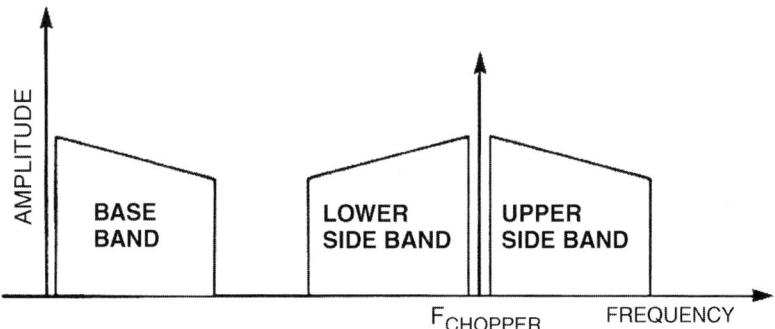

Fig. 5.4.8 Schematic to show the relation of the base band and the side bands due to chopper action.

between object and reference source. The proportionality factor, the instrument responsivity, needs to be determined in the calibration process.

Instead of a reflective chopper and deep space, blackened blades can be used. In this case the chopper serves as the reference. It is difficult, however, to obtain paints with high emissivities at all wavelengths. A small residual reflectivity exists, which may be objectionable in some cases. Furthermore, it is difficult to control the temperature of a black chopper. On a rotating chopper wheel it is even difficult to measure precisely the temperature of the blades. Power dissipation in the chopper motor or in the tuning fork may raise the blade temperature above that of the rest of the instrument. For these reasons it is preferable to use a reflective chopper and space or a cavity-type blackbody as a reference. The temperature rise of the chopper blades is then much less critical.

The chopper action may be viewed as a modulation process where the signal of interest, the frequencies generated by the scan motion, appear as side bands to a carrier, the chopper frequency (Fig. 5.4.8). The process is analogous to that of amplitude modulation of a radio transmitter. It is apparent from Fig. 5.4.8 that the chopper frequency must be at least twice the highest signal frequency, otherwise the high frequency end of the scan signal and the lower side band would overlap. As a rule of thumb, one should use a chopper frequency of at least four times the highest signal frequency. A reasonable degree of freedom exists then for the design of electrical filters, necessary to separate the carrier and the side bands from the base band.

The original signal frequencies are recovered by demodulation of the amplified signal. In the simplest case the demodulation consists of rectification and subsequent low-pass filtering. A better method uses a synchronous rectifier where the a.c. signal is multiplied by a sine or a square wave of constant amplitude derived from the chopper action. The phase of this reference frequency must be properly adjusted; the signal corresponding to the chopper open position must, for example, be multiplied by $+1$ and that corresponding to the closed position by -1. The multiplication may be carried out in analog form by four diodes in a ring modulator, or, as is generally preferred, in digital form, after the amplified signal has been sampled and quantized by an analog-to-digital converter. The quantization is conveniently carried out at twice the chopper frequency. Again, the phase relationship between chopper action and sampling is important. Synchronous demodulation provides an improvement factor of $2^{\frac{1}{2}}$ in signal-to-noise over nonsynchronous demodulation.

Only components of detector and amplifier noise within the side bands enter the system. Low frequency detector noise is unimportant. Also eliminated are all problems associated with d.c. or very low frequency amplification. Not eliminated are effects caused by the temperature dependence of detector responsivity or of optical transmission functions. The main disadvantage of chopping, besides the complexity of the chopper mechanism and associated circuitry, is the need for a faster detector in comparison with that for a d.c. radiometer of identical base band. In a chopping radiometer the detector must respond up to the frequency limit set by the upper side band.

In the chopper discussed so far, the detector is exposed to radiation from the object of interest only half the time while the other half is spent observing the reference source. The intrinsic efficiency of a chopped system with a single detector is, therefore, one half. It may appear that a d.c. radiometer is more efficient than an a.c. radiometer because it is exposed to the desired source for a long time; however, this is not the case. The time spent by the d.c. radiometer to view deep space must also be included in the efficiency budget. Therefore, both types of radiometer have the same observing efficiency of one half, assuming equal object and deep space observations. However, it is possible to increase this efficiency to unity with the help of a second detector as shown in Fig. 5.4.9. The a.c. signals at the identical detectors are $180°$ out of phase. After inverting the polarity of one detector channel, both signals may be added, or, without inverting, both channels may be fed to a

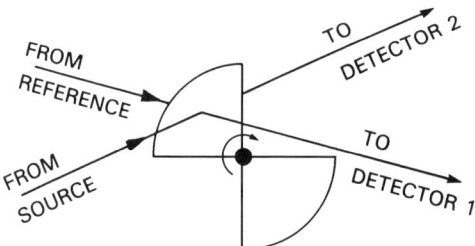

Fig. 5.4.9 A double sided, reflective chopper can be used to switch radiation from two sources alternately onto two detectors.

differential amplifier. Since the signal amplitude doubles and the noise increases only by the square root, the overall improvement in signal-to-noise over a single detector system is $2^{\frac{1}{2}}$; it is assumed that the noise sources of both detectors are uncorrelated. Whether such an improvement justifies the added complexity has to be judged for a particular application.

Chopping against an internal reference source may have advantages over chopping against deep space. The reference source, in most infrared applications a blackbody at a well-defined temperature, may be chosen so that the intensity from the source is of approximately the same magnitude as the mean intensity of the object to be measured. The detector responds to the difference between the intensities from the object and the reference source. The difference signal is expected to be smaller than the peak signal obtained while chopping against deep space. The dynamic range requirement of the instrument is thereby reduced. In the data processing task the measured intensity difference must then be added or subtracted from the known intensity of the reference source, depending on the polarity of the difference signal.

Implementation of a chopper as the first element of a radiometer is not always feasible. For a beam cross section larger than 10 cm, say, a conventional chopper mechanism becomes rather large; it is utterly impractical for ground-based astronomical telescopes of several meters in diameter. Following a suggestion by Low & Rieke (1974), it has become customary to oscillate the secondary mirror of the telescope between two stable tilt positions. The resulting angular displacement of the field of view between two positions in the sky is on the order of a few minutes of arc, that is, several diameters of the instantaneous field of view. The object, a planet for example, is then placed in one position, while the sky reference is placed sufficiently far away to clear the object, but not excessively far away to limit mirror motion. In addition, compensation for atmospheric emission is most effective when the reference channel is close to the object. For best results object and reference position are placed at identical elevations. Subsequently, planet and sky reference are interchanged to detect and eliminate a possible asymmetry between the responsivities of both positions. The angle over which the secondary mirror must be tilted is relatively small; even with heavy secondaries chopping frequencies in the low tens of hertz have been achieved. Care must be taken to prevent the beams from intercepting the rim of the primary mirror; in other words, a sufficiently small entrance pupil must be used, which must not be the primary. This technique of chopping by oscillating the secondary mirror of a telescope is only applicable to small angular displacements between both beams. It is, therefore, restricted to objects of small angular diameter, such as the planets seen from the ground or from Earth orbit, but it is not applicable to objects that subtend a large solid angle, such as a planet seen from a spacecraft in its vicinity.

184   *Instruments to measure the radiation field*

If neither chopping in front of the primary nor oscillating the secondary mirror is possible, the chopper can be placed at the focal point of the telescope, for example, as shown in Fig. 5.2.11. Placement at the aperture stop or at other positions is also possible, but may result in larger dimensions of the chopper blades. All parts in the optical path ahead of the chopper are being modulated along with the object under consideration. As with d.c. radiometers it becomes necessary to occasionally observe deep space and a warm blackbody to discriminate between emissions from the object of interest and the telescope. Instead of a warm blackbody, thermostating the telescope may suffice, as discussed in Section 5.13. Placing the chopper at the telescope focal point minimizes the number of elements in front of the chopper. Placing the chopper at the aperture stop has the advantage that all parts of the field of view are chopped simultaneously, even if the chopper blades take a finite time to traverse the cross section of the beam. If possible, the chopper should be ahead of apertures and optical filters.

The first chopped radiometer on a stabilized platform, functioning in a line by line scanning pattern perpendicular to the direction of spacecraft motion, was the High Resolution Infrared Radiometer (HRIR) on Nimbus II (1964) shown in Fig. 5.4.10. The purpose of the investigation was to obtain cloud images on the dark hemisphere

Fig. 5.4.10 The High Resolution Infrared Radiometer of Nimbus II. The radiative cooler is on the right (Nimbus Project, NASA).

for meteorological purposes by recording the intensity in the 3.5 μm atmospheric window. The chopper at the telescope focal plane had a large number of blades and chopped against the interior instrument cavity. An external blackbody and deep space were used for calibration. For the first time, a space radiator served to cool the detector, a lead selenide cell, to 190 K (Annable, 1970). The detector was thermally connected to a radiating surface emitting to space, but shielded from the Sun, the Earth, and warm parts of the spacecraft. An image recorded by the HRIR is shown in Fig. 5.4.11. Quite a number of scanning radiometers have been derived from this design.

c. Image motion compensation

To achieve a desired signal-to-noise ratio, it is often necessary to dwell at a particular scene and integrate the signal for a certain time, the exposure time. If within this time the apparent motion of the area under observation is appreciable, it may become necessary to adjust the pointing so that smear is minimized. This process is called image motion compensation. It can be implemented by rotating the whole spacecraft or by tilting a mirror inside an instrument. Both techniques are discussed.

The first case deals with images of Miranda, a satellite of Uranus. Pictures were taken by Voyager 2 on January 24, 1986. Normally one considers the shutter action of a Vidicon camera an instantaneous event, much as the shutter action in a conventional photographic camera. This is not so, however, for several reasons. The problem of taking pictures of Miranda can be compared to the task of taking photographs of a speeding car with a telephoto lens under low light conditions and with a slow film. The light level at Uranus (19.2 Astronomical Units from the Sun) is only 1/368 of that available at the distance of Earth, while the Vidicon sensitivity corresponds to an ASA film rating of two. These factors require an exposure time of 15 s. The spacecraft moved with a speed of $\sim$17 km s^{-1} through the Uranian system and the focal length of the narrow angle camera is 1.5 m. Substantial smear would have occurred if the camera pointing had not precisely followed Miranda during the exposure. Normally, stepping motors point the Voyager scan platform, which houses the camera and other instruments. However, these motors are inadequate for this purpose because they move only at specific rates and, moreover, they cause small jitter in the pointing during the motion. In normal operation this is of minor concern since exposure takes place only after the platform has come to rest. Here, another approach had to be taken. The desired pointing of the cameras was accomplished by rotating the gyros in the attitude control system at a predetermined rate. The whole spacecraft counter-rotated at a much slower rate to conserve angular momentum. During this time communication with Earth was interrupted because the spacecraft antenna did not point towards our planet; the pictures were recorded on magnetic tape. The procedure was carried out successfully; the picture of Miranda (Fig. 5.4.12)

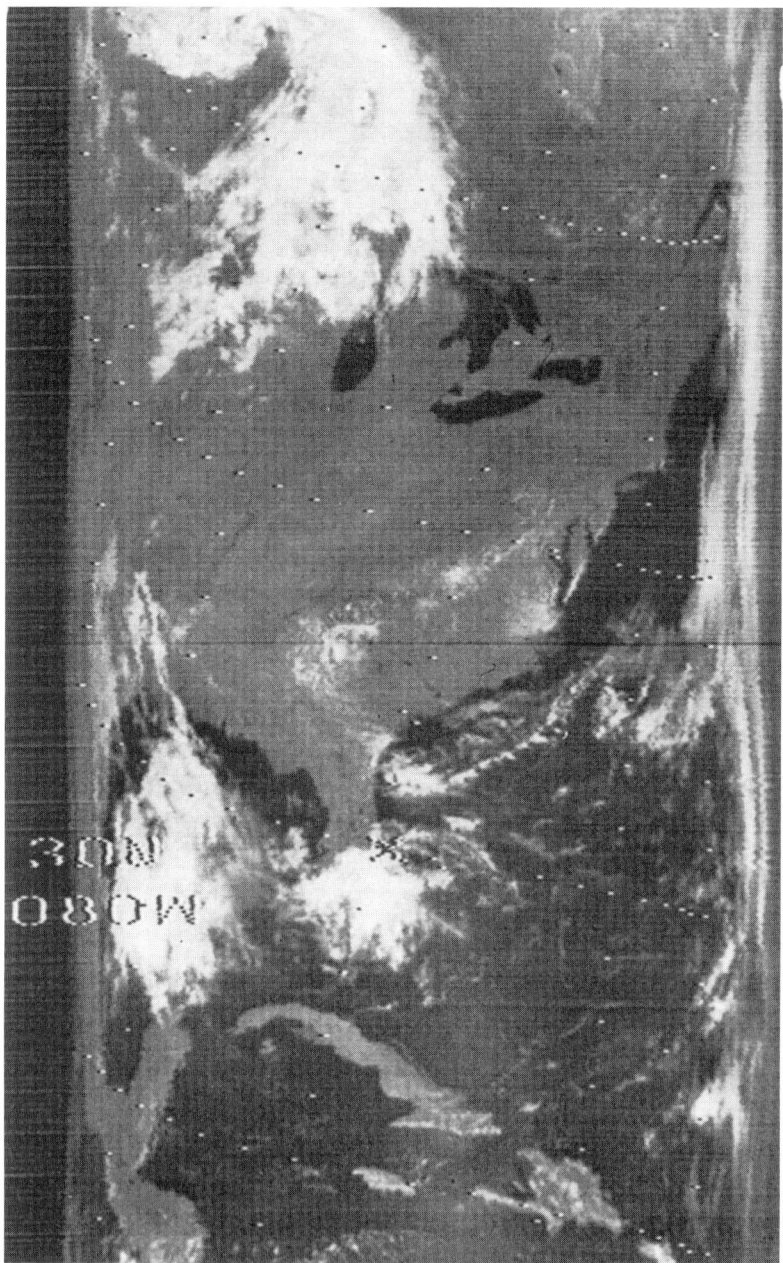

Fig. 5.4.11 Example of the continuous image strip generated by the High Resolution Infrared Radiometer on Nimbus 1. The east coast of North America is in the center. Strong distortion towards the horizons is apparent (Nimbus Project, NASA).

5.4 Chopping, scanning, and image motion compensation

Fig. 5.4.12 Miranda, innermost of Uranus' large satellites, is seen at close range in this Voyager 2 image, taken January 24, 1986. This clear-filter, narrow-angle image shows an area about 250 km across, at a resolution of about 800 m. Two distinct terrain types are visible: a rugged, higher-elevation terrain (right) and a lower, striated terrain. Numerous craters on the rugged, higher terrain indicate that it is older than the lower terrain. Several scarps, probably faults, cut the different terrains. The impact crater in the lower part of this image is about 25 km in diameter (Voyager Project, NASA).

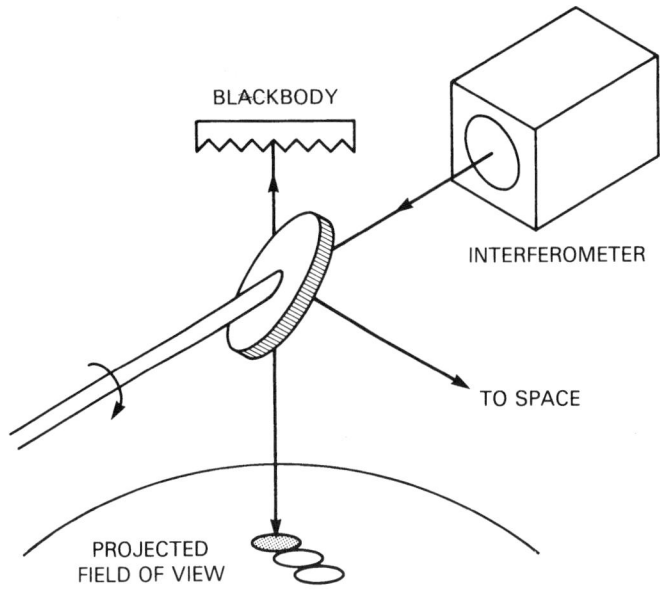

Fig. 5.4.13 A mirror mounted at 45° with respect to the rotating shaft permits radiation from a blackbody, deep space, and a planetary surface to be channeled in sequence towards the instrument. Slow rotation of the same mirror permits image motion compensation.

shows great detail, which would have been completely lost without image motion compensation (Smith *et al.*, 1986).

Rotation of the whole spacecraft is not always practical, but image motion compensation may be achieved by other means. An example is the pointing of the field of view of the infrared interferometers (IRIS) on Nimbus 3 and 4 and on Mariner 9. Dwell times of 11, 13, and 18 s, respectively, were used to record an interferogram. The Nimbus and Mariner 9 spacecraft circled their planets, Earth and Mars, in low-altitude, nearly circular and elliptical orbits, respectively. Image motion compensation was accomplished by slowly rotating a mirror, tilted at 45° to the instrument axis, in front of the entrance aperture (see Fig. 5.4.13). The field of view was pointed slightly ahead of the subspacecraft point at the beginning of an interferogram and slightly behind at the end of the frame. The same 45° mirror occasionally pointed the field towards deep space and towards a warm internal blackbody for calibration purposes.

5.5 Intrinsic material properties

a. Absorbing and reflecting filters

Transmission and reflection properties of certain materials form the basis of the oldest and simplest method of isolating spectral intervals. If one measures the transmission of polished plates of different thicknesses of an insulator, one may find spectral regions where the transmission is nearly independent of wavelength as well as of thickness. In this spectral range the material is transparent and attenuation is entirely due to reflection losses on both surfaces of the sample, called Fresnel losses. In other spectral regions the attenuation is larger and depends very much on the sample thickness. Vibrational transitions in the ion structure cause intrinsic absorption within the material.

Most glasses commonly used in the visible are also transparent in the near infrared up to about 2.6 μm. Quartz, in amorphous and in crystalline form, is transparent to at least 3.5 μm, and up to 4.5 μm in samples less than a few millimeters thick. Quartz becomes transparent again on the other side of a broad absorbing region, at wavelengths longer than 100 μm. Some crystals, spinel, rutile, and sapphire, are clear to about 6 μm and barium fluoride to 10 μm. Substances with longer cut-off wavelengths are silver chloride (AgCl, 20 μm), potassium bromide (KBr, 25 μm), and cesium iodide (CsI, 50 μm). Unfortunately, the halides are hygroscopic and relatively soft.

While a modest number of materials with a long wavelength cut-off exist, it is harder to find substances with a short wavelength cut-on. Some semiconductors are opaque in the visible, but become transparent at longer wavelengths. At short wavelengths the energy of the absorbed photons exceeds the energy gap width in the semiconductor and electrons are lifted from valence into conduction regions; strong intrinsic absorption at short wavelengths is the consequence. At long wavelengths the photon energy is insufficient to lift electrons and the material behaves as a transparent dielectric. Optically the most important semiconductors are silicon (Si), with a distinct turn-on at 1.1 μm, and germanium (Ge), which becomes transparent beyond 1.9 μm. The high refractive indices of Si and Ge (3.4 and 4.0, respectively) demand antireflection coatings to obtain reasonably good transmissions. Many optical and mechanical material properties, such as refractive indices, dispersion parameters, melting and softening temperatures, thermal conductivity and expansion coefficients, specific heat, and others can be found in *The Infrared Handbook* (Wolfe & Zissis, 1978). At cryogenic temperatures, some of the material properties may be different; for example, most substances are more transparent and have a lower heat capacity than at ambient conditions.

The importance of these materials goes beyond that of the design of broad-band transmission filters. The very same substances are also used in the construction of interference filters, prisms, Fabry–Perot etalons, beam dividers, dichroic mirrors, and sometimes as windows to seal parts of an instrument while permitting radiation to pass.

While glasses, natural and synthetic crystals, and semiconductors provide at least several choices for the design of optical components in the near and middle infrared, suitable bulk materials are more limited at longer wavelength. Quartz, diamond, and plastics, such as mylar and polyethylene, serve in the far infrared and millimeter range for the construction of lenses, transmission filters, and windows. It is also possible to construct filters by imbedding powders of magnesium oxide (MgO, cut-on ~50 μm), magnesium carbonate ($MgCO_3$ ~80 μm), and others into sheets of black polyethylene. Transmission data for a number of such filters can be found in the book by Bell (1972).

At long wavelengths reflection filters can be based on the residual ray effect (Hagen & Rubens, 1903). At certain frequencies the crystal lattice vibration is being excited and the crystal reflects well (Czerny, 1930; Strong, 1958).

Conductive grids may also serve as optical filters in the far infrared (Vogel & Genzel, 1964). A variety of techniques are used in their construction. Evaporated or electrochemically deposited metals on dielectric, transparent substrates (mylar, quartz) can be used, or simply a flat metal foil with patterns of small holes punched into it. Woven wire grids, such as commonly employed as sieves, can serve double duty as far infrared filters. With electroformed or vacuum deposited grids it is possible to construct two complementary versions, inductive and capacitive grid patterns (Smith *et al.*, 1972).

Wire mesh may also be used as a beam divider. At 45° and within a limited wavelength range of about two to four times the grid periodicity, reasonable efficiencies can be obtained. However, in view of the performance of linear wire polarizing grids in the Martin–Puplett interferometer (Subsection 5.8.c), wire mesh devices are less attractive. More promising is the use of metal grids in the construction of tuned, narrow-band filters. Several layers of mesh are stacked with precisely cut spacers to form interference filters for long wavelengths. A variety of narrow- and wide-band, as well as low-pass filters, have been constructed in this way (Manno & Ring, Chapter 5, 1972).

The detector characteristic may very well be included in the filter design. For example, an indium arsenide photovoltaic detector, operating at 195 K, has a very sharp cut-off at 3.6 μm. In combination with a thin germanium window, a well-defined 1.9–3.6 μm response function is obtained. However, with a limited number of substances available for the design of filters based on intrinsic absorption and reflection phenomena other methods must be found to construct filters where the transmission limits can be set by the scientific objectives and not so much by the absorption properties of available substances; such methods are based on the interference principle, to be discussed in Section 5.6, but first we deal with prism spectrometers, gas filters, and pressure modulation.

b. Prism spectrometers

Newton knew, as did others before him, that white light is separated into colors when it passes through a glass prism. Light rays of different colors are bent by

5.5 Intrinsic material properties

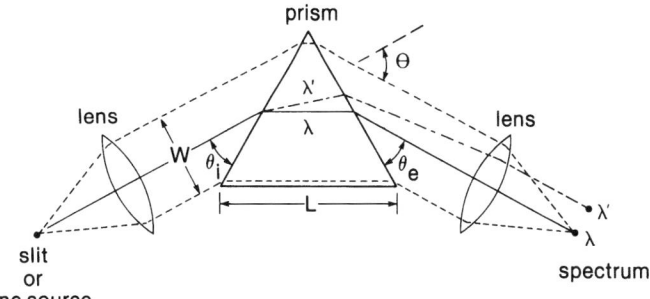

Fig. 5.5.1 Ray pattern in a prism spectrometer. The central and the outside rays are shown for wavelength λ. Only the central ray is shown for wavelength λ'.

different amounts; red is bent less than orange, with yellow, green, blue, and violet progressively bent more in succession. By using a second prism, Newton showed that individual colors could not be further divided, and by recombining the colors of the spectrum he produced white light again. Later, William Herschel (1800) used a prism to demonstrate the existence of infrared radiation by placing a thermometer at the position just below the red limit in the solar spectrum.

The prism separates radiation into a spectrum through refraction. In order to do so the index of refraction of the prism material must change with wavelength; adjacent wavelengths are diverted by different amounts in their paths through the prism. Figure 5.5.1 shows a schematic of a prism spectrometer. Radiation from a slit is collimated by a lens and passed through the prism. A second lens focuses the spectrum in the focal plane. When the incidence angle θ_i and emergence angle θ_e are equal the prism is adjusted for minimum deviation of the beam. In this position, the change in deviation angle Θ with wavelength λ is related to the change in index of refraction, n, by

$$\frac{d\Theta}{d\lambda} = \frac{L}{W}\frac{dn}{d\lambda}. \tag{5.5.1}$$

L is the base length of the prism and W the width of the beam. The resolving power, R, of the prism spectrometer can be found from Eq. (5.5.1),

$$R = \frac{\lambda}{\Delta\lambda} = \frac{\lambda}{\Delta\Theta}\frac{d\Theta}{d\lambda} = \frac{\lambda}{\Delta\Theta}\frac{L}{W}\frac{dn}{d\lambda}, \tag{5.5.2}$$

where $\Delta\lambda$ is the minimum resolvable wavelength interval. The diffraction limit given by the beam width W sets the minimum resolvable deviation angle

$\Delta\Theta = \lambda/W$. The resolving power is therefore

$$R = L\frac{dn}{d\lambda}. \qquad (5.5.3)$$

The derivative of refractive index with wavelength, $dn/d\lambda$, is different, and varies with wavelength, for different materials.

High resolving power requires large $dn/d\lambda$, and a prism of a specific material may be optimum over only a small portion of the spectrum. If a high resolving power is desired over a broad spectral region, several prisms may be needed. Information on refractive indices and on $dn/d\lambda$ for many materials of interest in the infrared are given in *The Infrared Handbook* (Wolfe & Zissis, 1978). If optimum resolving power is not required, the range between 1 and 50 μm can be adequately covered by prisms made of lithium fluoride (1–5 μm), sodium chloride (5–20 μm), and cesium iodide (20–50 μm).

To improve on the performance of a single prism element several such elements may be used in series. A cross section of the prism spectrometer of the McDonald Observatory, which uses two prism elements, is shown by Kuiper (1949). Alternatively, the same prism may be used twice by reflecting the radiation back at a slightly different angle. The exit slit is then located in the plane of the entrance slit, but slightly off-set. Until the 1950s most planetary infrared spectra were obtained with prism spectrometers.

c. Gas filter, selective chopper, and the pressure modulated radiometer

Besides solid substances infrared active gases may serve as filters. Many polyatomic gases have strong infrared absorption bands, each one consisting of numerous spectral lines. Absorption cells, charged with a suitable gas and hermetically sealed, are transparent, except at the narrow spectral intervals where the particular gas has absorption lines. Such filters are clearly not general purpose filters, but they are very useful for certain remote sensing tasks. The gas absorption cell differs in several aspects from a conventional narrow-band filter, say an interference filter. First, the gas cell is basically transparent over a reasonably wide spectral range, except at the line positions of the gas. In contrast, an interference filter is basically opaque over a wide spectral range, except in a narrow spectral interval where it is transparent. A second fundamental difference between the interference filter and the gas absorption cell concerns the details of the transmission function. In one case the transmission as a function of wavenumber is a smooth, normally bell-shaped curve, while in the other it mimics in every detail the very complex absorption characteristic of the gas in the cell. The degree of absorption depends, of course, on the chemical composition of the gas, on the strength of individual lines, on the path length, but also on the pressure and temperature within the cell.

In the study of atmospheric temperatures and composition one is often interested in the emission from a particular atmospheric constituent. For the analysis of the vertical temperature profile on Earth, Venus, and Mars, thermal emission from the CO_2 molecule can be used. If the same gas is contained in the absorption cell, the radiation of interest is being filtered out. How does one measure the radiation that has just been removed from the beam? This can be accomplished in several ways. For example, consider an absorption cell with two windows on opposing ends exposed to a beam of radiation. Wavenumbers outside the gas absorption band and in the transparent gaps between lines will penetrate the cell without a noticeable effect. Radiation within the width of strong lines will be absorbed and will cause a temperature rise in the gas. The corresponding pressure increase may be registered by a sensitive pressure transducer. The resulting infrared detector is sensitive only to radiation specifically tuned to the gas in the cell. Such detectors have been produced (the Patterson–Moos cell), but have, as far as we know, never been applied to planetary work.

Another way to measure the radiation absorbed by the gas in the cell is to construct two identical cells, one with and one without a gas inside, expose both cells to radiation from the object of interest, and measure the difference between the transmitted signals. This difference is just the radiation removed by absorption in the gas contained in the first cell. Indeed, the Selective Chopper Radiometer flown on Nimbus 4 is based precisely on this principle (Houghton & Smith, 1970). However, balancing radiation from both cells is difficult.

A third way is to use only one cell, but to remove the gas for the comparison measurement on the empty cell. One must not empty the cell completely. It is sufficient just to lower the gas pressure inside. The pressure modulation can be relatively rapid, ~40 Hz, similar to the chopping process in a conventional radiometer. Pressure modulation has another very important effect. The differential absorption between pressure extrema is simply the difference in the transmission functions of the gas at these pressures. If the minimum pressure is not too low, the line centers are still opaque and the modulation affects primarily the line wings. By selecting the path length, the mean pressure, and the pressure amplitude one may control, within limits, the position and width of the atmospheric weighting function. This function indicates from which atmospheric layer the modulated intensity originates. By selecting a spectral region with lines of nearly equal intensity, the width of the atmospheric layer probed by this technique can be reduced close to the theoretical minimum, applicable to monochromatic radiation. In spite of being sensitive to only a part of a spectral line, the method gains by summing the intensities from the shoulders of many lines, thereby increasing the signal with the number of usable lines in comparison with that from a measurement of only one line.

The method is not without complications. Infrared bands sometimes contain nearly overlapping lines, lines have different strength, thermal emission from the gas in the cell must be considered, and even the gas temperature changes due to adiabatic heating and cooling as a consequence of the pressure modulation. These effects can be taken into account either analytically or by empirical corrections. Good results have been obtained by the pressure modulation method. The Stratospheric and Mesospheric Sounder (SAMS) on Nimbus 7, the most advanced instrument of this type flown in space, contains seven individual pressure modulators and senses, besides atmospheric temperatures, a number of constituents: CO, NO, H_2O, N_2O, and CH_4 (Drummond *et al.*, 1980). Pressure modulated radiometers have also been combined with conventional grating radiometers to isolate emission from a wide range of atmospheric levels. The infrared instrument on the Pioneer Venus Orbiter (VORTEX) was such a combination (Taylor *et al.*, 1979a, b, 1980). The temperature field of the northern hemisphere of Venus was probed by VORTEX between about 55 and 150 km altitude. A recent summary of the pressure modulation technique, including results from measurements in the atmospheres of Earth and Venus, is given by Taylor (1983) and by Houghton *et al.* (1984).

5.6 Interference phenomena in thin films

A great variety of optical filters is needed in the design of remote sensing instruments. Only in a small number of cases can the demand be satisfied with filters based on intrinsic absorption properties of available substances. These cases concern mostly wide-band filters where neither the exact position nor the sharpness of the passband limits are of prime importance. More demanding requirements, such as for narrow-band filters, are best met with optical elements using interference phenomena in thin films and with diffraction gratings. Both subjects are treated in this and the following section. Interference phenomena leading to Fourier spectroscopy are discussed in Section 5.8.

In Subsection 5.6.a we review the theory of multilayer thin films. In Subsection 5.6.b this theory is applied to the design of antireflection coatings for infrared windows, lenses, and other components. The same theory is used again in Section 5.6.c to find suitable beam dividers of the free-standing, self-supporting type as well as of the type requiring a transparent substrate. Subsection 5.6.d deals with interference filters and Fabry–Perot interferometers.

a. Outline of thin film theory

The reflection and transmission properties of multiple layers of materials with different refractive indices can be treated either as a ray tracing or as a boundary value problem (e.g., Wolter, 1956; Born & Wolf, 1975). The ray tracing method leads to summations where it is sometimes difficult to follow the phase relations, especially if several layers are to be treated. We follow closely the boundary value method reviewed by Wolter (1956). In effect this method is a generalization of the one-interface boundary problem that led to the formulation of the Fresnel equations in Section 1.6.

Assume a stack of layers, as shown in Fig. 5.6.1. In each layer there exists a downward and an upward propagating field, indicated by the arrows A and R, respectively. The amplitude, phase, polarization, and direction of the arriving wave in the top layer, m, is assumed to be known as well as the material properties, μ_{rel} and $n_r + in_i$, of the substances forming the stack of plane parallel layers. We are interested in the amplitudes and phases of the reflected, R_m, and transmitted wave, A_0. In the lowest medium, 0, only a transmitted wave is postulated. All waves are assumed to be plane with the Poynting vectors in the x–y plane, that is, $S_z = 0$. To apply the boundary conditions to each interface the downward and the upward waves need to be calculated for each layer ($j = 1, 2, \ldots, m$). For layer j they are, respectively,

$$A_j = B_j e^{-ik_j(x \sin \phi_j - y \cos \phi_j)}, \tag{5.6.1}$$

$$R_j = C_j e^{-ik_j(x \sin \phi_j + y \cos \phi_j)}. \tag{5.6.2}$$

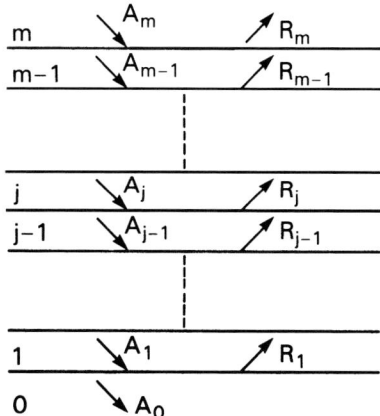

Fig. 5.6.1 Arrangement of many layers of transparent material. The incident and reflected radiation are A_m and R_m, respectively; the transmitted radiation in layer zero is A_0.

In general, each vector equation stands for six equations, one for each component of **E** and **H** in the three coordinates, but the number can be reduced by proper choice of the coordinate system. In setting all ϕ_j equal in the above equations we tacitly imply the validity of the reflection law. As in the discussion of the Fresnel equations we solve the problem for two orthogonal, linearly polarized waves. Other states of polarization may be represented by superposition of these solutions. For the transverse **E** wave (TE) the electric vector is perpendicular to the plane of incidence, and only the z-component of **E** exists, $E_z = E_\perp$. The same is true for the transverse **H** wave (TM), where only the $H_z = H_\perp$ component is present.

$$\text{For TE}: E_x = E_y = 0; \quad \text{for TM}: H_x = H_y = 0. \tag{5.6.3}$$

This is simply a consequence of the prudent choice of coordinates. In addition to the z-components, we need the other x-components of **H** and **E**, respectively, in order to apply the boundary conditions. They can be found from the z-components with the help of Maxwell's equations and the definition of the curl operator. For the TE and TM waves the tangential components in the x-direction, using the generalized form, are

$$(A_{jx})_{\text{TM}} = (g_j)_{\text{TE}}(A_{jz})_{\text{TE}}; \quad (R_{jx})_{\text{TM}} = -(g_j)_{\text{TE}}(R_{jz})_{\text{TE}}, \tag{5.6.4}$$

where

$$(g_j)_{\text{TE}} = \frac{n_j \cos \phi_j}{\mu_j}, \tag{5.6.5}$$

and

$$(A_{jx})_{\text{TE}} = -(g_j)_{\text{TM}}(A_{jz})_{\text{TM}}; \quad (R_{jx})_{\text{TE}} = (g_j)_{\text{TM}}(R_{jz})_{\text{TM}}, \tag{5.6.6}$$

but in this case,

$$(g_j)_{\text{TM}} = \frac{\mu_j \cos \phi_j}{n_j}. \tag{5.6.7}$$

In this section, μ_j is the relative permeability. At the boundary of layer j and $j-1$, the tangential components of **E** and **H** must be continuous,

$$A_{jz}(x, y_j, z) + R_{jz}(x, y_j, z) = A_{j-1,z}(x, y_{j-1}, z) + R_{j-1,z}(x, y_{j-1}, z), \tag{5.6.8}$$

5.6 Interference phenomena in thin films

and

$$A_{jx}(x, y_j, z) + R_{jx}(x, y_j, z) = A_{j-1,x}(x, y_{j-1}, z) + R_{j-1,x}(x, y_{j-1}, z). \quad (5.6.9)$$

If one introduces, for convenience's sake,

$$\rho_j = ik_j \cos\phi_j, \quad (5.6.10)$$

the boundary conditions may be expressed by

$$B_{jz}e^{\rho_j y_j} + C_{jz}e^{-\rho_j y_j} = B_{j-1,z}e^{\rho_{j-1} y_j} + C_{j-1,z}e^{-\rho_{j-1} y_j}, \quad (5.6.11)$$

and

$$g_j B_{jz}e^{\rho_j y_j} - g_j C_{jz}e^{-\rho_j y_j} = g_{j-1} B_{j-1,z}e^{\rho_{j-1} y_j} - g_{j-1} C_{j-1,z}e^{-\rho_{j-1} y_j}. \quad (5.6.12)$$

These equations are valid for TE as well as for TM waves, provided we substitute the appropriate values for g into the formulas. For TE waves one has to use g given by Eq. (5.6.5) and for TM waves by Eq. (5.6.7). By applying these equations to each boundary, $2m$ equations for $2m + 2$ quantities (also for layer 0) are obtained. However, the arriving wave in layer m is assumed to be known and the reflected wave in layer 0 is assumed to be nonexistent; therefore, all B and C can be determined, most conveniently by matrix inversion.

We test Eqs. (5.6.11) and (5.6.12) for one interface; they must reduce to the Fresnel equations. For $m = 1$; $j = 1$ and $j - 1 = 0$, and since $C_{0,z} = 0$ the equations can be solved for the normalized reflected and transmitted waves

$$R = \frac{C_1}{B_1}e^{-2\rho_1} = \frac{g_1 - g_0}{g_1 + g_0} \quad (5.6.13)$$

$$T = \frac{B_0}{B_1}e^{\rho_0 - \rho_1} = \frac{2g_1}{g_1 + g_0}. \quad (5.6.14)$$

The exponentials provide the phase information. Substituting the g values from Eqs. (5.6.5) and (5.6.7) yields $R_\perp$ or $R_\parallel$ and $T_\perp$ or $T_\parallel$ respectively, in agreement with the Fresnel equations derived previously [Eqs. (1.6.11) to (1.6.14)].

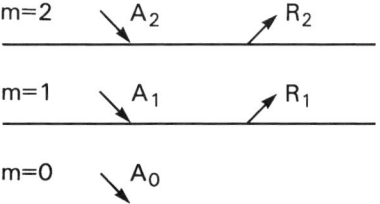

Fig. 5.6.2 Thin layer, $m = 1$, on a substrate, $m = 0$, exposed to a medium, $m = 2$. A_2 is the incident, R_2 the reflected, and A_0 the transmitted ray.

The Eqs. (5.6.13) and (5.6.14), in combination with the definition of the g values, are just another form of the Fresnel equations.

b. Antireflection coatings

To study the effect of one thin layer between two more extended media (Fig. 5.6.2), we use the general multilayer solution, Eqs. (5.6.11) and (5.6.12), for $m = 2$ (Wolter, 1956). The reflected amplitude R is

$$R = \frac{(g_2 - g_1)(g_1 + g_0)e^{\rho_1 d_1} + (g_2 + g_1)(g_1 - g_0)e^{-\rho_1 d_1}}{(g_2 + g_1)(g_1 + g_0)e^{\rho_1 d_1} + (g_2 - g_1)(g_1 - g_0)e^{-\rho_1 d_1}}. \quad (5.6.15)$$

In this equation d_1 has been substituted for $y_2 - y_1$, the thickness of layer 1. Medium 2 may be air or vacuum, and medium 0 may be glass, germanium, or another substance. The optical constants and dimensions of medium 1 shall be chosen so that the reflected amplitude R of the whole assembly becomes a minimum, which implies a transmission maximum if only nonabsorbing substances are admitted. To obtain zero reflectivity the numerator of Eq. (5.6.15) must be zero (and the denominator $\neq 0$), which requires

$$\frac{(g_2 - g_1)(g_1 + g_0)}{(g_2 + g_1)(g_1 - g_0)} = -e^{-2\rho_1 d_1}. \quad (5.6.16)$$

For nonabsorbing substances all refractive indices and, therefore, all g values are real quantities ($n_r \gg n_i$). The right side must, therefore, be real too; the imaginary component must vanish, which implies [see Eq. (5.6.10)],

$$2k_1 d_1 \cos \phi_1 = m\pi; \quad (5.6.17)$$

m is an integer, however, only odd values yield usable solutions. Since $k_1 = 2\pi/\lambda_1$, one obtains

$$d_1 = \frac{\lambda_1 m}{4 \cos \phi_1}. \tag{5.6.18}$$

The layer thickness must be an odd multiple of the wavelength in layer 1 divided by $4 \cos \phi_1$. For near normal incidence and $m = 1$ the layer should be a quarter wave thick. With condition (5.6.17) satisfied, Eq. (5.6.15) reduces to

$$R = \frac{g_2 g_0 - g_1^2}{g_2 g_0 + g_1^2}, \tag{5.6.19}$$

which is zero for

$$g_1 = (g_2 g_0)^{\frac{1}{2}}. \tag{5.6.20}$$

For nonmagnetic substances ($\mu_{\text{rel}} = 1$) the refractive index of the antireflection coating must be the geometric mean between the indices of the bordering substances.

To illustrate the effect of an antireflection coating, consider a germanium window ($n \sim 4$) in vacuum ($n = 1$) for near normal incidence. Without an antireflection treatment the reflected intensity, $r = |R|^2$, at the front surface would be 0.36. The refractive index of the antireflection layer should, according to Eq. (5.6.20), be 2. However, it is difficult to find a transparent material with precisely such an index that can be applied in a thin, uniform layer. Zinc sulfide (ZnS) with an index of 2.3 comes close, and the reflected intensity at the first layer reduces from 0.360 to 0.019, a substantial improvement. For a coating tuned to 10 μm, the optimum layer thickness, d, would be $10/(4 \times 2.3) = 1.087$ μm. Unfortunately, evaporated thin layers differ somewhat in their refractive index from that of the bulk material, and experimentation is often required (e.g., Heavens, 1955). Reflection nearly disappears only for one wavenumber, and for odd multiples thereof. The Ge window, coated for maximum transmission at 10 μm (1000 cm^{-1}) may be used over the whole 8–12 μm atmospheric window, but at 5 μm this particular antireflection coating is noneffective. The residual reflectivity of a germanium window, antireflection coated with ZnS ($n = 2.3$) or amorphous selenium ($n = 2.45$), is shown in Fig. 5.6.3. If antireflection performance must be achieved over a wider spectral range, it is necessary to use more than one intermediate film. Practical cases for visible wavelengths are discussed by Heavens (1955). It is even possible, by

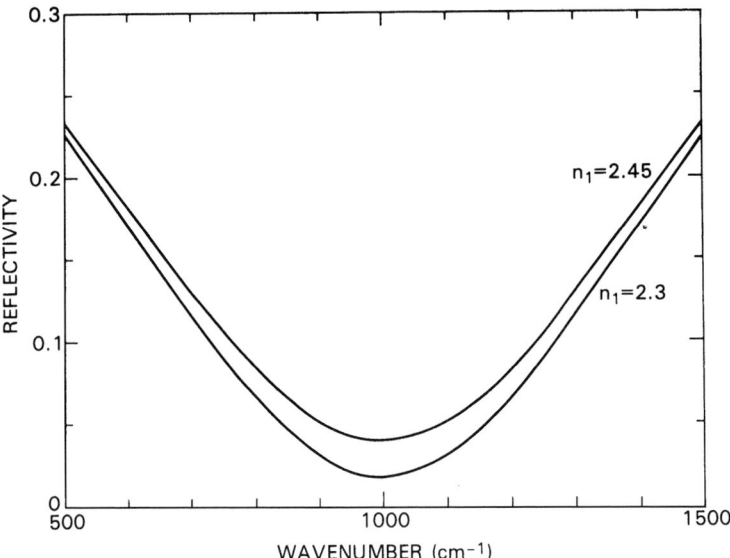

Fig. 5.6.3 Reflectivity of a germanium substrate is shown as a function of wavenumber. A single layer antireflection coating is illustrated, tuned to 1000 cm^{-1}. Two coating materials are shown, zinc sulfide ($n = 2.30$) and selenium ($n = 2.45$).

controlling the rate of deposition from two sources in the same evaporation chamber, to change the refractive index continuously between two values. Good antireflection properties may then be obtained over a wide wavelength range.

For large deviations from normal incidence, as they exist in fast lenses, the conditions of zero reflectivity cannot be met simultaneously for both TE and TM waves and a compromise must be chosen to minimize the total reflected intensity.

c. Beam dividers

Beam dividers, or beamsplitters, which divide reflected and transmitted waves nearly equally, are essential components in Michelson interferometers, but are also used in other instruments. Since they are based on thin film technology we discuss such dividers in this section. The solution for a two-interface (three-layer) stack, Eq. (5.6.15), is squared to derive the reflected intensity. After a straightforward, although tedious, calculation, one finds

$$r = \frac{\left(g_2^2 + g_1^2\right)\left(g_1^2 + g_0^2\right) - 4g_2 g_1^2 g_0 + \left(g_2^2 - g_1^2\right)\left(g_1^2 - g_0^2\right)\cos\left(4\pi \nu n_1 d_1 \cos\phi_1\right)}{\left(g_2^2 + g_1^2\right)\left(g_1^2 + g_0^2\right) + 4g_2 g_1^2 g_0 + \left(g_2^2 - g_1^2\right)\left(g_1^2 - g_0^2\right)\cos\left(4\pi \nu n_1 d_1 \cos\phi_1\right)}.$$

(5.6.21)

5.6 Interference phenomena in thin films

The solutions for the TE and TM waves are obtained by substituting the values for g as defined by Eqs. (5.6.5) and (5.6.7), respectively. With the assumption that the relative permeability of all layers is unity, the g values for both polarizations become

$$g_i(\text{TE}) = n_i \cos \phi_i; \quad g_i(\text{TM}) = \frac{1}{n_i} \cos \phi_i, \quad (5.6.22)$$

where i is the layer number.

With these preparations the one-layer beamsplitter can be treated ($n_0 = n_2 = 1$; $n_1 = n_1$). In the far infrared a self-supporting layer, such as a sheet of mylar stretched over a plane circular surface, is widely used. The intensity reflected from a 10 μm thick sheet of refractive index 1.85 (mylar) is shown in Fig. 5.6.4 for angles of incidence of 30° and 45°. The same figure gives the product $4rt = 4r(1-r)$ for all conditions. For an ideal beamsplitter this product would be unity. For the TE

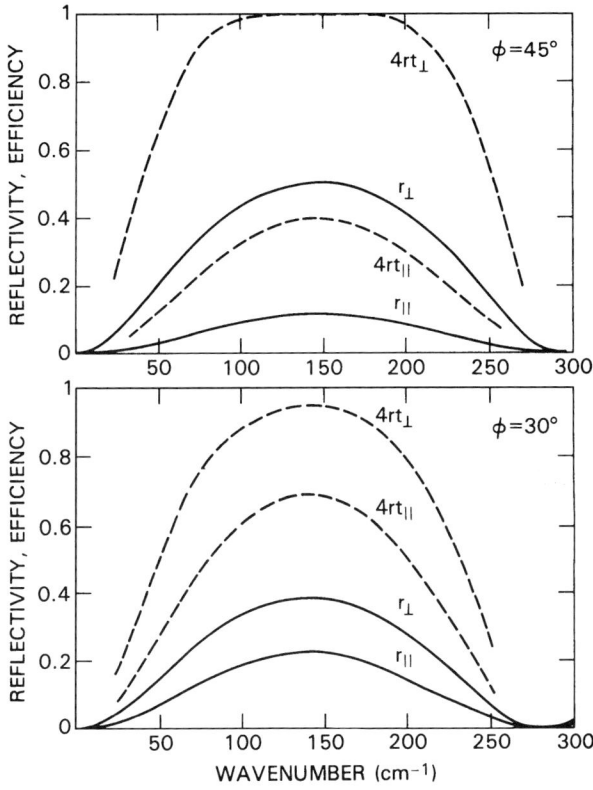

Fig. 5.6.4 Reflectivity (solid curve) and beamsplitter efficiency (dashed curve) of a 10 μm thick sheet of Mylar operated at 45° and 30°. Both planes of polarization are shown for both cases.

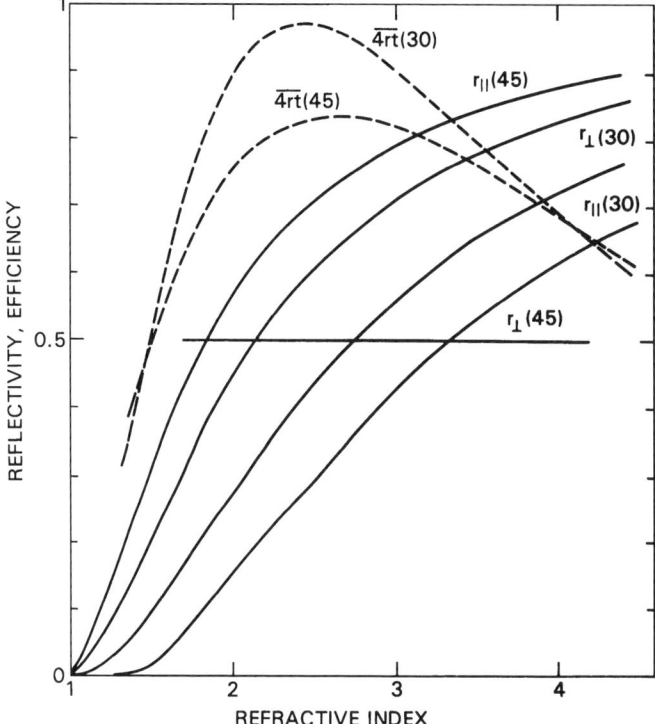

Fig. 5.6.5 Reflectivity (solid curve) and beamsplitter efficiency (dashed curve) of a single layer, self-supporting beamsplitter as a function of refractive index. Again, both polarizations are shown for 30° and 45° angles of incidence.

wave the mylar sheet at 45° is nearly a perfect beamsplitter at 150 cm^{-1}; however, the TM wave is poorly divided. At 30° both reflectivity curves are closer together and the maximum $4rt$ value, averaged over both polarizations, is better for 30° ($\sim$0.812) than for 45° (0.698). Despite the near perfect split of the TE wave with the 45° divider, it is preferable to operate mylar at 30°.

One may ask: what would be the optimum refractive index of a monolayer to be used as a beamsplitter? To answer this question we solve Eq. (5.6.21) for different values of n_1 for the maximum reflectivity, $\cos(4\pi \nu d_1 n_1 \cos\phi_1) = -1$. Again, both TE and TM values according to Eq. (5.6.22) have been used (Fig. 5.6.5). The optimum value of n_1 for 30° and the TE wave is just above 2 and for the TM wave about 2.8. Overall $n_1 \sim 2.4$ would be the best compromise for optimum conditions at the peak. However, it is desirable to use a larger n_1 if a broader wavenumber range is to be covered. For example, with $n_1 = 3$ one obtains double maxima in the $4rt$ curves, as shown in Fig. 5.6.6. A transparent material of refractive index 3 would be an attractive beam divider over a substantial wavenumber range; unfortunately

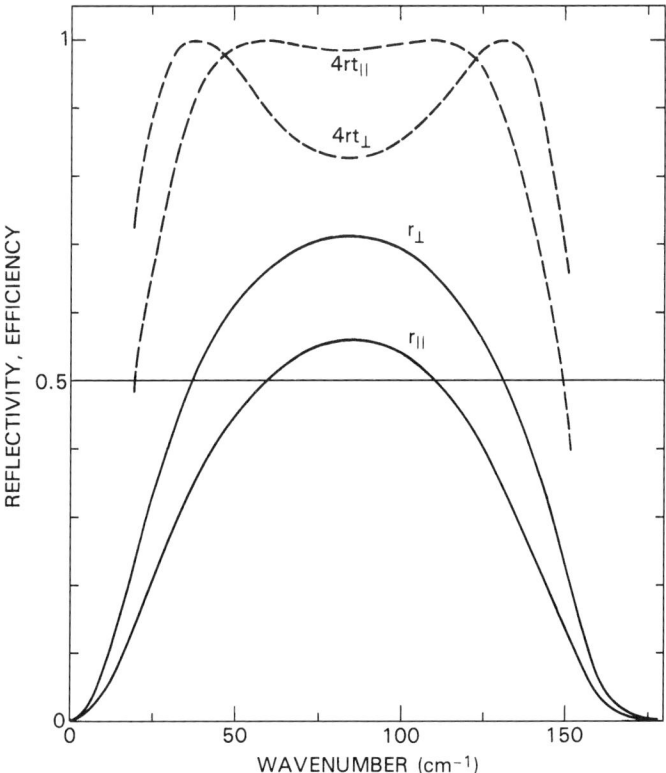

Fig. 5.6.6 Reflectivity (solid curve) and beamsplitter efficiency (dashed curve) for a self-supporting monolayer of refractive index 3 at 45°.

no suitable material of such an index is known that can also be manufactured in a suitable shape. Section 5.6.d shows that stacks of quarter wave layers of high and low refractive index materials may be used to generate equivalents to a layer of a certain index; however, this is hardly possible for self-supporting beamsplitters.

In the middle and near infrared it is common to construct beamsplitters by vacuum deposition of a thin film on transparent substrates. To analyze such a case we consider a 0.5 μm thick germanium ($n = 4$) layer on a potassium bromide ($n = 1.5$) substrate. Again, Eqs. (5.6.21) and (5.6.22) are solved for a beamsplitter at 45°; the result is shown in Fig. 5.6.7. The substrate cut-off below 400 cm^{-1} is omitted. Such a beamsplitter may be used between 400 and 2100 cm^{-1}; it shows an excellent efficiency (high $4rt$) for the TM wave between 700 and 1800 cm^{-1}, but considerably lower values for the TE wave over the same range. Overall, the average efficiency for unpolarized radiation is about 0.83 near 1300 cm^{-1}, but as high as 0.92 near 600 and 1900 cm^{-1}. Again, better performance may be obtained with additional layers.

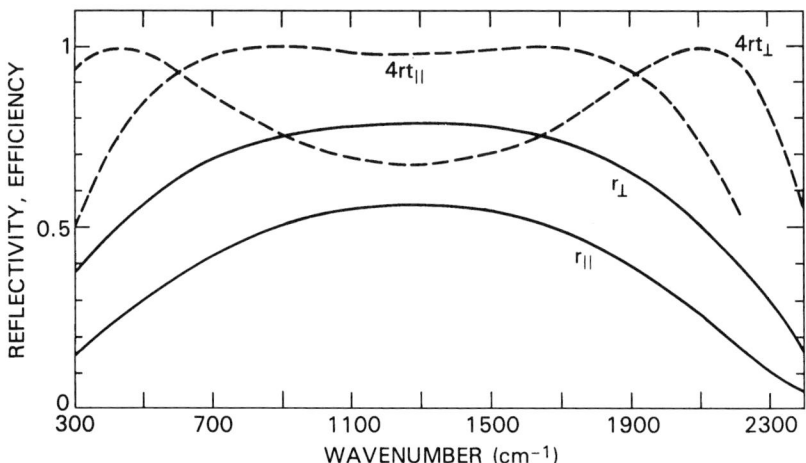

Fig. 5.6.7 Reflectivity (solid curve) and beamsplitter efficiency (dashed curve) of a germanium layer on a potassium bromide substrate.

The beamsplitter of the Nimbus IRIS, for example, used four main layers with refractive indices 4, 3.15, 2.47, 1.94 and the substrate, 1.53 (Hanel et al., 1970).

d. Interference filters and Fabry–Perot interferometers

The thin film theory developed in Subsection 5.6.a is directly applicable to the design of interference filters and Fabry–Perot interferometers. These devices are similar in concept and, therefore, are treated together. The normalized reflected amplitude for a three layer film of refractive indices n_2, n_1, and n_0 [see Eq. (5.6.21)] is applied first to the simplest case, a monolayer of index n_1 and thickness d_1 bordered by vacuum, $n_2 = n_0 = 1$. The calculated transmission, $t = 1 - r$, for normal angle of incidence, is shown in Fig. 5.6.8. Several refractive indices are illustrated, irrespective of the construction feasibility of such filters. In each case the layer thickness is chosen to produce a transmission peak at 1000 cm^{-1}. This peak, called the first-order peak, corresponds to a thickness of $d_1 = (2n_1 \nu_{\max})^{-1}$. The adjacent peak towards higher wavenumbers, called the second-order peak, occurs at 2000 cm^{-1} in this example, and so forth. The half width, $\Delta \nu$, of each peak is the same for all orders, but the resolving power, $\nu/\Delta\nu$, increases with order number. The germanium film shown in Fig. 5.6.8 has a thickness of 1.25 μm. Transmission peaks of thicker layers are correspondingly closer together in wavenumber space; for example, a 5 μm germanium layer has peaks at every 250 cm^{-1}. Based on this principle, interference filters can be constructed. In reality, it is not possible to manufacture a self-supporting layer of germanium ($n \sim 4.0$) or silicon ($n \sim 3.4$) with a thickness of a few micrometers and a diameter of a few centimeters; it is possible, however,

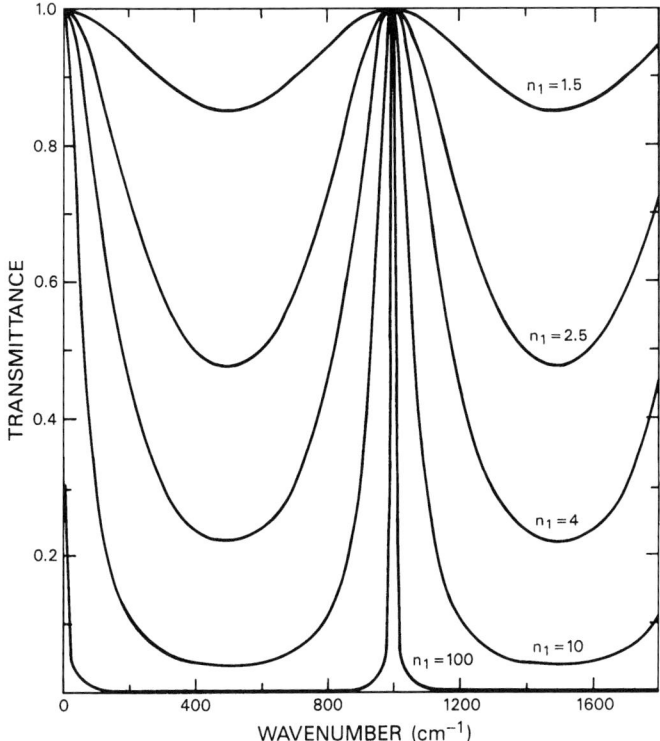

Fig. 5.6.8 Transmittance of a monolayer of material of different refractive indices. The layer thickness has been chosen to produce the first-order peak at 1000 cm^{-1}.

to produce silicon disks of 115 μm in thickness and to operate at an order number above 200. A tunable filter of this type has been flown on the Cryogenic Limb Array Etalon Spectrometer (CLAES) of the Upper Atmosphere Research Satellite (UARS), see also Reber (1990).

Now we return to a first-order germanium filter of 1.25 μm in thickness and the feasibility of realizing such a transmission function. Two approaches may be taken. In one case a 1.25 μm layer of germanium is evaporated onto a transparent, sufficiently flat substrate of low refractive index. For a potassium bromide substrate ($n \sim 1.5$) the peak transmission still occurs at 1000 cm^{-1}, but the peak is reduced from 1.00 for a monolayer to 0.96. At the same time the transmission minima at 500, 1500 cm^{-1}, etc., increase from 0.221 to 0.313, as shown in Fig. 5.6.9. The rejection ratio, that is the ratio of minimum to maximum transmission, increases from 0.221 to 0.326. Although the rejection ratio of the deposit is less desirable than that of the monolayer, the overall shape of the transmission function is not altered substantially by using a substrate of low refractive index. To improve the filter characteristic, that is, to reduce the rejection ratio and to narrow the half width

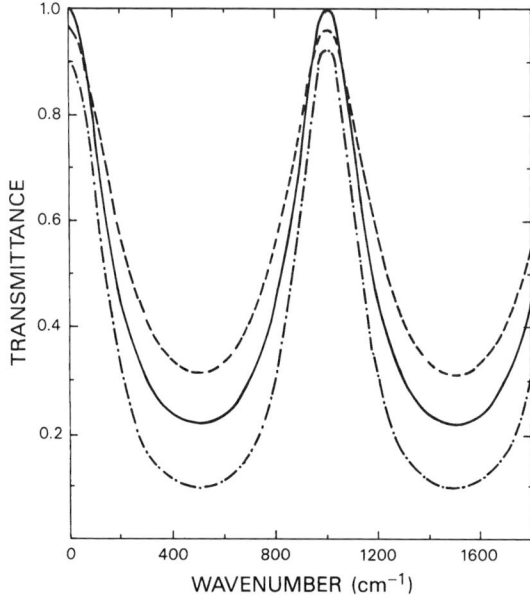

Fig. 5.6.9 Transmission of germanium films tuned to a maximum transmission at 1000 cm^{-1}. The solid curve is for a self-supporting monolayer, the dashed curve for a film on a potassium bromide substrate, and the dash-dot curve for two films, one on each side of the substrate.

of the transmission peak, one could apply an identical germanium layer on the other side of the substrate and obtain the product of the transmission functions; this product is also shown in Fig. 5.6.9. The rejection ratio of such a combination is 0.106. The substrate should be slightly wedged to prevent the filters on both substrate surfaces from being coupled, resulting in a ringing of the transmission function with a period of $(2nd)^{-1}$ wavenumbers; n and d are substrate parameters. The ringing, called channel spectrum, corresponds to interference between the parallel substrate boundaries.

Another approach to obtain a transmission function similar to that of a monolayer is to use a gap of $\lambda/2$ between two flat and parallel germanium plates; this requires a combination $n_2 = 4, n_1 = 1, n_0 = 4$, for example. At normal incidence this arrangement has a transmission function identical to the case $n_2 = 1, n_1 = 4, n_0 = 1$, provided the outside surfaces of the plates are antireflection coated. The thicknesses of the plates are immaterial since only their inside surfaces and the gap between them participate in the interference process. The location of the transmission peak depends only on the separation of the plates. Such a filter is easily constructed because the plates may be chosen thick enough to permit polishing to the required flatness of at least 1/20 of a wavelength. The spacer, which controls the gap width, is critical because it determines the location of the peak wavenumber. The spacer has to be made of a material with nearly zero thermal expansion coefficient or the peak wavenumber shifts with temperature. On the other hand the filter may easily be tuned to different wavenumbers by changing the gap width. Piezo electric crystals may be used as separators to remotely tune such a filter. In effect, this filter is a Fabry–Perot interferometer of low finesse (Fabry & Perot, 1899). Finesse is the ratio of the separation of adjacent fringes and the transmission half width; the germanium monolayer in Fig. 5.6.8 has a finesse of ~3.5.

For many applications a much larger finesse, that is, a much narrower filter function as well as a smaller rejection ratio, is needed than that obtained with two germanium plates. Clearly, a higher reflectivity than that provided with $n = 4$ and 1 would be desirable. A material with a refractive index of 100 or so would be attractive (Fig. 5.6.8). Unfortunately, such dielectric substances, which must also be transparent in the infrared, are unknown, but other possibilities exist. A very thin metal layer on a transparent substrate may serve the purpose. Such a layer, only a few hundredths of a micrometer thick, is semitransparent and can easily be produced by vacuum deposition. Traditionally silver coatings on two flat glass plates have been used to form Fabry–Perot interferometers for visible radiation. The same technique with infrared transmitting plates can be applied at longer wavelengths. Although the metal layers are very thin, they are noticeably absorbing due to the imaginary part of the refractive index. Thicker metal layers yield higher reflectivities and, therefore, narrower half widths, but also lower peak transmissions than thinner layers. Fabry–Perot interferometers with metal coatings and interference filters based on the same principle can be made with peak transmissions of not much more than 0.35. However, extremely narrow filter functions of a few percent in width and with excellent rejection ratios, as low as 10^{-4}, are possible with this technique.

An alternative method to obtain a high effective refractive index with non-metallic layers uses stacks of quarter wave layers of alternating high (H) and low (L) refractive indices. On each interface of such a HLHL etc. stack the reflected component is only modest, but the reflected waves from all interfaces are in phase, because the optical path length in each layer is just what is required to cause a wave reflected at one interface to constructively interfere with a wave reflected at an adjacent interface. The total reflection of the multilayer stack is then quite high, corresponding to that of a single layer of a very high refractive index material. Interference filters with stacks of as much as 15 layers on each side of a $\lambda(2n)^{-1}$ layer have been constructed, yielding peak transmissions as high as 0.8 and normalized half power widths, $\Delta \nu / \nu$, of only one or two percent (Smith *et al.*, 1972). The maximum number of layers in a stack is limited by the homogeneity and uniformity with which the layers can be deposited. Common materials for the low index layer are cryolite ($n \sim 1.35$), which is a sodium–aluminum fluoride compound, thorium fluoride ($n \sim 1.45$), or sodium chloride ($n \sim 1.54$). For the high index layer silicon ($n \sim 3.4$), germanium ($n \sim 4$), or lead telluride ($n \sim 5.1$), can be used. The larger the difference in the refractive indices between H and L layers the fewer layers are needed to achieve a certain finesse.

So far we have considered only cases of normal incidence. Inspection of Eq. (5.6.21) for the three-layer model indicates a wavenumber shift of the transmission peak due to the ϕ dependence. The transmission maximum (reflection minimum) occurs when

$$\cos(4\pi \nu_{\max} n_1 d_1 \cos\phi_1) = 1. \qquad (5.6.23)$$

Consequently, for the first-order peak,

$$\nu_{\max} = \frac{1}{2 n_1 d_1 \cos\phi_1}. \qquad (5.6.24)$$

The angle ϕ_1 is measured in material 1. For interference filters with a high index material as the $\lambda/2$ layer in the center, this angle is substantially smaller than the external angle. Filters with a high index material in the center have a much smaller wavenumber shift with external incidence angle than filters with a low index material or a Fabry–Perot interferometer with air or vacuum in the gap.

The angular dependence of the transmission peak has positive and negative aspects. A Fabry–Perot interferometer with a fixed spacer can be fine tuned within a substantial spectral range just by tilting the etalon. The disadvantage of the strong angular dependence of the Fabry–Perot interferometer is the limitation in the usable solid angle. Clearly, a Fabry–Perot interferometer with a narrow transmission peak of 5 cm^{-1} at 1000 cm^{-1} cannot be operated with a large solid angle. Even the 10° off-axis peak occurs at 1015 cm^{-1}, six times the half width of the filter. Fabry–Perot interferometers are best operated in a well-collimated beam. They have the same solid angle limit as the Michelson interferometer discussed in Subsection 5.8.a, Eq. (5.8.11).

All Fabry–Perot interferometers and interference filters of the type discussed so far have multiple peak transmissions. In most cases only one order is desired and other orders must be suppressed. A second interference filter tuned to the desired peak, but of different order, is then helpful, especially if the order number of the filters are not only different, but have no common factor. While the desired peaks coincide, other peaks do not occur at identical wavenumbers. Eventually, blocking filters with intrinsic absorption must be used. Often the substrate material, which supports the interference filters of different order on both surfaces, serves that purpose.

A continuously variable filter (CVF) may also be obtained by depositing the individual layers with a slightly varying thickness. On one end of the large substrate the filter is tuned to one wavenumber and towards the other end the layers gradually increase in thickness and, consequently, the interference peak shifts to lower wavenumbers. Such filters have been produced in disk shape. The active area is near the rim. The transmission peak can be shifted up to a factor of two by rotating the disk and exposing another part of the variable thickness filter. Such circular variable disk filters form spectral analyzers of great simplicity. The resolving power of commercially available filters is less than ~ 100. Hovis & Tobin (1967) and Hovis et al. (1967, 1968) have flown filter wedge spectrometers on airplanes, and Herr et al. (1972) on Mariner 6 and 7. A cryogenically cooled variable filter has been applied to rocket-borne measurements by Wyatt (1975).

Narrow-band interference filters have seen space environment in a large number of instruments, but they are not the only application of interference filters. Multilayer filters may be designed to provide band-pass as well as high-and low-pass filters in the infrared; such designs are discussed by Heavens (1955), Vasicek (1960), Hass (1963–71), and Smith et al. (1972).

The first space flight of a Fabry–Perot interferometer, in this case a spherical Fabry–Perot, was on OGO 6 (Blamont & Luton, 1972). The objective of that investigation was to derive thermospheric temperatures from a measurement of the shape of the 0.63 μm atomic oxygen line. A more standard Fabry–Perot, with flat mirrors but a high order number of 4×10^4, has been flown on the Dynamic Explorer (Hayes et al., 1981). The objective was to determine stratospheric temperatures and the line-of-sight components of winds. Again, the temperature information is derived from the line width and the wind information from the Doppler shift of the line, assuming the spacecraft velocity is known. In addition to several atomic oxygen (O) lines, lines of atomic nitrogen (N, 0.52 μm) and of sodium (Na, 0.5896 μm) have been observed. The same Fabry–Perot etalon, but with different order filters, has been used.

Several Fabry–Perot interferometers can also be operated in series. Each etalon is adjusted by piezo electric crystals to peak at the same wavenumber but at different orders. Up to three Fabry–Perot interferometers have been scanned together (Hegyi et al., 1972; Kurucz et al., 1977). Several Fabry–Perot interferometers were flown in November 1995 as part of the Short- and Long-Wave Spectrometers on the Infrared Space Observatory (ISO). A more detailed description of these instruments is given at the end of the next section (5.7).

5.7 Grating spectrometers

Spectral selection techniques discussed so far are based on bulk material properties and on interference phenomena. The spectral selection effect of gratings rests on diffraction phenomena. Diffraction theory has its roots in the Huygens (1690) principle, which considers every point of a wavefront as a source of a secondary disturbance. These disturbances propagate as an infinite number of spherical wavelets. Huygens viewed the forward envelope of all secondary disturbances as the primary propagating wave. Later, Fresnel (1816, 1819) postulated interference among those secondary waves and was highly successful in explaining phenomena that the corpuscular theory of light could not explain – for example, the penetration of light into the geometrical shadow behind an opaque disk. At that time (1818) the wave theory of light was fully accepted; the corpuscular theory was dismissed and was not revived until the discussion of the photoelectric effect by Einstein (1905a). Today, both the wave and particle nature of light are accepted as complementary views of electromagnetic radiation; light is sometimes better described as a wave, sometimes better as a particle.

A full treatment of diffraction requires solutions of the wave equations in the transparent medium (e.g., air or vacuum) and the opaque material that forms the barrier to the propagating beam of radiation (e.g., a dielectric or a metal). In each medium the appropriate values of the material constants (ε, μ, and σ) must be taken into account and the boundary conditions for the electric and magnetic vectors must be satisfied at all interfaces. Full solutions of the wave equation have been obtained for simple geometric shapes such as a long circular cylinder (Lord Rayleigh, 1881), a straight edge (Sommerfeld, 1986), and a sphere (Mie, 1908). For more complicated shapes the Kirchhoff (1883) diffraction theory provides good approximate solutions, as long as dimensions of obscurations or apertures exceed a few wavelengths. A detailed discussion of the Kirchhoff theory can be found in standard texts on optics, such as Born & Wolf (1959) or Sommerfeld (1954). Here we address directly the functioning of a grating spectrometer and the differences in comparison with prism and Fabry–Perot spectrometers.

Interference filters and Fabry–Perot etalons, discussed in the previous section, perform a spectral selection by passing only discrete, narrow wavelength bands. In contrast to this the diffraction grating disperses radiation into a spectrum by interference among many transmitted or reflected beams and by sending different wavelengths in different directions. In effect, this is similar to refraction in a prism, although refraction is not an interference phenomenon. Radiation from a narrow aperture is collimated and sent to a grating. The emerging radiation is refocused to form the spectrum.

In the traditional mode of operation, one exit slit and a single detector are used to scan the spectrum either by moving the exit slit in the focal plane or by rotating the grating while keeping the exit slit and the detector stationary. More recently, however, because the diffraction grating passes the entire spectrum to the focal plane and disperses it spatially, an array of detectors is placed in the focal plane to record all spectral elements simultaneously. Indeed the grating spectrometer has experienced renewed interest in infrared astronomy since the availability of detector arrays.

5.7 Grating spectrometers

A grating spectrometer with a detector array has a spectral multiplex advantage over a Fabry–Perot etalon. Compared with a Fourier transform spectrometer (discussed in the next section), a cryogenically cooled grating spectrometer with an array in its focal plane has an advantage in sensitivity in those applications where background and source radiation are the dominant causes of noise. In a Fourier transform spectrometer radiation from the entire bandpass falls on the detector. In a cooled grating spectrometer each detector in the array is irradiated with only the small spectral bandpass of one resolution element. The background noise in the grating spectrum is lower by $(N)^{\frac{1}{2}}$ where N is the number of resolution elements in the entire bandpass. This advantage is only realized when the entire bandpass of interest can be placed on the array. In practice this is limited by the availability of large arrays. Both a Fabry–Perot etalon and a Fourier transform spectrometer have an étendue or $A\Omega$ advantage over a grating spectrometer since they do not require the use of narrow slits.

The basic operation of a grating can be understood by considering a plane wave incident on a set of parallel narrow slits (Fig. 5.7.1). This corresponds to a transmission grating used at normal incidence. Infrared gratings are almost always reflection gratings used at high incidence angle, but this simple example serves to illustrate the basic operation of a grating. According to the Huygens principle the radiation beyond each slit travels in all directions. Radiation of a particular wavelength, λ, will constructively interfere in those directions that correspond to an integer number, m, of wavelengths between adjacent slits. The grating equation for

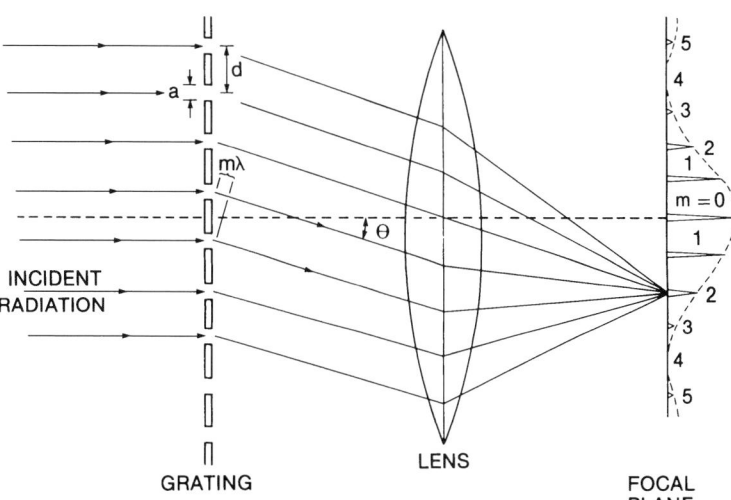

Fig. 5.7.1 Diffraction of a plane wave at a set of parallel narrow slits. Radiation of wavelength λ interferes constructively with itself at particular angles where a constant difference, $m\lambda$, exists between adjacent slits. A spectrum forms at the focal plane of the lens.

normal incidence is

$$m\lambda = d \sin\theta_d, \quad (5.7.1)$$

where d is the slit spacing and θ_d is the diffraction angle. If a lens is placed after the grating, all rays of wavelength λ (for a given value of m) will arrive at the same location in the focal plane at an angle θ_d through the lens center (see Fig. 5.7.1). Radiation of slightly different wavelengths will be focused at slightly shifted positions, so that a spectrum is formed in the focal plane. The integer m is called the order number. A change in the value of m for a fixed λ changes the diffraction angle θ_d and the position in the focal plane. Therefore, the radiation is divided into several orders in the focal plane. Spectral analysis is impossible where spectra of different orders overlap, and a spectral order filter must be used to isolate a specific order.

Not all orders of a given wavelength appear equally strong in the focal plane, however. The single-slit diffraction pattern corresponding to the width a of each slit forms an envelope in the focal plane that determines the positions of the orders of significant intensity. The intensity pattern in the focal plane, as derived in standard texts for Fraunhofer diffraction by multiple slits, is

$$I = I_0 \left(\frac{\sin\alpha}{\alpha}\right)^2 \left(\frac{\sin N\beta}{N \sin\beta}\right)^2, \quad (5.7.2)$$

where $\alpha = (\pi/\lambda) a \sin\theta_d$, and $\beta = (\pi/\lambda) d \sin\theta_d$. The $(\sin\alpha/\alpha)^2$ term is the diffraction pattern for a single slit of width a, and the $(\sin N\beta/N \sin\beta)^2$ term is the grating diffraction pattern, which has maxima given by Eq. (5.7.1), one for each order m. Only orders that fall within the single-slit envelope show strong intensities.

Infrared gratings are most commonly of the reflection rather than transmission type. A reflection grating is a mirror with parallel rulings on its surface instead of transparent gaps. The radiation is diffracted upon reflection. To improve the spectral resolution reflection gratings are often used at high rather than normal incidence angle. If the direction of the incident radiation is at an angle θ_i with respect to the grating normal, radiation of wavelength λ will be diffracted in the direction θ_d (see Fig. 5.7.2) according to the relation

$$m\lambda = d(\sin\theta_i + \sin\theta_d), \quad (5.7.3)$$

where d is the spacing between rulings. A special case is the 'Littrow' configuration, where the incidence and reflection angles are nearly identical, so that Eq. (5.7.3) reduces to

$$m\lambda = 2d \sin\theta. \quad (5.7.4)$$

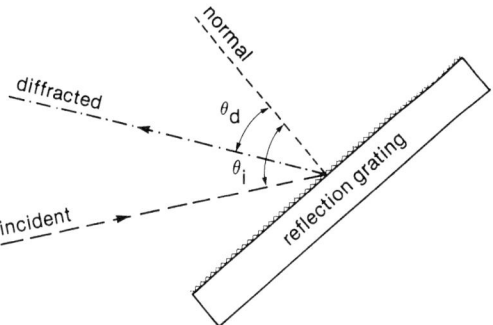

Fig. 5.7.2 Definitions of angles of incidence, θ_i, and diffraction, θ_d, at a reflection grating.

This Littrow grating equation is generally useful because θ_i and θ_d can be regarded as approximately equal in many applications of reflection gratings, even when the optical configuration is not strictly Littrow.

With a single detector in the focal plane, the spectrum can be scanned by rotating the grating. If the grating is turned through a small angle $\Delta\theta$, Eq. (5.7.4) gives the relation to change in wavelength $\Delta\lambda$,

$$\frac{\Delta\lambda}{\Delta\theta} = \frac{2d}{m} \cos\theta, \qquad (5.7.5)$$

for the Littrow case. Before the availability of detector arrays, grating rotation was the common technique used to record infrared spectra. With arrays it is possible to record an entire region of the spectrum simultaneously. The same multiplex property is obtained with photographic and CCD grating spectrometers in the visible. In an array spectrometer the grating angle is fixed, and radiation at all wavelengths is incident at the same angle. The diffracted radiation varies in angle with wavelength according to Eq. (5.7.3), with a dependence of $\Delta\lambda$ on diffraction angle given by

$$\frac{\Delta\lambda}{\Delta\theta_d} = \frac{d}{m} \cos\theta_d. \qquad (5.7.6)$$

Note that $\Delta\lambda$ changes twice as much for a change in grating angle θ (Eq. 5.7.5) as it does for a change in diffraction angle θ_d (Eq. 5.7.6). The wavelength dispersion decreases in higher orders and at higher diffraction angles. It is often more convenient to write the wavenumber dispersion; since $\Delta\nu/\nu = -\Delta\lambda/\lambda$ Eqs. (5.7.4) and (5.7.6) give for the Littrow configuration,

$$\frac{\Delta\nu}{\Delta\theta_d} = -\frac{\nu}{2} \cot\theta. \qquad (5.7.7)$$

In this form the wavenumber dispersion is only a function of wavenumber and angle, but still depends implicitly on the spacing of the grating ruling and the order number.

The ability of a grating to resolve two closely spaced features in a spectrum depends on the dispersion of the grating and the angular interval in the focal plane over which a single wavelength is spread. The smallest angle that can be subtended by a single wavelength is given by the diffraction limit,

$$\Delta\theta_{\min} \sim \frac{\lambda}{D} = \frac{1}{\nu D}, \tag{5.7.8}$$

where D is the diameter of the radiation beam at the grating. The grating must be wide enough to intercept the entire beam when used at angle θ, which means that its minimum width is $W = D/\cos\theta$. Substituting Eq. (5.7.8) in Eq. (5.7.7) yields

$$\Delta\nu_{\min} = \frac{1}{2D\tan\theta} = \frac{1}{2W\sin\theta}. \tag{5.7.9}$$

The minus sign in Eq. (5.7.7) has been deleted. The denominator is just the difference in path length between rays at the two edges of the grating; this difference is called the 'optical retardation' of the radiation.

An alternative parameter for specifying the spectral discrimination capability of a grating, which is often used in astronomy, is the resolving power, defined as $R = \lambda/\Delta\lambda$. The resolving power has a simple relationship to the order m and the total number of rulings on the grating N,

$$R = \lambda/\Delta\lambda = mN, \tag{5.7.10}$$

which can be derived from Eq. (5.7.9) and $\lambda/\Delta\lambda_{\min} = -\nu/\Delta\nu_{\min}$. However, this relation can sometimes be misleading. The resolving power and resolution ($\Delta\nu = \nu/R$) cannot be increased by making the total number of rulings arbitrarily large with a fixed grating width. The ruling spacing will eventually become too small to operate at the desired wavelength. From a practical standpoint, Eq. (5.7.9) is easier to apply, because the operating angle θ and the grating width W are chosen to give the desired resolution. The ruling spacing, d, is then determined, using Eq. (5.7.4), by the order m in which the desired wavelength is to be observed.

As Eq. (5.7.9) implies, high resolution is achieved by using a large grating at steep angles. The largest available gratings are about 20×40 cm, and are operated at angles around $60°$. Grating rulings are usually blazed, that is, shaped to produce high efficiency at the desired working angle. If the grating is to be used at a high angle, the rulings often have a step-shaped cross section, and care is taken during

5.7 Grating spectrometers

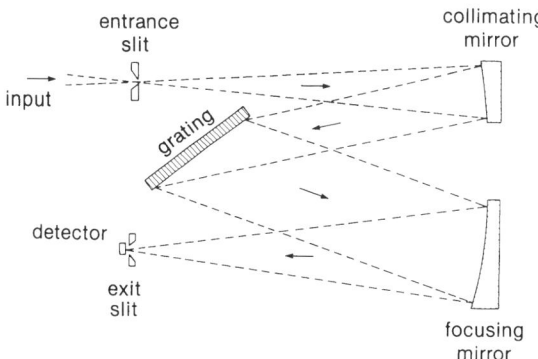

Fig. 5.7.3 Schematic of a grating spectrometer. Radiation from the entrance slit is collimated and sent to the grating. After diffraction, the radiation is focused at the detector (possibly through an exit slit).

manufacture to produce a good mirror surface on the facet of each step that faces the radiation. Gratings of this type, called echelles, are common where high spectral resolution is needed.

A grating spectrometer normally consists of an entrance slit, a collimating mirror, the grating, a focusing mirror, an exit slit, and a detector (Figure 5.7.3). The entrance slit limits the field on the source and, therefore, the $A\Omega$. The collimating mirror produces plane waves at the grating. The focusing mirror displays the spectrum in the focal plane, and a portion of that display, containing a small spectral interval, passes through the exit slit and falls on the detector. Depending on the application, pinhole apertures may replace the slits. An array may be used in place of the exit slit and the detector. There are several configurations of plane grating spectrometers used in the infrared. Here we give a brief discussion of two common types; the Littrow, mentioned above, and the Ebert–Fastie.

In the Littrow version the incident and diffracted beams occupy the same section of the collimating mirror. The collimator, usually an off-axis parabola, places the focal plane next to the grating. Because the beams overlap and only one mirror is needed, the Littrow design allows the spectrometer to be housed in a small volume. An additional advantage of the Littrow configuration is that grating efficiencies are normally higher when the incident and diffracted beams are similar. In the Littrow design the exit image of a straight entrance slit is curved, and the curvature changes with grating angle. A curved exit slit can only be optimized for a small range of grating angles. However, if the grating is used in high order near its blaze angle the range of angles can, indeed, be kept small.

The Ebert–Fastie configuration uses two separate sections of a common mirror for collimating and focusing. The entrance and exit slits can be placed on opposite sides, or above and below the grating. The primary advantage of the Ebert–Fastie configuration is that, although the mirrors are used off-axis, the aberrations caused by the collimating mirror are largely corrected by the focusing mirror. If the incidence and diffraction angles at the grating are minimized, the Ebert–Fastie spectrometer will have less coma than the Littrow spectrometer. Because of this, long curved slits can be used to increase throughput without loss of resolution. The basic Ebert–Fastie design has several variants. The most common version is the Czerny–Turner configuration, which uses two separate mirrors for collimating and focusing. The reader is refered to standard texts, such as James & Sternberg (1969), for complete discussions and diagrams of various designs of grating spectrometers.

In order for a grating spectrometer to achieve its ultimate resolution it must be used at the diffraction limit set by Eq. (5.7.8). This requires that the range of incidence angles at the grating be very small, which in turn limits the width of the entrance slit and, consequently, the angular size of the object that can be observed. Because planets and many other objects are extended sources, it is often necessary to observe fields-of-view that are larger than the diffraction limit. Under these conditions the resolution is degraded by the ratio of the size of the actual field-of-view or slit width to the diffraction limit.

An example of a multichannel radiometer using a grating was the satellite infrared spectrometer on Nimbus 3, SIRS (Fig. 5.7.4). The spectral resolution was determined by the width of the exit slit, which was formed by a germanium wedge with a spherical surface, and a combination of an immersion lens and a conical reflector (Dreyfus & Hilleary, 1962). The SIRS on Nimbus 3 had eight channels, seven in the 667 cm^{-1} CO_2 band and one centered at 900 cm^{-1}, a relatively transparent part of the spectrum. The purpose of the CO_2 channels was to obtain temperatures at different atmospheric levels and that of the 900 cm^{-1} channel was to obtain the surface temperature (Wark & Hilleary, 1969). An improved version of SIRS flown on Nimbus 4 had six additional channels located in the water vapor rotation band. The concept of multichannel radiometers became the design pattern of a large number of instruments serving as remote temperature sensors for numerous weather satellites.

Another example of a multichannel radiometer using a diffraction grating was the water vapor detector on the Viking Orbiter (Farmer & LaPorte, 1972; Farmer *et al.*, 1977). This instrument measured absorption of reflected solar radiation in and near the water vapor band at 7200 cm^{-1}.

The Galileo mission to Jupiter carries a Near Infrared Mapping Spectrometer (NIMS). This imaging spectrometer is used to characterize the geological and mineral content of Jupiter's satellites and to investigate the composition and temperature of the atmosphere of Jupiter. NIMS is a grating spectrometer operating in the 0.7 to 5.2 μm spectral range with a resolving power, $\lambda/\Delta\lambda$, of 50 to 200. The focal plane is cooled by radiation to 70–80 K and consists of a linear array of two silicon and

5.7 Grating spectrometers

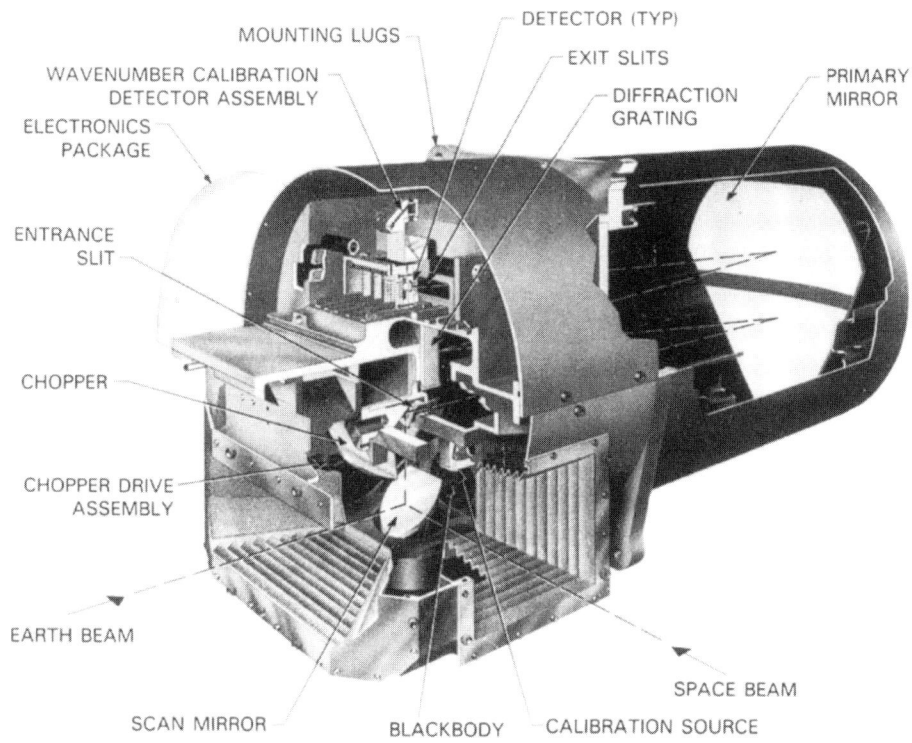

Fig. 5.7.4 Schematic view of the Satellite Infrared Spectrometer (SIRS) on Nimbus 3 (Nimbus Project, NASA).

seventeen indium antimonide (InSb) detectors. Spectral information is dispersed along the array axis. The spatial scene is constructed in whisk-broom mode, with a wobbling secondary mirror producing cross-track and the motion of the spacecraft producing along-track scanning. The angular resolution of an image element is 0.5×0.5 mrad, and the cross-track field of view is 10 mrad (Aptaker, 1987; Carlson et al., 1992). All instruments on the Galileo spacecraft and probe are discussed in *Space Science Reviews*, **60**, 1–4, 1992 and in a collection of reprints published in book form, edited by Russel (1992). Scientific results obtained by the NIMS instrument are discussed in Subsection 6.5.b.

An instrument similar to NIMS (called VIMS) is on board the Cassini spacecraft scheduled to reach Saturn in July 2004. It consists of two grating spectrometers. VIMS-V covers the spectral range from 0.3 to 1.05 μm and VIMS-IR from 0.85 to 5.1 μm. The resolving power is the same as in the Galileo instrument, 50–200. VIMS-V samples both spectral and spatial information using a two-dimensional array of detectors. The nominal field of view is 32×32 mrad with a 0.5 mrad pixel size. In VIMS-IR the spectrum is detected with a linear InSb array, cooled by radiation to 70–80 K. The image of the scene is generated by rastering the optics in two dimensions. VIMS-IR is a modified version of the NIMS on the Galileo spacecraft. The scientific objectives of VIMS are to study the atmospheres of Saturn and Titan, and the composition of the rings and satellites of Saturn (Jurgens *et al.*, 1990).

Between January and March 1989, an imaging spectrometer (ISM) on board the Phobos-2 spacecraft obtained infrared spectra (0.73–3.15 μm) of Phobos and of the equatorial region of Mars. The spectrometer used the first- and second-order of a grating. In the focal plane four groups of 32 (total of 128) cooled lead sulfide detectors measured each spatial pixel simultaneously. Thus far, for the most part only two groups have been used. The instantaneous field of view of the instrument is 12×12 arc minutes. At a spacecraft altitude of 6300 km, this corresponds to a surface resolution of 22×22 km. A subset of the data was taken at a lower spacecraft altitude, yielding a 7×7 km resolution. The signal-to-noise ratio is very good, about 500 over most of the range, but somewhat lower at the longer wavelength end. A scanning mirror provides cross-track and the spacecraft motion along-track imaging. The instrument is described by Puget *et al.* (1987), and results are mentioned in Section 6.2. An advanced version of the ISM instrument, called Omega, is scheduled to fly in 2003 on the Mars Express Mission.

Two infrared spectrometers, an imaging photo-polarimeter, and a camera flew in Earth's orbit on the Infrared Space Observatory (ISO), a project of the European Space Agency. The spacecraft contained a cryostat with 2300 liters of super-fluid helium to cool the instruments and the telescope to between 2 and 8 K, (Kessler *et al.*, 1996). The Short-Wave Spectrometer covers the spectral range from 2.5 to 45 μm (de Graauw *et al.*, 1996). The Long-Wave Spectrometer operates between 43 and 196.9 μm (Clegg *et al.*, 1996). Both instruments are grating spectrometers with detector arrays. Two orders of each grating can be used with or without additional sets of Fabry–Perot interferometers. A schematic of the far-infrared spectrometer is shown in Fig. 5.7.5.

The grating of the Long-Wave Spectrometer is ruled with 7.9 lines per mm and can be tilted over $\pm 7°$. In the neutral position, the grating is illuminated at an incident angle of $60°$. One beryllium-doped and five gallium-doped germanium detectors cover the range from 45 to 90 μm using the second-order of the grating. Four stressed gallium-doped germanium detectors cover the range from 90 to almost 197 μm, using the first-order of the grating. The design permits sufficient overlap in the spectral range assigned to each tilt angle to tolerate the failure of one detector without loss of a particular spectral interval. The spectral resolution in the second-order of the grating (45–90 μm) is 0.29 μm, and in the first-order (90–197 μm) it is 0.60 μm.

To achieve a much higher spectral resolution two Fabry–Perot etalons can be rotated into the collimated optical path between the fore-optic and the grating.

5.7 Grating spectrometers

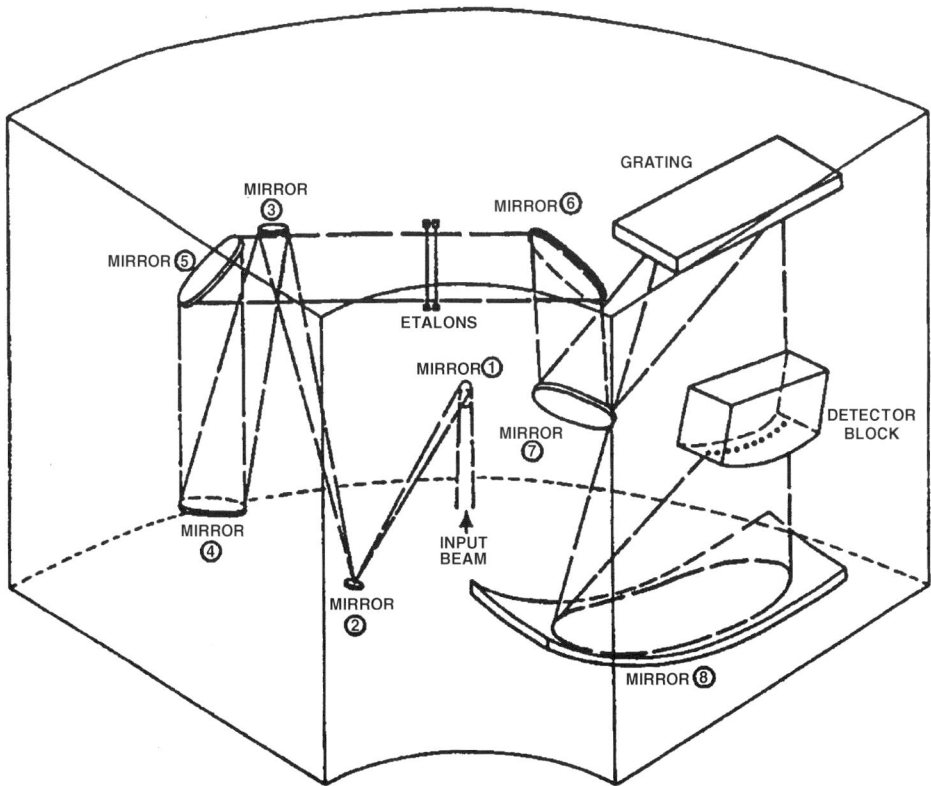

Fig. 5.7.5 Optical train of the Long-Wave Spectrometer of the Infrared Space Observatory (ISO). Mirrors 1, 2, and 6 are folding mirrors. Mirror 2 is at the focal point of the telescope, mirror 3 collimates the beam for the benefit of the Fabry–Perot etalons. The etalons are mounted on a rotating wheel together with a transparent opening and an opaque disk (Clegg, 1992; Clegg et al., 1996).

The resolving power achieved with the Fabry–Perot ranges from 6800 to 9700. The Fabry–Perot plates consist of rectangular grids of nickel meshes, 3 μm in thickness. The widths of the individual stripes of the grid are 6 μm and the periodicities are 15.5 μm for the short-wave (45–90 μm) and 19 μm for the long-wave (90–180 μm) etalons. The moving plate, containing one grid, was supported by leaf springs. An identical grid was mounted on a stationary structure. Electric coils in magnetic fields provide the driving force, and capacitive sensors control the motion of the carriage which must be kept parallel to the stationary grid. Figure 5.7.6 shows the etalon (Clegg et al., 1996; Davis et al., 1995).

The calibration of the Long-Wave Spectrometer is discussed by Swinyard et al. (1996) and by Burgdorf et al. (1998). A description of the Short-Wave Spectrometer is given by de Graauw et al. (1996). Flux calibration has been discussed by Schaeidt et al. (1996) and wavelength calibration by Valentijn et al. (1996).

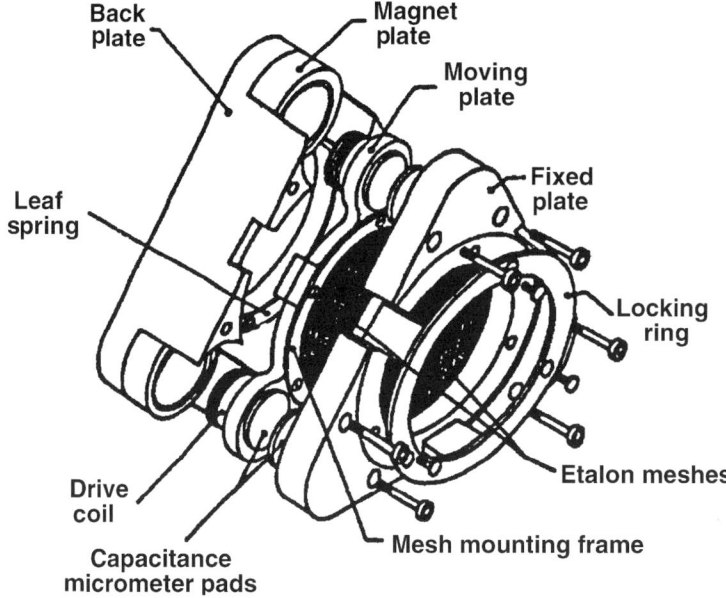

Fig. 5.7.6 The moving plate is suspended on leaf springs between the back plate and the fixed plate. A moving coil in a magnetic field provides the force for the displacement. Three capacitive position sensors in servo-loops adjust the current to the coils and thereby control the motion (Clegg, 1996).

The achieved NEP of the detectors was on the order of 5×10^{-18} W Hz$^{-1/2}$ as measured in the laboratory, but was somewhat higher in orbit due to the cosmic ray background. The very low noise of the detectors and the cryogenic temperature of the whole device contributed greatly to the extraordinary performance of the instrument. Results obtained by ISO are discussed in Chapters 6 and 8.

5.8 Fourier transform spectrometers

a. Michelson interferometer

Albert A. Michelson (1881) constructed his two-beam interferometer more than a century ago. His prime motivation was to find a technique precise enough to measure the motion of the Earth relative to the 'ether'. The negative result of the Michelson–Morley (1887) experiment required major revisions of the classical space–time concepts. To gain spectral information Michelson recorded the 'visibility curve' using his eye as a detector, while gradually increasing the path in one arm of his interferometer. Applying a Fourier transformation he was able to generate crude spectra for simple cases. With this method he identified the red cadmium emission as a single, narrow line and confirmed the yellow sodium emission as a doublet. To perform the Fourier transformation Michelson & Stratten (1898) constructed a mechanical Fourier analyser, and Michelson (1898) applied it to the investigation of magnetically broadened spectral lines. Although the principles of Fourier transform spectroscopy were clearly demonstrated, Michelson could not have used his interferometer as a spectrometer in the modern sense without the availability of electronic and computing technology.

A few years later Rubens & Wood (1911) performed a spectral analysis with the Michelson interferometer by guessing the number and strength of a few lines in a narrow spectral range and synthesizing an interferogram by Fourier technique. Finally, they compared the computed envelope to the observed visibility curve. After that Fourier transform spectroscopy lay dormant for several decades until it was revived in the late 1940s and 1950s. Only since then has the potential of the Michelson interferometer as a powerful spectrometer been recognized. Jacquinot & Dufour (1948) called attention to the high energy-gathering ability of the interferometer, which under comparable conditions may be as high as several hundred times that of a conventional grating instrument. The high étendue, also called throughput, or $A\Omega$-advantage, has been fully realized with space-based interferometers observing the planets, but it may not always be possible to exploit it in other applications. In stellar or even in low resolution planetary spectroscopy from the ground, $A\Omega$ is limited by the small angular size of the object and, in spite of large telescope diameters, the high light-grasp of the interferometer cannot always be exploited.

The second major advantage of the Michelson interferometer, the multiplex advantage, was first pointed out by Fellgett (1951, 1958). Within a given range the interferometer measures all spectral intervals simultaneously, while a conventional grating instrument with a single detector must record the intervals sequentially. For equal observing time an improvement factor follows in signal-to-noise equal to the square root of the number of resolved spectral intervals. The improvement is most impressive in high resolution, wide-band observations. Although space-borne interferometers have fully benefited from the multiplex advantage, conditions exist where it cannot be realized. The multiplex advantage holds as long as the system noise is independent of the signal level; if the system is dominated by source noise, for which the noise level is proportional to the signal, the multiplex advantage does not exist; it even becomes a disadvantage. Background-noise-limited operation, that is, operation where statistical fluctuations in the arriving photons are the dominant noise source, may be reached with wide-band, cooled interferometers and modern cryogenic detectors. One method to overcome this limitation is the post-dispersion technique discussed in Section 5.8.b.

A third advantage of the Michelson interferometer, the ability to have a very precise wavenumber calibration by comparing wavenumbers of the planetary spectra to that of a monochromatic reference source, is discussed further below. With the $A\Omega$- and multiplex-advantages understood, the stage was set in the late 1950s and early 1960s for the development of Michelson-type spectrometers and Fourier transform spectroscopy (FTS) in general. Many of the early design principles were established by Janine Connes (1961) in her thesis and with Pierre Connes (e.g., 1966; see also Aspen Conference, Vanasse et al., 1971). Strong (1954, 1957), Mertz (1965), Loewenstein (1966), Vanasse & Sakai (1967), Bell (1972), and others made significant contributions to and provided reviews of this quickly evolving field. The rapid advances of FTS in the 1960s and the application to planetary research have been helped by simultaneous progress in detector and computer technology.

The first to transform an interferogram of an astronomical object was Fellgett (1951, 1958, 1971). In Michelson's time such a transformation would have been time consuming, even for only a few hundred data points. Today, with modern computers and the Cooley–Tukey (1965) algorithm, which Forman (1966) introduced to the practitioners of FTS, transformation of a million data points is not uncommon.

The Michelson two-beam interferometer records the autocovariance function of the observed radiation, the interferogram, as a function of optical path difference (delay) between both beams. The spectrum is obtained by Fourier analysis of the interferogram. The operation of a Michelson interferometer as an infrared spectrometer is discussed with the help of Fig. 5.8.1. The essential part of the instrument is the beamsplitter, which divides the incoming radiation into two beams of nearly equal intensity. After reflection from the stationary and the movable mirrors, the beams recombine at the beamsplitter. The phase difference between the beams is proportional to their optical path difference, including a phase shift due to the

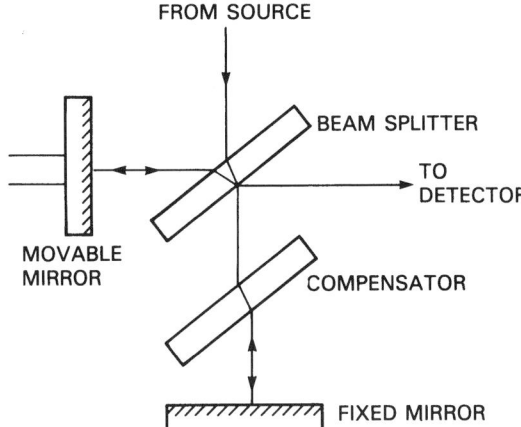

Fig. 5.8.1 Basic configuration of a Michelson interferometer. The beamsplitting surface is on the lower-right side of the beamsplitter substrate. The compensating plate is of equal thickness and of the same material as the beamsplitter substrate.

difference between internal and external reflections at the beamsplitter (e.g., Born & Wolf, 1959). Suppose a collimated beam of monochromatic radiation strikes the interferometer while the movable mirror is set at the balanced position where both arms have equal length. In a nonabsorbing beamsplitter the phase difference between internal and external reflection is 180°. Consequently, both beams interfere destructively as seen from the detector; the central fringe is dark. At the same time, the interference is constructive as seen from the entrance port. The incoming energy is reflected back towards the source; the interferometer acts as a mirror. Now suppose the movable mirror is displaced by a quarter of a wavelength. The phase of the beam reflected from that mirror changes by half a wavelength or 180°. Detector and source sides experience now constructive and destructive interference, respectively. The interferometer as a whole becomes transparent at that wavelength, and the incoming radiation strikes the detector. The interferometer may be viewed as a sinusoidal modulator, which alternately switches the arriving beam between detector and source, depending on the position of the movable mirror. Emission from the detector experiences the same fate, since the interferometer is symmetric as far as source and detector are concerned. As in a chopped radiometer, the intrinsic efficiency of the interferometer configuration shown in Fig. 5.8.1 is one half. Later, more efficient configurations are discussed.

Michelson interferometers can be operated in two modes. In one mode the movable mirror, sometimes called the Michelson mirror, is stepped in small, equal increments. Sampling theory requires an optical step size of less than one half of the shortest wavelength to be measured; otherwise aliasing may occur. However, if we know that the signal consists only of a narrow band of frequencies (because we are interested only in a particular narrow spectral interval and we have inserted corresponding optical and electrical filters), sampling may occur at larger intervals without risking confusion due to aliasing. The number of sample points can be reduced considerably by this technique. Alternatively, in a method sometimes employed in the continuous mode, the interferogram may be oversampled and the reduction to the minimum number of necessary data points may be accomplished by numerically filtering after digitization.

In the stepping mode, the signal is integrated at each rest position for a certain time, the dwell time. After that, the Michelson mirror advances to the following position and the next point is recorded. As in the case of an a.c. radiometer an external chopper may be used to modulate the incoming radiation. The stepping technique has been perfected by J. and P. Connes.

In the continuous mode the Michelson mirror advances at a constant speed and the signal is sampled at small, equal intervals. At one time the misconception existed that this mode is less efficient than the discrete step mode because the time spent in taking the sample is small in comparison with the time between samples, but

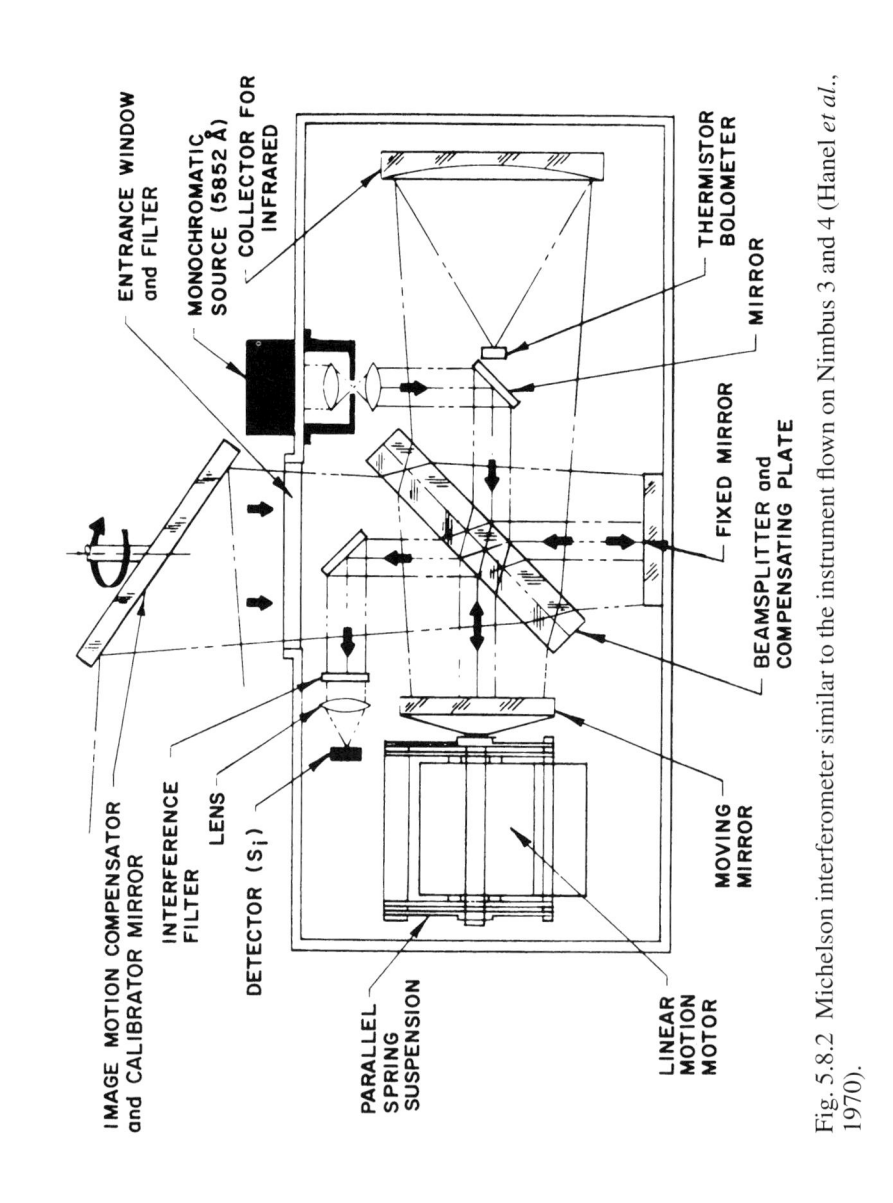

Fig. 5.8.2 Michelson interferometer similar to the instrument flown on Nimbus 3 and 4 (Hanel et al., 1970).

this is not the case. Integration between samples takes place in the electrical filter used to limit the frequency range before the sampling process. On the contrary, the stepping mode is slightly less efficient, because the time spent while moving the mirror between measurement points is lost.

In the constant speed mode the detector signal is modulated sinusoidally at a frequency equal to the product of optical mirror speed, v, and wavenumber, ν,

$$f = v\nu. \qquad (5.8.1)$$

For constant v the wavenumber spectrum is directly mapped into an electrical frequency spectrum. In the configuration shown in Fig. 5.8.1, the optical speed is twice the actual mirror speed; a factor of 2 appears in Eq. (5.8.1) if v is interpreted as the mechanical mirror speed. For double passing, to be discussed later, another factor of 2 appears on the right side of Eq. (5.8.1). The interferometers flown in space (IRIS, ATMOS, FIRAS, TES, and CIRS) are all operated in the continuous scan mode. Interferometers with a constant velocity drive, in principle, may also be operated with an external chopper to accommodate a fast detector with strong $1/f$ noise. In general, however, in this case one prefers to scan more rapidly and thereby move the passband of interest towards higher frequencies (Mertz, 1965). Instruments without an external chopper, such as IRIS, must occasionally be aimed at deep space to discriminate between planetary and instrument emission. The situation is analogous to that of an a.c. radiometer with the chopper not being the first element (see Subsection 5.3.a). In other interferometer configurations, with complimentary output, the instrument emission is canceled otherwise.

The Mariner IRIS (similar to the Nimbus schematic shown in Fig. 5.8.2) has a coated cesium iodide window to permit purging of the instrument with dry nitrogen to protect the hygroscopic beamsplitter, also of cesium iodide, from water vapor damage. The purging was continued up to lift-off. This window reflects a small fraction of the outgoing interferogram back to the interferometer. The sinusoidal signals of the autocovariance function generate terms with twice the frequency, which appear as small ghosts in the Fourier transformed spectra. For example, the sharp Q-branch of CO_2 at 667 cm^{-1} has a weak signature at 1334 cm^{-1}. The effect is most noticeable in spectra of low temperature scenes. Planetary and calibration spectra are affected. However, a correction term eliminates most of the problem (Formisano et al., 2000).

The mirror position in the stepping mode, and the sampling point in the constant velocity mode, must be accurate to a small fraction of a wavelength. To achieve this accuracy it is customary to employ a second interferometer with a monochromatic source of a wavenumber considerably higher than that of the radiation to be measured. The reference interferometer shares part of the main infrared interferometer

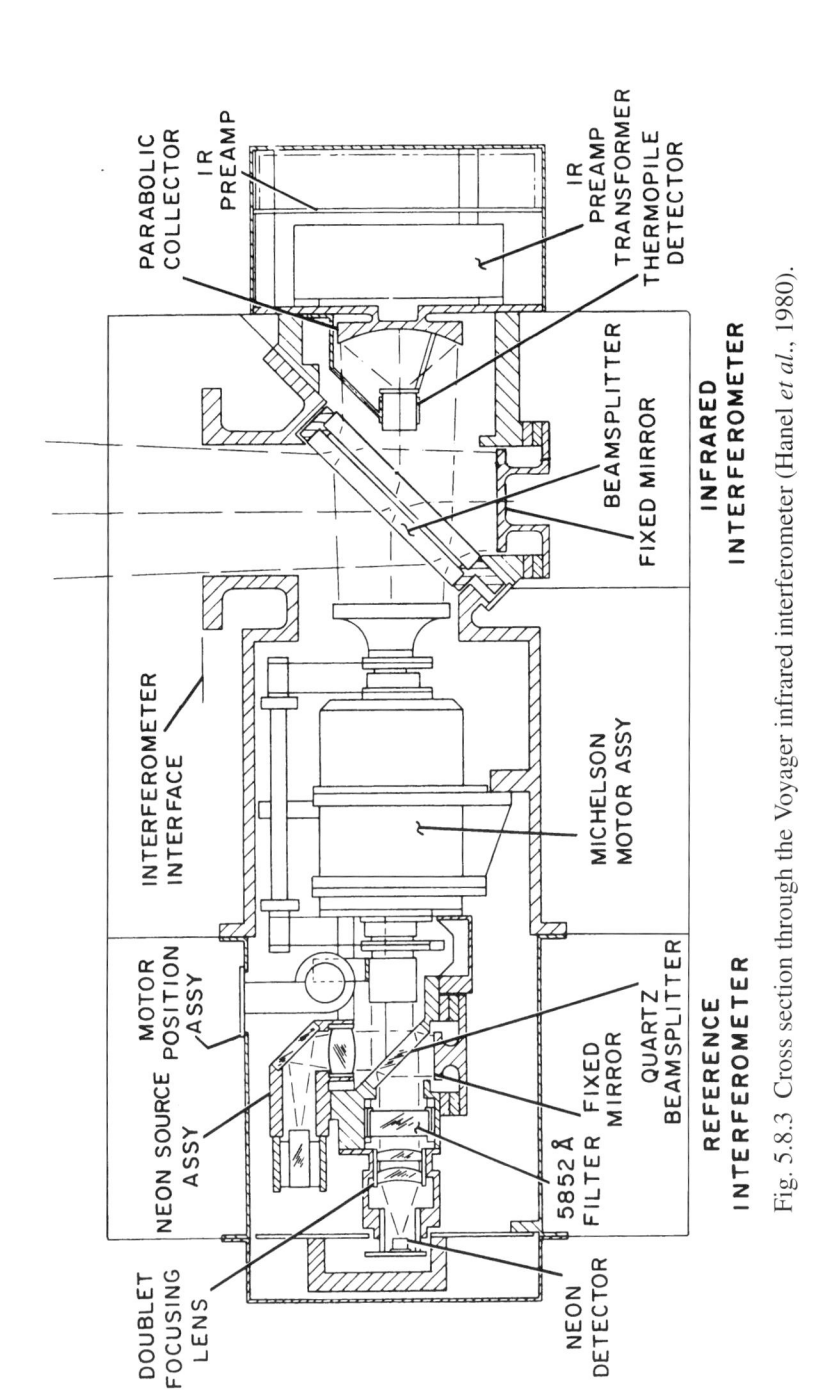

Fig. 5.8.3 Cross section through the Voyager infrared interferometer (Hanel et al., 1980).

or must be rigidly connected to it. In the Nimbus and Mariner IRIS's (Fig. 5.8.2), the reference interferometer occupies the center of the optical beam. In the Voyager design (Fig. 5.8.3), the reference interfcrometer is directly coupled to the mirror drive and uses a quartz beamsplitter, which was expected to be less susceptible to scintillation caused by Jovian high-energy particles than the CsI substrates of the main unit. A spectral line of a neon discharge lamp serves in all IRIS as a reference source. When power is not limited a He–Ne laser is often used as a reference. The Cassini interferometer (CIRS), to be discussed further below, uses a solid-state laser. Such a source is strong and very efficient, but the signal frequency is somewhat sensitive to the temperature of the device. The stepping interval or the sampling point is derived by counting zero-crossings in the reference signal. The reference frequency of IRIS is also used to phase lock the mirror motion to the spacecraft clock. The bit stream (the sequence of digital signals) in the interferometer data channel is then fully synchronized with the spacecraft data system. The use of a stable reference line provides an absolute wavenumber scale to the measured spectra. This is the third major advantage of that type of spectrometer.

Planetary radiation consists of many spectral lines and a radiation continuum. The detector signal is then the superposition of many individual interference patterns (Fig. 5.8.4). In a well-balanced and well-aligned constant-velocity instrument the phase characteristic of the whole analog chain (detector, amplifier, filter) is linear, and the derivative of the phase is constant. Thus

$$\frac{d\phi(\omega)}{d\omega} = \tau = \text{constant}; \quad \omega = 2\pi f. \tag{5.8.2}$$

All sinusoidal signals are in phase at one particular mirror position and the interferogram is symmetrical with respect to that point. This mirror position corresponds to the crossing of the neutral position of the mirror if the detector, amplifier, and filter have nearly infinite electrical bandwidth; then the delay time, τ, is zero. In reality, one has to limit the bandwidth to avoid aliasing of noise into the passband. An electrical filter with a linear phase characteristic must be introduced, which results in a finite delay between events in the optical domain and the electrical signal at the point of digitization. If the overall phase characteristic is not linear the interferogram is not symmetrical and different frequencies show different delays. If the interferometer is to be insensitive to small variations in mirror speed, which may occur due to external vibration, the sampling action must be delayed by the same amount τ as the main interferometer signal (Hanel et al., 1972d; Logan, 1979). Only then is the signal sampled by the correct sampling pulse. If the mirror velocity is truly constant, then it does not matter that signal and sampling pulse

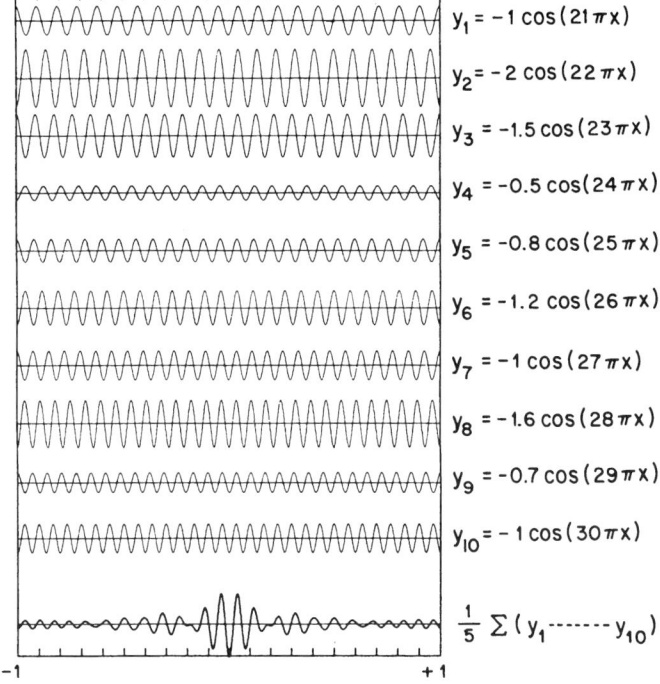

Fig. 5.8.4 Superposition of many frequencies of different amplitudes produces the interferogram shown in the lowest line.

do not correspond to the same mirror position. However, even under phase locked conditions, the mirror speed varies within the dynamic range of the phase locked loop, and much more so in systems where the speed is not so firmly controlled. In the stepping mode this problem does not arise as long as the dwell time is long in comparison with the signal delay.

For a perfectly symmetrical interferogram it would be adequate to record only from the center to one maximum mirror displacement. Only a cosine transformation would need to be performed, because a sine transformation of a symmetrical function is zero. In reality, the interferometer may not be fully compensated, the phase linearity may not be perfect, detector noise is different on both sides, and the sampling points may not be precisely symmetrical with respect to the center. Therefore, real interferograms deviate somewhat from symmetry. The amplitude of the sine transformation reveals the magnitude of this deviation. In the presence of a non-negligible sine amplitude, the power spectrum (the square root of the sum of the squares of the cosine and sine terms of a two-sided interferogram) may be taken as the measured amplitude. The preferred method is to record only one side and a small part of the interferogram at the other side of the center and to obtain a low resolution phase spectrum from the short two-sided portion of the interferogram. In data such as obtained from Voyager the phase relation between the individual frequency components does not change rapidly with wavenumber and a low resolution phase spectrum is, therefore, adequate. Phases in the full, one-sided interferogram are then shifted with respect to the assumed center by a numerical convolution process. The interferogram is thereby symmetrized and a cosine transformation can be applied (Forman, 1966). The advantage of using the cosine over the power spectrum can be as much as $2^{\frac{1}{2}}$ in signal-to-noise.

Consider only the alternating component of a symmetrized interferogram, that is, the component amplified by an a.c. amplifier:

$$i(\delta) = \int_{v_1}^{v_2} r_v[I_v - B_v(T_i)] \cos 2\pi v \delta \, dv. \tag{5.8.3}$$

In this equation $i(\delta)$ is the interferometer amplitude at mirror position δ; r_v is the responsivity of the instrument; I_v is the radiance of the object under investigation, and $B_v(T_i)$ is the Planck function corresponding to the instrument temperature. The interferogram at mirror position δ is the summation over all wavenumbers of the modulated net flux at the detector. The constant term, although not transmitted, may play an undesired role when the radiation source is not truly constant. Source fluctuations may occur due to telescope vibration and pointing errors, or due to atmospheric scintillations in ground-based measurements. Fluctuations in the 'constant' term enter the passband of the amplified signal and affect the interferogram adversely. Moreover, the multiplicative nature of such modulation causes nonlinear effects and the formation of side bands to strong spectral features. Configurations with double output cancel the constant term. They prove to be superior to single output configurations, particularly for ground-based astronomical observations. In Eq. (5.8.3) the responsivity, r_v, lumps together all instrumental parameters such as modulation depth, detector response, optical efficiency of components, and others. The amplitude, C_v, is recovered by a Fourier transformation of the interferogram; C_v and $i(\delta)$ are Fourier pairs:

$$C_v = r_v(I_v - B_v) = \int_{-\infty}^{+\infty} i(\delta) A(\delta) \cos 2\pi v \delta \, d\delta. \tag{5.8.4}$$

This equation holds strictly only for the apodization function $A(\delta) = 1$. However, the interferogram can only be recorded over a finite interval, from $-\delta_{\max}$ to $+\delta_{\max}$, and not beyond these limits. The function $A(\delta)$ is introduced and set equal to one for $|\delta| \leq \delta_{\max}$ and equal to zero for $|\delta| > \delta_{\max}$. The reconstructed spectrum is, therefore, not the true spectrum but the convolution of the true spectrum with the Fourier transform of the rectangular weighting function $A(\delta)$. The Fourier transform of the

230   *Instruments to measure the radiation field*

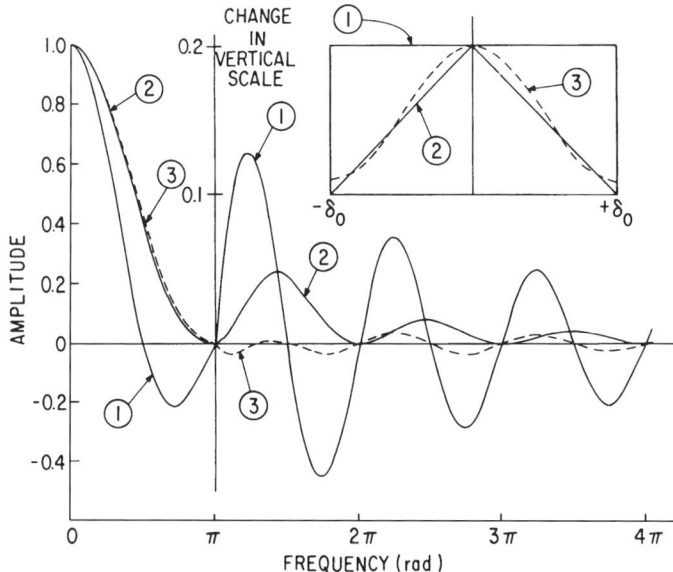

Fig. 5.8.5 Fourier transforms of several apodization functions shown in the insert. (1) refers to the box-car function, (2) to triangular apodization, and (3) to the Hamming function. (Hanel, 1983)

product of two functions is the convolution of the transforms of both functions. The Fourier transform of the rectangular, box-car function is the sinc function

$$\text{sinc}\,(2\pi\nu\delta) = \frac{\sin 2\pi\nu\delta}{2\pi\nu\delta}, \tag{5.8.5}$$

shown in Fig. 5.8.5. Other weighting functions may also be applied, a process called apodization (Jacquinot & Roizer-Dossier, 1964), but all apodization functions must be zero above $\delta_{\max}$. The sinc function applies to a weighting of unity and yields the highest spectral resolution for a given $\delta_{\max}$, but it has relatively strong side lobes. A weighting function

$$A(\delta) = 0.54 + 0.46\cos\pi\delta/\delta_{\max}, \tag{5.8.6}$$

called the Hamming function, yields somewhat lower resolution, but much reduced side lobes [Fig. 5.8.5; see also Blackman & Tukey (1958)].

To extract I_ν from Eq. (5.8.4), one must determine for each resolved wavenumber interval the unknown quantities r_ν and B_ν. This is conveniently achieved by measuring two known sources in addition to the object of interest. Deep space may be one convenient reference and a warm blackbody may serve as the other. In case of an isothermal instrument temperature, $B_\nu(T_i)$ is the Planck function corresponding

5.8 Fourier transform spectrometers

to that temperature. The process of deriving I_ν from the measured amplitude is discussed further in Section 5.13.

The spectral resolution, $\Delta \nu$, of an interferometer may be derived by considering the number of fringes in the interference pattern between $-\delta_{max}$ and $+\delta_{max}$ of two adjacent wavenumbers, ν and $\nu + \Delta \nu$ (e.g., Loewenstein, 1971). Two adjacent wavenumbers are considered resolved when their respective numbers of wavelengths, n and $n + 1$, differ by one count,

$$2\delta_{max} = \frac{n}{\nu} = \frac{n+1}{\nu + \Delta \nu}, \qquad (5.8.7)$$

which yields

$$\Delta \nu = \frac{1}{2\delta_{max}}. \qquad (5.8.8)$$

This definition of spectral resolution is equivalent to the Rayleigh criterion. As a rule of thumb, apodization doubles $\Delta \nu$, although the precise value depends on the particular apodization function used (Bell, 1972).

The maximum permissible solid angle in an interferometer can be determined from the size of the central fringe in the interference pattern at the detector. The detector size and, therefore, the solid angle in the interferometer must be kept smaller than the first null in the interference pattern generated by the highest wavenumber of interest at the maximum path difference. A larger detector would receive out of phase components from the next fringe, which would tend to reduce the signal. In other words, only the central part, the Airy disk, is spatially coherent.

The phase difference between the central ray and one inclined at an angle α is

$$\Delta \phi = 2\pi \nu \delta (1 - \cos \alpha) \sim \pi \nu \delta \alpha^2. \qquad (5.8.9)$$

For the maximum permissible α, that is, for the first zero in the pattern, $\Delta \phi$ equals π, and

$$\pi \alpha_{max}^2 = \Omega = \frac{\pi}{\nu \delta_{max}}, \qquad (5.8.10)$$

or with the help of Eq. (5.8.8)

$$\Omega = \frac{2\pi \Delta \nu}{\nu} = \frac{2\pi}{R}, \qquad (5.8.11)$$

where R is the resolving power. The maximum solid angle is inversely proportional to the resolving power, a fundamental relationship for the design of interferometers. Even for detectors slightly smaller than the first null of the interference fringes, the interferogram amplitudes are attenuated at mirror displacements approaching $\delta_{\max}$. Apodization has the same effect and one speaks of natural apodization due to the finite value of Ω.

The signal at the detector is proportional to

$$S_\nu \sim \eta_1 \eta_2 A\Omega \Delta\nu \sqrt{\tau} I_\nu, \qquad (5.8.12)$$

and the noise is

$$N_\nu \sim \mathrm{NEP}_\nu \left[\mathrm{W\,Hz}^{-\frac{1}{2}} \right]. \qquad (5.8.13)$$

We call the signal for which the signal-to-noise ratio is unity the Noise-Equivalent-Spectral-Radiance,

$$\mathrm{NESR}_\nu = \frac{\mathrm{NEP}_\nu}{\eta_1 \eta_2 A\Omega \Delta\nu \sqrt{\tau}}, \qquad (5.8.14)$$

where NEP_ν is the Noise-Equivalent-Power of the detector (Section 5.10), η_1 the system efficiency, η_2 the optical efficiency, $A\Omega$ the étendue, $\Delta\nu$ the width of the resolved spectral element, and τ the duration of the interferogram. The system efficiency takes the configuration of the interferometer into account and factors necessary to convert all values to rms terms.

It is possible to combine the spectrometric performance of the Michelson interferometer with simultaneous imaging capabilities. To achieve this, the object of interest must be imaged at the focal plane of the interferometer and, within the permissible solid angle, the single detector must be replaced by a two-dimensional array of detectors. Obviously, each element in the array receives only a fraction of the energy falling on the total array, or that falling on the single detector in the strictly spectrometric mode of operation. In the imaging mode each element records an interferogram, which, after transformation, produces a spectrum. An image may be formed by assembling one or a group of spectral intervals from each interferogram into a two-dimensional pattern. Such a process would place large demands on the data transmission or on the on-board processing facility, if high spectral resolution and a large number of pixels are to be combined; a pixel is the smallest element, a dot, in a two-dimensional image. But with modest numbers such an approach is feasible.

An interferometer, operating between 200 and 1600 cm^{-1}, with a spectral resolution of $\sim$5 or 10 cm^{-1} and an array of 6 detectors in a 2 × 3 pattern was flown on the

5.8 Fourier transform spectrometers

Mars Global Surveyor in 1997. The Thermal Emission Spectrometer (TES) contains two broad-band radiometers and a Michelson interferometer with flat mirrors. The interferometer has a cesium iodide beamsplitter and operates in the spectral range from 200 to 1600 cm^{-1} with a selectable, nominal resolution of 5 or 10 cm^{-1}. Six uncooled, deuterated triglycine sulfide (DTGC) pyroelectric elements in a 2 × 3 array with a NEP of 3×10^{-11} W Hz$^{-\frac{1}{2}}$ serve as detectors for the interferometer. The field of view of the instrument is 16.6 mrad in the down-track and 24.9 mrad in the cross-track direction. The radiometers have similar detectors but slightly better NEP of 2×10^{-11} W Hz$^{-\frac{1}{2}}$. The fields of view of the radiometer detectors are coincident with those of the interferometer. An entrance mirror at 45° permits pointing of all fields of view towards the forward and aft horizons, at any downward-looking angle including nadir, at an internal blackbody, and at space for calibration purposes. The signal of the interferometer detectors can be Fourier transformed, calibrated, and formatted in the instrument. The scientific objectives of the TES investigation are to measure the composition and distribution of minerals on the surface, to determine the structure of the atmosphere and aspects of its circulation, and to determine the temporal and spatial distribution, abundance, sources and sinks of volatile material, and of the dust over a seasonal cycle (Christensen *et al.*, 1992; Christensen, 1998). Scientific results are discussed in Sections 6.2 and 9.2.

The simple configuration of a Michelson interferometer with a flat mirror (Fig. 5.8.2) has a high optical efficiency, but, at the same time, a rather high sensitivity to errors in optical alignment. The Michelson mirror must be translated while maintaining the optical wavefront flat and aligned to a small fraction of a wavelength over the beam cross section. To maintain the alignment over the mirror displacement, rather tight design tolerances for the drive motor are required. The IRIS motor uses a set of 12 matched, symmetrically oriented springs (Hanel *et al.*, 1970). For rocket flight and a larger displacement, a motor with ball bearings, designed by J. Pritchard of Idealab, has been used by Stair *et al.* (1983) and by Murcray *et al.* (1984). The problem becomes more severe for high spectral resolution, that is, for large mirror displacement, and for operation at short wavelengths. One way to avoid the difficult motor design is to servo control the optical alignment. Henry Buijs of Bomem, Inc. (Quebec, Canada) has designed an interferometer with flat mirrors, which uses the split signals from a single-mode laser to actively control the alignment over the full travel of up to several tens of centimeters (Buijs, 1979).

Another way to free the mirror carriage from the high precision requirement is to use retroreflectors instead of flat mirrors. Then a small tilt of the reflector does not matter. The original instrument of Fellgett used corner cubes. High precision requirements are shifted from the motor design to that of the reflector, where they are easier to handle, but are still not trivial. The Cassini spacecraft (to arrive at Saturn in July 2004) and the Mars Express spacecraft (to arrive at Mars in December

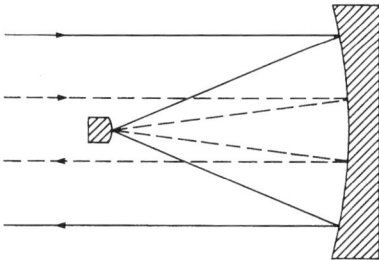

Fig. 5.8.6 Cat's eye reflector. Each return ray is parallel to the incoming ray and independent of a small tilt of the reflector.

2003), carry Michelson interferometers. Both instruments, to be discussed further below, use corner cubes. An interesting configuration with two retroreflectors on the same rotating structure and moving simultaneously is described by Burkert et al. (1983) and by Jaacks & Rippel (1989). Both retroreflectors tilt slightly as they move towards and away from the beamsplitter, but this does not matter for corner cubes. If dynamically balanced this configuration is rather insensitive to external vibration. A third method uses cat's eyes (Fig. 5.8.6). The precision is in the quality of the optical elements and in maintaining the distance between the primary and secondary mirrors (Beer & Marjaniemi, 1966).

The lateral displacement of the return beam, in the corner cube as well as in the cat's eye, permits separation of both inputs and both outputs. The signals at the output terminals are 180° out of phase, while each detector responds to the difference between inputs. The same effect can be achieved with flat mirrors and division of the solid angle inside the interferometer (Hanel et al., 1969). The differential nature of the interferometer is advantageous for many applications. In ground-based astronomical observations it is common practice to place the object of interest, a planet for example, in one input beam and the sky at the same elevation right next to the object into the other. This procedure reduces the dynamic range requirement and, more importantly, tends to cancel fluctuations in the thermal emission of the Earth's atmosphere. The differential output removes the constant component in the interferogram, which, if not perfectly constant, can be a disturbance (Martin, 1982).

An alternate method to remove the constant component in an interferogram is to modulate the path difference. Following a suggestion by Mertz (1965), Connes et al. (1967a, b) have demonstrated the merit of this technique. It can be implemented by rapidly translating the small mirror of a cat's eye. The amplitude of the oscillation must be large enough to produce a reasonable signal, but it must be small in comparison with the shortest wavelength. In effect, the modulation of the pathlength corresponds to a differentiation of the interferogram, which is now antisymmetric. Instead of a cosine, a sine transformation is required. With this method

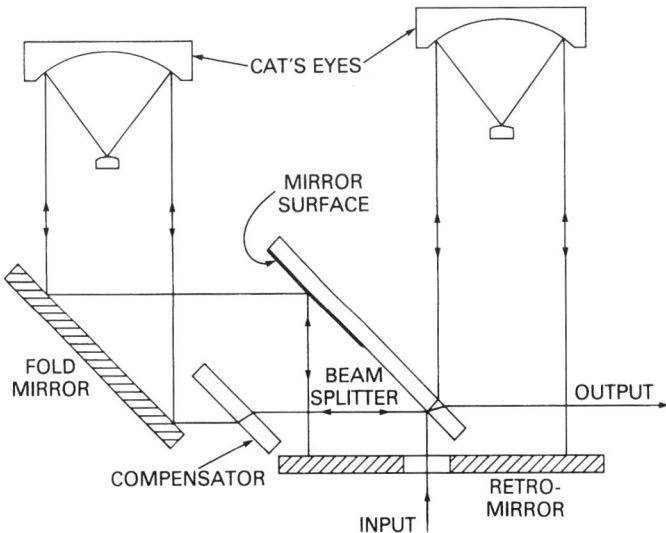

Fig. 5.8.7 Interferometer with double passing and cat's eyes is compensated for small errors in tilt of optical components as well as lateral shift of the carriage motion. The ATMOS interferometer has a similar design.

Connes *et al.* (1967b) have obtained excellent planetary spectra, apparently reducing sky noise to a large degree. The method becomes inefficient when a wide spectral range has to be covered (Chamberlain, 1971).

The cat's eye was used by Fizeau, later by Michelson (1927), and more recently by Connes & Connes (1966), Pinard (1969), and many others. The cat's eye produces systems insensitive to small degrees of reflector tilt. To also gain immunity to small amounts of lateral displacement, double passing can be used as shown in Fig. 5.8.7. The Atmospheric Trace Molecule Spectroscopy (ATMOS) interferometer used a variation of this configuration to completely compensate for all errors in carriage motion and tilts of optical components (Farmer & Raper, 1986). As can be seen, the price for this degree of freedom is a large number of reflections in each interferometer arm. Apparently, this is not critical while observing a strong radiation source. ATMOS measures stratospheric absorption in an occultation mode while observing the rising or setting Sun. The field of view must be kept much smaller than the solar disk, and an interferogram must be obtained within a few seconds, resulting in a high data burst twice each orbit at sunrise and sunset. A sample spectrum obtained by ATMOS is shown in Fig. 5.8.8. The objective of the ATMOS investigation was to obtain the stratospheric concentration and vertical distribution of major and minor constituents. Many of the gases of interest have strong infrared bands. Photochemical production and destruction of natural and man-made compounds, even at a dilution of a few parts in 10^9, may affect the delicate stratospheric equilibrium. For

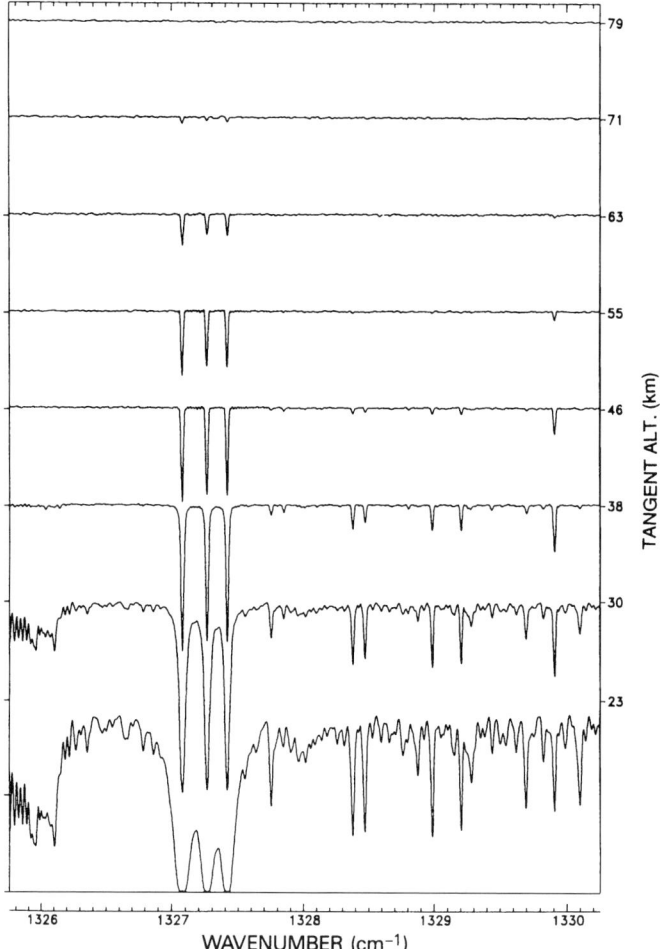

Fig. 5.8.8 Absorption spectrum recorded from Spacelab 3 by the ATMOS interferometer while observing the Sun near the Earth's limb. The tangent altitude in km above the Earth's surface is indicated on the right side. The three prominent lines near 1327 cm^{-1} are due to methane (Farmer & Norton, 1989); see also Fig. 5.8.10.

example, of great concern is the effect of man-made constituents on the destruction of the ozone layer. Tropospheric chemistry is equally important; all atmospheric layers and the Earth's surface are interacting in many ways. Obviously, some of the processes are of great interest to the well-being of man. Therefore, abundance measurements of minor constituents will become an increasingly important tool of atmospheric research.

Absorption measurements from spacecraft take advantage of the long path provided by limb observations of the Sun and are a powerful observation technique. The Sun is an extremely strong source, which permits the recording of high resolution spectra within a short time. However, this method has limitations as well. All observations must necessarily occur at local morning or evening. Only two sets of measurements are obtained per spacecraft orbit, and, depending on that orbit, the measurements tend to concentrate at certain latitudes. Moreover, solar observations on the Earth's limb yield average atmospheric conditions along the direction of the solar rays – they smear in the direction of local time. Many of the compounds that participate in photochemical reactions show large diurnal fluctuations and rapid changes in concentration just at sunrise and sunset. A study of the concentration with respect to local time requires spectroscopic observations of thermal emission; they can be carried out normal to the Sun direction, that is, at a distinct local time. The energy available for emission measurements is many orders of magnitude less than that available for solar absorption studies. However, the task can be accomplished with modern Fourier transform spectrometers and cryogenically cooled detectors. Murcray *et al.* (1984) have demonstrated the feasibility of such measurements from balloon altitude. Kunde *et al.* (1987) have flown a Michelson interferometer cooled to the temperature of liquid nitrogen for the same purpose. A stratospheric emission spectrum obtained by the latter instrument from a high altitude balloon flight over Texas is shown in Figs. 5.8.9 and 5.8.10.

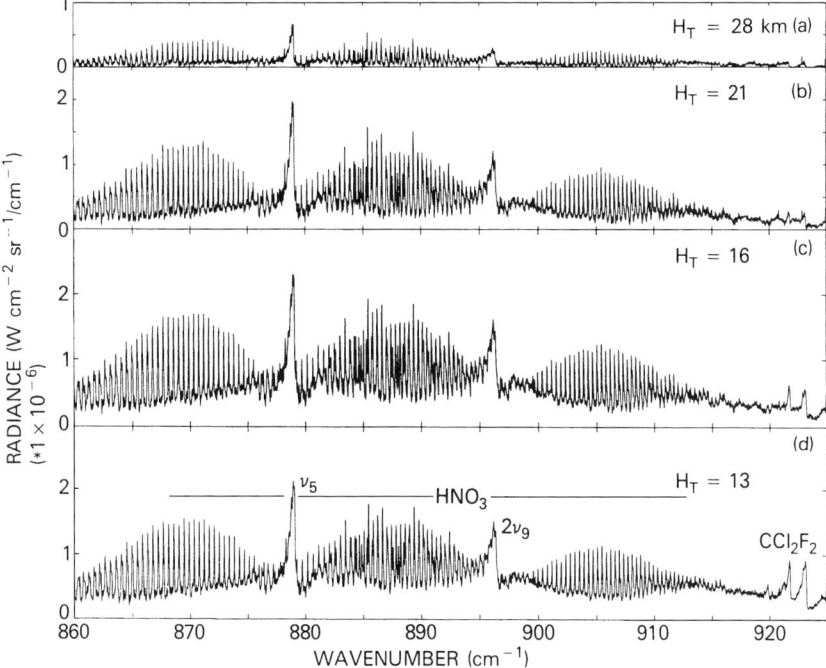

Fig. 5.8.9 Thermal emission spectrum between 860 and 920 cm^{-1} of the Earth's stratosphere recorded by a cryogenically cooled Michelson interferometer on a balloon. H_T gives the height of the center of the field of view above the surface at the horizon. Several stratospheric gases are identified (Kunde *et al.*, 1987).

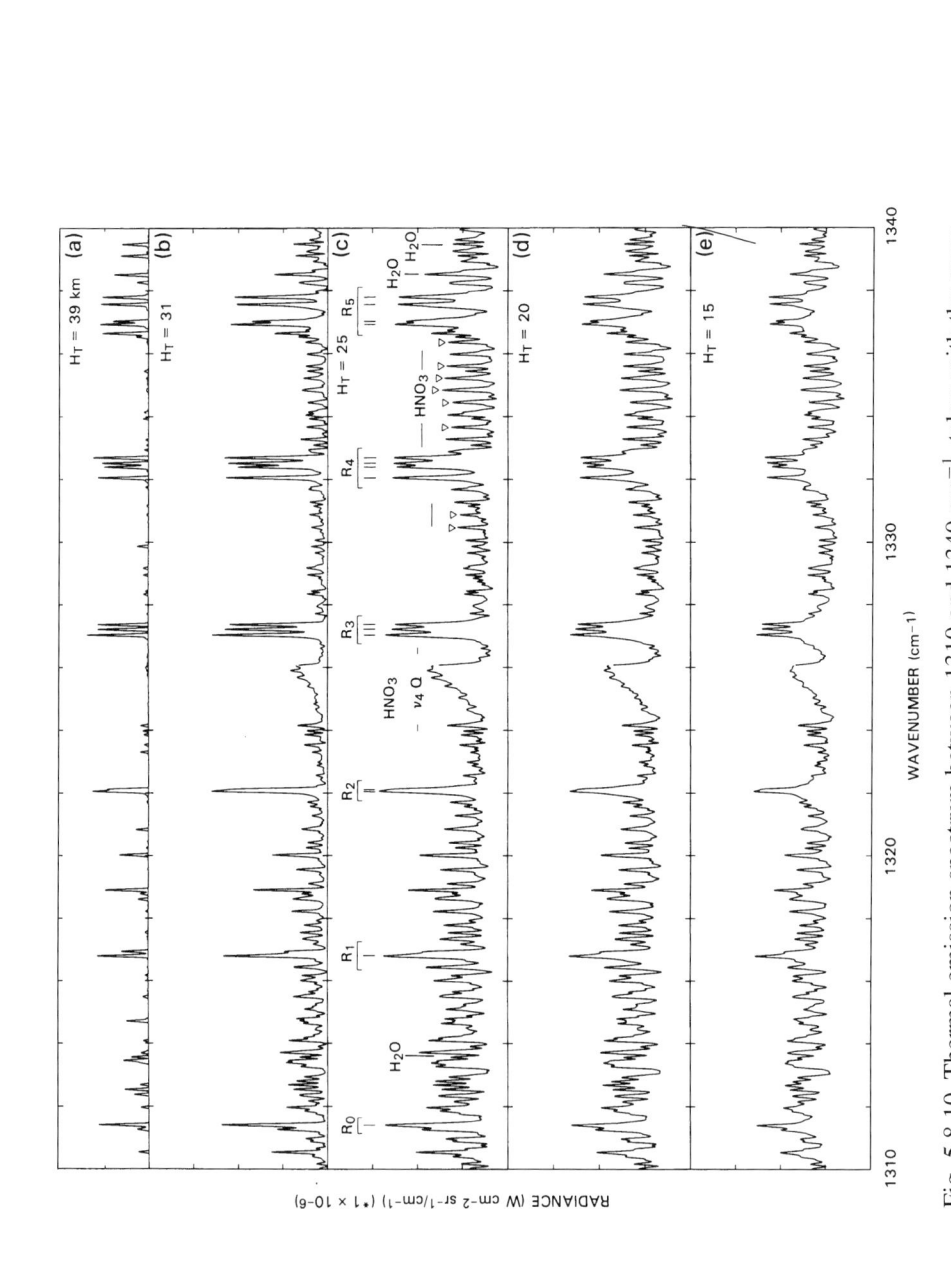

Fig. 5.8.10 Thermal emission spectrum between 1310 and 1340 cm^{-1}, taken with the same instrument as the spectrum of Fig. 5.8.9 (Kunde *et al.*, 1987).

5.8 Fourier transform spectrometers

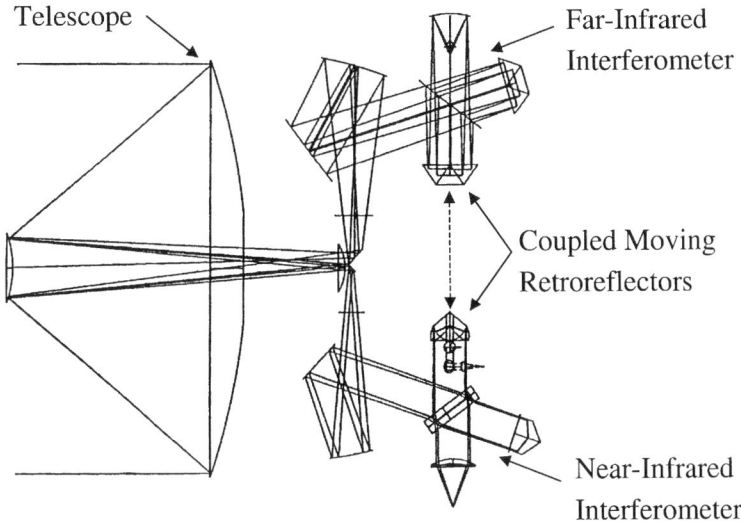

Fig. 5.8.11 Optical layout of the Composite Infrared Spectrometer, flown on the Cassini mission to Saturn (Kunde *et al.*, 1996).

At this point, it is also appropriate to mention the Composite Infrared Spectrometer (CIRS) on the Cassini spacecraft, that flew past Jupiter in December 2000 and will arrive at Saturn in July 2004, and the Planetary Fourier Spectrometer (PFS) on the Mars Express spacecraft, to arrive at Mars in December 2003. Both spectrometers are dual channel instruments consisting of short- and long-wave interferometers. First, we discuss CIRS.

The optical design of CIRS is shown in Fig. 5.8.11. The short-wave channel of CIRS operates with a potassium bromide beamsplitter and corner cubes. The spectrum measured by this instrument ranges from about 600 to 1400 cm^{-1}. This interferometer has two 10-element, linear mercury cadmium telluride (HgCdTe) arrays, covering adjacent spatial ranges. Each element in the arrays has a square field of view of 0.3 mrad. The long wavelengths are covered by a polarization interferometer with roof reflectors, see also Subsection 5.8.c. The spectrum of this channel ranges from 10 to 600 cm^{-1}. The polarization interferometer uses two thermopile detectors in a common 3.9 mrad field of view. The two interferometers share a linear motor and a telescope with a diameter of 50 cm. The apodized spectral resolution can be selected within the limits of 0.5 and 20 cm^{-1}. CIRS operates at 170 K. The HgCdTe arrays are cooled by radiation to 80 K (Maymon, *et al.*, 1993; Kunde *et al.*, 1996).

The PFS instrument consists of two double-pendulum interferometers with corner reflectors. The near infrared interferometer operates from 2000 to 8000 cm^{-1} with a calcium fluoride (CaF$_2$) beamsplitter and a 2° field of view. The short-wave detector, a lead selenide photoconductor, has a NEP of 1×10^{-12} W Hz$^{-\frac{1}{2}}$ and operates at a temperature of 220 K. The far infrared instrument covers the range from 230 to 2000 cm^{-1} using a cesium iodide beamsplitter and a 4° field. The long-wavelength detector, a lithium tantalum oxide (LiTaO$_3$) pyroelectric element, achieves a NEP of 4×10^{-10} W Hz$^{-\frac{1}{2}}$ at an operating temperature of 290 K. Both devices have an apodized resolution of 2 cm^{-1} and share a laser diode source in the reference interferometer. The calibration source for the short-wave instrument is a Lambertian screen, which must be illuminated by the Sun during calibration. The long-wave instrument uses an internal blackbody and deep space (Formisano, 1999).

b. Post-dispersion

As mentioned in the previous section, the multiplex advantage of the Michelson interferometer turns into a disadvantage when background noise exceeds the signal-independent noise of the detector. Part of the background noise may originate from the warm instrument and the telescope. That part can, at least in principle, be reduced by cooling. Another part of the background noise originates from emission of the object under study. It is obviously not subject to modification. To reduce background noise one could reduce $A\Omega$, but this is an undesirable choice since it reduces the signal simultaneously. The only reasonable approach to reduce the NEP due to background noise is to decrease the bandpass of the interferometer. The reduction in bandwidth can be accomplished by inserting a narrow-band optical filter at cryogenic temperatures directly in front of the detector. This procedure is entirely appropriate as long as the scientific objective can be satisfied by recording the spectrum just within a narrow band, say 2 cm^{-1} wide. With a spectral resolution of the interferometer of 0.02 cm^{-1}, for example, about 100 spectral elements are available for interpretation. In that case the multiplex advantage would be reduced to $\sim 10^2$, but as long as one is only interested in detecting one or two particular lines of a suspected gas within the 2 cm^{-1} interval, this multiplex advantage is adequate. For this particular observation one may consider the use of a Fabry–Perot instead of a Michelson interferometer. The Fabry–Perot has the same high étendue as the Michelson, and the multiplex advantage of the Michelson may not be that significant in view of overall optical efficiency, complexity, and other considerations.

For many applications a 2 cm^{-1} passband may not be adequate. One may wish to see a larger number of lines of that particular gas, or one may want to search for other, possibly unknown features at the same time. One solution is to use a cryogenically cooled, low resolution grating spectrometer with a linear array of detectors at the exit of a high resolution Michelson interferometer (Fig. 5.8.12). Each detector in the array receives only radiation within a 2 cm^{-1} wide interval given by the spectral resolution of the grating. With 100 detectors in the array the full

5.8 Fourier transform spectrometers

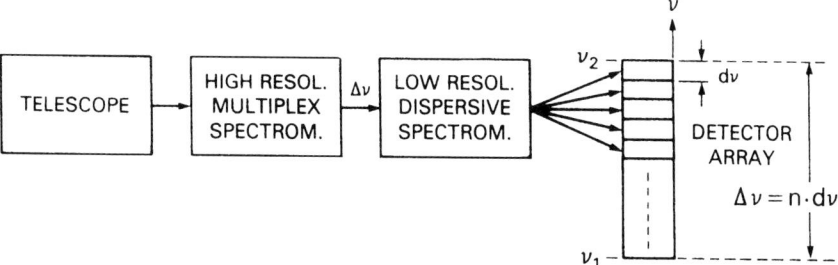

Fig. 5.8.12 The post-dispersion principle uses a low resolution grating and a detector array at the exit of a high resolution Michelson interferometer (Wiedemann *et al.*, 1989).

passband of 200 cm^{-1} can be covered. Each detector generates an interferogram, which is transformed, and the 2 cm^{-1} wide spectral segments are assembled to form the full spectrum of 200 cm^{-1}. The multiplex factor is $\sim 10^4$. The advantage of post-dispersion is demonstrated in the following example.

Consider an Earth orbiting telescope of 2.7 m diameter, equipped with a Michelson interferometer operating between 300 and 500 cm^{-1} with a spectral resolution of 0.02 cm^{-1}. This is an entirely fictitious case and does not refer to an existing or planned project. Of course, as a useful observation, the hydrogen dimer feature at 356 cm^{-1} could be examined on Jupiter (Frommhold *et al.*, 1984) and, at the same time, one could search for hydrocarbon emissions from the Jovian stratosphere. Consider also observations with a 40 arcsecond field of view. The $A\Omega$ of the telescope is 1.69×10^{-3} cm^2 sr. With an interferometer beam of 4 cm diameter the system étendue is still telescope limited. It is assumed that the interferometer is equipped with a cryogenically cooled bolometer with a NEP of 1×10^{-14} W Hz$^{-\frac{1}{2}}$. To calculate the contributions from the interferometer, telescope, and planet to the background noise at the detector, we evaluate Eq. (5.11.33) for several temperatures and a passband of 200 cm^{-1} as well as 2 cm^{-1}. Then we use Eq. (5.11.21) to find the individual contributions of the different background-noise sources to the systems noise (Table 5.8.1).

In part (a) of Table 5.8.1, the squares of the individual noise components are shown for different temperatures. The contribution of the detector without external flux (1×10^{-28}) and the Jovian contribution (1.6×10^{-28}), taken for simplicity equal to that of a blackbody of 120 K, are kept constant. The interferometer and the telescope are permitted to assume several temperatures. In the computations the interferometer transmission was taken to be 0.4; consequently, the emissivity is 0.6. The telescope transmission was assumed to be 0.85 and its emissivity 0.15. To prevent the interferometer from contributing substantially to the overall background noise it is necessary to cool it to 80 K. Acceptable values are underlined

Table 5.8.1 *Background-noise sources*

(a) 300–500 cm^{-1}; NEP$_{det}$ = 1×10^{-14} W Hz$^{-\frac{1}{2}}$

T(K)	(NEP$_{det}$)^2	(NEP$_{Instr}$)^2	(NEP$_{Tele}$)^2	(NEP$_{Jup}$)^2
280	—	2.5×10^{-26}	2.5×10^{-28}	—
200	—	9.3×10^{-27}	9.3×10^{-29}	—
120	—	5.1×10^{-28}	5.1×10^{-30}	1.6×10^{-28}
80	—	1.2×10^{-28}	1.2×10^{-30}	—
2	1×10^{-28}	—	—	—

NEP$_{sys}$ = 2.17×10^{-14} W Hz$^{-\frac{1}{2}}$

(b) 300–302 cm^{-1}; NEP$_{det}$ = 1×10^{-14} W Hz$^{-\frac{1}{2}}$

T(K)	(NEP$_{det}$)^2	(NEP$_{Instr}$)^2	(NEP$_{Tele}$)^2	(NEP$_{Jup}$)^2
280	—	1.7×10^{-28}	1.7×10^{-30}	—
200	—	7.1×10^{-29}	7.1×10^{-31}	—
120	—	1.4×10^{-29}	1.4×10^{-31}	4.4×10^{-30}
80	—	2.2×10^{-30}	2.2×10^{-32}	—
2	1×10^{-28}	—	—	—

NEP$_{sys}$ = 1.04×10^{-14} W Hz$^{-\frac{1}{2}}$

in Table 5.8.1. Similar considerations require a telescope temperature as low as 200 K. Jovian emission contributes the most to the system's NEP, which is more than 2×10^{-14}, twice as much as the detector NEP without background radiation. The system is background-noise limited. Cooling the interferometer or the telescope below 80 and 200 K, respectively, would not help much because of the strong Jovian contribution.

For a spectral width of 2 cm^{-1}, as applicable to a post-dispersion system, the situation is quite different [see Table 5.8.1(b)]. Only the case 300–302 cm^{-1} is shown. At higher wavenumbers the system's NEP is even lower than 1.04×10^{-14} W Hz$^{-\frac{1}{2}}$. The Jovian contribution is now considerably less than that of the detector, and, most remarkably, it is now possible to use a telescope of 280 K, although the interferometer would still have to be cooled.

A cross section of a post-dispersion grating, mounted in a slightly taller, but otherwise conventional detector dewar, is shown in Fig. 5.8.13. This dewar fits directly into the detector position of a ground-based, astronomical interferometer. Ground-based infrared observations with warm telescopes and residual emission from the warm Earth atmosphere benefit equally from the post-dispersion use of an array. Such a system has been successfully used by G. Wiedemann *et al.* (1989) in the 750–1250 cm^{-1} atmospheric window, although with only a small number

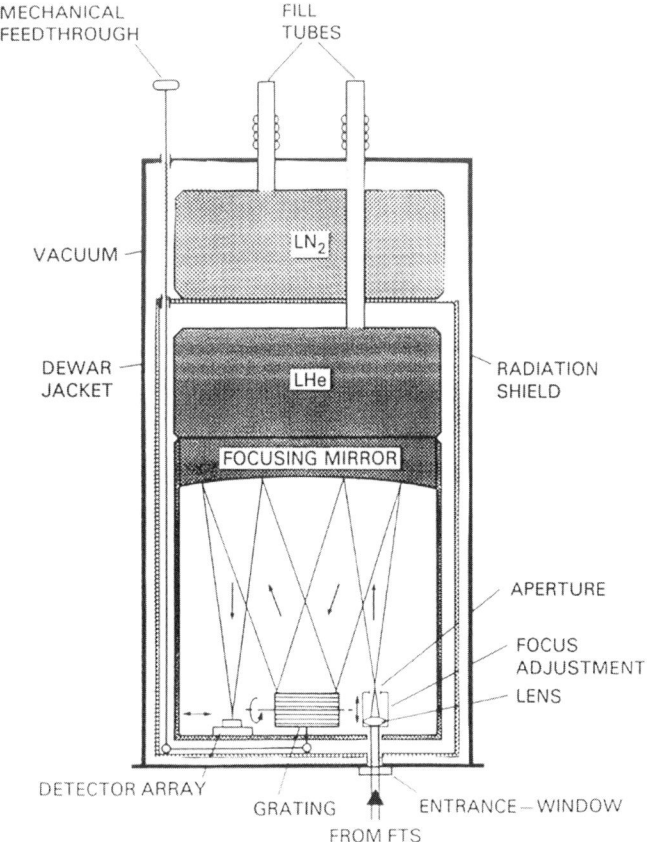

Fig. 5.8.13 Cross section through a post-dispersion detector. Radiation from a Fourier transform spectrometer enters from below (Wiedemann *et al.*, 1989).

of detectors instead of a full array. The ISO spectrometers are examples of post-dispersion arrays in conjunction with gratings and Fabry–Perot interferometers.

c. Martin–Puplett interferometer

The Martin–Puplett (1969) interferometer is a polarizing interferometer that operates over a wide spectral range in the far infrared. The success of this type of interferometer rests to a large degree on the highly efficient polarizing properties of wire grids. An array of metal wires stretched uniformly on a flat, rigid frame provides an efficient polarizer over a wide wavenumber range. For example, tungsten wires of a diameter of 10 μm work well up to 100 cm^{-1} (Ade *et al.*, 1979; Martin, 1982). The upper wavenumber limit is given by the wire dimensions and spacing. The lower useful limit of $\sim$1 cm^{-1} is given by the size of components, which must

244 *Instruments to measure the radiation field*

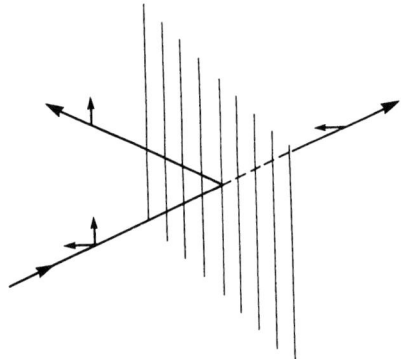

Fig. 5.8.14 The unpolarized radiation impinging on the vertical wire grid is split into two components. The component with the electric vector parallel to the wires is reflected and the orthogonally polarized component is transmitted.

be at least several wavelengths across, or considerations based on geometric optics do not apply. The construction of wire grids for the Far Infrared Absolute Spectrometer (FIRAS) of the Cosmic Background Explorer (COBE) has been discussed by Eichhorn & Magner (1986).

A wire grid is a good reflector for incident radiation polarized in the same direction as the orientation of the wires, that is, with the electric field vector parallel to the conductors. The induced current flows as freely along the wires as in a solid metal surface. At the same time, radiation polarized orthogonally, that is, with the magnetic field vector parallel to the wires, passes the grid with little attenuation (Fig. 5.8.14). Wire polarizers work uniformly well over a wide range of wavenumbers; in the above example the range is from 1 to 100 cm^{-1}. If one accepts less than near perfect performance, the same wire grid may be used up to about 150 cm^{-1}. Smaller diameter wires have been placed at closer centers to construct polarizers for higher wavenumbers, but it is difficult to obtain good performance with free standing wires much above 200 cm^{-1}. Apparently, it is not possible to achieve the required uniformity and flatness. Evaporated metal grids have been produced by photolithography and sputteretching on 4 μm thick mylar substrates (Mok *et al.*, 1979; Challener *et al.*, 1980). The polarizers in CIRS are photolithographically deposited wires spaced by 2 μm on a 1.5 μm mylar substrate. These polarizers work up to 700 cm^{-1}, but show absorption as well as reflection effects due to the substrate. A grid with the wires placed at 45° to the electric (and magnetic) vector provides an excellent beamsplitter. Nearly half of the incident polarized radiation is reflected, with the other half transmitted.

Another important ingredient of the Martin–Puplett interferometer is a reflector that rotates the plane of polarization of the returning beam by 90° with respect to the incident beam. Two metal mirrors, set at 90° with respect to each other, can rotate

5.8 Fourier transform spectrometers

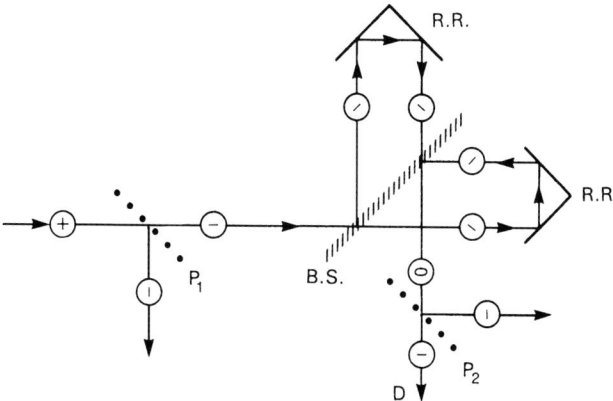

Fig. 5.8.15 Simple Martin–Puplett polarization interferometer. P_1 and P_2 are polarizers with the wires normal to the plane of the paper. B.S. is a beamsplitter with the wires rotated at 45° with respect to that plane. R.R. are roof reflectors that rotate the plane of polarization by 90°. Symbols in the circles indicate the plane of polarization.

the doubly reflected beam by that angle (Martin & Puplett, 1969). This so-called roof-top reflector is insensitive to misalignment in one axis, that is, against tilt around the ridge of the roof, but not in other directions. It is a one-dimensional corner reflector. The important property of the roof reflector is not the sensitivity or insensitivity to misalignment, but the horizontal displacement of the beam and the 90° rotation of the plane of polarization. In the far infrared, alignment is generally not as important an issue as it is in the near infrared.

With wire grids and roof-top reflectors, one may construct a variety of polarizing interferometers (Martin, 1982). Unpolarized collimated radiation may strike the polarizer, P_1, set with the wires in the vertical direction (Fig. 5.8.15). The components of the radiation with the electric vector horizontal pass the polarizer with little attenuation while the components of the radiation with the electric vector normal to the paper are reflected downward and discarded in this example. Another wire grid, this time with the wires set at 45°, acts as a beam divider. The horizontally polarized radiation is split into a reflected component polarized +45°, and a transmitted component polarized −45°. Both roof reflectors, R.R., rotate the planes of polarization by 90°. The returning radiation recombines at the beam divider set at 45°. The combined beam is now elliptically polarized. As the path difference changes the polarization switches between horizontal and vertical. The polarizer P_2 modulates the intensity at the detector, D. Another detector may be used to register the vertically polarized component reflected at P_2. By translating one of the roof reflectors the path difference between both arms is varied and the detector records an interferogram as with a conventional Michelson interferometer.

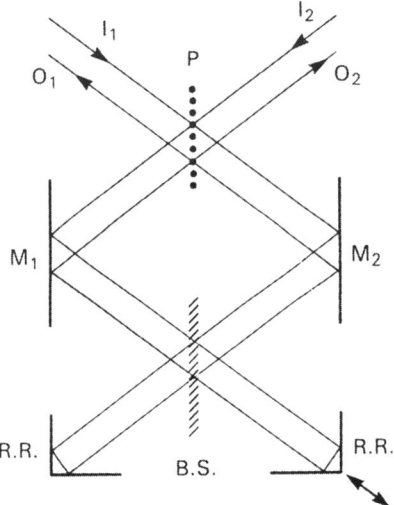

Fig. 5.8.16 Completely symmetrical Martin–Puplett interferometer with two outputs and two inputs for difference measurements.

Obviously, this configuration has deficiencies. Half of the incoming radiation is lost at the first polarizer, and half of the remainder is lost at the second, yielding an overall efficiency of only 0.25. The configuration shown in Fig. 5.8.16 corrects both shortcomings. It also provides access to both inputs and outputs. Input beams one and two both strike the polarizer at 45° from opposite sides. The wires of the polarizer are oriented vertically. The vertical component of I_1 and the horizontal component of I_2 both proceed to mirror M_1 and to one side of the beamsplitter. The orthogonally polarized components of I_1 and I_2 propagate towards M_2 and the other side of the beamsplitter, set at 45°. The arriving components are divided in equal amounts. After further rotation by 90° and a horizontal displacement on the roof reflectors, both beams recombine at the beamsplitter and work their way back via the mirrors M_1 and M_2 towards the polarizer and outputs O_1 and O_2. The interferometer is completely symmetrical and input and output ports could be interchanged. Detectors placed in the output beams register 180° out of phase interferograms. The amplitudes are proportional to the differences between both inputs as in other four-terminal Michelson configurations. The interferometer shown in Fig. 5.8.16 uses both planes of polarization from both inputs yielding a system efficiency of unity. Only reflecting surfaces and wire grids are needed; transmitting optical components for beamsplitters and substrates are unnecessary. This accounts for the wide spectral range of the Martin–Puplett interferometer, limited only by geometric considerations. It also makes this interferometer capable of operation over a wide range of temperatures. Polarizing interferometers have been flown on balloons by Woody *et al.* (1975) and by Carli *et al.* (1984).

The polarizing interferometer (FIRAS) of the Cosmic Background Explorer (COBE) had the same configuration as that of Fig. 5.8.16, except that the mirrors M_1 and M_2 were oriented nearly normally with respect to the incident beams. The beam divider was located on top of the polarizer. The whole interferometer was operated at 2 K. Furthermore, M_1 and M_2 had a small curvature and imaged the input aperture onto the output ports. An extremely compact and efficient configuration was obtained (Mather & Kelsall, 1980). The COBE spacecraft was launched in 1989 and operated cryogenically for 10 months until the liquid helium reservoir was exhausted. FIRAS measured the cosmic background to be a blackbody at 2.735 ± 0.060 K (Mather & Boslough, 1996).

d. Lamellar grating interferometer

The lamellar grating interferometer divides wavefronts, in contrast to the Michelson interferometer that splits amplitudes. John Strong (1954, 1957) invented the lamellar grating instrument and constructed the first working model with the help of his student (Strong & Vanasse, 1958, 1960). A spherical lamellar grating was designed by Hansen & Strong (1971). Schematic cross sections of both versions are shown in Fig. 5.8.17. The plane version consists of a metal block with rectangular slits of the same width as the solid ridges in between the movable set of metal tongues, which fit into the slits. In the zero-path-difference position the top surfaces of the ridges and of the tongues form a single, flat mirror surface. In that position, the grating is fully reflective and all of the incoming radiation is passed on to the detector. As the movable tongues retreat into the slits (or extrude), radiation reflected on the tongues is delayed (or advanced) in comparison with that reflected on the stationary ridges. The delay is proportional to twice the depth of the grooves. The two sets of reflected wavefronts interfere at the exit aperture, and an interferogram is recorded at the detector as a function of tongue displacement, just as in the case of mirror displacement in the Michelson interferometer. A detailed analysis is given by Bell (1972).

The lowest useful wavenumber of the lamellar grating is determined by the cavity effect; $\nu_{min} \sim 3.2/a$, where a is the grating constant. At lower wavenumbers the corresponding wavelength becomes larger than 0.6 times the width of the groove, the waves have difficulties entering the cavity between the ridges, and phase errors occur. The upper limit of the spectral range is given by the point where the first order of the grating, considered as a diffraction grating, reaches the detector. Only the zero order is used in the lamellar grating interferometer. The first order interferes and reduces the efficiency of the instrument. The breakpoint occurs when $\nu_{max} = f/(a \cdot s)$; f is the focal length of the collimating mirror, a the grating constant, and s the diameter of the entrance and exit apertures. For large displacements the beam must be well collimated or shadowing effects will lead to natural apodization.

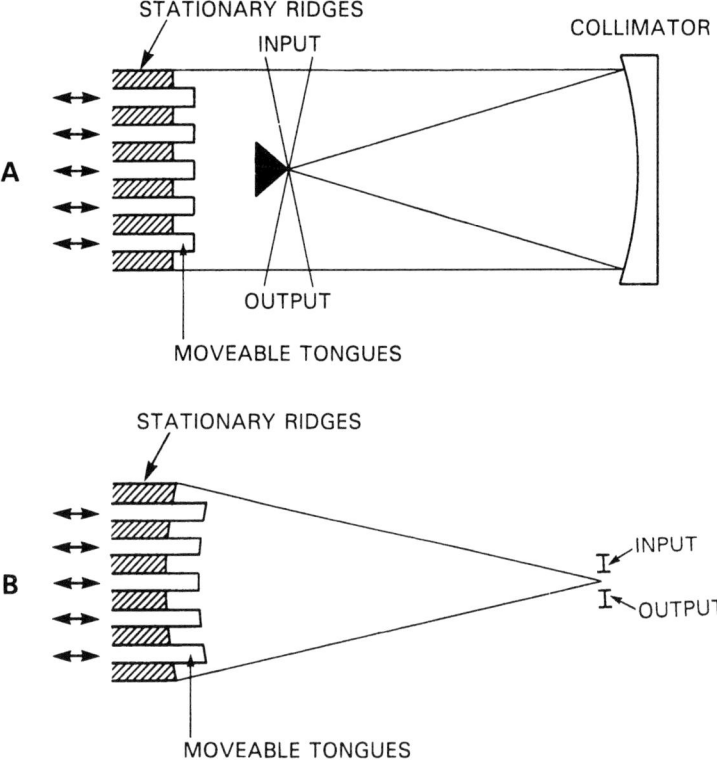

Fig. 5.8.17 Lamellar grating interferometers. The upper panel (A) shows the plane version. The lower panel (B) illustrates the spherical configuration.

In the spherical lamellar grating the surface at zero path difference has not a plane but a concave shape. This interferometer operates with a noncollimated beam and is, therefore, limited to lower spectral resolution than the plane version; otherwise the spherical configuration is the most simple form of a lamellar grating instrument. Only a source and a detector need to be added adjacent to the center of curvature to form a working far infrared interferometer.

The main advantages of the lamellar grating instruments are the high optical efficiency, the absence of a beamsplitter, and the wide spectral range. The main application of such gratings is in the far infrared. In spite of these attractive properties the lamellar grating has not been used in space, probably because of the relatively heavy parts required to form the grating and the large dimensions due to the low f-number collimating mirror. However, an instrument operating over the range from 50 to 500 cm^{-1} with a spectral resolution of 0.05 cm^{-1} has been developed by the Max-Planck-Institute for Physics and Astrophysics for balloon operation (Hofmann et al., 1977). A similar lamellar grating was constructed at the Paris Observatory (Meudon, France) but never flew in space.

5.9 Heterodyne detection

Heterodyne spectrometers achieve extremely high spectral resolution of about 10^{-3} to 10^{-4} cm^{-1} in relatively narrow spectral intervals. The technique of heterodyning in the infrared resembles that used in heterodyne radio receivers. Radiation from an infrared source is superimposed by a beamsplitter with radiation from a local oscillator, a laser. The combined signals are then mixed in a nonlinear detector. The superimposed electric field at the detector is

$$E(t) = E_S \, e^{i[\omega_S t + \phi(t)]} + E_0 \, e^{i\omega_0 t}, \quad (5.9.1)$$

where the first term on the right represents the field of the infrared source and the second term the field of the local oscillator. E_S and E_0 are the amplitudes of these fields, while ω_S and ω_0 are the corresponding frequencies. The phase, $\phi(t)$, is random due to the temporal incoherence of radiation from a point source. If the local oscillator is polarized, only the component of the source with the same polarization will be detected. The infrared detector responds to the intensity of the radiation,

$$I(t) = EE^* = E_S^2 + E_0^2 + 2E_S E_0 \cos[(\omega_S - \omega_0)t + \phi(t)], \quad (5.9.2)$$

where E^* is the complex conjugate of E.

At the output of the detector, a signal appears with the beat frequency (the intermediate frequency, or IF), $\Delta\omega = \omega_S - \omega_0$, in addition to the constant intensities $I_S = E_S^2$ and $I_0 = E_0^2$. The IF can be amplified using radio frequency (RF) techniques. The radio frequency circuits following the detector produce a time averaged signal proportional to the power in the oscillating term in Eq. (5.9.2),

$$S \sim I_S I_0. \quad (5.9.3)$$

Phase information is no longer present in the output, and the recorded signal, S, is linearly proportional to the source intensity. Since each source frequency, ω_S, is associated with an intensity, a spectrum, $I(\omega_S)$, is produced by scanning the beat frequencies (difference between signal and local oscillator frequencies), or by using several fixed channels to detect all beat frequencies simultaneously.

While the basic process is common to all heterodyne systems, differences exist between radio and infrared heterodyning. In radio receivers the local oscillator frequency is variable, permitting tuning of the receiver over a wide range, while the IF filters are set at a fixed frequency. In the infrared heterodyne spectrometer the typical local oscillator, a gas laser, operates at a fixed frequency. However, a small portion of the spectrum around each laser frequency can be reached with a radio frequency filter bank covering the intermediate frequencies. A source intensity,

$I(\omega_S)$, can be measured if it lies near one of the many laser frequencies, and the beat frequency falls within the response bandwidth of the detector. Because infrared mixers, often diodes of mercury cadmium telluride, have bandwidths of no more than a few gigahertz (~ 0.1 cm^{-1}), observations are limited to small spectral intervals. However, by using all transitions and all isotopes of a lasing gas, a large fraction of the spectrum can be covered in discreet segments. For example, the local oscillator most commonly used in infrared astronomy is the carbon dioxide (CO_2) gas laser operating in the range from 9 to 12 μm. By employing all isotopic variants of CO_2, that is, ^{13}C, ^{14}C, ^{16}O, ^{17}O, or ^{18}O, approximately 30% of this spectral range is accessible. The range of a CO_2 laser heterodyne spectrometer overlaps spectral bands of several molecules of interest and has been used to observe Doppler shifted CO_2 lines on Mars and Venus (Betz *et al.*, 1976; Mumma *et al.*, 1981; Deming *et al.*, 1982; Käufl *et al.*, 1984), and C_2H_6 (ethane) lines on Jupiter (Kostiuk *et al.*, 1983, 1987). Figure 5.9.1 shows an 11 μm Jovian ethane spectrum recorded with a heterodyne spectrometer.

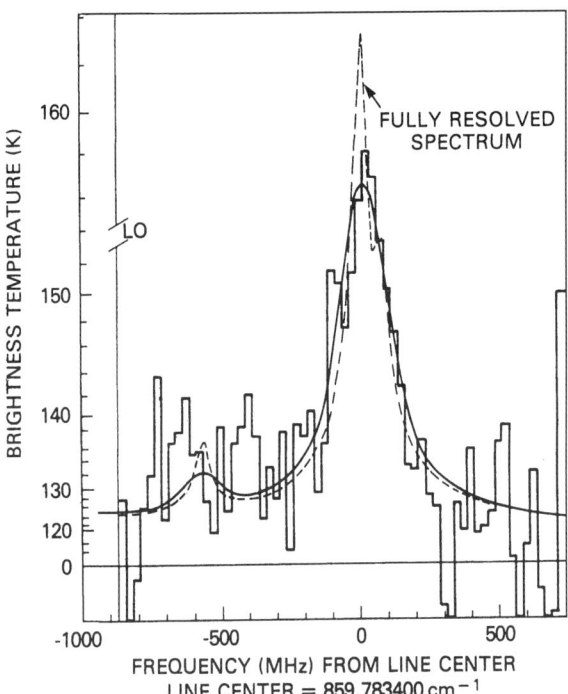

Fig. 5.9.1 A spectrum of the R(11) line of ethane on Jupiter recorded with a CO_2 laser infrared heterodyne spectrometer. 'LO' marks the frequency of the laser local oscillator. The observed 25 MHz spectrum (histogram) was detected with a 64-channel radio frequency receiver at the output of the infrared detector mixer. The dashed and solid curves are calculations of the line shape (Kostiuk *et al.*, 1987).

5.9 Heterodyne detection

The power spectrum can be recorded only as a function of the absolute value of $\Delta\omega$. Since the cosine function is symmetric, the intensities at $\omega_0 + \Delta\omega$ and $\omega_0 - \Delta\omega$ are indistinguishable from each other and both are summed in the detected spectrum at the same difference frequency, $\Delta\omega$, generating a double side band spectrum. This folding of the spectrum with respect to the laser frequency, ω_0, (LO in Fig. 5.9.1) causes an ambiguity in the position of detected spectral lines, and care must be taken to determine whether a line belongs to the upper or lower side band. Also, the spectrum will usually have a continuum, which changes only gradually and which is nearly identical in both side bands. Because both side bands are added in the heterodyne spectrum, the measured continuum level is twice the true level.

The resolution of an infrared heterodyne spectrometer is ultimately limited by the width of the laser line. For gas lasers this width is typically less than 100 kHz (3×10^{-6} cm^{-1}). In practice the resolution is limited by the width of the microwave filters used in the detection circuit, typically 3–30 MHz (10^{-4}–10^{-3} cm^{-1}). In either case, the instrument resolution is narrower than the width of molecular lines found in planetary spectra, and the technique is well suited for studying details of line shapes, especially near a line center. Since the core of a line contains information on gases at high altitudes, the high spectral resolution achieved in heterodyning allows probing to higher atmospheric levels than are accessible with lower resolution. The accuracy of the local oscillator frequency also permits highly precise measurements of line positions, making it possible to determine wind velocities, for example.

Because the detected radiation must all be of a common phase in order to interfere constructively with the local oscillator, spatial coherence is required. The field of view must be limited in diameter to the central interference fringe formed at the focal plane of the telescope. For a telescope with an unobstructed, circular, primary mirror this is the angular radius of the Airy disk, $\theta = 1.22\lambda/D$, as discussed in Section 5.3; λ is the wavelength and D the telescope diameter. Only within the Airy disk is the source radiation spatially coherent. A telescope of diameter 1 m operating near 10 μm has a diffraction limited field of view of 2.5 arc seconds. Larger telecopes give proportionally smaller fields. Although spatial resolution is improved by observing with a larger telescope, no additional radiation can be collected if the source is already larger than the field of view. For objects that are smaller than the diffraction limit, however, larger telescopes offer higher signal levels up to the point where atmospheric fluctuations and telescope aberrations begin to limit the spatial coherence of the arriving wavefronts. The étendue of heterodyne receivers is given by $A\Omega \sim \lambda^2$, where A is the aperture area and Ω the solid angle.

The sensitivity of heterodyne detection is limited by photon statistics at the infrared detector. The noise caused by the discrete, random arrival of photons is referred to as shot noise. The laser power is usually raised to a level where shot noise from the laser dominates all other noise sources. This is called the quantum detection limit. In this case the ideal spectrometer is able to detect one photon in each spectral and time resolution element of the system. The spectral resolution element is equivalent to the electronic bandwidth, β, of each channel and the time resolution element is β^{-1}. The minimum detectable power in a single polarization in bandwidth β after τ seconds is

$$P_{\min} = \frac{\hbar\omega\beta}{2\pi}(\beta\tau)^{-\frac{1}{2}} = \frac{\hbar\omega}{2\pi}(\beta/\tau)^{\frac{1}{2}}, \tag{5.9.4}$$

where $\hbar\omega$ is the photon energy. The factor $(\beta/\tau)^{\frac{1}{2}}$ indicates the improvement in sensitivity with integration time. The signal-to-noise ratio for a source delivering power P_S in bandwidth β is, therefore,

$$\frac{S}{N} = \frac{2\pi P_S}{\Delta \hbar\omega}\left(\frac{\tau}{\beta}\right)^{\frac{1}{2}} = \frac{2\pi \rho_S}{\Delta \hbar\omega}(\beta\tau)^{\frac{1}{2}}, \tag{5.9.5}$$

where ρ_S is the source power per unit bandwidth, P_S/β. In practice the degradation factor Δ must be applied to the signal-to-noise ratio to account for loss of one source polarization component, chopping of the source, mixer efficiency, and optical losses in the telescope and the spectrometer. The ratio, $2\pi\rho_S/\hbar\omega = F_S$, is just the number of photons per second per unit bandwidth. In terms of this flux the signal-to-noise ratio becomes

$$\frac{S}{N} = \frac{F_S}{\Delta}(\beta\tau)^{\frac{1}{2}}. \tag{5.9.6}$$

For $S/N = 1$ the Noise Equivalent Flux is

$$\text{NEF} = \Delta(\beta\tau)^{-\frac{1}{2}}[\text{photons s}^{-1}\text{ Hz}^{-1}]. \tag{5.9.7}$$

Typical values for Δ are about 10. Other factors that contribute to system degradation are amplifier noise, shot noise from the source and background (usually negligible), and insufficient laser power. A more complete treatment of infrared heterodyne spectroscopy is presented by Abbas *et al.* (1976), Kingston (1978), and Kostiuk & Mumma (1983).

5.10 Infrared detectors in general

Frequencies associated with infrared radiation, that is, from approximately 3×10^{11} to 3×10^{14} Hz, cannot be detected directly by radio frequency techniques. One exception exists in the heterodyne method discussed in Section 5.9, but, in general, infrared frequencies are too high for a direct application of microwave technology. On the other hand, the frequencies and, therefore, the energies of infrared photons are too low to liberate electrons from surfaces by the photoelectric effect. Consequently, standard photomultipliers and all devices based on photoelectric phenomena are also unsuitable as detectors above about 1 μm. For all practical purposes only two classes of phenomena provide infrared detection mechanisms.

One class depends on the temperature rise in the detector element caused by the absorption of infrared energy. The resulting temperature difference between detector element and the surrounding heat sink is then registered by electronic means. Detectors based on this principle are called thermal detectors; they are more or less sophisticated thermometers. William Herschel (1800) discovered the infrared by observing with a blackened mercury-glass-type thermometer the heating effect of different colors apparent in a dispersed solar spectrum. He found that the heating was not limited to the visible colors, but continued strongly below the red limit; hence the term 'infrared'.

The other class of detectors is based on the effect infrared photons exert directly on electrons in semiconducting materials; such detectors are called photon or quantum detectors, a somewhat unfortunate name because thermal detectors absorb photons or quanta as well. The energy of an absorbed infrared photon may not be high enough to cause emission of an electron by the photoelectric effect, but it is sometimes sufficient to lift an electron from a valence band into a conduction band, thereby altering the macroscopic properties of the material. The change in the electrical resistance (in photoconductors) or in the electrical potential (in photovoltaic elements) may then be sensed electrically.

One characteristic parameter of a detector is the responsivity, expressed in volt watt^{-1} or in ampere watt^{-1}, depending on open or short-circuit operation. To avoid repetition of equations for both cases we limit the discussion to open circuit terminology, but we keep operation into a low impedance amplifier in mind. The responsivity of a detector is defined as the change in voltage, ΔV, measured across the output leads, while the detector is exposed to a change in radiative power, ΔW,

$$r = \frac{\Delta V}{\Delta W} [\text{V W}^{-1}]. \tag{5.10.1}$$

The responsivity may be a function of wavenumber (or wavelength); then r_ν (or r_λ)

is called the spectral responsivity, applicable to a voltage change due to a power increase within the spectral range ν and $\nu + \Delta\nu$ (or λ and $\lambda + \Delta\lambda$).

Besides the responsivity, the noise properties of a detector are of fundamental importance. The noise is measured in volts (rms) within a specified electrical bandwidth. Since noise in one frequency interval is statistically independent of that in others, the noise power increases linearly with bandwidth and the noise voltage with the square root of the electrical passband. Therefore, the noise characteristic of a detector is expressed in units of $V\,Hz^{-\frac{1}{2}}$. More instructive than the noise voltage *per se* is the noise normalized to the responsivity. Neither a high responsivity detector with excessive noise nor a low noise element that lacks responsivity is of interest. The noise normalized to the responsivity [$V\,Hz^{-\frac{1}{2}}/V\,W^{-1}$] is expressed in $W\,Hz^{-\frac{1}{2}}$, and is called the Noise-Equivalent-Power (NEP) per root hertz. In the literature the term NEP (which has units of power, e.g., W) is often applied to the NEP per root hertz (which has units of $W\,s^{\frac{1}{2}}$). This inconsistency is deeply embedded in the literature and we also use the term NEP for both, but we state units where needed to avoid confusion. As the responsivity, the NEP may be a function of wavenumber; the term spectral NEP, NEP_ν or NEP_λ is then appropriate. The NEP [watt] can also be understood as the signal power for a signal-to-noise ratio of unity. The NEP [watt] is generally a small number, the smaller the value the better the detector. The inverse of the NEP [watt] is called the detectivity, D (Jones, 1952),

$$D[W^{-1}] = 1/NEP. \qquad (5.10.2)$$

For certain types of detectors, D is inversely proportional to the root of the detector area and the term D^* is more characteristic of this class of detectors; it allows comparison of detector quality independently of size (Jones, 1959). Expressing it in units of root hertz,

$$D^* = (A\Delta f)^{\frac{1}{2}} D[W^{-1}\,Hz^{\frac{1}{2}}\,cm]. \qquad (5.10.3)$$

A is the detector area and Δf is the electrical bandwidth. In some cases, when the detector is limited by background noise, D^* is proportional to the square root of the solid angle of illumination; then

$$D^{**} = D^* \sin\theta [W^{-1}\,Hz^{\frac{1}{2}}\,cm\,sr^{\frac{1}{2}}] \qquad (5.10.4)$$

has been suggested as a figure of merit, where θ is the half cone angle of the illuminating beam. Unfortunately, no one figure of merit (NEP, D, D^*, D^{**}) fits all types of detectors under all operating conditions. Intercomparison of different types of detectors is, therefore, not always straightforward.

In addition to the responsivity and the NEP, other parameters are of concern in selecting a detector for a particular application. The time constant or the response to signal modulation frequency is important, and so is the linearity or at least the reproducibility. In a more practical sense mechanical integrity, insensitivity to high-energy particle radiation, convenience in matching preamplifier characteristics and that of a bias power supply, temperature range, and other subtleties need to be considered in a detector selection process.

In Section 5.11, we discuss thermal detectors, their responsivity, their noise characteristics, and we provide examples of energy conversion mechanisms as well as a few data on actual detector performance. Then, in Section 5.12 we discuss photodetectors and we provide similar information on these types of transducers.

5.11 Thermal detectors

a. Temperature change

The responsivity of a thermal detector may be expressed as a product:

$$\frac{\Delta V}{\Delta W} = \frac{\Delta V}{\Delta T} \frac{\Delta T}{\Delta W}. \qquad (5.11.1)$$

The first factor on the right side describes the conversion from temperature change to voltage change, while the second factor concerns the temperature rise as a result of radiative power dissipation. This latter factor applies to all types of thermal detectors and is discussed first.

Consider a detector element of area A and of heat capacity $\mathscr{C}$ (Fig. 5.11.1); script letters denote thermal quantities. The detector is embedded in a housing of uniform temperature, T_0, except for the solid angle, Ω, where it views the outside world through an optical filter, also at temperature T_0. If the external blackbody, shown in Fig. 5.11.1, is at the same temperature as the housing, $T_1 = T_0$, after a while the detector will reach equilbrium at that temperature and the net flux of heat will be zero through the thermal conductor that connects the detector element with the surrounding housing. The heat conduction path, with thermal resistance $\mathscr{R}$, consists of mechanical structure to hold the detector in place, of leads to sense the temperature of the element electrically, and possibly of a heat conducting link installed deliberately to obtain a short thermal time constant. The inverse of the thermal resistance $\mathscr{R}$ is the thermal conductance $\mathscr{G}$.

Within the solid angle and within the passband of the filter the specific intensity of the incident radiation supposedly increases by ΔI. An excess power, ΔW, will

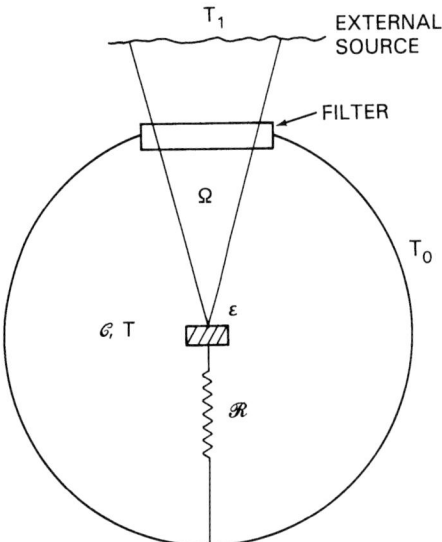

Fig. 5.11.1 A detector of emissivity ε, temperature, T, and heat capacity, $\mathscr{C}$, is imbedded in a housing of temperature, T_0, and exposed to an external source of brightness temperature, T_1, through an aperture that limits the solid angle, Ω. The aperture contains a spectral filter.

be absorbed by the detector,

$$\Delta W = \Delta I \varepsilon A \Omega. \tag{5.11.2}$$

The emissivity of the absorbing side of the detector element is ε. If the power ΔW is constant, after a while, the detector will reach a new equilibrium temperature, T, which will exceed T_0 by ΔT. The temperature increase, ΔT, will be proportional to the absorbed power and the thermal resistance. Thus

$$\Delta T = \Delta W_0 \mathscr{R} \quad \text{or} \quad \Delta W_0 = \mathscr{G} \Delta T. \tag{5.11.3}$$

ΔW_0 is the steady state value of ΔW. In analogy to the displacement current in Maxwell's equations [Eq. (1.1.1)] the variable component of ΔW, called ΔW_1, has to be a time derivative. Thus

$$\Delta W_1 = \frac{d(\Delta Q)}{dt}, \tag{5.11.4}$$

where ΔQ is the increment of heat (joules) deposited onto the detector. With the definition of heat capacity,

$$\mathscr{C} = \frac{\Delta Q}{\Delta T}, \tag{5.11.5}$$

one obtains, assuming $\mathscr{C}$ to be constant,

$$\Delta W_1 = \mathscr{C}\frac{\mathrm{d}(\Delta T)}{\mathrm{d}t}. \tag{5.11.6}$$

Letting $\Delta W(t) = \Delta W_0 + \Delta W_1$ yields

$$\Delta W(t) = \mathscr{G}\Delta T + \mathscr{C}\frac{\mathrm{d}(\Delta T)}{\mathrm{d}t}. \tag{5.11.7}$$

This differential equation describes relaxation phenomena; $\Delta W(t)$ is a time-dependent perturbation function. The homogeneous solution ($\Delta W(t) = 0$) is

$$\Delta T_{\mathrm{hom}} = (\Delta T)_{t-0}\,\mathrm{e}^{-t/\tau}, \tag{5.11.8}$$

where $\tau = \mathscr{R}\mathscr{C}$ is the thermal time constant. The element would approach a new equilibrium exponentially. However, such initial transients are of minor importance. What is of great importance is to find a particular integral of Eq. (5.11.7) when $\Delta W(t)$ is periodic. If a chopper is used to interrupt the radiation falling on the detector, $\Delta W(t)$ is a periodic function with a dominant frequency ω_n, the chopper frequency. Side bands and, depending on the nature of the chopper, higher harmonics of ω_n may also appear. The solution of Eq. (5.11.7) for one frequency, ω_n, of amplitude ΔW_n is

$$\Delta T_n = \frac{\mathscr{R}\Delta W_n \mathrm{e}^{\mathrm{i}\omega_n t}}{1 + \mathrm{i}\omega_n \tau}. \tag{5.11.9}$$

The absolute value of $\Delta T_n / \Delta W_n$ is

$$\left|\frac{\Delta T_n}{\Delta W_n}\right| = \frac{\mathscr{R}}{\left(1 + \omega_n^2 \tau^2\right)^{\frac{1}{2}}}. \tag{5.11.10}$$

With such a detector a flat frequency response is achieved by operating below the critical frequency, $\omega_c = (\mathscr{R}\mathscr{C})^{-1}$. To obtain a large ΔT for a given ΔW the heat capacity of the detector element must be made as small as possible and, at the same time, $\mathscr{R}$ must be chosen as large as possible; however, $\mathscr{R}$ is often limited by the need to achieve a desired time constant.

The similarity between thermal and electrical phenomena in circuits containing resistors and capacitors is apparent. Equation (5.11.7) can equally be applied to

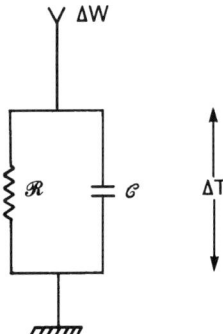

Fig. 5.11.2 Electrical analog circuit of thermal detector shown in Fig. 5.11.1.

electrical circuits by making the following substitutions:

$$\begin{aligned} \text{heat flow rate} &- \text{electric current,} \\ \text{temperature difference} &- \text{voltage difference,} \\ \text{thermal resistance} &- \text{electrical resistance,} \\ \text{thermal conductance} &- \text{electrical conductance,} \\ \text{thermal capacity} &- \text{electrical capacity, etc.} \end{aligned}$$

These thermal–electrical analogies provide a powerful tool; one may construct analog computers to simulate heat flow problems. The electrical analog circuit to the detector arrangement of Fig. 5.11.1 is shown in Fig. 5.11.2. The conditions required to maximize the voltage, ΔT, for a given ω and a certain current, ΔW, are now clear. The difficulties in making $\mathscr{C}$ small and $\mathscr{R}$ about $\tau/\mathscr{C}$ are in the practical implementation, however.

Heat flow through the thermal resistance, $\mathscr{R}$, representing conduction in the support structure and in electrical wires, is not the only form of heat exchange between a detector and its environment. Exchange by thermal radiation also takes place. To analyze radiative exchange, again consider Fig. 5.11.1. The energy emitted from both sides of the detector toward the housing is

$$W_{(\text{elem})} = \pi A \int_0^\infty (\varepsilon_1 + \varepsilon_2) B(\nu, T) \, d\nu. \tag{5.11.11}$$

The emissivities of the front and back sides are $\varepsilon_1(\nu)$ and $\varepsilon_2(\nu)$, respectively. The detector of cross section A is assumed to be thin in comparison with its horizontal dimensions, so that radiative exchange from the edges can be neglected. If the surrounding housing has an emissivity of one, or if it is much larger than the detector

5.11 Thermal detectors

(when its inner surface emissivity does not matter, as in the case of an isothermal cavity), the energy emitted by the housing and absorbed by the detector is

$$W_{(\text{housing})} = \pi A \int_0^\infty (\varepsilon_1 + \varepsilon_2) B(\nu, T_0) \, d\nu. \tag{5.11.12}$$

Tacitly, we assume the passband of the optical filter to be insignificant in comparison with the total thermal emission. If ΔT is small, the difference between the Planck functions is

$$B(\nu, T) - B(\nu, T_0) = \frac{dB(\nu, T_0)}{dT} \Delta T \tag{5.11.13}$$

and the net exchange between element and housing is:

$$\Delta W_{\text{rad}} = \pi A \int_0^\infty (\varepsilon_1 + \varepsilon_2) \frac{dB(\nu, T_0)}{dT} \Delta T \, d\nu. \tag{5.11.14}$$

If the emissivities ε_1 and ε_2 do not depend strongly on ν they may be taken outside the integral, and since

$$\pi \int_0^\infty B(\nu, T) \, d\nu = \sigma T^4; \qquad \pi \int_0^\infty \frac{dB}{dT} d\nu = 4\sigma T^3, \tag{5.11.15}$$

$$\Delta W_{\text{rad}} = 4A(\varepsilon_1 + \varepsilon_2) \sigma T_0^3 \Delta T, \tag{5.11.16}$$

where σ is the Stefan–Boltzmann constant. This heat flow may be represented in the analog circuit by a conductance, $\mathscr{G}_{\text{rad}}$, in parallel to that representing heat flow in solid matter. Thus

$$\mathscr{G}_{\text{rad}} = \frac{\Delta W_{\text{rad}}}{\Delta T} = 4A(\varepsilon_1 + \varepsilon_2) \sigma T_0^3. \tag{5.11.17}$$

If the emissivity of the far side of the element is small, that is, if $\varepsilon_2 \ll \varepsilon_1 = \varepsilon$, $\mathscr{G}_{\text{rad}}$ approaches

$$\mathscr{G}_{\text{rad}} \sim 4\varepsilon A \sigma T_0^3. \tag{5.11.18}$$

Thermal detectors operating in air or in a gas have an additional term associated with thermal conduction and convection in the gas. We omit this case because all high sensitivity detectors are operated in vacuum.

b. Noise in thermal detectors

The precision of every radiometric instrument is ultimately limited by fundamental disturbances, which have a random character and which are predictable only in a statistical sense. These disturbances, called noise, originate from several physical mechanisms. In this section we consider noise in detectors only. Amplifier noise can be significant for modern cryogenic detectors and may influence the choice of the operating parameters or even the type of detector. We will address fundamental limits and not disturbances caused by poor electrical contacts, improper shielding, or faulty grounding, which we assume have all been eliminated.

The recognition that the precision of physical measurements is ultimately limited by random fluctuations materialized only gradually, following the development of the kinetic theory of gases and of statistical mechanics, although indicators to that effect existed before that time. For example, the botanist Robert Brown (1828) observed under high magnification small, inanimate particles in suspension in a liquid and noted their erratic behavior, now called Brownian motion. Later, the kinetic gas theory provided an explanation (Einstein, 1905b, 1906a; Smoluchowski, 1906). Each particle is constantly bombarded by surrounding molecules of the liquid and the net impulse on the particle changes constantly. The kinetic energy of the molecules in the liquid is partially transferred to the particles and causes the observed motion as well as a small, constantly changing heating effect. The heat is returned to the liquid, mostly by conduction. The constant exchange and dissipation of energy on a molecular or atomic scale is characteristic of noise generating mechanisms. Einstein also realized that in an electrical conductor free electrons would also participate in random motion and as a consequence would cause a continuous transport of space charge. A few years later, de Haas-Lorentz (1913) recognized that such random transport of space charge must manifest itself in a randomly fluctuating electrical potential. This voltage fluctuation could even be measured with recently invented vacuum tube amplifiers. Johnson (1925) explained the fluctuations in terms of the shot-effect and Nyquist (1928) provided an elegant explanation in terms of eigenvalues of a transmission line. In a thought experiment a loss-less line is terminated at both ends by resistors matching the wave impedance of the line. In a state of equilibrium the noise power produced by each resistor is transmitted by the line and dissipated by both resistors. At one moment, both ends of the line are short circuited. The signals in transit are then reflected at both ends and form standing waves with nodes at the ends. An infinite number of discrete waves (eigenvalues) exists. Each eigenvalue constitutes a degree of freedom and may be considered to be a harmonic oscillator with a total energy (kinetic and potential) of $hf[\exp(hf/kT) - 1]^{-1}$. A similarity exists with the derivation of the Planck formula, where blackbody radiation can be expressed by summation of eigenvalues

5.11 Thermal detectors

in a three-dimensional cavity. To obtain the energy stored in the line (which was originally supplied by the resistors) within a frequency interval Δf one must sum the eigenvalues in that interval, which yields

$$P_{\Delta f} = kT\Delta f \frac{hf/kT}{e^{hf/kT} - 1}. \tag{5.11.19}$$

Except for ultrahigh frequencies the factor $hf/kT[\exp(hf/kT) - 1]^{-1}$ is unity. For all practical purposes, electrical noise is in the 'Rayleigh–Jeans' limit. Then the power $P_{\Delta f}$ corresponds to a mean rms noise voltage of

$$U_{\Delta f} = (4kTR\Delta f)^{\frac{1}{2}}, \tag{5.11.20}$$

which is the Nyquist formula. For example, an electrical resistor of 1 MΩ at room temperature ($T = 300$ K) produces 4 μV within an electrical band of 100 Hz.

In order to express all noise sources in common terms, it is desirable to refer the Johnson noise produced by the electrical resistance of the infrared detector to radiative units and express it in terms of NEP [W Hz$^{-\frac{1}{2}}$]. The total NEP of the system can then be found by quadratically adding the NEPs of all individual noise sources. Thus

$$\text{NEP}_{\text{system}} = \left[\sum_i (\text{NEP}_i)^2\right]^{\frac{1}{2}}. \tag{5.11.21}$$

The NEP due to Johnson noise alone,

$$\text{NEP}_{\text{Johnson}}[\text{W Hz}^{-\frac{1}{2}}] = \frac{(4kTR)^{\frac{1}{2}}[\text{V Hz}^{-\frac{1}{2}}]}{\Delta V/\Delta W[\text{V W}^{-1}]}, \tag{5.11.22}$$

can be found by dividing the Johnson noise voltage per root bandwidth by the detector responsivity, as is apparent from an examination of the dimensions of the quantities involved. Johnson – or Nyquist – noise is present in all electrical resistors. In electrical networks it is represented by the real part of the impedance; the imaginary part does not contribute to that noise.

Johnson noise is not the only noise source present in detectors. In a thermal detector temperature changes are converted into electrical signals. Consequently, random fluctuations in the temperature of the detector translate directly into random noise at the output terminals. This noise, called temperature noise, is inherent to all thermal detectors. Statistical mechanics shows that the mean square fluctuation of the energy in a small volume is (e.g., Sommerfeld, 1964)

$$(\Delta E)^2 = kT^2 \mathscr{C}. \tag{5.11.23}$$

To obtain the expression for the temperature noise in terms of the NEP [W Hz$^{-\frac{1}{2}}$] one must divide $[(\Delta E)^2]^{\frac{1}{2}}$ [J] by the square root of time [s$^{\frac{1}{2}}$]. This time is the thermal time constant of the detector. In analogy to Eq. (5.11.22)

$$\text{NEP}_{\text{temp}} = \left(\frac{4kT^2\mathscr{C}}{\tau}\right)^{\frac{1}{2}} = (4kT^2\mathscr{G})^{\frac{1}{2}}. \tag{5.11.24}$$

The NEP due to temperature noise is proportional to T and the square root of the thermal conductance. It represents noise due to the exchange of phonons, that is, due to quantized energy exchange by heat conduction between detector and surroundings.

As discussed before, heat exchange between detector and housing also leads to a radiative conductance [Eq. (5.11.18)], which also contributes to noise. The temperature noise due to radiative exchange is obtained by substituting the expression for radiative conductance, Eq. (5.11.18), into Eq. (5.11.24), yielding

$$\text{NEP}_{\text{rad}} = (16\varepsilon Ak\sigma T^5)^{\frac{1}{2}}. \tag{5.11.25}$$

The NEP due to radiative coupling is proportional to $T^{\frac{5}{2}}$ and represents noise due to the photon exchange between detector and housing.

The last, but not the least important detector noise sources are fluctuations due to radiative exchange between detector and external sources that illuminate the detector within the solid angle Ω. If the filter shown in Fig. 5.11.1 is transparent at all wavenumbers, Eq. (5.11.25) applies just within the solid angle Ω. In this case

$$\text{NEP}_\Omega [\text{W Hz}^{-\frac{1}{2}}] = \left(16\varepsilon Ak\sigma T_1^5 \frac{\Omega}{2\pi}\right)^{\frac{1}{2}}. \tag{5.11.26}$$

With the filter being only partially transparent, the filter function has to be weighted according to dB/dT and one obtains

$$\text{NEP}_\Omega = \left[\frac{8}{\pi}\varepsilon A\Omega k\sigma T_1^5 \frac{\int_0^\infty F(\nu)\frac{dB(\nu,T)}{dT}d\nu}{\int_0^\infty \frac{dB(\nu,T)}{dT}d\nu}\right]^{\frac{1}{2}}. \tag{5.11.27}$$

5.11 Thermal detectors

Since

$$B(\nu, T) = 2hc^2\nu^3 \frac{1}{e^x - 1}; \quad x = \frac{hc\nu}{kT}, \qquad (5.11.28)$$

the Planck function derivative with respect to temperature is

$$\frac{dB}{dT} = \frac{2k^3 T^2}{h^2 c} \frac{e^x x^4}{(e^x - 1)^2}. \qquad (5.11.29)$$

We approximate the filter effect by a rectangular transmission function. Between ν_1 and ν_2 the filter shall be transparent [$F(\nu) = 1$], and outside this range it shall be opaque [$F(\nu) = 0$]. Then the filter can be taken into account by setting the integration limits accordingly, yielding

$$\text{NEP}_\Omega = \left[\frac{8}{\pi} \varepsilon A \Omega k \sigma T_1^5 \frac{\int_{x_1}^{x_2} \frac{e^x x^4}{(e^x - 1)^2} dx}{\int_0^\infty \frac{e^x x^4}{(e^x - 1)^2} dx} \right]^{\frac{1}{2}}. \qquad (5.11.30)$$

The integral in the denominator is (e.g., Groebner & Hofreiter, 1957)

$$\int_0^\infty \frac{e^x x^4}{(e^x - 1)^2} dx = \frac{4\pi^4}{15}. \qquad (5.11.31)$$

Consequently,

$$\text{NEP}_\Omega = \left[\frac{30}{\pi^5} \varepsilon A \Omega k \sigma T^5 \int_{x_1}^{x_2} \frac{e^x x^4}{(e^x - 1)^2} dx \right]^{\frac{1}{2}}, \qquad (5.11.32)$$

or after substituting numerical values for the constants,

$$\text{NEP}_\Omega = 2.77 \times 10^{-18} \left[\varepsilon A \Omega T^5 \int_{x_1}^{x_2} \frac{e^x x^4}{(e^x - 1)^2} dx \right]^{\frac{1}{2}}, \qquad (5.11.33)$$

which is the equation for background noise (Lewis, 1947). The emissivity in the above expressions is that of the detector. If the background source is not a blackbody, the emissivity of that source must also be factored under the square root. For example, the effective emissivity of the residual atmospheric water vapor above an

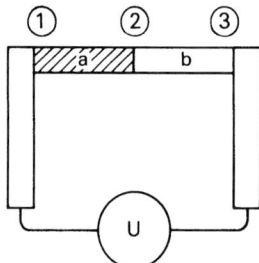

Fig. 5.11.3 Thermoelectric junction of two metals, a and b.

observing site may have to be considered in the background noise, although foreground noise would be a better term for ground-based astronomical observations. If the emissivities are strongly wavenumber dependent and cannot be parameterized, it may be necessary to include them under the integral.

Background noise and temperature noise due to radiative and conductive coupling between detector and housing are all manifestations of energy fluctuations due to the exchange of photons and phonons between the detector and its environment.

In detectors that need a direct current for bias, such as bolometers, a so-called current noise is sometimes noticeable. This noise source, sometimes also present in preamplifiers, has in general a $1/f$ frequency dependence and, therefore, can be minimized by operating with a sufficiently high chopping rate. The origin of the $1/f$ noise is not well understood. In recent years the effect has been reduced substantially by controlling the manufacturing process of attaching the electrical leads to the detector element and by better controlling impurities. It also helps to operate at a small bias current. However, in cryogenically cooled detectors, where other noise sources are reduced to small values, current noise is sometimes of concern (Low & Hoffman, 1963; Zwerdling et al., 1968).

This concludes our discussion of the second factor (i.e., $\Delta T/\Delta W$) in Eq. (5.11.1), and the associated noise sources of a thermal detector due to Johnson noise (fluctuations of electrons) and temperature noise (fluctuations of photons and phonons). We now return to the first factor of Eq. (5.11.1), $\Delta V/\Delta T$. This factor deals with individual conversion mechanisms, and specific classes of thermal detectors.

c. Temperature to voltage conversion

The oldest conversion mechanism, the thermoelectric effect, was discovered by Seebeck (1826). If two dissimilar conductors form part of an electrical circuit, as shown in Fig. 5.11.3, and junction 2 is heated while the rest of the circuit is kept at a constant temperature, a voltage, U, can be measured at the terminals. The polarity of U changes if junction 2 is cooled instead. To heat the junction by radiation, one could blacken the thin wires or one could attach to the junction a thin blackened metal foil to serve as the infrared absorber. The first such detector was constructed by Nobili (1831); Melloni (1833) made the first thermopile by connecting several thermocouples in series. The electric potential is in first order proportional to the temperature difference, ΔT, between junction 2 and junctions 1 and 3. The proportionality factor depends primarily on the nature of the materials, a and b, used in the construction of the element and to a lesser degree on the ambient operating temperature. At low temperatures the thermoelectric effect diminishes, although thermocouples have been described for operation at temperatures as low as that of liquid nitrogen (Cartwright, 1933; Pearson, 1954). However, with few exceptions, thermoelectric detectors are used at room temperature. Different pairs of material show quite different magnitudes of the thermoelectric effect, depending on their electrochemical potential. Most suitable are alloys of bismuth and tellurium with antimony, sulfur, and several other trace ingredients. Such couples show a thermoelectric voltage as high as 650 μV K^{-1}.

Shortly after the discovery of the thermoelectric effect by Seebeck, Peltier (1834) found the inverse phenomenon. Suppose all junctions shown in Fig. 5.11.3 are at ambient temperature. Then an external current source is connected to the terminals. This current causes Joule heating, $i^2 R$, as expected, but, in addition, one junction experiences cooling while the other shows excess heating. The Peltier effect has been exploited to construct small refrigerators. Thermoelectric coolers, specifically designed for the operation of infrared detectors, are commercially available. They are able to cool a small detector up to 50 or even more degrees below ambient (Wolfe & Zissis, 1978).

A third thermoelectric effect, discovered by W. Thomson (1843), later Lord Kelvin, is also related to the Seebeck and Peltier effects. Thomson found that even in a conductor made of one substance, but with a temperature gradient, heat can be removed or added depending on whether the electric current and the temperature gradient coincide or point in opposite directions.

Thermocouples, also called thermopiles, in which several couples are combined electrically, played an important role as infrared detectors in the last decades of the nineteenth and the first half of the twentieth century. Thermocouples provide low impedance energy converters, well-suited to drive sensitive galvanometers. Before the use of electronic amplification this was of great value. Even today many laboratory spectrometers still use thermopiles, but, in nearly all cases, with electronic amplifiers and recorders. An excellent discussion of various construction techniques and test results of thermoelectric detectors is given by Smith *et al.*, (1957).

With the advent of cryogenic detectors thermoelectric elements have lost most of their importance, except for applications where the use of cryogens is not necessary or not possible. The latter was the case for the Voyager spacecraft, where the use of cryogens or of a refrigeration cooler was ruled out by the long duration of the mission and limitations in available power and weight. On Voyager a thermopile serves as the detector for the infrared spectrometer. This detector consists of a cluster of four Schwarz-type thermocouples with a time constant of about 12 ms.

As an example we calculate the magnitudes of different noise sources for the Voyager thermopile. One particular detector of the set constructed for Voyager has a responsivity of 6 V W^{-1} at the operating temperature of 200 K, a resistance of 62 Ω, and a sensitivity of $\sim$600 μV per degree K for each of the four junctions. The NEP measured in the laboratory was $\sim 1.4 \times 10^{-10}$ W Hz$^{-\frac{1}{2}}$. Inserting the appropriate values in Eq. (5.11.22) yields a NEP of 1.38×10^{-10} W Hz$^{-\frac{1}{2}}$ for Johnson noise. The NEP for thermal conduction [Eq. (5.11.24)] is almost an order of magnitude lower, 3×10^{-11} W Hz$^{-\frac{1}{2}}$, and that for radiative coupling to the environment [Eq. (5.11.25)], including radiation from a 200 K blackbody in the field of view, is only 3.8×10^{-12} W Hz$^{-\frac{1}{2}}$. The Voyager thermopile is in essence limited by Johnson noise.

Another type of thermal detector, called the bolometer, was first developed by Langley (1881); it takes advantage of the temperature dependence of certain electrical conductors to produce a voltage change, ΔV, from a temperature change, ΔT. A variety of materials can be used to construct bolometers. Most metals have a temperature coefficient, $\alpha = dR/RdT$, of approximately $1/T$. Langley employed a platinum strip with $\alpha = 0.0038$ K^{-1} because platinum can be worked into very thin foils; others preferred nickel because of the higher temperature coefficient, $\alpha = 0.006$ K^{-1}. The active side of the metal strip must be coated to absorb radiation, but the other side is left uncoated to take advantage of the low emissivity of bare metal. Gold black, produced by slowly evaporating gold in a low pressure, oxidizing atmosphere, yields a good infrared absorber of low heat capacity. In the far infrared, thicker layers must be applied. The bolometer is often placed in one branch of a bridge circuit and an identical element in the other branch to compensate for changes in ambient temperature. The bolometer, with resistance R, is connected to a load resistor, with resistance R_L. The latter element is identical to the former, except that it is shielded from radiation. Two constant voltage sources provide the potential, U (see Fig. 5.11.4). Application of Ohm's law yields

$$I(R + R_L) = U = \text{const}; \quad (\Delta I + \delta I)(R + R_L) + I(\Delta R + \delta R + \Delta R_L) = 0, \tag{5.11.34}$$

and

$$IR = V; \quad (\Delta I + \delta I)R + I(\Delta R + \delta R) = \Delta V + \delta V. \tag{5.11.35}$$

5.11 Thermal detectors

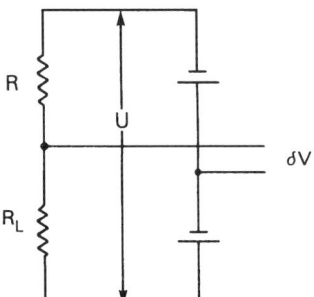

Fig. 5.11.4 Electric circuit of a bolometer, R, and a load resistor, R_L.

The Δ symbol applies to changes due to heating by the bias current and the δ symbol due to the much smaller heating by the absorbed radiation, δW. Since

$$R = R_0[1 + \alpha(T + \delta T - T_0)] \qquad (5.11.36)$$

and

$$R_L = R_0[1 + \alpha(T - T_0)], \qquad (5.11.37)$$

it follows that

$$R - R_L = \delta R = R_0 \alpha \delta T. \qquad (5.11.38)$$

The voltage change δV due to the temperature change δT is

$$\frac{\delta V}{\delta T} = \frac{U \alpha R_0}{2(R + R_L)}. \qquad (5.11.39)$$

The power dissipated in R must be conducted away. Thus

$$\frac{U^2}{4R} = (T - T_0)\mathcal{G}. \qquad (5.11.40)$$

Substituting for R, R_L and U [Eqs. (5.11.36) (5.11.37), and (5.11.40)] into Eq. (5.11.39) yields

$$\frac{\delta V}{\delta T} \approx \frac{\alpha[R\mathcal{G}(T - T_0)]^{\frac{1}{2}}}{2[1 + \alpha(T - T_0)]}. \qquad (5.11.41)$$

The objective must be to maximize the signal-to-noise ratios S/N. Since Johnson noise is the dominant noise source in metal bolometers, S/N is proportional to $(\delta V/\delta T)(T)^{-\frac{1}{2}}$, which has a broad maximum at

$$T = \tfrac{1}{4}\bigl[3T_0 \pm (T_0^2 + 8T_0/\alpha)^{\frac{1}{2}}\bigr]. \qquad (5.11.42)$$

This implies an optimum operating temperature of about 137 K above ambient for a platinum and 100 K above ambient for a nickel strip bolometer. The maximum is rather broad and one loses only 7% in signal-to-noise ratio by operating at half the temperature difference; nevertheless, the high temperatures lead to relatively high noise levels. For this reason, and probably also for the inconvenience of requiring a strong, low-noise bias supply, metal bolometers with direct current bias have seen little application recently. Operation with a short readout pulse avoids strong heating as long as the pulse is of modest amplitude.

In contrast, thermistor bolometers have been used frequently in space-borne instruments, particularly in the first decade of space exploration. For years thermistor bolometers were the only infrared detectors with a short time constant, about 1 ms, and a sensitivity to tens of micrometers, suitable for operation at ambient temperatures. Thermistor bolometers, originally developed by Brattain & Becker (1946), are fabricated from oxides of nickel, manganese, and cobalt. Flakes of several micrometers thickness are cemented on a substrate. For a square flake the electrical resistance at room temperature is 2.5 or 0.25 MΩ, depending on the particular composition, referred to as material type 1 or 2. Both types have a temperature coefficient α of ~ -0.04. The negative value of α can lead to a runaway condition for an operating temperature of only 30 K above ambient. For that reason $T - T_0$ is normally limited to not more than about 5 K. In comparison with thermocouples, thermistor bolometers have the advantage of lower time constants, but, in general, they do not achieve NEP values as low as good thermopiles. Another drawback of thermistors is the relatively high bias voltage required, which can reach several hundred volts; the optimum bias depends on the material and the operating temperature. In thermistor bolometers the dominant noise source is Johnson noise. For a typical element size of 1×1 mm, a resistance of 2.5 MΩ at 300 K, and a responsivity of 700 V W^{-1} the achievable NEP would be 3×10^{-10} W Hz$^{-\frac{1}{2}}$; however, such low values have rarely been achieved.

Another temperature to voltage conversion mechanism, used in pneumatic detectors, is based on the gas thermometer. The best known representative of this type of detector is the Golay cell. The infrared absorber is imbedded in a small volume of gas contained in a sealed cell. Radiation heats the absorber and the surrounding gas. The associated increase in pressure causes a small displacement of a thin membrane. In the Golay cell this displacement is sensed by the deflection of an external light beam and registered by an appropriate detector (Golay 1947;

Hickey & Daniels, 1969). In a more recent version of a pneumatic detector, the displacement of the membrane is sensed electrostatically as in a condenser microphone (Chatanier & Gauffre, 1971). Good pneumatic detectors have obtained NEP values of 2 to 3×10^{-10} W Hz$^{-\frac{1}{2}}$. Golay cells are still in use in many laboratory spectrometers, but have not been applied to space-borne observations, probably because of their weight, complexity, and lower reliability compared to thermopiles and pyroelectric detectors, which are discussed next.

For room temperature operation, a most attractive thermal detector is the pyroelectric element. It is a small capacitor with a dielectric material that possesses a temperature sensitive dipole moment. So far, the most successful dielectric is triglycine phosphate (TGS), particularly if doped with L-alanine. Its Curie point is at 49 °C and, consequently, it must be operated below that temperature. (Above the Curie point, these dielectrics lose their pyroelectric properties.) Other suitable materials include lithium tantalate and strontium barium niobate. The voltage across a capacitor of charge Q is

$$V = \frac{Q}{C}, \tag{5.11.43}$$

where the capacity, C, is

$$C = \varepsilon \frac{A}{s}. \tag{5.11.44}$$

The dielectric constant, the element area, and the thickness are ε, A, and s, respectively. The change in voltage due to a small temperature change is

$$\frac{\Delta V}{\Delta T} = \frac{s}{A} \frac{d}{dT}\left(\frac{Q}{\varepsilon}\right) = \frac{s}{A}\left(\frac{1}{\varepsilon}\frac{dQ}{dT} - \frac{Q}{\varepsilon^2}\frac{d\varepsilon}{dT}\right). \tag{5.11.45}$$

The first term in the right-hand parentheses describes the pyroelectric effect, a change in charge due to a change in spontaneous polarization (Chynoweth, 1956a, b; Cooper, 1962). The second term describes the effect on which the dielectric bolometer is based, a change in voltage due to a change in the dielectric constant (Moon & Steinhardt, 1938; Hanel, 1961c; Maserjian, 1970). In a capacitive element both effects are present; so far, the pyroelectrical bolometer has led to more practical implementations. If the variation of the dielectric effect with temperature can be neglected,

$$\frac{dV}{dT} = \frac{A}{C}\frac{dP}{dT}, \tag{5.11.46}$$

where the polarization $P = Q/A$. In good pyroelectric elements the equivalent resistor parallel to the capacitive element, representing dielectric losses and leakage, is large and the capacity small. In practice both are of the same order as the input parameters of a typical FET (field-effect transistor) preamplifier, $\sim 10^{10}$ Ω and 10 pF. To keep stray capacity small, the FET is often mounted in the same evacuated housing as the detector.

Considering modulated radiation of frequency ω (chopper frequency), calling the total capacity (detector and preamplifier) C, the total resistance (detector resistance and amplifier input resistance in parallel) R, and the change in polarization $dP/dT = \alpha$, the voltage change per temperature change becomes

$$\frac{\Delta V}{\Delta T} = \omega \alpha A R \mathscr{R} \left(1 + \omega^2 \tau_E^2\right)^{-\frac{1}{2}}. \tag{5.11.47}$$

The electrical time constant τ_E is the product of R and C. To obtain the responsivity one has to multiply Eq. (5.11.47) by Eq. (5.11.10), and considering the emissivity of the element yields

$$r = \frac{\Delta V}{\Delta W} = \varepsilon \omega \alpha A R \mathscr{R} \left(1 + \omega^2 \tau_E^2\right)^{-\frac{1}{2}} \left(1 + \omega^2 \tau_T^2\right)^{-\frac{1}{2}}, \tag{5.11.48}$$

where τ_T is the thermal time constant of the element. Depending on the actual value of electrical and thermal time constants three different regimes may be distinguished. Typical values for the time constants are 0.1 and 1 s, respectively. The low frequency ($\omega \tau_E \ll 1; \omega \tau_T \ll 1$) and the intermediate frequency cases ($\omega \tau_E < 1; \omega \tau_T > 1$) are of minor importance. By far the most common operating domain is that for high frequencies ($\omega \tau_E \gg 1; \omega \tau_T \gg 1$). In this case the responsivity becomes

$$r_{\text{high}} = \varepsilon \alpha A C^{-1} \mathscr{C}^{-1} \omega^{-1}. \tag{5.11.49}$$

The responsivity falls off as the reciprocal of ω; to obtain a flat frequency response, amplification proportional to ω has to be provided.

Since Johnson noise is not present in the imaginary part of an electrical impedance, only noise due to the loss tangent of the capacitor, the load resistor, and the amplifier input resistor contributes to Johnson noise. However, the basically capacitive nature of the element shunts Johnson noise to some degree. As a consequence the NEP [W Hz$^{-\frac{1}{2}}$] of pyroelectrical detectors due to Johnson noise is only proportional to the square root of frequency (Putley, 1977). Pyroelectric detectors can, therefore, be used to relatively high modulation frequencies. Temperature noise, due to thermal conduction and radiative exchange, is also present, as in all thermal detectors. In pyroelectric elements, Johnson noise, temperature noise, and amplifier noise are approximately of the same order of magnitude (Putley, 1977). NEP values of 5×10^{-11} W Hz$^{-\frac{1}{2}}$ have been obtained.

The temperature to voltage conversion mechanisms discussed so far are best suited for operation at room temperature. As mentioned in the section on detector noise (Subsection 5.11.b), Johnson noise and all forms of temperature noise are substantially reduced by operating the detector and its environment at low temperatures. Besides the reduction of all noise sources other advantages exist in cryogenic operation of thermal detectors. For most substances the specific heat and consequently the thermal capacity is much reduced in comparison with that at ambient temperature. Furthermore, infrared absorbing substances with high temperature coefficients of the electrical resistance are available. For cryogenic operation only resistive bolometers have seen wide application; the thermoelectric effect is greatly diminished at cryogenic temperatures.

A convenient way to maintain a detector at a low and constant temperature is to thermally couple the detector housing to a cryostat filled with liquid helium. At one bar the boiling point of helium is 4.2 K and less at lower pressures. For operation down to 0.3 K the isotope helium 3 can be used instead of the normally available helium, which is dominantly helium 4. Operation at even lower temperatures (~ 0.1 K) is possible using demagnetization effects in certain paramagnetic salts; ferric ammonium sulfate is an example (Castles, 1980; Serlemitsos et al., 1990).

One attempt to construct a cryogenic bolometer centered on operation at a temperature just where the element becomes superconductive (e.g., Martin & Bloor, 1961; Gallinaro & Varone, 1975). NEP values of 10^{-13} W Hz$^{-\frac{1}{2}}$ have been achieved, but superconducting bolometers have not found wide application. The main drawback seems to be the need for a very high degree ($\Delta T/T \sim 10^{-5}$) of temperature control. Furthermore, when superconducting ($\sigma = \infty$) the element becomes highly reflective; even before superconduction occurs the infrared absorption characteristic becomes poor, especially in the far infrared. However, with the recent discovery of materials superconducting at higher temperatures reexamination of superconducting bolometers may be worthwhile.

Carbon bolometers have been made of slices of material used in the construction of electrical resistors (Boyle & Rodgers, 1959); however, far better results have been achieved with doped germanium elements (Low, 1961). Single crystals of germanium, doped with gallium or with a combination of gallium (9×10^{15} cm^{-3}) and antimony (1×10^{15} cm^{-3}), have been used (Zwerdling et al., 1968). The doping level is chosen to provide a good infrared absorption characteristic of the bulk material. However, without a black or an antireflection coating the high refractive index of germanium causes strong reflection and such elements have a relatively low emissivity (see Fig. 1.6.1). For that reason, such elements are often placed in the center of a highly reflective sphere with a relatively small feed cone (~15°) so that radiation reflected by the element is refocused onto the detector by the reflective walls. After several reflections most of the photons entering the cavity are absorbed by the bolometer. The effective emissivity of such a cavity type detector arrangement is quite high, approaching unity.

Cryogenic bolometers can also be made of silicon (Beerman, 1975; Kinch, 1971; Chanin, 1972). Silicon has the advantage of an eight times lower heat capacity than germanium, but so far, silicon bolometers have not demonstrated NEP values lower than good germanium bolometers (Dall'Oglio *et al.*, 1974). More recently, composite bolometers have been constructed where the functions of infrared absorption and of temperature measurement have been separated (Coron *et al.*, 1972). Such detectors consist of an absorbing layer of metal deposited on a good heat conducting substrate (sapphire or diamond) and a small semiconducting crystal (germanium or silicon) attached to the thin substrate. Very thin layers of gold, silver, or bismuth can be made with an impedance nearly matching that of free space. Such films are good absorbers (50% absorption) over a wide spectral range in the infrared and at millimeter wavelengths (Nishioka *et al.*, 1978; Dragovan & Moseley, 1984). The small semiconducting crystal can now be optimized as a temperature sensor irrespective of infrared absorption characteristic. Such composite bolometers show great promise, particularly for the far infrared.

5.12 Photon detectors

Unlike thermal detectors, which sense the power of the absorbed radiation, photon detectors respond to the number of photons arriving per unit time. Photon as well as thermal detectors are incoherent transducers, which means that the detection process is independent of the wave properties of the incident radiation field. Incoherent detectors produce an electrical signal proportional to the intensity of the radiation. In contrast, coherent detectors, such as the nonlinear elements in heterodyne receivers discussed in Section 5.9, register the amplitude and phase of the electric field associated with the absorbed radiation. Due to the simultaneous measurement of amplitude and phase, coherent detection is subject to a fundamental noise limit that has its origin in the quantum mechanical uncertainty principle. Incoherent detectors are free of this particular limit. However, as we shall see, they are subject to other noise sources.

In a photon detector a sufficiently energetic photon produces a charge separation in the detector material. The charge can be measured at the output leads of the detector as a voltage or current depending on the input impedance of the preamplifier and load resistor compared with that of the detector. A photon detector produces the same signal for a photon at 10 μm as for one at 5 μm, while a thermal detector gives a greater signal for the more energetic 5 μm photon. Conversely, if the radiant power is the same at two wavelengths, the photon detector will produce a smaller signal at the shorter wavelength, where there are fewer photons.

a. *Intrinsic and extrinsic semiconductors*

Photon detectors are commonly manufactured from semiconducting materials. The semiconductor may be intrinsic when charge carriers arise within the bulk material, or extrinsic when charge carriers are formed at impurity sites doped into the bulk material. Mercury cadmium telluride (HgCdTe) and indium antimonide (InSb) are examples of intrinsic semiconductors, while arsenic-doped silicon (Si:As) and copper-doped germanium (Ge:Cu) are examples of extrinsic devices. Intrinsic detectors are, generally, most sensitive at wavelengths shorter than 5 μm; however, some operate at wavelengths as long as 20 μm. Extrinsic detectors are best at wavelengths longer than 5 μm; some operate up to a few hundred micrometers. The wavelength range of intrinsic detectors is determined by choosing the type of material, and of extrinsic detectors by selecting an appropriate dopant for silicon or germanium.

Figure 5.12.1 shows the generation mechanisms of charge carriers in both kinds of semiconductors. In an intrinsic semiconductor an incident photon creates an electron–hole pair that decreases the detector resistance, resulting in a voltage change across the element when a constant current bias is maintained. The positively charged hole is placed in the valence band and the electron is placed in the conduction band. However, this is possible only if the photon energy (frequency) is high enough to overcome the band gap of the material. The wavelength corresponding to the band gap energy is the long wavelength cut-off of the material.

In an extrinsic semiconductor the incident photon does not have sufficient energy to form a free electron–hole pair, but instead produces a free charge and an immobile, ionized dopant-impurity. If the material is *p*-type, the photon forms a free hole in the valence band, while in *n*-type material a free electron is formed in the conduction band. For extrinsic materials the long wavelength cut-off corresponds to the photon energy at which the impurity ionization energy is exceeded. The impurity ionization energy is the energy between the valence band and the acceptor states in *p*-type material and between the donor states and the conduction band in *n*-type material.

The long wavelength cut-off of a photon detector is given by

$$\lambda_C(\mu m) = \frac{1.24}{E(\text{eV})}, \tag{5.12.1}$$

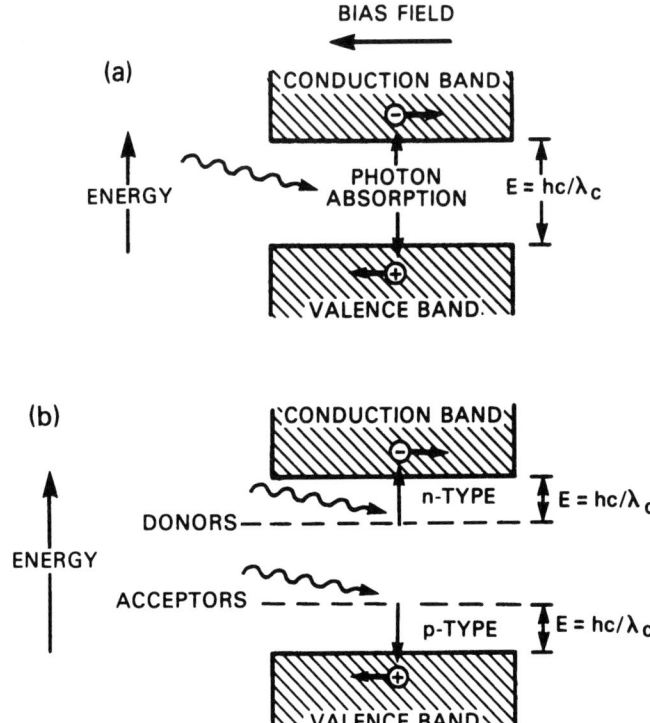

Fig. 5.12.1 Charge carrier generation mechanisms. (a) In intrinsic semiconductors an incident photon creates an electron–hole pair. (b) In extrinsic semiconductors an immobile ion and a free charge (electron or hole) are created by the photon.

where λ_C is expressed in micrometers and E in electron-volts. Typical values of band gap energies for intrinsic materials are 1.1 eV for Si, 0.23 eV for InSb, and 0.1 eV for HgCdTe. The wavelength cut-offs for these materials are 1.1, 5.4, and 12 μm, respectively. Examples of impurity ionization energies in extrinsic materials are 0.054 eV for Si:As and 0.04 eV for Ge:Cu, with cut-offs at 23 and 30 μm, respectively. Tables of band gap and ionization energies are listed for the most common photoconductor materials by Keyes (1977), for example.

b. Photoconductors and photodiodes

Photon detectors can be classified as either photoconductive (PC) or photovoltaic (PV) devices. In photoconductors the resistance of the material is lowered when impinging photons generate free charge carriers. The signal is measured as a change in voltage across the detector when it is connected to a constant current source or as a change in current when it is connected to a constant voltage source. Photoconductors

5.12 Photon detectors

can be made from either intrinsic or extrinsic material. In intrinsic detectors photoconductivity can be generated in bulk semiconductor material, or the material may be fabricated in layers to enhance the performance of the detector. Fabrication techniques for extrinsic photoconductors have improved dramatically with the development of blocked-impurity band (BIB) detectors (Petroff & Stapelbroek, 1980; Wiedemann et al., 1989). The BIB detector is fabricated with a thin layer of undoped silicon separating the doped infrared-active layer and the positive electrode. This prevents spurious electrons in the active layer from reaching the electrode. The dopant concentration can then be much higher than in conventional photoconductors, producing much higher detection efficiencies. This technology has also permitted the development of solid-state photomultipliers for the 1–28 μm region (Petroff et al., 1987). These infrared photomultipliers are operated under high bias voltage, so that a single photon causes an avalanche of carriers and, therefore, high gain within the detector.

In photovoltaic detectors, also called photodiodes, electrons and holes generated by photons are separated by an electric field formed at a potential barrier within the device. This field is formed at a specifically introduced interface in the material, for instance a p–n junction. Photodiodes used in astronomy are almost always made of intrinsic material. When an electron–hole pair is created at the junction by a photon the electron drifts to the n-region and the hole to the p-region. The separation of charge causes a voltage to be generated across the detector terminals, which can be sensed directly or, alternatively, a current can be measured when the circuit is completed.

Because a PV detector is a diode, it has a very high resistance in one direction and a very low resistance in the other. Frequently, photodiodes are reverse-biased to enhance their response by placing a voltage source and a load resistor in series with the detector; the positive voltage is applied to the n-side of the junction. The signal measured across the load resistor is proportional to the current generated by photons at the p–n junction. Reverse-biasing enhances the field in the junction region and improves the response time of the detector. However, detector noise also increases with reverse-biasing, and the bias voltage must be adjusted to optimize the signal-to-noise ratio.

c. Responsivities

The sensitivities of several common PC and PV detectors are shown in Figure 5.12.2. The cut-off wavelengths are readily seen, as are the wavelength dependent decreases due to fewer photons per watt at shorter wavelengths. In the figure the sensitivity is given in terms of D^* [see also Eq. (5.10.3)], which is

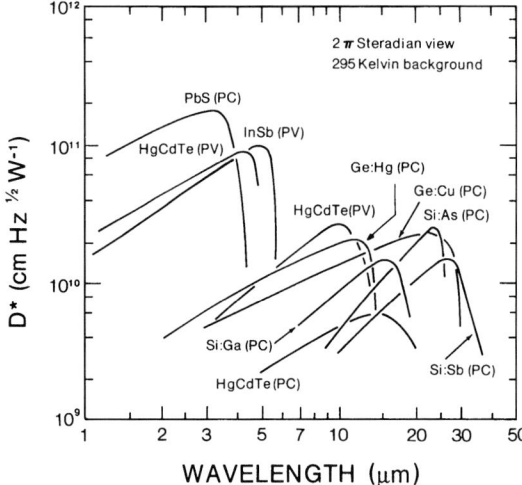

Fig. 5.12.2 Detectivities, D^*, of several commonly used infrared photoconductive (PC) and photovoltaic (PV) detectors.

defined by

$$D^* = \frac{(A\Delta f)^{\frac{1}{2}}}{\text{NEP}}. \qquad (5.12.2)$$

Here A is the detector area, Δf the electrical bandwidth, and NEP the radiant power at the detector required to produce a signal voltage equal to the noise voltage. To be of practical use, a detector must produce a signal greater than the noise level of the detector and the preamplifier. In particular, the incident photons must be able to excite electron–hole pairs, while these pairs must not be thermally generated at the operating temperature of the detector. The response of the detector is determined by its quantum efficiency, i.e., the probability that an incident photon will generate an electron–hole pair. Quantum efficiencies vary from a few to about 70 percent, depending on the type of detector and the fabrication process.

When the load resistance greatly exceeds the detector resistance and the signal frequency is low, the signal voltage of a PC detector is

$$V_S = Q_{PC}\eta\lambda V_b P(\lambda), \qquad (5.12.3)$$

where Q_{PC} is a coefficient that depends on the number and lifetime of the charge carriers, η is the quantum efficiency (number of charge carriers generated per photon), λ the wavelength, V_b the bias voltage, and $P(\lambda)$ the incident radiant power as a function of wavelength. For a PV detector operated without bias the expression is

$$V_S = Q_{PV}\eta\lambda P(\lambda), \qquad (5.12.4)$$

where Q_{PV} depends on the impedence of the detector. The sensitivity of PC and PV detectors varies with modulation frequency, generally decreasing at higher frequencies. This behavior is due to properties of the detector itself and the electronic circuit to which it is connected. In PC detectors the charge carrier lifetime in the detector material determines the cut-off frequency if the lifetime is greater than the electric time constant of the circuit. Typical cut-off frequencies for PV detectors are 10^8 Hz, much higher than for PC detectors, which typically operate up to about 10^3 Hz.

d. Noise in photon detectors

Noise arises in semiconductor detectors from several mechanisms. Johnson noise is found in all resistive elements. It has already been discussed in connection with thermal detectors [see Subsection 5.11.b and Eq. (5.11.20)]. If the load resistance in the circuit is larger than the detector resistance, the Johnson noise of the detector element dominates because load and detector act electrically in parallel as far as the noise properties are concerned.

Another type of noise, often dominating at low modulation frequency, is called $1/f$ noise because the noise power is inversely proportional to frequency. Its physical cause is not well-understood. In PV detectors it is proportional to the direct current flowing through the detector, and can be greatly reduced by operating the detector with a small d.c. bias. Because PC detectors must operate with a d.c. current, $1/f$ noise is always present, particularly at low modulation frequencies. With advanced fabrication techniques, the level of this noise can be reduced below other, more dominant noise sources.

Generation–recombination (G–R) noise is found in PC detectors, and is caused by random fluctuations of charge carriers. These fluctuations can be due to thermal excitation within the semiconductor. Sometimes G–R noise is also defined to include random arrival of photons at the detector. PC detectors are normally operated at temperatures low enough to reduce the thermal generation of carriers well below that of all other noise sources. The residual G–R noise is then the photon noise already discussed in Subsection 5.11.b [see Eq. (5.11.33)]. Photon noise can also be understood in terms of the Poisson probability of n photons arriving during a given time interval,

$$P(n) = \frac{N^n e^{-N}}{n!}, \qquad (5.12.5)$$

where N is the mean number. For large arrival rates the probability distribution becomes Gaussian, with a standard deviation

$$\sigma = \sqrt{N}. \qquad (5.12.6)$$

The photon noise is simply due to the uncertainty, σ, in the number of photons per second, i.e., the detected signal level. In the photon noise limit the signal-to-noise ratio is proportional to the square root of the photon arrival rate. When the measured noise is the photon noise in the background radiation, the detection is said to have reached background limited performance (BLIP).

The total noise voltage can be found from the individual noise voltages by adding the squares of the individual rms noise voltages [see Eq. (5.11.21)]. Whenever possible, detectors are operated with sufficient incident radiation to achieve BLIP performance, that is, where the background noise is larger than the noise from all other sources combined. For ground-based observations the noise is then due only to the radiation from the source, sky, and warm optical elements in the field of view of the detector.

Noise production mechanisms strongly depend on the detector temperature. Therefore, photon detectors operating at infrared wavelengths must usually be cooled to eliminate noise caused by thermal generation of charge carriers. In materials with small band gaps or impurity ionization energies, particularly at room temperature, it is likely that thermal excitation will overcome the potential barrier and create more free charges than absorbed photons do. For this reason lead sulphide (PbS) and lead selenide (PbSe) detectors are generally operated below 200 K, while InSb and HgCdTe detectors perform best at liquid nitrogen temperature (77 K) or lower. Doped silicon and germanium detectors require operation near liquid helium temperature (<10 K). Low temperatures are usually required for detectors with cut-offs at wavelengths greater than 3 μm. By contrast, intrinsic silicon, with a cut-off at 1 μm, performs well at room temperature (295 K).

e. Circuits for photon detectors

Figure 5.12.3 shows circuits most commonly used with photon detectors. The bias circuit illustrated in (a) is well-suited for photoconductors. The signal voltage is measured across a load resistor. The simple circuit in (b) is appropriate for a PV detector without an external bias. The more complex circuit in (c) is for a reverse-biased PV detector. In this circuit, which resembles (a) to some extent, an external voltage is applied across the detector and a load resistor arranged in series.

Figure 5.12.4 shows a modification of the circuits illustrated in Fig. 5.12.3. The trans-impedance amplifier (TIA) is used primarily with cooled detectors. The Johnson noise of the feedback resistor decreases as the square root of the resistor temperature [see Eq. (5.11.20)]. Therefore, in cases where the Johnson noise would

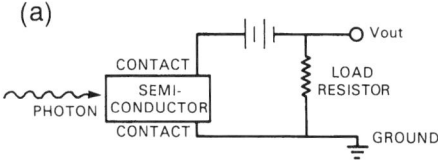
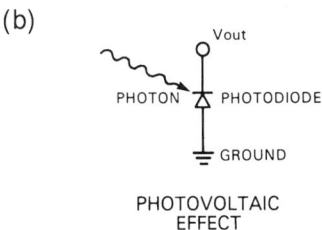
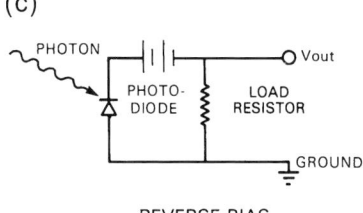

Fig. 5.12.3 Simple circuits used with photon detectors. Circuit (a) is used with photoconductors. Circuits (b) and (c) are both used with photodiodes.

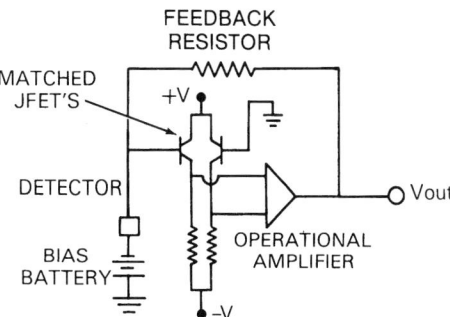

Fig. 5.12.4 The trans-impedance amplifier (TIA) often used with cooled detectors. The feedback resistor and Johnson field-effect transistors (JFETs) are placed on the cold stage with the detector.

dominate other noise sources, it is necessary to also cool the feedback resistor by placing it on the cryogenic mount along with the detector.

The operational amplifier causes the current in the feedback resistor to equal that in the detector, and the gain in the system is the ratio of feedback and detector resistances. The matched pair of Johnson field-effect transistors (JFETs) provide a high impedance input to the operational amplifier. The JFETs are usually placed close to the detector and feedback resistor to minimize the lengths of wire leads, thereby reducing the capacitance between the signal wire and ground. A large capacitance would greatly limit the maximum modulation frequency of the system. With very high impedances even a small stray capacitance can have a noticeable effect. This circuit can also be used with photovoltaic detectors, with or without bias.

f. Detector arrays

During the past decades infrared detectors have been developed in array format, with thousands or millions of detector elements arranged in two-dimensional configurations. Applications of arrays have allowed significant advances in infrared astronomy similar to those experienced in visible astronomy when Charge Coupled Devices (CCD) in array format were introduced. Infrared arrays of InSb, HgCdTe, Si:As, and other materials are available in a variety of formats ranging from 128×128 up to 2048×2048 elements. Individual elements are typically tens of micrometers in size.

All large format arrays are monolithic (made as a single unit) and are bonded to multiplexer integrated circuits. Multiplexers normally use a small number of preamplifiers that are each switched among detector elements in a subsection of the array without loss of integration time. This is generally done by allowing charge to accumulate at each detector (or capacitor attached to each detector) and reading the charge levels in sequence. A major advantage of using charge accumulation is that very long exposures are possible between read-outs when the source intensity is weak. A disadvantage is that under high intensity conditions it is sometimes difficult to read the charge levels at a sufficiently high rate to avoid saturation. Charges of 10^5 to 10^8 electrons may be accumulated between readings.

Infrared arrays are particularly advantageous for imaging. High quality, two-dimensional images can be taken of planets and other astrophysical objects that could not be obtained with scanning radiometers containing a single detector. The image of Jupiter shown in Fig. 5.12.5 was taken with a 256×256 array from a ground-based telescope (Harrington *et al.*, 2000). A similar advantage exists for arrays in dispersive spectrometers, where all spectral elements can be recorded simultaneously while the second dimension of the array is used for spatial coverage. Infrared array technology has received impetus from use on the Hubble Space Telescope, the Space Infrared Telescope Facility, and the planned Next Generation Space Telescope. For more complete discussions of photon detectors and arrays, the reader is referred to reviews by Keyes (1977), Richards & Greenberg (1982),

Fig. 5.12.5 Image of Jupiter recorded with a 256 × 256 detector array and a filter limiting the spectral response to a narrow region centered at the ν_4 band of methane at 1304 cm^{-1}. The image was obtained by Harrington *et al.* (2000) using the MIRLIN infrared camera at the Infrared Telescope Facility on Mauna Kea, Hawaii.

Dereniak & Crowe (1984), Fowler (1993), Dereniak & Boreman (1996), and more detailed papers listed in Dereniak & Sampson (1998).

5.13 Calibration

a. Concepts

A physical measurement is of little value unless accompanied by a reasonable estimate of its uncertainty. For remote sensing investigations intensity calibration is of prime concern, but wavenumber calibration also is very important. In all measurements one must distinguish between precision and accuracy. Precision refers to the reproducibility of a measurement in the presence of random errors, while accuracy indicates the deviation of the measured value from the true value. The level of precision may be established by repeating a measurement several times. Accuracy involves a judgment of systematic errors, which are generally not well known. Therefore, a good estimate of the accuracy of a measurement is sometimes difficult to obtain.

The important distinction between precision and accuracy can be illustrated by an example. Consider an infrared radiometer with a high étendue ($A\Omega$), a low-noise detector, and a good communication channel, but without a means of on-board calibration. Presumably, the instrument has been calibrated several months before launch in the laboratory. Suppose also that the intensities measured after launch appear reasonable; they fall within wide limits set by expectation. What confidence level can one assign to the measurements? How can one estimate random and systematic errors? To find the random error is relatively straightforward. One observes repeatedly the same scene under identical conditions and derives the degree of reproducibility of the results. If nearly the same numerical value appears every time the radiometer faces the same planetary area or deep space, one concludes that the random error in the measured intensity is small. This is expressed by estimating the standard deviation,

$$s = \left(\frac{\sum_{i=1}^{k}(x_i - \bar{x})^2}{k-1} \right)^{\frac{1}{2}}. \tag{5.13.1}$$

Here, the individual radiometer readings are x_i and their mean value is $\bar{x}$; the number of measurements taken into consideration is k. When this number is large, at least 10 (or, better yet 100), then s approaches the standard deviation, σ, which is the limit of s for large k. It is assumed that the individual measurements are statistically independent of each other. The probable error, which can be assigned to the large average, is then

$$s(\bar{x}) = \frac{s}{\sqrt{k}}. \tag{5.13.2}$$

By repeating the same measurement and averaging the results the random uncertainty in the average is reduced by the square root of k; of course, the time required to perform these measurements increases proportionally to k.

From a low standard deviation one may conclude that the measuring system has good precision, but one cannot judge the accuracy of the data. The mean value, $\bar{x}$, may differ systematically from the true value for a variety of reasons. Some may be rather obvious; for example, the instrument temperature in space may differ

5.13 Calibration

from that at the time of the laboratory calibration. Some may be more subtle; the detector response may have changed because of the high energy particle radiation, or the optical alignment may have shifted because of a rough launch environment. The experimenter who knows the instrument well may be in a position to estimate the magnitude of some of these systematic errors, but will rarely be able to assess all possible effects. The radiometer considered in this discussion may have a high precision, but it has an unknown accuracy.

Good accuracy can be established with an on-board calibration. The ability to calibrate before, during, and after collection of scientific data, is a great advantage. Changes in the instrument performance as a consequence of launch environment, exposure to space, and operation as part of the spacecraft system can all be taken into account in the on-board calibration process. The radiometer, the spacecraft electronics, and the data transmission channel have to be included in the calibration. Each link in this long chain may affect the precision and accuracy of the information.

Intensity calibration of a linear system requires measurements of at least two stable reference sources. It is desirable to have one source at the lower and the other at the upper end of the measurement range. Planetary data may then be scaled to these known sources by interpolation, which is always more desirable than extrapolation. Measurements of additional known sources are required for a verification of system linearity. Long term instrumental drift and all other changes in instrument performance are, thereby, removed from the data. In contrast to the radiometer discussed above, an instrument with well-designed calibration sources, but with a noisy detector or with severe truncation in the digital transmission channel (not enough bits in the digital words), would be an example of a device with good accuracy, but with poor precision. An increase in observation time improves precision, but not accuracy.

In a discussion of calibration methods one should not assume that all radiometers or spectrometers need to be calibrated. The need depends entirely on the scientific objective. If this objective can be reached by an examination of two-dimensional patterns generated by a scanning radiometer, for example, then calibration may not be necessary. Cloud patterns may be recognized sufficiently well for weather forecasting purposes without an absolute radiance scale. Good precision and frequency response are more important than accuracy in this case. On the other hand, if data from that radiometer are to be used to estimate the altitude of dense clouds, an absolute calibration of the instrument is of vital importance. The same holds true for spectrometers. If the only objective is to detect spectral features of atmospheric constituents, good precision is more important than high accuracy. What is required in this case is a capability to identify even weak spectral features at the right wavenumbers. High accuracy in the intensity scale is then not necessary, although it is required in the wavenumber scale. Such an identification of constituents in the atmosphere of Mars was the objective of the infrared spectrometer (IRS) investigation on Mariner 6 and 7. The infrared spectra obtained have been presented with relative scales only, but, in spite of the absence of an intensity calibration, the presences and upper limits of several gases have been derived from these spectra (Herr & Pimentel, 1969, 1970; Herr et al., 1972; Horn et al., 1972). Recently, Kirkland (1999) succeeded in the calibration of the IRS spectra with respect to wavelength and intensity for the long-wave channel (5 to 14 μm). The calibrated spectra are much more useful than those previously published and demonstrate the low noise characteristic of the data, particularly in the 5 to 10 μm region. If one wishes to obtain from the spectral measurements surface or gas temperatures at the same time, or more precise gas abundance values, a calibration capability is required. Such a capability was part of the infrared spectrometer on Mariner 9 (Hanel et al., 1972a, 1972b; Conrath et al., 1973; Maguire, 1977).

b. Middle and far infrared calibration

The most convenient calibration sources for the middle and far infrared are blackbodies at constant and well-known temperatures. The range of blackbody temperatures should match approximately the range of expected brightness temperatures of the planetary atmospheres and surfaces under investigation. Of course, calibration sources are also subject to systematic errors; the temperature sensor of a blackbody may be systematically off, or the actual emissivity of the device may not be as close to unity as assumed. However, these types of error generally can be kept small in number and in magnitude in comparison with systematic changes of the instrument response.

To examine the calibration issue more closely consider an infrared radiometer with a spectral response, $r(\nu)$, observing a planetary area of intensity $I(\nu)$. The objective of the measurement is to determine that intensity. The amplitude, in volts or in digital numbers registered by the radiometer while facing the planet, is

$$A_1 = \int_\nu r(\nu)[I(\nu) - B(\nu, T_{\text{eff}})] \, d\nu \qquad (5.13.3)$$

The responsivity includes all instrumental properties, such as the transmission characteristics of optical filters, the detector response, amplifier gain, etc. The term $B(\nu, T_{\text{eff}})$ is the Planck function corresponding to the effective instrument temperature. If the instrument and the detector are at the same temperature, T_i, then that temperature is the effective temperature. However, the detector and the rest of the instrument are often at different temperatures; for example, the detector may be cryogenically cooled or the power dissipation in the detector may elevate its temperature above that of the instrument. In these cases the effective temperature must be somewhere between those of the instrument proper and the detector. It can be defined by

$$B(T_{\text{eff}}) = \alpha B(T_{\text{i}}) + (1 - \alpha) B(T_{\text{d}}). \tag{5.13.4}$$

The parameter α may assume values between zero and one. If instrument and detector are at the same temperature, clearly α is one. In general, α deviates from unity and may vary also with wavenumber. For simplicity, we have assumed that the instrument, including the telescope, baffles, and all other elements within the optical path, is at one well-defined temperature, T_{i}; this is not absolutely necessary, however. What is absolutely necessary is that the temperatures and temperature gradients that exist within the instrument are maintained between the times of planetary measurement and calibration. In view of different viewing directions and changing incident solar radiation with that direction it is often very difficult to maintain not only the temperatures but also temperature gradients. One way to satisfy these requirements is by thermostating the instrument. Temperatures are then assured of staying constant between measurements and calibrations, and internal temperature gradients are minimized. If thermostating cannot be provided to the required degree, absolute calibration may be very difficult, especially if a low temperature object must be measured with a relatively warm instrument.

The measured amplitude must be zero if one exposes the radiometer to a blackbody at the effective instrument temperature. It is also evident that the radiometer output changes polarity between observations of scenes colder and warmer than the effective instrument temperature. To establish absolute calibration we expose the field of view of the instrument first to a blackbody of temperature T_2, significantly higher than T_{eff}. The temperature of that blackbody must be measured accurately, for example, by a resistance thermometer. The radiometer reading is then

$$A_2 = \int_\nu r(\nu)[B(\nu, T_2) - B(\nu, T_{\text{eff}})] \, d\nu. \tag{5.13.5}$$

Subsequently, a reading is obtained while the radiometer faces another blackbody at a temperature T_3, significantly lower than T_{eff},

$$A_3 = \int_\nu r(\nu)[B(\nu, T_3) - B(\nu, T_{\text{eff}})] \, d\nu. \tag{5.13.6}$$

If this second calibration source is deep space, then $B(v, T_3)$ equals zero for our purpose. The cosmic background and emissions from stellar and galactic objects are usually negligible in comparison with instrument noise and systematic errors. Furthermore, we assume that the instrument is stable, that is, $r(v)$ and $B(v, T_{\text{eff}})$ do not change between calibrations and planetary measurements. Maintaining constant instrument and detector temperatures is, therefore, very important. The calibration objective is to find the intensity, I, in absolute units, from the readings A_1, A_2, and A_3, provided we know the temperatures of the calibration sources. First, we eliminate the term $B(v, T_{\text{eff}})$ by forming the differences $A_1 - A_3$ and $A_2 - A_3$; then we calculate the ratio,

$$\frac{A_1 - A_3}{A_2 - A_3} = \frac{\int r(v)[I(v) - B((v, T_3)]\, dv}{\int r(v)[B(v, T_2) - B(v, T_3)]\, dv}. \tag{5.13.7}$$

If the third observation is deep space, then $B(v, T_3) = 0$ and the expression simplifies, but the use of deep space is not a requirement. Calibration with a blackbody at a temperature corresponding to the lowest expected brightness temperature has advantages. With T_2 set at the maximum expected brightness temperature, T_3 set at the minimum, and the instrument temperature chosen at the midrange of the corresponding intensities, the required dynamic range is a minimum.

Holding a calibration blackbody at a temperature higher than that of the instrument is easily achieved in practice by thermally isolating and electrically heating that blackbody in a thermostatic circuit. It is also possible and highly advisable to thermostat the whole instrument. To thermostat a blackbody at a temperature substantially lower than that of the instrument may involve active cooling and is inconvenient. Therefore, deep space has often been used as a cold calibration source for space-borne infrared instruments; to simplify the discussion we also assume $B(v, T_3) = 0$. Then

$$\frac{A_1 - A_3}{A_2 - A_3} = \frac{\int_0^\infty r(v)I(v)\, dv}{\int_0^\infty r(v)B(v, T_2)\, dv}. \tag{5.13.8}$$

To solve for $I(v)$ we make the special assumption that the radiometer response is constant between wavenumbers v_1 and v_2 and zero outside this spectral range. The spectral integration can then be limited to the range between v_1 and v_2 and r cancels, yielding

$$\int_{v_1}^{v_2} I(v)\, dv = \frac{A_1 - A_3}{A_2 - A_3} \int_{v_1}^{v_2} B(v, T_2)\, dv. \tag{5.13.9}$$

The mean value of $I(v)$ within the spectral response of the radiometer is thereby determined in absolute units. Realistic filter functions often deviate substantially from the ideal rectangular response function considered above, so that Eq. (5.13.9) is then only a more or less valid approximation.

A related case for which Eq. (5.13.8) can be evaluated is a spectral response function for which the spectral width $v_2 - v_1 = \Delta v$ is at most a few wavenumbers. Within such a narrow interval the Planck function changes little, and may be taken outside the integral, so that

$$\int_{\Delta v} r(v) B(v, T_2)\, dv \sim B(\bar{v}, T_2) \int_{\Delta v} r(v)\, dv, \tag{5.13.10}$$

where $\bar{v}$ is the mean wavenumber within Δv. Consequently,

$$\frac{\int_{\Delta v} r(v) I(v)\, dv}{\int_{\Delta v} r(v)\, dv} = \frac{A_1 - A_3}{A_2 - A_3} B(\bar{v}, T_2). \tag{5.13.11}$$

On the left is the mean value of $I(v)$, within the interval Δv, weighted by the instrument function, which may have a nonrectangular shape in this case. Narrow-band interference filters are often bell-shaped; grating spectrometers and Michelson interferometers with triangular apodization have a $(\sin x/x)^2$ response function. If one defines $I(\bar{v})$ as the spectrum weighted by the narrow instrument function one obtains

$$I(\bar{v}) = \frac{A_1 - A_3}{A_2 - A_3} B(\bar{v}, T_2). \tag{5.13.12}$$

For a spectrometer, each of the resolved spectral intervals yields a set of $A_i(v)$ for each object. In effect each spectral interval is calibrated independently of all others.

Another special case invokes the assumption that the spectral shape of the planetary emission is known, except for a constant factor C, that is,

$$I(v) = C I'(v). \tag{5.13.13}$$

Then Eq. (5.13.8) can be evaluated to find C,

$$C = \frac{A_1 - A_3}{A_2 - A_3} \frac{\int r(\nu) B(\nu, T_2) \, d\nu}{\int r(\nu) I'(\nu) \, d\nu}. \tag{5.13.14}$$

We use the radiometer measurements only to adjust the known spectrum, I', as required by the factor C. This method is used often, since planetary spectra are relatively well-known by now, or they can be estimated by the radiative transfer calculations assuming an atmospheric composition and a vertical temperature profile.

In addition to providing the measured spectrum in absolute units, one may also determine the responsivity, the effective temperature, and the parameter α from the calibration measurements. Again, for a spectrometer the responsivity may be found from

$$r(\nu) = \frac{A_2(\nu) - A_3(\nu)}{B(\nu, T_2)}. \tag{5.13.15}$$

Knowing the responsivity as well as the temperatures of the instrument and the detector one can solve the calibration equations for α,

$$\alpha = \frac{1}{B(T_i) - B(T_d)} \left[\frac{A_3}{A_3 - A_2} B(T_2) - B(T_d) \right]. \tag{5.13.16}$$

The Planck functions, the parameters A_i, and α are functions of ν.

To be truly effective the aperture of the calibration blackbody must be large enough to accommodate the full cross section of the entrance pupil of the radiometric instrument. If this cross section is only a few centimeters, a true blackbody, that is, an aperture leading into a large isothermal cavity, can be constructed. The surface emissivity of the internal material of the cavity should be high, but this is not so critical if the aperture hole is small in comparison with the surface area of the cavity interior. The effective emissivity of an aperture in a spherical cavity with an internal surface emissivity of ε is shown in Fig. 5.13.1. For small apertures the effective emissivity approaches unity (Hanel, 1961b). Matters are not substantially different for cylindrical shapes, which are easier to manufacture than spherical cavities. With either shape, however, such a blackbody would be prohibitively large for instruments with large collecting areas. For that reason many radiometer designs have replaced the warm blackbody with a flat, heated plate. A high surface emissivity is achieved with special paints. However, even good paints rarely have infrared

5.13 Calibration

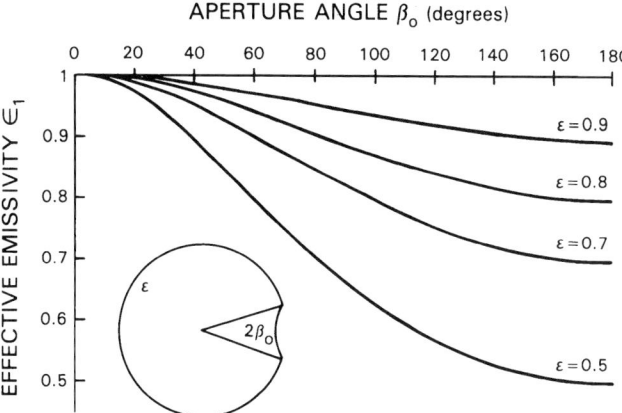

Fig. 5.13.1 Effective emissivity of an aperture in a spherical cavity. The half cone angle of the aperture is β_0. The parameter ε refers to the inside surface emissivity of the cavity.

emissivities higher than 0.97; in the far infrared it is even difficult to obtain a value as high as this. One can improve the emissivity of a plate by cutting deep, V-shaped grooves or by drilling many holes into it before applying the paint. In effect, a large number of small blackbodies are created in this way. With this method emissivities higher than 0.99 can be obtained. For many remote sensing applications this is adequate. Plates with deep concentric grooves served as calibration sources for a large number of infrared instruments. The first scanning radiometers on Nimbus, nearly all infrared radiometers flown on meteorological satellites and planetary spacecraft, as well as the infrared interferometers on Nimbus and Mariner 9, used such grooved plates as the warm calibration source. The cold source has, without exception, been deep space since it is readily available to space-borne instruments.

The calibration of the infrared spectrometer on Voyager presented a special problem. This interferometer, shown in Fig. 5.2.10, has a telescope of 50 cm diameter. Calibration in the telescope focal plane, where the beam cross section is less than 2 cm, would have been possible, but could not provide the desired accuracy, since the whole instrument must be calibrated, including the telescope. Use of a true blackbody with a cavity large enough to accommodate a 50 cm beam diameter was out of the question. Even a grooved metal plate of such a diameter was considered impractical. An entirely different approach was taken. The whole instrument was designed to be as isothermal as possible. Three independent thermostats, one for the interferometer, one for the primary, and one for the secondary mirror hold the instrument at 200 K within a small fraction of a degree. The instrument is thermally insulated from the spacecraft and wrapped in multilayer, thermal blankets, except for the telescope aperture, the mounting foot, and the radiating surface. Without the small thermostatic heaters the instrument would cool to temperatures below

200 K by radiation to space. With thermostatic control the thermopile detector and all optical elements are accurately held at the same temperature.

Suppose that this isothermal instrument views a 200 K blackbody. The detector sees only objects of its own temperature. The motion of the interferometer mirror inside this perfectly isothermal enclosure cannot affect the net flux at the detector; the interferogram must be zero. Consequently, the amplitude $A_2(\nu)$, which is the Fourier component for wavenumber ν of the interferogram, must also be zero and Eq. (5.13.12) simplifies to

$$I(\nu) = \frac{A_1(\nu) - A_3(\nu)}{-A_3(\nu)} B(\nu, T_i). \qquad (5.13.17)$$

The bar over ν has been omitted; it is understood that the measured intensity, $I(\nu)$, is reproduced only with the spectral resolution of the instrument. In reality, the measurement with the 200 K blackbody is not carried out in space; however, it was done in a test chamber on the ground to verify the applicability of Eq. (5.13.17). Therefore, in addition to the planetary measurement, A_1, only a measurement of deep space, A_3, and a knowledge of T_i are needed to achieve absolute calibration (Hanel *et al.*, 1980).

To find the responsivity we use Eq. (5.13.15) with $A_2(\nu) = 0$ and the blackbody temperature equal to that of the isothermal instrument. We find

$$r(\nu) = \frac{-A_3(\nu)}{B(\nu, T_i)}. \qquad (5.13.18)$$

The negative sign compensates for the negative value of A_3, which results from our polarity convention. Again, a deep space measurement and a knowledge of the instrument temperature are sufficient to permit computation of the spectral responsivity. The responsivity is monitored to look for secular changes of the instrument performance.

Repeated measurements of the responsivity also reveal random errors in the system. These errors are expressed for each resolved spectral interval by the standard deviation, $s(r(\nu))$. The relative error is then $s(r(\nu))/r(\nu)$. The Noise-Equivalent-Spectral-Radiance, NESR, is defined by

$$\text{NESR}(\nu) = \frac{s(r(\nu))}{r(\nu)} B(\nu, T_i). \qquad (5.13.19)$$

The NESR is a measure of the random errors of the instrument, expressed in radiometric units. It represents the one-sigma uncertainty in an individual spectrum. This fully quantifies the random errors of the spectrometer. Examples of the responsivity and the NESR of the Voyager interferometer are shown by Hanel *et al.* (1980). For that instrument sources of systematic errors are relatively small; it is mainly errors in the measurement of the instrument temperature that enter in a systematic way [Eq. (5.13.14)]. To provide redundancy and to verify the absence of temperature gradients, the instrument has several temperature sensors, which have been calibrated at the US National Bureau of Standards.

c. Near infrared calibration

In the near infrared and visible part of the spectrum, planetary spectra are dominated by reflected solar radiation. Blackbodies that approximate the spectral distribution of the radiation to be measured would require much higher temperatures than can be conveniently produced in a spacecraft. A calibration lamp with a filament temperature of 2000 to 3000 K can be used; however, calibration lamps, such as normal laboratory standards, are large and power consuming (Hickey, 1970). Smaller lamps may suffice for the detection of changes in the instrument response; this is still useful, even if it does not provide a true radiometric calibration. The most readily available calibration source for the visible and near infrared is the Sun. For many planetary investigations the ratio of reflected to incident solar radiation is of prime interest; therefore, it is logical to use the Sun as a calibration source. The effect of Fraunhofer lines, for example, tends to cancel in the ratio. However, the Sun is such a strong source in comparison with reflected planetary radiation that instruments, designed to be sensitive enough for planetary measurements, cannot be pointed directly at the Sun. Solar radiation must be attenuated to approximately the expected planetary levels. Neutral density filters, opaque disks with small holes, integrating spheres, or diffuser plates may be used to attenuate solar radiation. Calibration of the whole system over the total aperture area becomes more difficult for instruments with large size telescopes. As an example of such a calibration we discuss the on-board calibration method of the IRIS radiometer on Voyager.

The Voyager spacecraft has a large plate mounted on the main bus so that the instruments on the scan platform can view the plate almost normally. The surface of the aluminum plate is chemically etched and scatters light in all directions when illuminated by the Sun. From the visible to ~ 20 μm, the scattering characteristic is that of a near perfect Lambertian diffuser, as verified by laboratory measurements one or two years before launch. Exposure of the calibration plate to sunlight requires a complex maneuver of the spacecraft. Since the telemetry antenna does not point in the Earth direction at that time, such calibration sequences cannot be performed frequently. Calibrations have been carried out a month or so before and after each planetary encounter, except following the Saturn encounter by Voyager 2, where problems with the scan platform articulation mechanism caused cancellation of the Sun calibration sequence.

The advantage of calibration with a diffuser plate is clear. The plate attenuates solar radiation to the level suitable for the dynamic range of the radiometer. The reflectivity of a planetary surface is directly compared with the reflectivity of the calibration plate, which was carefully measured before launch.

The signal, S_p, in volts or in digital numbers (DN), registered by the radiometer, is proportional to the energy reflected from the planetary surface element within the field of view,

$$S_p = C A_i \Omega_i \int_0^\infty r(\lambda) I_p(\lambda) \, d\lambda. \tag{5.13.20}$$

The constant C must be determined in the calibration process; $A_i \Omega_i$ is the étendue and $r(\lambda)$ the normalized spectral response function of the radiometer. The intensity, $I_p(\lambda)$, is the spectral radiance emanating from the planetary surface within the field of view.

When observing the diffuser plate an expression similar to that of Eq. (5.13.20) is obtained,

$$S_d = C A_i \Omega_i \int_0^\infty r(\lambda) I_d(\lambda) \, d\lambda. \tag{5.13.21}$$

In the ratio

$$\frac{S_p}{S_d} = \frac{\int r(\lambda) I_p(\lambda) \, d\lambda}{\int r(\lambda) I_d(\lambda) \, d\lambda} \tag{5.13.22}$$

the calibration constant and the étendue cancel. Tacitly, we assume that the radiometer is linear and that the planet and the diffuser plate are larger than the projected field of view and the instrument aperture, respectively. The form of Eq. (5.13.22) is similar to that of Eq. (5.13.8) used in the discussion of the thermal infrared calibration; the special assumptions made there are equally applicable to the near infrared region.

Eventually, the goal is to find I_p, and from that cloud or surface properties or the planetary albedo, provided we know the instrument response, the intensity of the solar radiation reflected from the diffuser, I_d, and, of course, the measured values S_p and S_d. The task of finding the planetary albedo from such measurements is discussed in Section 8.6. The disadvantages of calibration with a diffusor plate are the uncertainties in the changes of the reflectivity of the plate before and after launch, which may differ from that of the witness sample used in the laboratory calibration. Indeed, differences in the derived albedo measurements between Voyager 1 and 2 point to different degradation factors of the plates on the two spacecraft (see Pearl *et al.*, 1990; Pearl & Conrath, 1991).

d. Wavenumber calibration

The main thrust of the calibration of radiometric instruments concerns the establishment of an absolute intensity scale. However, generation of a trustworthy wavenumber or wavelength scale must not be overlooked. Clearly, in the identification of an unknown spectral feature an accurate knowledge of the wavenumber is crucial. Without a good wavenumber scale the identifications of the nitrile, hydrocarbon, and carbon dioxide features discovered in the Titan spectrum would have been difficult (see Section 6.4).

Broad-band radiometers, where the spectral range is determined by intrinsic material properties, have generally not been calibrated in orbit with respect to wavenumber. In most cases the cut-off wavelengths of optical materials are predictable functions of temperature, but they are assumed to be relatively stable, otherwise. A major shift in the wavenumber range of such a radiometer is not likely, and the scientific objectives of broad-band measurements rarely depend on the cut-off wavenumbers in a critical manner.

In contrast, results from instruments with narrow-band filters, such as interference filters or Fabry–Perot interferometers, may depend strongly on a precise knowledge of the position of the peak wavenumber. An unnoticed shift of the instrument response away from the position of a spectral line may drastically alter the scientific conclusion. Poorly constructed interference filters are sometimes subject to deterioration. For example, individual layers may partially separate from each other or from their substrate due to thermal stresses. Not only does the peak position change in such a case, but so does the maximum transmission. A good intensity calibration, which indicates that the overall transmission of the filter has not changed since the laboratory calibration, is a necessary, but not a sufficient indicator that the position of the peak wavenumber is still correct. To measure the position of the peak wavenumber in orbit one could introduce, sequentially, two low resolution filters, such as the one layer interference filters shown in Fig. 5.6.9. One of these broad filters is tuned to a wavenumber slightly above and the other one to a wavenumber slightly below that of the narrow-band filter for which the peak wavenumber must be checked. While observing the same strong source, the ratio of intensities, measured first with one and then with the other of the low resolution auxiliary filters in the optical path, depends strongly on the position of the peak wavenumber of the narrow-band filter to be checked. It is tacitly assumed that the simple auxiliary filters are more stable than the complex, multilayer filter to be tested; generally, this is a valid assumption.

For tunable filters, whether they are tilted interference filters, Fabry–Perot etalons, or scanning grating spectrometers, the appearance of known lines in the measured spectra is the best means of wavenumber calibration. The tuning range of the instrument must be sufficiently wide to include at least two, preferably several lines of a well-known gas. The gas may exist on the planet or it may be contained in a special calibration cell inside the instrument. For ground-based observations these lines may very well be formed by gases in the Earth atmosphere; they are not Doppler shifted in this case. Planetary spectra are rich in spectral lines that may serve as fiducial marks for wavenumber calibration. There are, however, regions devoid of sharp spectral features, for example, in the domain of the collision-induced hydrogen transitions on the giant planets. A good wavenumber calibration then requires a sufficiently wide tuning range or an on-board calibration cell.

Fourier transform spectrometers, with a reference interferometer for location of sampling positions, have a built-in wavenumber calibration. The laser frequency or the particular line of the gas discharge tube of the reference interferometer is the wavenumber standard against which the planetary feature is compared. As discussed in Section 5.8.a, this property is a distinct advantage of FTS. However, small systematic differences may still occur between true and measured positions of spectral lines. The solid angles of the main and the reference interferometer may not be the same. The apparent foreshortening of the off-axis rays may be slightly different in both interferometers, leading to a small deviation in the wavenumber calibration. Such deviations are linear with wavenumber and generally smaller than the spectral resolution of the instruments (Hanel *et al.*, 1968; Maguire, 1977). They are routinely removed from the spectra in the Voyager data reduction process.

5.14 Choice of measurement techniques

a. Scientific objectives

The merit of observations, especially those made from spacecraft, must be judged in terms of the potential increase in knowledge and the probability of success of the measurement. If present theory is at a crossroad, that is, if a hypothesis can be accepted or rejected on the basis of an experiment, then the particular measurement should be considered. However, the potential value of an investigation and the probability of success are often difficult to predict. Discoveries simply cannot be scheduled. Furthermore, information theory permits specification of the amount of information, but it is silent when the quality of information is to be quantified.

5.14 Choice of measurement techniques

For example, the Voyager 1 picture that showed, for the first time, the image of a volcanic plume on Io must be assigned a much higher information content than all other, subsequent pictures of volcanic eruptions on that relatively small satellite of Jupiter. Strictly from a quantitative point of view all Io frames contain the same amount of information, that is, all contain the same number of bits. Without a clear numerical measure of the significance of information it appears to be difficult to assign a value to a particular measurement. After the fact, however, it is generally quite possible to assign such a value, at least in a relative sense. The picture that permitted the discovery of the volcanic eruption on Io must clearly be considered much more significant than any other picture of Io. Before the fact, such judgement would have been impossible.

How should one then plan an investigation and how should one select a particular instrument approach? For the sake of discussion, we divide scientific objectives into two groups. The first group searches for unknown phenomena. The second group increases the present knowledge by improving on the precision, accuracy, spatial or temporal resolution of quantities that are already known to exist or, at least, can be predicted by theory in a noncontroversial manner. New and significant discoveries may be made within the framework of the second group as well. The first group involves survey-type, general-purpose instruments, such as cameras, mapping radiometers, and spectrometers with good spectral resolution and wide spectral coverage. Discoveries by space-borne cameras include the existence of large, extinct volcanos on Mars by Mariner 9, of rings around Jupiter, of active volcanos on Io, and of the bizarre landscapes of Miranda and Triton by Voyager, just to name a few. The detection of sulfur dioxide in a volcanic plume on Io and of carbon dioxide in the reducing atmosphere of Titan are examples of discoveries made by the Voyager infrared spectrometer. Characteristic of all these discoveries is that none could have been predicted at the time when these instruments were designed.

However, many spacecraft measurements fall into the second group. Meteorological measurements from Earth orbit are typical representatives of this group. The temperature in the Earth's atmosphere can be predicted from climatological data for any place and time to within a few degrees K. The significance of space-borne measurements of atmospheric temperatures is in knowing the deviation from the predicted value, the improvement in precision, and the achieved global converage. The measurement of the energy balances of the giant planets and their atmospheric helium concentrations are further examples where the increase in knowledge comes from improvements in the accuracy of already known facts. Rough estimates on the energy balance of Jupiter existed from ground-based and Pioneer observations before Voyager arrived there. Helium was expected to be present in the atmospheres of all massive planets, simply from the cosmic abundances of elements. The significance of the Voyager measurements of these quantities lies in the achieved accuracy, although further improvements in that direction are desirable.

Examples where a high spatial resolution contributed to a better understanding of conditions are numerous in the short history of space exploration. Great details of topography and surface structure have been revealed by space-borne cameras, photometers, and scanning radiometers. The complexity and fine structure of the rings of Saturn and the other giant planets have been revealed by the cameras, the photopolarimeter, and the radio science investigation of Voyager (Smith *et al.*, 1981; Lane *et al.*, 1982; Tyler *et al.*, 1981).

An example of a measurement where time has played an important role is the surface pressure measurement performed by the Viking landers (Hess *et al.*, 1977, 1979). Seasonal pressure variations were found that correspond to the deposition of atmospheric carbon dioxide on the polar caps. In general, however, discoveries of entirely new phenomena have been made primarily by general-purpose instruments, while single-purpose instrumentation contributed primarily to improvements concerning already known quantities. In this process exciting scientific insight may be gained with single-purpose instruments as well.

b. Instrument parameters

A choice of instrumentation always involves trade-off considerations between signal-to-noise ratio, spectral and spatial resolution, telescope aperture, detector characteristics, frequency response, bit rate, and many other parameters. Clearly, the overall scientific objective should dictate the instrumental approach, but, in reality, the chosen solution must also fit within the framework of available resources such as volume, weight, power consumption, time required to complete the design, and cost. Within reason, these limitations are generally also subject to trade-off considerations. To limit ourselves in the presence of a multitude of instrumental approaches consider the Noise-Equivalent-Spectral-Radiance, NESR, of a radiometric instrument, Eq. (5.8.14),

$$\mathrm{NESR}(\nu) = \frac{\mathrm{NEP}(\nu)}{\eta_1 \eta_2(\nu) \Delta \nu A \Omega \sqrt{\tau}}. \qquad (5.14.1)$$

The Noise-Equivalent-Power, NEP, of the detector is often a function of wavenumber (see Sections 5.9 to 5.11). The efficiency is expressed by the product $\eta_1 \eta_2$; η_1 is the system efficiency, which includes the chopper efficiency, for example, and η_2 is the optical efficiency, which includes the transmission characteristics of optical elements. The spectral bandwidth is expressed by the equivalent width, $\Delta \nu$. For spectrometers $\Delta\nu$ is the spectral resolution. The étendue, $A\Omega$, is the product of collecting areas times solid angle. The telescope obscuration, due to the shadowing effect of the secondary mirror in a Cassegrain configuration, for example, must be subtracted from the total area of the primary mirror. Alternatively, one may calculate the collecting area simply from $D^2\pi/4$, where D is the diameter of the primary, and take the obscuration factor into account in the calculation of the optical efficiency. The observation time, τ, refers to a particular measurement. For the duration τ the field of view is exposed to the same spot on the planet; sometimes τ is called the dwell time.

Although the NESR can be measured without great difficulties, it is often instructive to consider the signal-to-noise ratio, S/N, which is

$$\frac{S}{N} = \frac{I(\nu)}{\text{NESR}(\nu)} = \frac{\eta_1\eta_2(\nu)\Delta\nu A\Omega\sqrt{\tau}I(\nu)}{\text{NEP}(\nu)}. \tag{5.14.2}$$

An estimate of the specific intensity, $I(\nu)$, of the object to be measured is thereby introduced; clearly, it must play a role in optimizing the measurement system. Equation (5.14.2) may be rearranged,

$$\frac{S/N}{\Delta\nu\Omega\sqrt{\tau}} = \frac{\eta_1\eta_2 AI}{\text{NEP}}. \tag{5.14.3}$$

For a particular class of instruments the right side of the above equation is more or less constant. For example, the system efficiency is determined by selecting the chopping mode or a single detector Michelson interferometer. It is 0.35 for both cases; 0.7 accounts for the conversion of the zero to peak value of the signal to the rms value of the noise and 0.5 accounts for the chopper action or the reflection in a standard Michelson interferometer. The S/N ratio is defined as the rms signal over the rms noise level. The optical efficiency will, in all likelihood, be between 0.1 and 0.9, depending on the number and type of optical components; needless to say, one strives to maximize it. The collecting area, A, depends on the telescope size. Volume and weight scale as the third power of the telescope diameter. The value of $I(\nu)$ is given by the planet to be observed and the spectral range needed to reach the scientific objective. The NEP depends on the type and the operating temperature of the detector. For thermal detectors, operating at or near room temperature, the NEP will be close to 10^{-10} W Hz$^{-\frac{1}{2}}$; for cryogenically cooled detectors it will be substantially lower.

However, after one has chosen a reasonable telescope aperture, the type and degree of cooling of the detector, and the lowest intensity for which a reasonable

S/N ratio must be obtained, it would be difficult to increase the value of the right-hand side by even as much as a factor of two. True, one can always increase the collecting area; however, in practice one quickly reaches a limit. This limit is most likely set by the maximum size or weight that can be accommodated on a spacecraft or by the largest telescope available to a ground-based or air-borne observer. Within the boundaries discussed, the right side of Eq. (5.14.3) is a constant, characterizing the class of instrument, but not so much the type.

The terms on the left side of Eq. (5.14.3) are much more flexible and can be traded against each other, as long as the equation balances. For example, enlarging the solid angle by a factor of two doubles S/N, provided the object is large enough to fill the field of view, and $\Delta \nu$ and τ are kept constant; S/N may be traded for spatial resolution. The S/N ratio may also be doubled by increasing the observation time by four. It may appear that an increase in observation time can always be counted on to improve the S/N ratio or to increase spectral or spatial resolution in a particular design, but τ is a valuable commodity and should not be given away lightly. In a low, circular orbit the spacecraft speed sets limits to τ, especially if a number of measurements must be performed while the craft is over a certain area. On a planetary fly-by the time available to view the planet with a certain spatial resolution, say a resolution of better than $\frac{1}{10}$ in diameter, is again limited.

In a discussion of trade-off options it is desirable to consider the S/N ratio per square root of time. The three parameters $1/\Omega$, $S/N\, \tau^{-\frac{1}{2}}$, and $1/\Delta \nu$ may be viewed as the axes of a three-dimensional coordinate system, Fig. 5.14.1. A point within

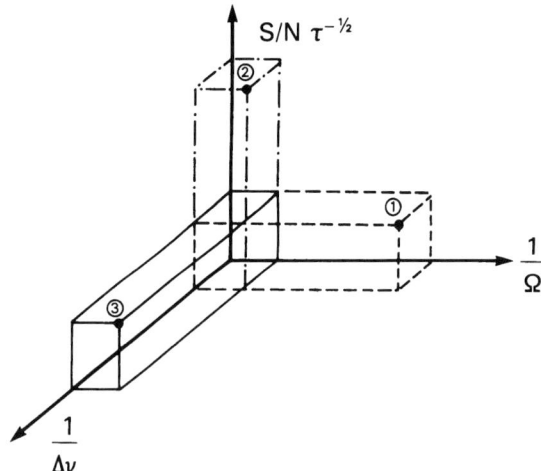

Fig. 5.14.1 Three-dimensional space representing spatial resolution, spectral resolution, and signal-to-noise ratio per square root of time. All radiometric instruments are characterized by a volume in this space.

5.14 Choice of measurement techniques

this system defines a volume, xyz, which equals the more or less constant right side of Eq. (5.14.3). One can emphasize one particular direction. For example, if one chooses to emphasize the direction of $1/\Omega$ (point 1 in Fig. 5.14.1) at the expense of the other directions, one obtains a narrow beam radiometer, capable of spatially resolving great details, but with limited spectral resolution and restrained signal-to-noise. Of course, diffraction sets a limit on how far one may expand in the $1/\Omega$ direction. If one chooses to stretch the volume so that S/N $\tau^{-\frac{1}{2}}$ is emphasized (point 2 in the figure), one obtains a photometer with high precision, but with modest spatial and spectral resolution. If one chooses to design the system to have a narrow spectral interval, then the available volume is stretched in the $1/\Delta\nu$ direction (point 3 in the figure). Clearly, one would like to optimize all three properties and construct an instrument with good S/N, high spectral and spatial resolution, but that is not always possible. It would require a larger constant on the right side – in other words, a larger telescope or a better detector. The design of a remote sensing instrument involves carefully balancing the three parameters and choosing a point in the space shown in Fig. 5.14.1.

The trade-off between the three axes of Fig. 5.14.1 is an important, but not the only consideration in the design of a radiometric instrument. The electrical bandwidth needed to transmit the information may become a significant parameter for scanning radiometers or for spectrometers. So far, the discussion has been limited to single-channel instruments. If one uses an array of detectors or an instrument with a multiplex property, such as the Michelson interferometer, Eq. (5.14.3) still applies to each detector element or to each spectrally resolved interval, but the electrical bandwidth or the bit rate, if the information is converted into digital format, increases according to the number of elements in the array, or the multiplex factor, both expressed by M. The required bit rate is proportional to

$$\text{Bit rate} \sim \frac{M}{\tau} \log_2(\text{S/N}). \qquad (5.14.4)$$

For high resolution mapping instruments, such as the Thematic Mapper, with a number of individual channels and a short observation time per resolved spatial element, the bit rate becomes large, 84.9 Mbit s^{-1}. The same is true for high resolution spectrometers, such as the ATMOS, where the multiplex factor, M, is large and τ is small (15.76 Mbit s^{-1}). For missions to the outer planets the bit rate is limited by the large communication distance; consequently, the telemetry rate becomes an important design parameter.

The overall design of a radiometric instrument must be an iterative process. After the initial concept has been chosen to satisfy the scientific objective, Eq. (5.14.3) must be optimized. Then one must assure that the design does not violate given boundaries such as weight, power, etc. If a conflict is noted either the design has to be modified or the boundaries have to be renegotiated. After these steps are complete one must ask again; does the design still meet the scientific objective? In some cases it may be necessary to scale this objective down. Finally, the whole design sequence should be iterated, probably several times, until one arrives at the optimum solution.

6
Measured radiation from planetary objects up to Neptune

In this chapter we provide a partial survey of infrared measurements of a number of objects in our Solar System. In view of the large quantity of available data, we have to be very selective in choosing examples. As in other parts of the book, we give space-borne spectrometry preference over broad-band radiometry. Results from lower resolution data can always be simulated by smoothing higher resolution spectra with a corresponding instrument function. To provide a more balanced view of the potential of infrared techniques we occasionally include examples of spectrometric, polarimetric, and radiometric results obtained from ground-based telescopes in addition to spectrometric data from spacecraft.

In Section 6.1 we show the effects of finite spectral resolution and other instrument characteristics on the recording of the emerging radiation field. Infrared data from the terrestrial planets, that is Venus, Earth, and Mars, are treated in a comparative way in Section 6.2. Emphasis is given to an understanding of the physical principles that cause the structure in the measured spectra. The spectra of the giant planets are discussed in Section 6.3, again in a comparative manner. Section 6.4 is devoted to Titan; as a satellite with a deep atmosphere it is in a class by itself. The last section in this chapter (6.5) is concerned with astronomical bodies without substantial atmospheres. Mercury, the Moon, and Io are most interesting examples of this class of objects. The numerical treatment of information retrieval is postponed until Chapter 8.

6.1 Instrument effects

Interpretation of planetary measurements is a difficult task, even when emission and reflection spectra are precisely known. In reality, the task is even more complicated because physical parameters of planetary atmospheres and surfaces must be retrieved from data recorded by real instruments. Such devices do not faithfully reproduce planetary spectra, such as the full resolution spectra discussed in

Chapter 4. Instead, instruments modify the true spectral radiances in several ways. The modifications can be understood as filtering processes, which reflect characteristic properties of the instrument involved. The filtering effect must be fully recognized in the analysis and interpretation of the data.

The most significant instrumental effects are the limitations in spectral resolution and signal-to-noise ratio. Within the chosen spectral region radiometric instruments reproduce the spectrum only with a finite spectral resolution. This resolution is an important parameter, which is chosen in the instrument optimization process discussed in Section 5.14. For spectrometers the resolution depends on the type and design of the instrument. For example, maximum mirror displacement determines the spectral resolution of a Michelson interferometer, while the bandwidths of individual channels provide the resolution of a multichannel radiometer. In either case, only the radiances convolved with the instrument function as recorded by the radiometric device are available for interpretation. In addition, all radiometric measurements contain a certain level of random noise, which is commonly expressed as the Noise-Equivalent-Spectral-Radiance (NESR). The ratio of the radiance, $I(\nu)$, of the object under investigation to the NESR(ν) of the instrument is the signal-to-noise ratio, $S/N(\nu)$. A low value of S/N may have a significant effect on the quality of the data and on the uncertainties of the deduced parameters.

Besides the main instrumental effects, finite spectral resolution, and random noise, other modifications of the true spectrum occur. For example, the recorded data cover only certain, often very small parts of the total spectrum. Other spectral regions may offer the opportunity to provide redundant information. Simultaneous analysis of redundant regions is often desirable in order to gain confidence in the deduced conclusions. Radiometric instruments are also subject to systematic errors. Depending on the quality of the on-board calibration system, such errors can be kept within tolerable limits; however, the true magnitude of systematic errors is often difficult to estimate.

The numerical output of a remote sensing instrument refers to the average radiance within the field of view. Inhomogeneous conditions, such as may be produced by a partial cloud cover, generate a single set of data applicable to a weighted average of clear and cloudy conditions. Lack of homogeneity is often difficult to recognize. Images, if recorded simultaneously and at a higher spatial resolution than the infrared measurements, may be helpful in judging the uniformity within the field of view. Finally, when the diameter of the field of view projected onto the object of interest is not small in comparison with the diameter of that object, the emission angle and the solar incidence angle may assume wide ranges of values within the field of view; the recorded spectrum represents an average over these angles as well.

The most significant instrumental parameters that affect radiometric measurements are spectral resolution and the signal-to-noise ratio. Other effects can be

6.1 Instrument effects

minimized by choosing the proper spectral range, by providing an adequate onboard calibration, and by taking data when the field of view is small in comparison with the scale of gross inhomogeneities on the target object. Of course, these conditions cannot always be met; limitations and their effects on the results must then be examined for a particular set of measurements. As an example, we discuss the effect of spectral resolution on the information content of a measurement.

For this analysis we select the Martian emission spectrum between 500 and 800 cm^{-1} because the 667 cm^{-1} CO_2 band is well isolated and nearly free of lines from other constituents. The Martian spectrum has been measured by the infrared spectrometer on Mariner 9 with a spectral resolution of 2.4 cm^{-1} and a good signal-to-noise ratio. Random errors are not noticeable in the large average of spectra shown in Fig. 6.1.1. The set of emission spectra displayed in Fig. 6.1.2 have been calculated with different spectral resolutions for a pure CO_2 atmosphere with a surface pressure of 7 mbar and a surface temperature of 275 K. A temperature profile similar to that deduced from the measured spectrum, shown in Fig. 6.1.1, served in the radiative transfer calculations. The computed high resolution emission spectrum has been convolved with $(\sin 2\pi \Delta \nu / 2\pi \Delta \nu)^2$ instrument functions

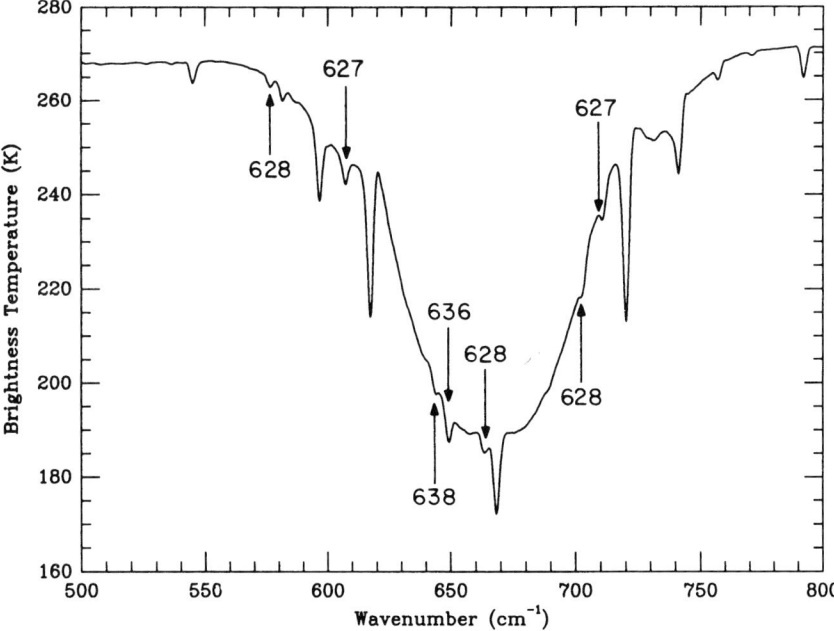

Fig. 6.1.1 Average thermal emission spectrum of Mars between 500 and 800 cm^{-1} recorded by Mariner 9 at a spectral resolution of 2.4 cm^{-1}. The arrows point to isotopic features of CO_2. The number 628 refers to $^{16}O^{12}C^{18}O$, for example, and 638 indicates $^{16}O^{13}C^{18}O$. All unmarked features are due to $^{16}O^{12}C^{16}O$. (See also Maguire, 1977.)

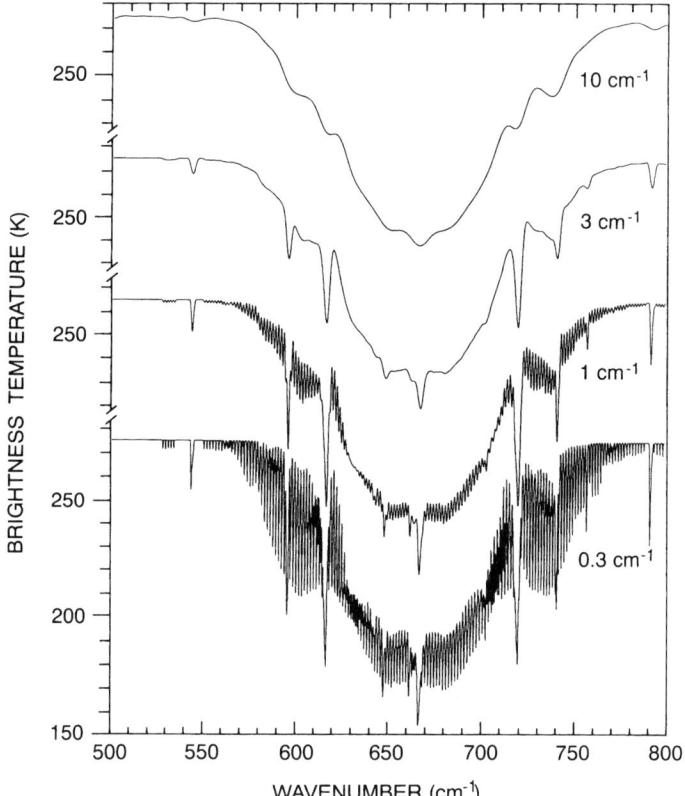

Fig. 6.1.2 Calculated emission spectra assuming a CO_2 atmosphere with a surface pressure of 7 mbar, a surface temperature of 275 K, and a typical Martian temperature profile. Different spectral resolutions have been used in the calculations as indicated on the right side.

of different half width, $\Delta \nu$, as indicated in the figure. This is the instrument function of a grating spectrometer and of a Michelson interferometer with triangular apodization. The spectrum calculated with a resolution of 3 cm^{-1} is closest to the 2.4 cm^{-1} resolution of the Mariner 9 spectrum displayed in Fig. 6.1.1.

An examination of the spectra shown in Fig. 6.1.2 permits a qualitative judgment of the order of magnitude of spectral resolution necessary to achieve a certain objective. For example, an analysis of the isotopic ratio $^{16}O^{13}C^{16}O/^{16}O^{12}C^{16}O$ from the strength of the feature near 648 cm^{-1} can be accomplished with a resolution of 3 cm^{-1}, but not with a resolution of 10 cm^{-1}; a 1 cm^{-1} resolution would be preferable for that purpose. The final judgment on the optimum resolution required must come from an examination of the end product of the spectral measurement, which may be the isotopic ratio or the vertical temperature profile, for example. In the temperature analysis the vertical resolution, as well as the range in altitude over which the temperature profile can be retrieved, is affected by the spectral resolution. Random noise in the spectral measurement can have a strong effect on the probable error of a derived temperature profile. The effects of spectral resolution and random noise on the derived quantities are discussed further in Chapter 8, where we deal with numerical retrieval methods. However, many qualitative statements can be made just from a visual examination of the spectra.

6.2 The terrestrial planets

In this section we compare in a qualitative way thermal emission spectra of Venus, Earth, and Mars. As noted in Appendix 3, Venus and Earth are nearly equal in size; Mars is somewhat smaller. Earth and Mars rotate at almost the same rate, while Venus is turning very slowly. The atmospheres of Venus and Mars consist mostly of carbon dioxide, while that of Earth is predominately nitrogen and oxygen with only a trace of carbon dioxide. The planets differ substantially in their mean distance from the Sun, atmospheric pressure at the surface, and mean surface temperature.

The atmosphere of Mars is fairly transparent, except for several CO_2 bands, weak water vapor lines, spectral features of occasional water ice clouds, and signatures of dust periodically stirred up by strong surface winds. Emission from the solid surface can be observed over wide spectral ranges. The atmosphere of Earth is much more opaque. Absorption by strong bands of H_2O, CO_2, and O_3 (ozone) block infrared transmission from the surface over a significant part of the spectrum. In the thermal infrared the surface can be observed only in certain spectral windows (780–1000 cm^{-1}, 1080–1240 cm^{-1}, and 2500–2800 cm^{-1}), and there only under cloud-free conditions. The surface of Venus is completely hidden by several cloud layers, largely of sulfuric acid droplets, and at lower altitudes by far-wing pressure broadening and collision-induced absorption of gaseous carbon dioxide. Only in the microwave region is the atmosphere of Venus transparent enough to permit surface observation.

Because of these dissimilarities we would expect to see great differences in the emission spectra of these three planets. Indeed, large differences do exist in some spectral regions, but in others striking similarities are apparent, as shown in Fig. 6.2.1. The Venus spectrum was recorded by Venera 15 at a resolution of about 5 cm^{-1} in 1983 (Spänkuch *et al.*, 1984; Zasova *et al.*, 1985; Moroz *et al.*, 1986). The Earth spectrum was measured by Nimbus 4 at a resolution of 2.8 cm^{-1} in 1970 (Hanel *et al.*, 1971, 1972c; Conrath *et al.*, 1971; Kunde *et al.*, 1974). The Mars spectrum was observed by Mariner 9 at a spectral resolution of 2.4 cm^{-1} in 1971 (Hanel *et al.*, 1972a, b; Conrath *et al.*, 1973; Maguire, 1977). The spectra shown are typical of mid-latitude conditions. Particularly on Earth and Mars great variabilities exist, mostly between low latitudes and polar regions, but also between

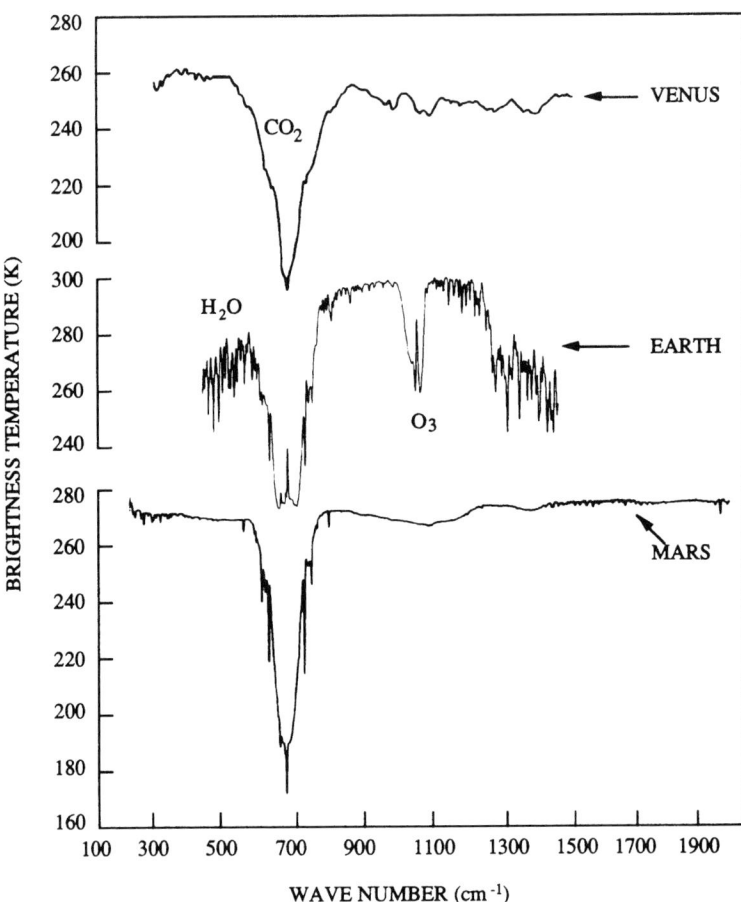

Fig. 6.2.1 Thermal emission spectra of Venus, Earth, and Mars, all measured by Michelson interferometers. The Venus spectrum was recorded from Venera 15; the Earth and Mars spectra are from Nimbus 4 and Mariner 9, respectively. See also Moroz et al. (1986) and Hanel et al. (1972c and 1972a).

clear and cloudy areas. Despite many differences among the spectra shown in Fig. 6.2.1, certain intervals, such as the region between 550 cm^{-1} and 750 cm^{-1}, display great similarity. The spectral signature of this domain is characteristic of the 667 cm^{-1} system of CO_2 bands. This is not surprising for Venus and Mars since their atmospheres consist predominately of carbon dioxide, but on Earth CO_2 is only a minor constituent. However, the major gases of the Earth's atmosphere, nitrogen and oxygen, are homopolar molecules without dipole moments and, therefore, without infrared vibration or rotation spectra. Collision-induced absorption also is insignificant at terrestrial pressures and path lengths. The third most abundant atmospheric constituent, argon, is a monatomic gas and also lacks an infrared spectrum. Therefore, the emission spectrum of the Earth is dominated by signatures from minor atmospheric constituents and to a lesser extent by surface emission.

The absorption-type shapes of the 667 cm^{-1} CO_2 band in all three spectra indicate decreasing atmospheric temperatures with increasing altitude on all three planets. An exception is the reversal of the Q-branch in the spectrum of Earth at the 667 cm^{-1} band center. This reversal signals warmer atmospheric layers above a temperature minimum (compare the measured Earth spectrum of Fig. 6.2.1 with the computed spectrum of Fig. 4.3.2; see also Fig. 4.2.5). The temperature rise in the Earth's stratosphere is caused by absorption of solar ultraviolet radiation in the ozone layer. The lower brightness temperatures in the Q-branch at 667 cm^{-1} on Mars and Venus indicate the absence of such a warm layer and a continuous reduction of atmospheric temperatures with increasing height, at least at mid-latitudes and to pressure levels where the contribution function corresponding to the center of the Q-branch has its maximum [see Figs. 8.2.2 and 9.1.1 (below 100 km)]. For Mars and a resolution of 2.4 cm^{-1} this corresponds to a pressure of about 0.1 mbar or an altitude of about 40 km above the surface.

As this example shows, even without going through a formal retrieval process, we can draw certain conclusions on the temperature structure, just by applying the knowledge gained in Chapter 4 to a visual inspection of the emission spectrum. The same can be said about the presence of gases, if their signatures can be identified at the proper spectral positions. Of course, it is implied that we have *a priori* knowledge of the spectral characteristics of potentially present gases. If we want to determine the temperature versus pressure precisely, or to find the concentration of a particular constituent in terms of the molar abundance, the more quantitative methods discussed in Chapter 8 must be applied.

By inspecting the emission spectrum we can also reach a preliminary judgment on the suitability of a spectral region for retrieving the temperature profile. Since the CO_2 concentration is known on all three planets and is nearly constant with altitude over the range of interest, the strong CO_2 band may be a good candidate for a numerical retrieval analysis, provided this spectral region is free of absorption from other atmospheric gases or particulates. On Mars this is generally the case. Between the most absorbing Q-branch at 667 cm^{-1} and the atmospheric window region near 800 cm^{-1} the CO_2 absorption coefficient changes over many orders of magnitude and the contribution function shifts from high in the atmosphere down to the surface. Absorption by silicate dust and occasionally by clouds of water ice are the prime sources of interference for temperature retrieval on Mars.

On Earth the same spectral interval (from 667 cm^{-1} to 800 cm^{-1}) is suitable for temperature retrieval in cloud-free areas (Kaplan, 1959). Indeed, this spectral region served in the first derivations of the vertical temperature profile from the Nimbus 3 meteorological satellite, and initiated a new era in weather forecasting

(Hanel & Conrath, 1969; Wark & Hilleary, 1969). Nimbus 3 was launched into Earth orbit on 14 April, 1969. In this spectral region some interference by weak O_3 absorption must be considered in precise analyses. On Venus, the use of the 667 cm^{-1} CO_2 band is restricted to altitudes above the opaque cloud deck at approximately 65 km above the solid surface.

In all spectra shown in Fig. 6.2.1, several other features can be attributed to CO_2 besides the 667 cm^{-1} band. These features, only weakly present in the Martian spectrum, include the bands at 961 cm^{-1} and 1064 cm^{-1}; the distinct Q-branches near 545 cm^{-1}, 791 cm^{-1}, 830 cm^{-1}, and 865 cm^{-1}; and the strong features at 1918 cm^{-1} and 1932 cm^{-1}. For comparison, a ground-based spectrum of Venus (Fig. 6.2.2) shows the CO_2 bands between 750 cm^{-1} and 1000 cm^{-1} at a much higher spectral resolution of 0.2 cm^{-1} (Kunde et al., 1977). If the spectral display of Fig. 6.2.1 extended to higher wavenumbers, then the strong ν_3-band of CO_2 would appear at 2349 cm^{-1}. This is another spectral region where temperature sounding is feasible. Temperature retrieval in the atmosphere of the Earth is also

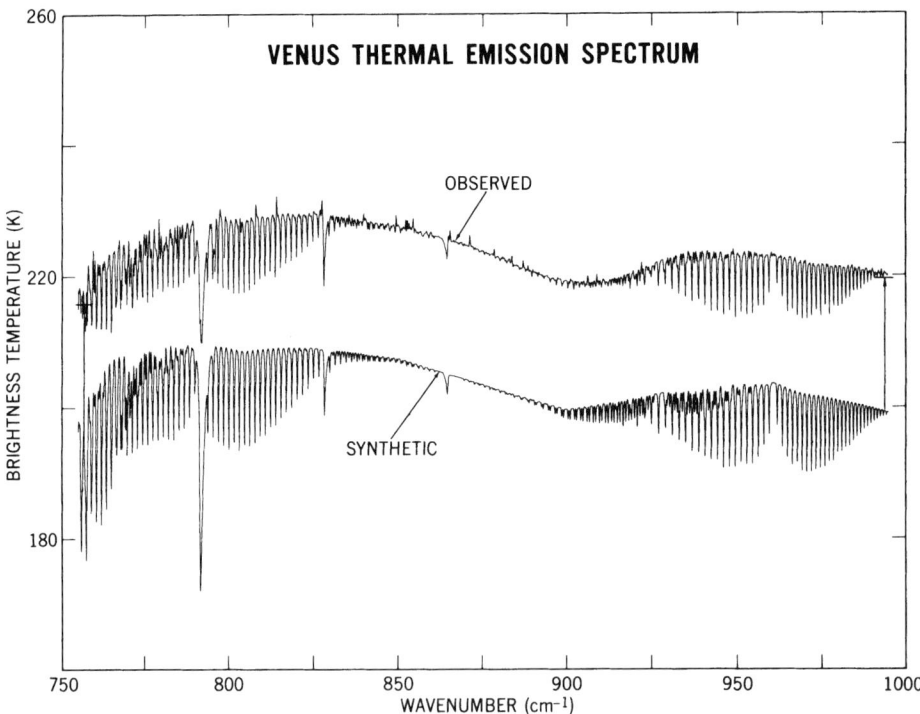

Fig. 6.2.2 Measured and calculated thermal emission spectra of Venus. The observed full disk spectrum was recorded by a Fourier transform spectrometer from the MacDonald observatory in Texas. The spectral resolution is 0.2 cm^{-1}. The upward spikes in the measured spectrum are due to overcompensation of strong water vapor lines in the Earth's atmosphere. (Kunde et al., 1977.)

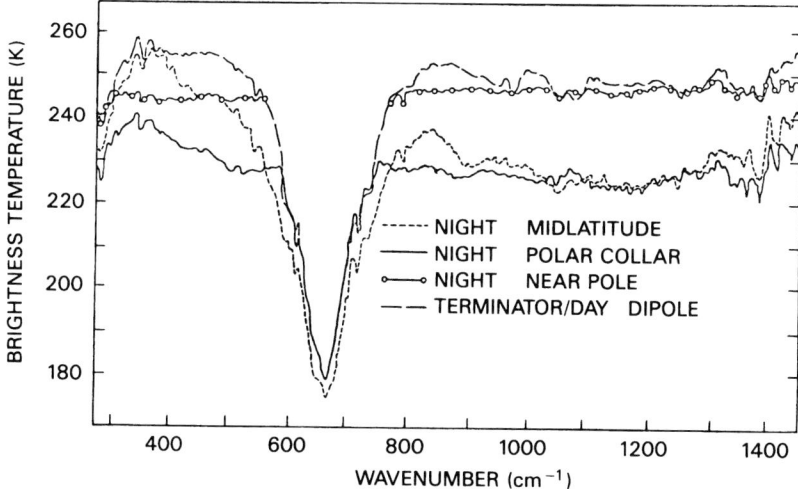

Fig. 6.2.3 Venus emission spectra from different latitudes recorded by Venera 15 (Spänkuch et al., 1984).

possible in the microwave region, using the strong magnetic oxygen line at 57.3 GHz (1.914 cm^{-1}).

Signatures of water vapor are scattered throughout the terrestrial spectrum shown in Fig. 6.2.1; below 500 cm^{-1} and between 1300 cm^{-1} and 2000 cm^{-1} the atmosphere is nearly opaque at low latitudes. Water vapor features also appear in the spectra of Mars and Venus, but are reduced in strength. The Venus spectrum also shows the broad absorption features of liquid sulfuric acid (H_2SO_4) clouds near 900 cm^{-1} (Fig. 6.2.2) and possibly of sulfur dioxide (SO_2) gas near 1360 cm^{-1} (Fig. 6.2.3). In the Earth spectrum the strong ν_3-band of ozone (O_3) at 1042 cm^{-1} and the ν_4-band of methane (CH_4) at 1306 cm^{-1} can easily be identified.

Over wide intervals the spectra of Venus show differences in brightness temperature of several tens of degrees Kelvin. The contrast is most pronounced between the warm, north polar area and the cooler, circumpolar band near 70° North, known as the polar collar, see Fig. 6.2.3 (Spänkuch et al., 1984). The apparent temperature differences reflect a combination of genuine differences in kinetic temperatures between observed areas, and differences in vertical concentrations and droplet sizes of cloud particles (see Subsection 4.1.b). It is difficult to evaluate cloud effects by casual examination of the spectra; methods discussed in Chapter 8 are needed for a thorough interpretation. Actual temperature gradients provide the driving forces for strong circumpolar winds. Latitudinal temperature differences, as well as a dipole shaped polar pattern have been observed by the pressure modulated radiometer on Pioneer Venus (Taylor et al., 1980). Earlier, ground-based thermal maps of Venus already indicated a circumpolar disturbance (Murray et al., 1963). Ground-based

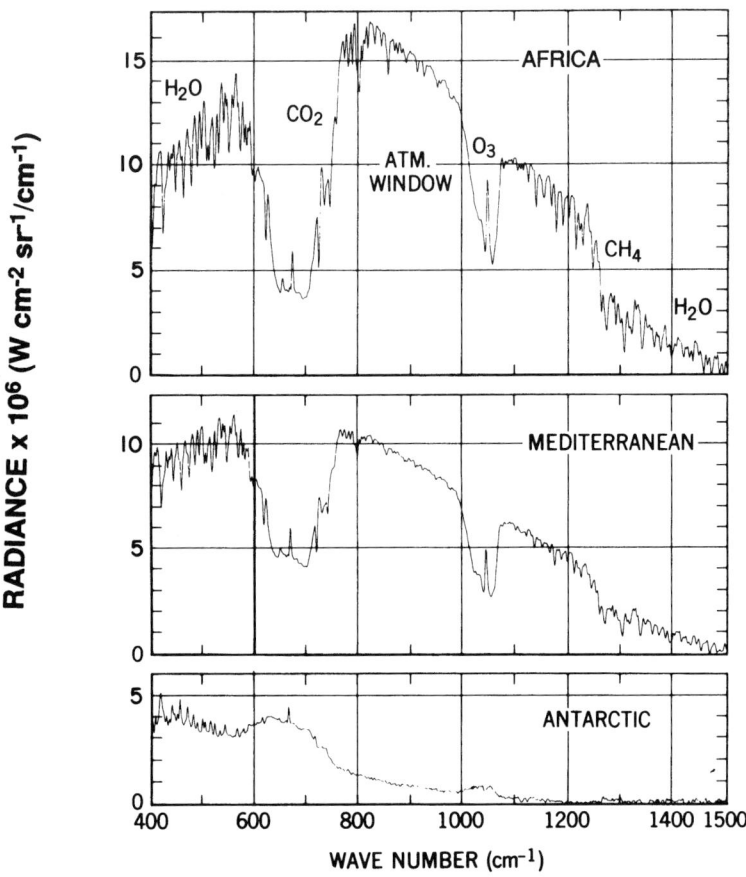

Fig. 6.2.4 Typical terrestrial emission spectra of cloud-free areas recorded over Central Africa, the Mediterranean Sea, and the Antarctic from Nimbus 4 (Hanel et al., 1972c).

spectroscopic measurements in the near infrared have discovered traces of hydrogen chloride (HCl) and hydrogen fluoride (HF) (Connes et al., 1967c) and of carbon monoxide (CO) (Connes et al., 1968).

The Earth spectra, recorded by Nimbus 4 in 1970, show strong contrast between tropical and polar latitudes (Fig. 6.2.4). In the Antarctic spectrum the whole 667 cm^{-1} band of CO_2, the water vapor lines below 550 cm^{-1}, and the O_3 band at 1042 cm^{-1} appear in emission. This indicates warmer atmospheric levels above a cold surface. The 1042 cm^{-1} O_3 band is hardly visible because stratospheric and surface temperatures are nearly the same. In the African spectrum surface emission, emerging in the atmospheric window, is strong. The high atmospheric lapse rates cause the large contrast in the CO_2 band. The ozone band appears in absorption because temperatures in the stratosphere, where most of the ozone resides, are lower than at the surface.

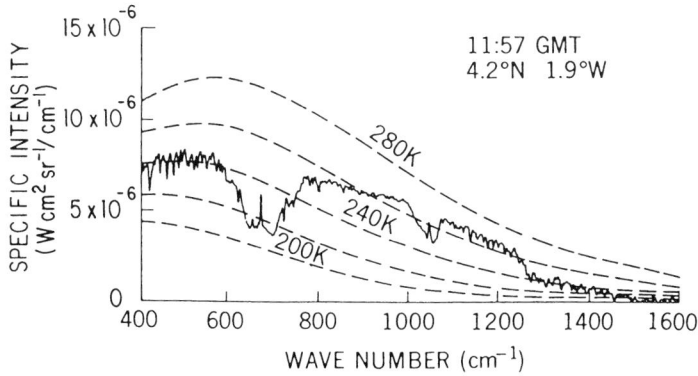

Fig. 6.2.5 Emission spectrum from a cloudy area (Hanel *et al.*, 1972c).

The spectrum of Fig. 6.2.5 is distinctly different from the tropical spectrum of Fig. 6.2.4, although both were recorded not too far apart. The former is from a densely clouded area while the latter is from a clear region. Clouds affect not only the overall radiance level, but also the shape of the spectrum. The slope in the brightness temperature between 400 cm^{-1} and 1000 cm^{-1} and the nearly constant level between 1100 cm^{-1} and 1250 cm^{-1} are characteristic of dense, high clouds.

Surface effects can be observed in Fig. 6.2.6, where a Sahara spectrum is compared to one recorded over the densely forested region of central Africa. Substantial differences appear in the atmospheric window between 1100 cm^{-1} and 1250 cm^{-1}, while the low wavenumber parts of the window between 800 cm^{-1} and 1000 cm^{-1} are very much alike. Strong water vapor bands are evident in all cases. The substantially lower brightness temperatures in the 1100 cm^{-1} to 1250 cm^{-1} window of the Sahara spectrum are caused by a lower emissivity of the exposed silicate sand in that region, in contrast to the near blackbody behavior of dense vegetation. Differences in the brightness temperatures between both atmospheric window regions indicate the presence of silicate sand. Prabhakara & Dalu (1976) used this phenomenon to map the global distribution of desert areas. The low wavenumber part of the same spectral window also permits measurement of sea surface temperatures, land temperatures, and the thermal inertia of the surface.

The spectra of Mars (Fig. 6.2.7) display a north and two south polar spectra recorded during the dust storm of 1971. The north polar spectrum (a) shows atmospheric features in emission, resembling the Antarctic spectrum of the Earth (Fig. 6.2.4). The south polar spectrum (b) shows the whole 667 cm^{-1} CO$_2$ band, water vapor lines below 400 cm^{-1}, and broad features due to suspended dust particles (most noticeable near 480 cm^{-1} and 1100 cm^{-1}), appear stronger than the background. The background, indicated by the smooth solid line in the lowest spectrum, (c), deviates from a blackbody spectrum shown in the middle, (b); however,

312   *Measured radiation from planetary objects up to Neptune*

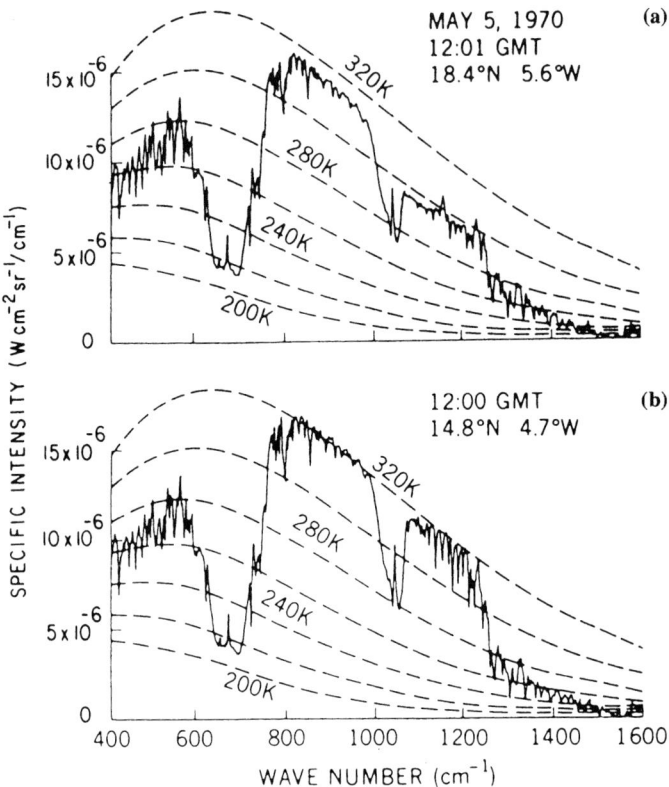

Fig. 6.2.6 (a) Emission spectra from a desert area showing the effect of low emissivity between 1100 and 1250 cm^{-1} caused by residual rays in quartz sand. (b) The comparison spectrum from an area covered by vegetation shows nearly the same brightness temperature on both sides of the ozone band at 1042 cm^{-1} (Hanel *et al.*, 1972c).

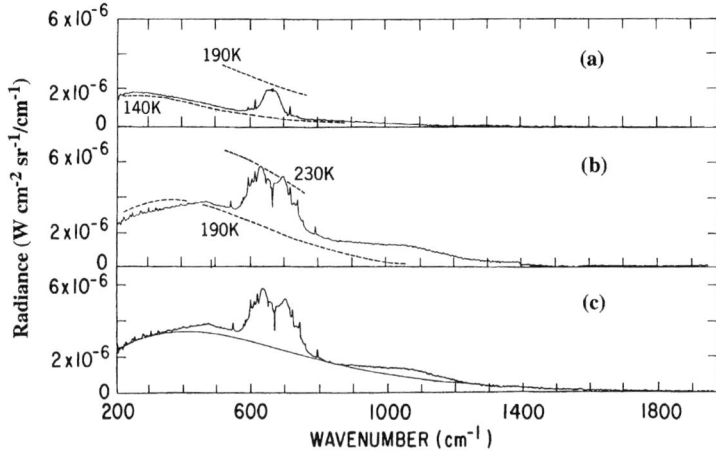

Fig. 6.2.7 (a) North and (b) and (c) two identical south polar spectra of Mars recorded by Mariner 9. Spectrum (b) also shows a blackbody spectrum of 190 K for comparison, while the smooth curve in spectrum (c) is the sum of two blackbody spectra, see text.

it is fit reasonably well by the sum of two blackbody spectra, one at 140 K covering 65% of the field of view, and the other at 235 K covering the rest (Hanel *et al.*, 1972*b*). At the time of the observation, it was late spring in the southern hemisphere. The polar cap was evaporating rapidly, and about one-third of the field of view was already ice-free. The lower blackbody temperature of 140 K can be identified with the equilibrium temperature of CO_2 ice at the Martian surface pressure, and the higher temperature of 235 K with the radiative equilibrium temperature of the ice-free surface areas. The surface had warmed rapidly as soon as the ice disappeared. The presence of water vapor lines in emission in the same south polar spectrum is consistent with this model; with the complete evaporation of the polar CO_2 cap in several areas, small amounts of water are freed, which had been frozen below and in the CO_2 ice deposit in the previous fall. At the time of the Mariner 9 arrival the north polar area was in shadow and the north polar cap was forming (Fig. 6.2.7). The surface is approximately at 145 K, low enough to form CO_2 ice at a pressure of about 8 mbar.

The mid-latitude spectra (Fig. 6.2.8) have been recorded (a) at the height of, (b) in the latter part of, and (c) after the dust storm. The upper two spectra show strong atmospheric dust features in absorption, particularly between 900 and 1200 cm^{-1}. In addition, the 667 cm^{-1} CO_2 band is relatively subdued, especially in Fig. 6.2.8 (a). The spectrum in Fig. 6.2.8 (c) was recorded in July 1972, after most of the dust had settled to the surface; it shows a pronounced contrast in the CO_2 band and hardly any dust features.

Figure 6.2.8 demonstrates a perfect example of a 'laboratory experiment' showing the effects of aerosols in a planetary atmosphere; in this case the laboratory is Mars. During dusty conditions the suspended dust absorbs sunlight and heats the atmospheric gas by conduction. At the same time the surface is partially shielded from direct sunlight and does not reach as high a temperature as it would under dust-free conditions. As a result the temperature contrast between atmosphere and surface is lowered, and the atmospheric temperature profile becomes more nearly isothermal; consequently, the 667 cm^{-1} CO_2 band shows relatively weak contrast. Later, after the dust had settled, the surface temperature had risen and the atmospheric temperature had dropped, thereby enhancing the apparent strength of the CO_2 absorption band relative to the adjacent continuum.

None of the laboratory spectra of fine-grain minerals that were available at the time fitted the absorption maximum at 1100 cm^{-1} (9 μm) in the IRIS data very well. This made it difficult to associate a specific mineral with the dust composition. Hanel *et al.* (1972*b*) suggested the SiO_2 content of the dust to be about 60 ± 10%, indicating differentiation of Mars into a silicate-rich mantel and an iron-rich core. The mineral composition of the dust was further studied by Toon *et al.* (1977) using a radiative transfer model. They concluded that a small size range of particles (2.7 μm mean radius) such as montmorillonite basalt, as previously suggested by

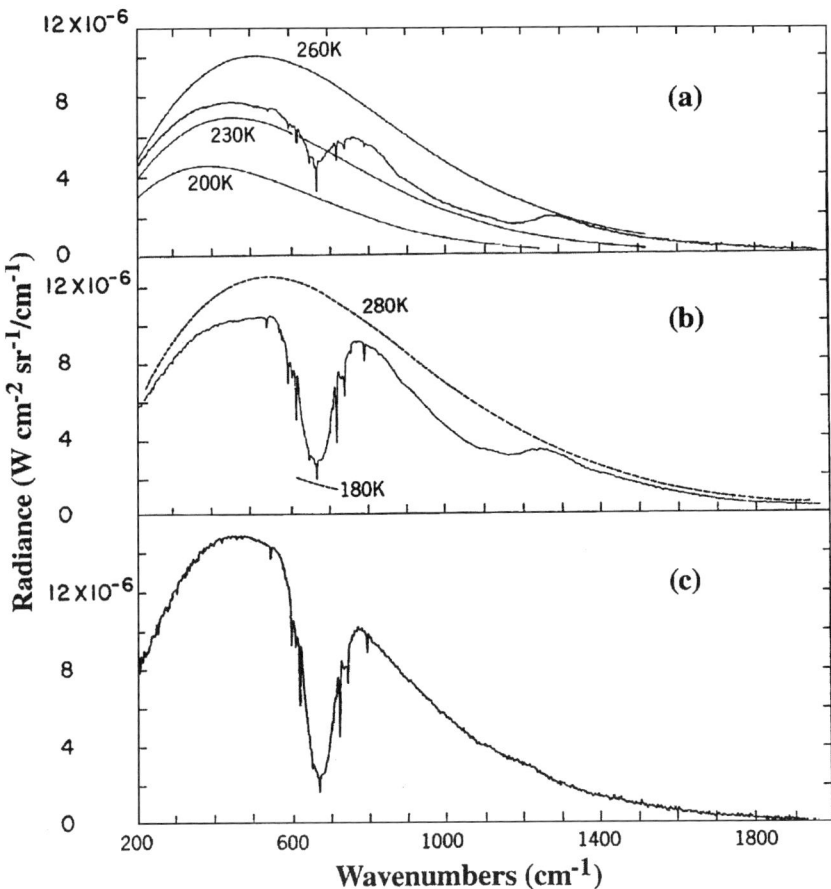

Fig. 6.2.8 Martian mid-latitude spectra recorded by Mariner 9 (a) during the dust storm of 1971, (b) towards the latter part of the storm, and (c) after the storm ended.

Salisbury (see Hunt et al., 1973), produced an excellent fit to the broad absorption region around 9 μm in the IRIS spectrum. Clancy et al. (1995) reexamined that problem and found that a palagonite-like composition with a wider size range and much smaller particles produced a better overall fit. This took into consideration the long wavelength (17 to 30 μm) range of the IRIS spectra as well as the visible-to-9 μm dust opacity and the ultraviolet albedo of Mars. However, the 7 to 8 μm region was less well reproduced. The recently discovered calibration error of IRIS (a ghost of the 15 μm band at half the wavelength; see Formisano et al., 2000) may account for the discrepancy, at least in part.

An example of the spectral signatures of water ice clouds is shown in Fig. 6.2.9 (Curran et al., 1973). As confirmed by the cameras of Mariner 9, the field of view of the infrared instrument contained several white clouds at the time the Tharsis Ridge

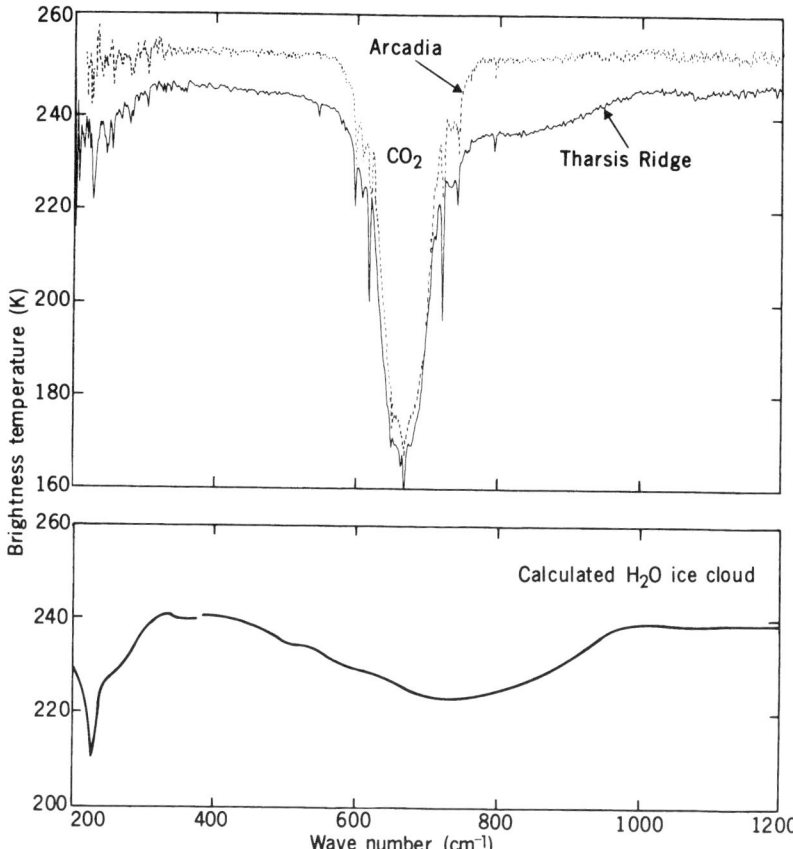

Fig. 6.2.9 Martian spectra over the Tharsis Ridge and over Arcadia. The Tharsis Ridge area contained water ice clouds. The lower spectrum was calculated with such clouds (Curran *et al.*, 1973).

spectrum was taken. The adjacent spectrum of Arcadia was nearly cloud free. The lower panel of the same figure shows a calculated spectrum of small ice crystals in a model atmosphere of Mars. The Tharsis Ridge spectrum compares favorably with that of the ice cloud, including the strong feature at 229 cm^{-1}. Water ice clouds in the Earth atmosphere have been studied by Prabhakara *et al.* (1988) using Nimbus 4 IRIS data.

The Viking Orbiters carried an Infrared Thermal Mapper (IRTM) with broadband channels centered at 7, 9, 11, 15, and 20 μm, as well as one channel covering the visible and near infrared (0.3 to 3 μm), designed to measure the albedo. A description of the instrument and scientific results have been published by Chase *et al.* (1978) and by Kieffer *et al.* (1977). Christensen (1998) compared the results of the Viking IRTM and the Mariner 9 IRIS in order to gain information on the surface composition of Mars. This was in preparation for the interpretation of the TES spectra obtained by the Mars Global Surveyor.

The near infrared imaging spectrometer (ISM) on Phobos used the 2 μm CO_2 band to provide the depth of the atmosphere and, therefore, a topographic map. Bands of CO and H_2O resulted in maps of these gases, respectively. The average water vapor content was found to be about 11 precipitable-μm. From aerosol spectra a mean particle size of 1.25 μm and a size distribution was derived; both the mean particle size and the width of the size distribution are about one half of that given by Toon et al. (1977). For the bright regions a surface composition of poorly crystalline iron oxides was suggested, and for the dark regions two-pyroxene basalt. Results have been reported by Bibring et al. (1989), Drossart et al. (1991), Rosenqvist et al. (1992), Murchie et al. (1993), Erard & Celvin (1997), and others (see references therein).

A description of the Pathfinder Mission and excellent pictures of the Martian surface can be found in the articles by Golombek et al. (1997) and by Golombek (1998).

Mars Global Surveyor was launched in November, 1996. Shortly thereafter, on its way to Mars, the Thermal Emission Spectrometer (TES) observed the Earth. Although our planet filled only a fraction of the field of view, a reasonable spectrum of the Earth was obtained, demonstrating the proper performance of the instrument (Christensen & Pearl, 1997). First Martian results from the Global Surveyor Mission were reported in *Science* (1998, **279**, 1671–98). More in-depth analyses can be found in the *Journal of Geophysical Research* of April 25, 2000, pages 9507–739. Using a radiative transfer model (and a deconvolution model) TES results suggest a surface composition about 53% (43%) plagioclase feldspar, 19% (26%) clinopyroxine, 12% (12%) olivine, 11% (15%) sheet silicate, and possibly very small amounts of quartz and gypsum (Christensen et al., 2000b). Crystalline hematite was discovered in the Sinus Meridiani region but nowhere else (Christensen et al., 2000a). Limb observations in the Northern Hemisphere reveal low-lying dust hazes and detached water ice clouds up to an altitude of 55 km (Christensen et al., 1998). A dust storm was observed in the Noachis Terra region. Examples of atmospheric temperatures, derived zonal wind velocities, and dust opacities are shown. Atmospheric temperatures during the aerobraking and science phasing of the mission have been discussed by Conrath et al. (2000), while M. D. Smith et al. (2000) have presented dust opacity results. Samples of nadir and limb spectra are displayed in Fig. 6.2.10.

Malin & Edgett (2000) have obtained MGS MOC images that they interpret as evidence for sedimentary layering, which would imply the existence of liquid water in the early history of Mars. The global topography of Mars was measured by the Laser Altimeter (MOLA), also on board of the Mars Global Surveyor (D. E. Smith et al., 1999). This altimeter investigation has produced the best topographic maps of Mars so far.

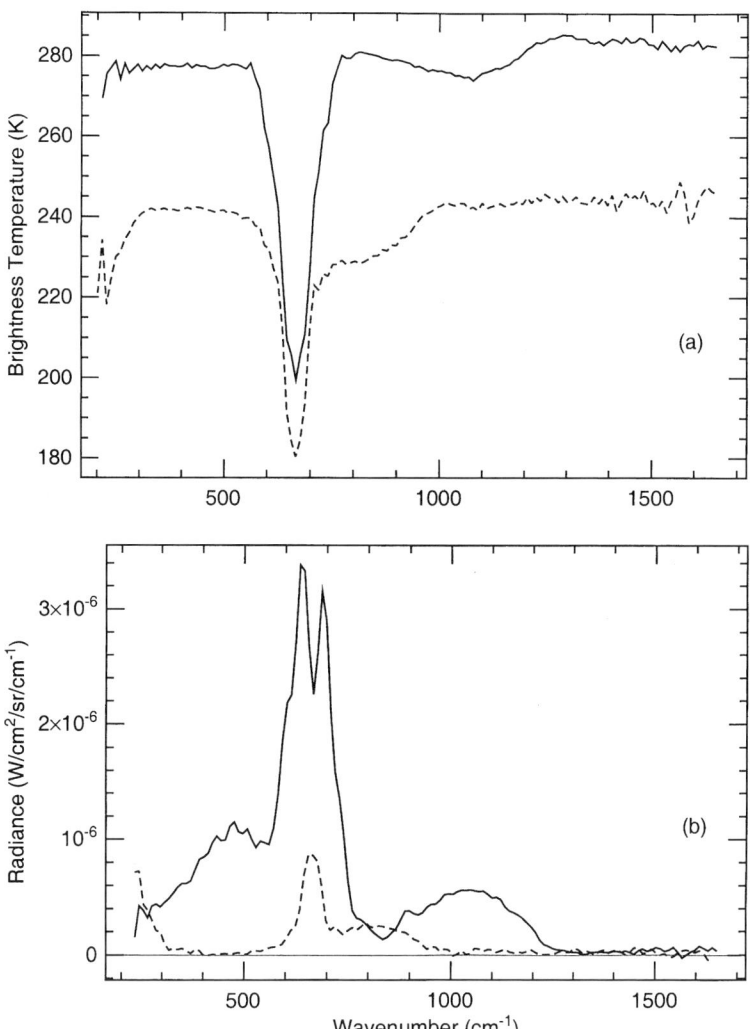

Fig. 6.2.10 Representative TES spectra showing dust and water ice features. Spectra dominated by dust features are shown as solid lines, while those dominated by ice features are dashed. Signatures of gaseous CO_2 (strongest near 667 cm^{-1}), water ice clouds (strongest near 230 and 825 cm^{-1}), and dust aerosol (strongest near 1075 cm^{-1}) are apparent. (a) Brightness temperature spectra for nadir viewing. (b) Radiance spectra for limb observations. (Pearl *et al.*, 2001.)

6.3 The giant planets

The giant planets, Jupiter, Saturn, Uranus, and Neptune, occupy the 5th through 8th planetary orbits, counting outward from the Sun. They seem to form pairs; Jupiter and Saturn are similar in size and other properties, and so are Uranus and Neptune (see Appendix 3). All giant planets have deep atmospheres, composed mainly of

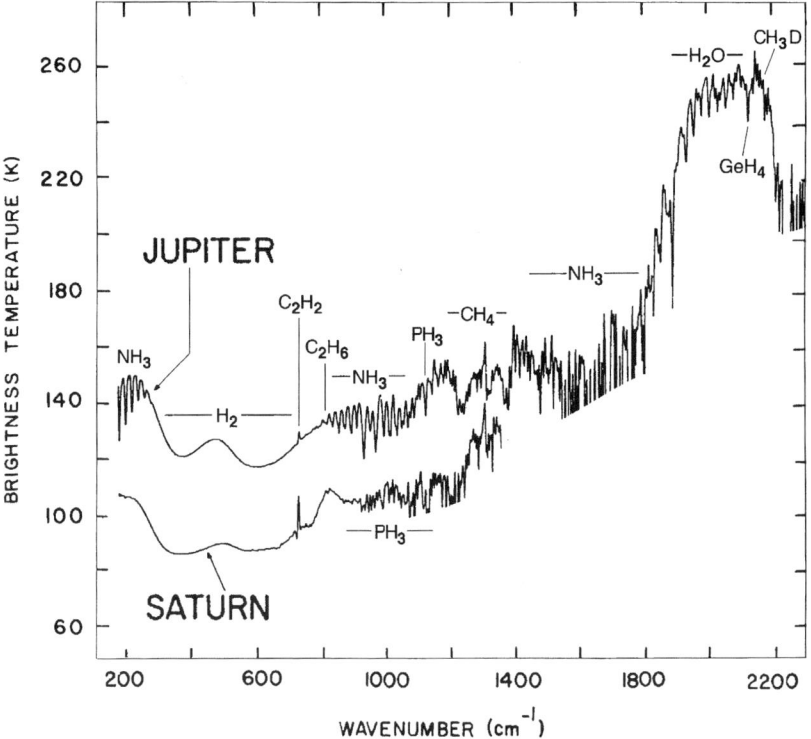

Fig. 6.3.1 Spectra of Jupiter and Saturn recorded with the infrared interferometer of Voyager.

molecular hydrogen (H_2) and helium; all spin relatively rapidly. Uranus is unique in the orientation of its spin axis, which lies nearly in the orbital plane. The axes of the others, as well as of the terrestrial planets, are more or less normal to the ecliptic plane. Infrared emission of the giant planets originates entirely from atmospheric gases and aerosols with surface emission completely absent. It is not even known whether Jupiter has a solid surface; it may be gaseous and fluid throughout. The emission spectra of Jupiter and Saturn, shown in Fig. 6.3.1, were recorded in 1979 and 1981, respectively, by both Voyager infrared instruments. A spectrum of Jupiter recorded with the Composite Infrared Spectrometer (CIRS) on Cassini during its 2000–2001 fly-by is shown in Fig. 6.3.2. The spectra of Uranus and Neptune (Fig. 6.3.3), were taken in 1986 and 1989 by the Voyager 2 IRIS (Hanel et al., 1986; Conrath et al., 1989a).

As seen in the figures the brightness temperature levels of the spectra decrease from Jupiter (5.2 AU), to Saturn (9.5 AU), and to Uranus (19 AU). Neptune (30 AU), however, is slightly warmer than Uranus. As will be shown later, the effective temperatures, that is, the temperatures of blackbodies that emit the same spectrally integrated energy as the planets, are 124.4 K, 95.0 K, 59.1 K, and 59.3 K, respectively. Although Neptune is still further away from the Sun, it contains a substantial

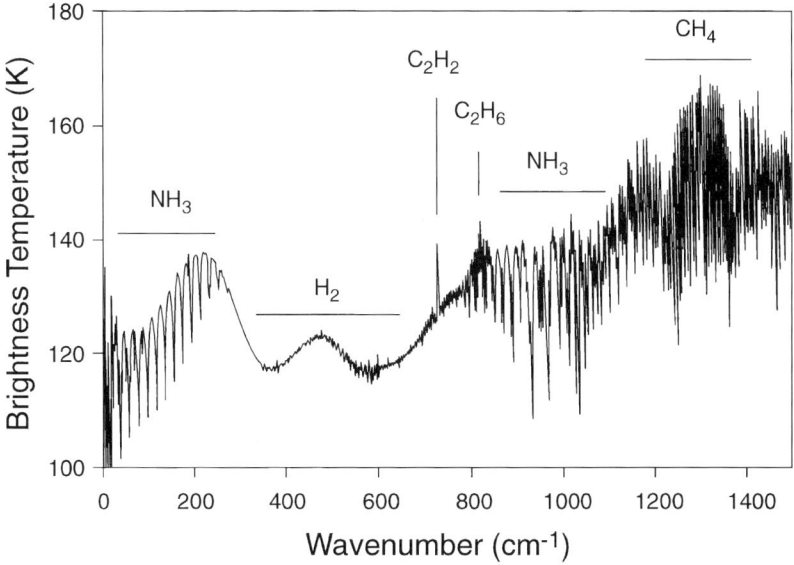

Fig. 6.3.2 Spectrum of Jupiter recorded with the Composite Infrared Spectrometer (CIRS) on Cassini.

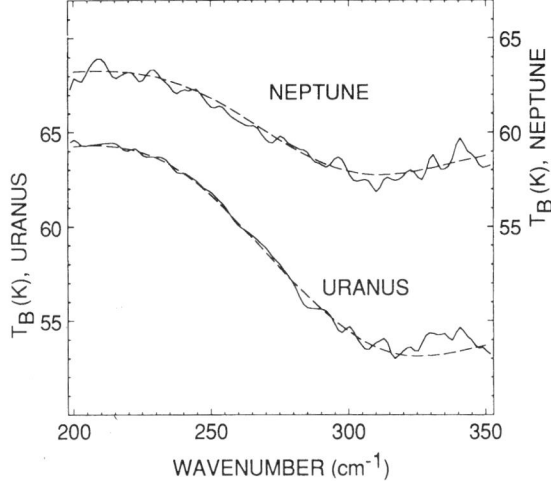

Fig. 6.3.3 Thermal emission spectra of Uranus and Neptune obtained with the Michelson interferometer (IRIS) carried on Voyager 2. The solid curves represent averages of 125 individual spectra from Uranus and 157 spectra from Neptune. For clarity the spectra have been offset from one another by 5 K. The superposed broken curves are model spectra calculated using temperature profiles obtained by inversion of the measured spectra.

internal energy source, in contrast to Uranus, which has at most a relatively small internal heat flux. The lower spectral intensities of the colder objects, in conjunction with the Noise-Equivalent-Spectral-Radiance of the Voyager spectrometer, set limits to the useful spectral range. This explains the strong curtailment of the displayed spectra towards high wavenumbers. The limitation is particularly severe in the spectra of Uranus and Neptune. All Voyager spectra are limited at the lower wavenumber end to about 200 cm^{-1} by absorption in the cesium iodide (CsI) beamsplitter substrate of the Fourier transform spectrometer.

The spectra of Jupiter and Saturn (Figs. 6.3.1 and 6.3.2) and Uranus and Neptune (Fig. 6.3.3) differ fundamentally from those of Venus, Earth, and Mars (Fig. 6.2.1). The 667 cm^{-1} band of CO_2, which is so prominent in the inner planets' spectra, is very weak or absent in those of the outer planets. Only with much higher spectral resolution has CO_2 become detectable in the outer planets; on Jupiter Saturn, and Neptune with the Infrared Space Observatory (ISO) and on Jupiter with Cassini (de Graauw *et al.*, 1997; Feuchtgruber *et al.*, 1997; CIRS team, 2001, private communication). Instead, two highly broadened lines are apparent, centered at 354 cm^{-1} and 602 cm^{-1}. Only a part of the low wavenumber range of the broad 354 cm^{-1} line appears in Fig. 6.3.3. These features are collision-induced lines of molecular hydrogen. At relatively high pressures and long path lengths collision-induced absorption becomes significant (see Subsection 3.3.d); otherwise, molecular hydrogen should not have infrared rotation or vibration lines since it is a homopolar molecule without a permanent electric dipole moment. Obviously, the atmospheres of the outer planets contain large quantities of hydrogen. Their large masses and low exospheric temperatures have prevented the escape of significant amounts of hydrogen over the age of the Solar System, while on Venus, for example, large amounts of atomic hydrogen (H) could have escaped without difficulty.

The spectra of Uranus and Neptune have been measured by Voyager 2 over only a small spectral range (Fig. 6.3.3). Comparison with the spectra of Jupiter and Saturn identifies the measured feature as part of the broad S(0) line of molecular hydrogen. With the data analysis methods discussed later, the atmospheric temperature profiles were obtained. However, little can be deduced from a visual inspection of the spectra shown in Fig. 6.3.3, except that large quantities of hydrogen must be present and that the temperatures increase with pressure in the tropospheres of both planets. Voyager results from Uranus and Neptune systems can be found in the books edited by Bergstralh *et al.* (1991) and Cruikshank (1995), respectively.

The rotational lines of ammonia (NH_3) below 250 cm^{-1} can easily be identified in the Jovian spectrum, but they are less prominent on Saturn, and completely absent on Uranus and Neptune. A comparison of the vapor pressure curve of NH_3 with the ambient temperatures on these planets indicates that the atmospheres of Uranus and Neptune are just too cold to contain much NH_3 in gaseous form at the pressure levels pertinent to these measurements, that is, at pressures up to about one bar. If NH_3 were pushed up with a pocket of gas from lower levels, it would first supercool and, eventually, form small ice crystals. Earth-based measurements of the microwave emission of Uranus indicate that NH_3 is abundant at much deeper levels (de Pater *et al.*, 1989; Hofstadter & Muhlman, 1989). The same process occurs at Saturn, but there the condensation level is high enough in the atmosphere to permit the existence of NH_3 gas within the pressure region observed by the spectrometer. The Jovian spectrum also shows strong NH_3 features in the spectral ranges 850–1200 cm^{-1} and 1700–1900 cm^{-1}. On Saturn the 850 cm^{-1} to 1200 cm^{-1} interval is dominated by signatures of phosphine (PH_3). This gas has a lower vapor pressure than NH_3 and, therefore, is not condensed at the corresponding pressure–temperature levels. Incidentally, the Jovian spectrum also shows a Q-branch of PH_3 at 1118 cm^{-1}; other PH_3 lines are masked, primarily by NH_3.

Both Jupiter and Saturn show the ν_4-band of CH_4 near 1304 cm^{-1}. The center of the band, that is, the most strongly absorbing part, appears in emission, indicating the presence of this gas in relatively warm stratospheres. The temperature minima near the tropopause are sufficiently high to prevent condensation of CH_4 there. In contrast, on Uranus, Neptune, or Titan, CH_4 haze or clouds are possible. Because CH_4 does not condense on Jupiter and Saturn, the CH_4 vapor abundances are fairly constant with height, at least up to the levels where photochemical destruction takes place, but this is generally above the level probed by infrared spectroscopy. Therefore, the CH_4 band may be used for vertical temperature sounding, provided we know the CH_4 concentration. The collision-induced H_2 lines are another spectral region where temperature sounding is possible.

The spectra of Jupiter and Saturn also show signatures of deuterated methane (CH_3D); they appear much weaker and are slightly shifted towards lower wavenumbers than the features of normal methane (CH_4). The hydrocarbons, acetylene (C_2H_2) and ethane (C_2H_6), can be identified in emission at 729 cm^{-1} and in a broader feature centered at 882 cm^{-1}, respectively. These and other hydrocarbons are produced photochemically from methane in the upper stratosphere and above from where they diffuse down to regions where they can be detected by infrared spectroscopy.

The hot spot spectrum of Jupiter around 2000 cm^{-1}, enlarged in Fig. 6.3.4, permits identification of water vapor lines and a Q-branch of germane (GeH_4) at 2111 cm^{-1}, in addition to the lines of CH_3D and NH_3. Hot spots occur in the Jovian belt system. Absorption by the dominant, infrared-active gases (H_2, CH_4, NH_3, PH_3) is relatively weak in that spectral region and, in the absence of dense clouds, radiation can escape from lower levels (Bjoraker *et al.*, 1986). In that atmospheric window, brightness temperatures up to 260 K have been measured. Hot spots on Jupiter were discovered from Earth-based measurements by Westphal (1969). Terrile & Beebe (1979) observed the distribution of hot spots from the ground, simultaneously with the Voyager 1 observations. Fink *et al.* (1978) identified the GeH_4 feature and Larson *et al.* (1975) the water vapor lines in spectra

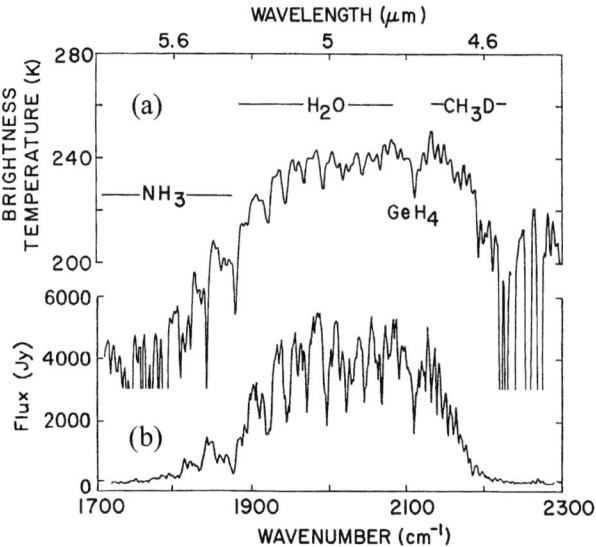

Fig. 6.3.4 The 5 μm spectrum of Jupiter recorded by IRIS (Hanel *et al.*, 1979) and by the Infrared Space Observatory (Fig. 2 of Encrenaz *et al.*, 1996). To facilitate comparison, the ISO spectrum (which is the spectrum with the higher resolution) is plotted with increasing wavenumbers. (a) Voyager IRIS and (b) ISO spectrum.

recorded from the C141 airborne observatory. Beer & Taylor (1973) first recognized the CH_3D lines in the hot spot spectrum from ground-based spectroscopic measurements.

The ISO spectrum shown in Fig. 6.3.4 gives the disk average, but in that spectral range only the 'hot spots' contribute to the radiance significantly. The spectrum was obtained by the Short-Wave Spectrometer at a spectral resolution of 1500 (1.3 cm^{-1} at 2000 cm^{-1}) with an exposure time of 110 minutes. Both spectra, IRIS (a) and ISO (b) are shown linear in wavenumber. Although of lower spatial resolution, ISO had higher spectral resolution and better signal-to-noise ratio than Voyager.

In contrast to the terrestrial atmospheres, differences between cloudy and less cloudy areas or between low and high latitudes are much less pronounced on the giant planets. An example of how clouds affect the spectrum of Jupiter is shown in Fig. 6.3.5. The two observed spectra are almost identical in the 300 to 700 cm^{-1} interval and have comparable emission angles, indicating that the vertical temperature structures of the upper tropospheres of the south polar and north temperate regions are almost identical. Downward extrapolations of these temperature structures should lead to very similar temperature profiles for these two regions at the

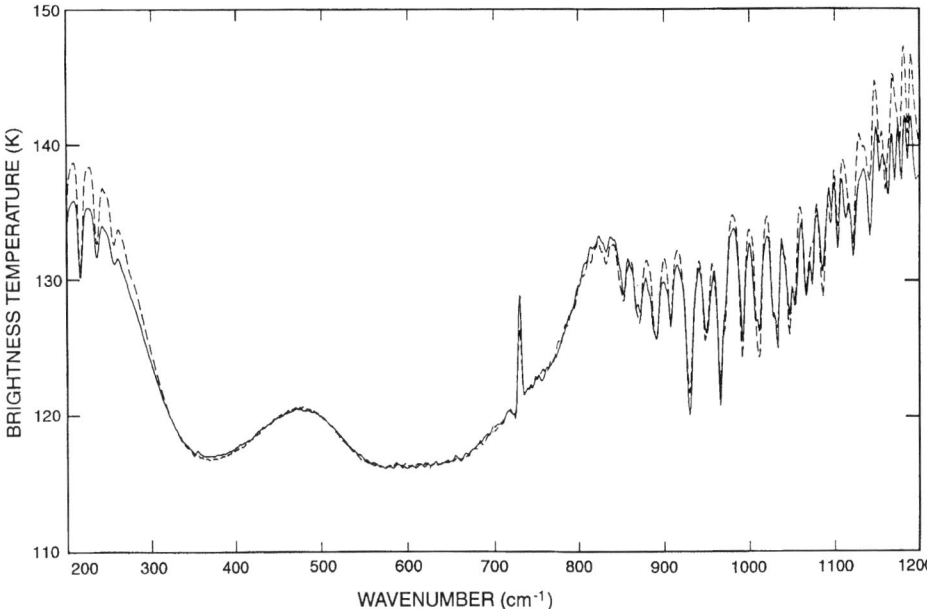

Fig. 6.3.5 Averages of 55 Voyager 1 IRIS spectra from Jupiter's south polar region (dashed curve) and 46 spectra from the north temperate zone (solid curve).

lower levels also, since these lower levels are in regions of convective transport where the lapse rates are adiabatic. Hence the observed differences between the two spectra over the intervals 200–300 cm^{-1} and 1100–1200 cm^{-1} are due to different amounts of cloud cover blocking radiation from below. The gaseous atmosphere is relatively transparent in these spectral intervals, and the mostly cloud-free south polar region emits radiation from an effectively deeper, hotter level than the completely cloudy north temperate zone.

Chronologically between Voyager and ISO, the Galileo Orbiter Mission began exploring the Jovian system. Results from Jupiter and its satellites have been published in special sections of *Science* (1996, **272**, 837–60, and **274**, 377–413), and in a special issue of *Icarus* (1998, **135**, No. 1). Atmospheric results are also discussed by Atreya *et al.* (1997). A summary of the Galileo Mission with spectacular pictures of Jupiter and the Galilean satellites has been presented by Johnson (2000). For our discussion, the most important results are the probe measurements. The helium concentration (von Zahn and Hunten, 1996; von Zahn *et al.*, 1998) and gas abundance data by the Mass Spectrometer (Niemann *et al.*, 1996, 1998) are discussed in Section 8.3. The results from the Galilean satellites are treated in Subsection 6.5.b.

The extraordinary quality of the ISO spectra is demonstrated by the disk spectra of Jupiter, Saturn, and Neptune shown in Fig. 6.3.6. On Jupiter, emission from the

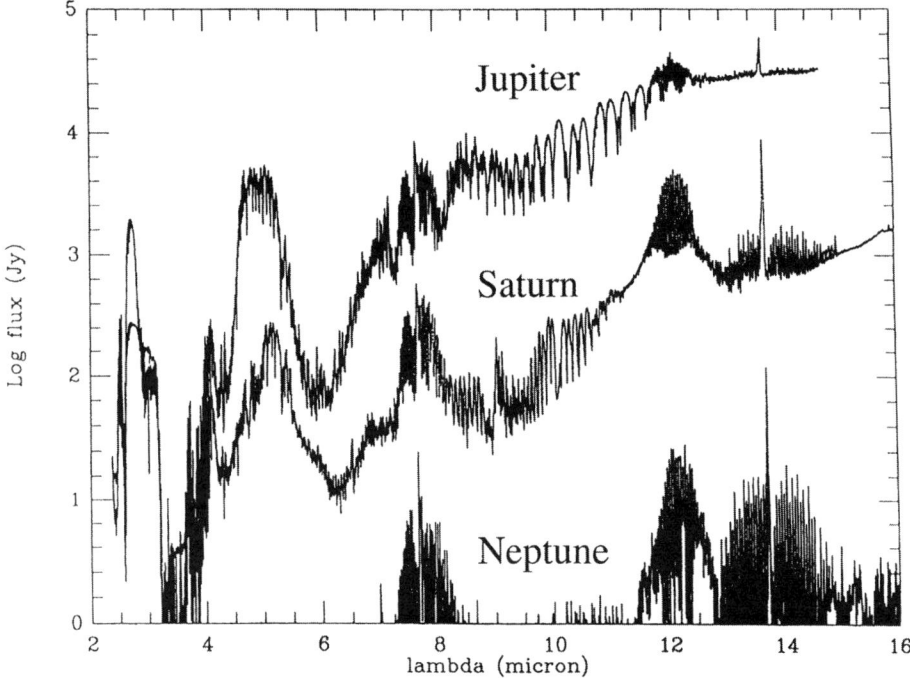

Fig. 6.3.6 Spectra of Jupiter, Saturn, and Neptune between 2 and 16 μm by the ISO Short-Wave Spectrometer (Fig. 1 of Encrenaz *et al.*, 1997*b*).

hot spots at 5 μm, from methane near 8 μm, from ethane near 12 μm, and from acetylene near 14 μm, are quite apparent. Absorption by ammonia dominates the 9 to 11.5 μm region. The same features can be identified on Saturn, except that the 9 to 11 μm region is dominated by phosphine. Ambient temperatures are too low for ammonia to exist there in gaseous form. On Neptune, emissions from methane, ethane, and acetylene are clearly shown. At 15 μm the Q-branch of CO_2 may be seen on Saturn and Neptune.

De Graauw *et al.* (1997) discovered the signatures of CO_2, CH_3C_2H, C_4H_2, and tropospheric H_2O on Saturn, although CH_3C_2H had been tentatively identified previously from IRIS data (Hanel *et al.*, 1981*a*). The existence of CO_2, an oxygen containing molecule, was surprising at the time it was detected in the reducing stratosphere of Titan (Samuelson *et al.*, 1983). ISO found H_2O in the stratospheres of all four giant planets and CO_2 on Jupiter, Saturn, and Neptune, but not on Uranus. As Y. Yung suggested in Samuelson *et al.* (1983), the source of oxygen may have been water ice crystals external to the atmosphere. The same may be the case for the giant planets (Feuchtgruber *et al.*, 1997).

Although the ISO measurements are in general full disk measurements, it was possible to obtain crude spatial coverage on Jupiter, separating the northern and southern hemispheres. In the north CO_2 was not detected, while in the south CO_2 was clearly present. Possibly, the impact at 45°S of the Comet Shoemaker–Levy 9 may have been responsible for bringing H_2O and CO into the Jovian stratosphere. Subsequently, CO and the OH radical produced CO_2 (Moses, 1996; 1997). Encrenaz et al. (1997a) found water vapor features in the Galileo–NIMS spectra. Although from a great distance, Galileo instruments also observed the impact area of the Comet Shoemaker–Levy 9.

ISO has also obtained measurements of the D/H ratios on all giant planets and on the Comets Halley and Hale–Bopp (Meier et al., 1998; Bézard et al., 1999a). The D/H ratio of close to 2×10^{-5} (Griffin et al., 1996) for Jupiter and Saturn is in agreement with Galileo mass spectrometer results and with ground-based determinations. The D/H ratios derived from ISO spectra of Uranus and Neptune are higher, about 6×10^{-5} (Feuchtgruber et al., 1999), in agreement with present models of the formation of these planets. The D/H value derived for the comets Halley and Hale–Bopp is still higher, 3×10^{-3}, which has implications for models of comet formation (Bockelee-Morvan et al., 1998). Delsemme (1999) discussed the D/H enrichment in comets in comparison with the D/H measurements in sea water on Earth. In addition, ISO spectra revealed the presence of the CH_3 radicals on Saturn (Bézard et al., 1998) and on Neptune (Bézard et al., 1999b). The far infrared spectra have been examined by Davis et al. (1996).

6.4 Titan

With a radius of 2575 km, Titan is the second largest satellite in the Solar System. Only Jupiter's Ganymede (2631) is larger, while the planet Mercury (2421 km) is smaller than Titan. However, it is not the size, but the thick nitrogen atmosphere that makes this satellite of Saturn so unique. The Voyager radio science investigation found a surface pressure of 1.5 bar, and an atmospheric thermal structure in which the temperature decreases with increasing altitude between 0 and ~ 42 km (the tropopause), and increases with increasing altitude between ~ 42 and 200 km (lower and middle stratosphere). These results were obtained from an occultation by Titan of the spacecraft as seen by radio antennas on Earth. Radio signals transmitted from Voyager 1 at wavelengths of 13 cm (S-band) and 3.6 cm (X-band) were used to derive the atmospheric height profiles of the gas refractivity, from which the vertical thermal structure (apart from a normalization factor provided later by the thermal infrared spectra) was derived (Lindal et al., 1983; see also Samuelson et al., 1981, for the normalization procedure).

The visible appearance of Titan is disappointing. Images recorded by the Voyager cameras revealed distinct haze layers at the limb, but no trace of surface features (Smith et al., 1981). Apparently, Titan is completely shrouded by clouds and haze layers, presenting a nearly uniform disk of dark-orange or brownish color. Earth-based images taken more recently show surface features in the near infrared between methane absorption bands, indicating the atmospheric haze is more transparent in those spectral regions (e.g., Smith *et al.*, 1996; Combes *et al.*, 1997).

Titan's infrared spectrum is distinctly different from those of the terrestrial or the giant planets. Figure 6.4.1 shows spectra of the disk, the north polar region, and the north limb all recorded in 1981 by Voyager 1. The spectra display a relatively smooth continuum with a number of strong emission peaks. Features already familiar from the planetary spectra discussed earlier include the v_4-band of methane (CH_4), the Q-branch of the v_5-band of acetylene (C_2H_2), and the broader v_6-band of ethane (C_2H_6). In addition, emission bands of ethylene (C_2H_4), methyl acetylene (C_3H_4), propane (C_3H_8), diacetylene (C_4H_2), carbon dioxide (CO_2), hydrogen cyanide (HCN), cyanogen (C_2N_2), and cyanoacetylene (HC_3N) are also evident in Titan's spectra, although the last two nitriles are associated only with high northern latitudes. By analogy with Fig. 4.2.2, these features are formed in the stratosphere, above 42 km. The large amounts of nitrogen and methane in Titan's atmosphere are sources for the chemistry that takes place to produce this exotic suite of organic compounds (Yung *et al.*, 1984; Hunten *et al.*, 1984; Toublanc *et al.*, 1995; Lara *et al.*, 1996). Laboratory spectra were the principal means by which the emission bands shown in Fig. 6.4.1 were identified. Examples of such comparisons are shown in Figs. 6.4.2 and 6.4.3 (Maguire *et al.*, 1981; Kunde *et al.*, 1981). A list of gases identified in this way is shown in Table 6.4.1. The precise wavenumber calibration of the Voyager spectrometer was essential for these identifications.

The existence of CO_2 was quite unexpected. The question arose as to what the source of oxygen in such a highly reduced atmosphere could be. Y. Yung (in Samuelson *et al.*, 1983) suggested that a meteoritic influx of ice particles from the surrounding ice-rich environment of the Saturn system might be responsible. Quantitative confirmation of this idea was developed following the discovery of CO from ground-based observations (Lutz *et al.*, 1983) and H_2O from the Infrared Space Observatory (Coustenis *et al.*, 1998).

Prior to the Voyager 1 encounter, Danielson *et al.* (1973) and Caldwell (1977) had developed a simple model for Titan's atmospheric thermal structure and haze that fit the ground-based far infrared data available at that time. In this model, an isothermal 160 K atmosphere containing a haze with an emissivity proportional to wavenumber overlay a surface of 78 K. The observed 1304 cm^{-1} v_4-band of CH_4 and the 821 cm^{-1} v_9-band of C_2H_6 were adequately reproduced, as was the continuum between 300 and 600 cm^{-1} (see Fig. 6.4.1a). Based on the Voyager 1 occultation temperature profile, it was natural to assume that the surface in the Danielson–Caldwell model corresponded to opaque methane clouds near the tropopause, and

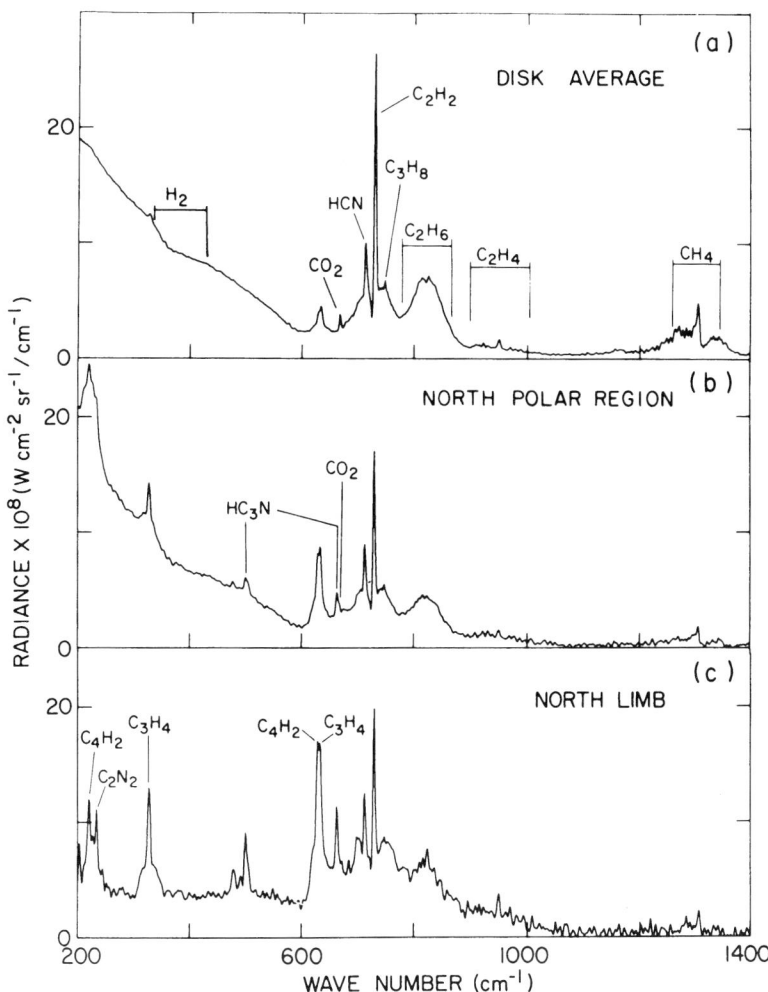

Fig. 6.4.1 Averages of (a) 346 low latitude, (b) 30 north polar region, and (c) three north limb spectra of Titan. Identification of various atmospheric gases are indicated (Samuelson et al., 1983).

that any extant broad spectral feature associated with the haze would be seen in emission above the clouds. However, efforts to infer the spectral properties of such emission features were unsuccessful.

The reason for this failure soon became apparent. Two spectral averages of high spatial resolution IRIS data were obtained near the same latitude as the radio occultation ingress point. One average consisted of spectra taken near the center of the apparent disk (low emission angle), while the other was taken near the limb (high emission angle). According to Section 4.2, the contribution function for the limb

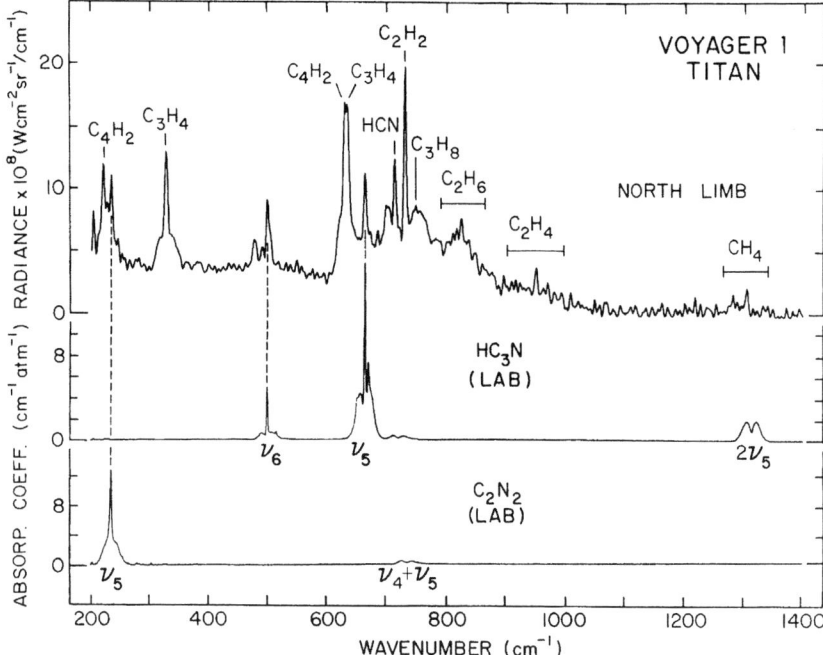

Fig. 6.4.2 Spectrum of the north limb of Titan recorded by Voyager 1 and laboratory spectra of cyanoacetylene (HC_3N) and cyanogen (C_2N_2) (Kunde et al., 1981).

spectrum would peak at higher altitude than those of the disk-center spectrum. The limb spectrum should appear more intense than the disk-center spectrum at wavenumbers for which contribution functions peak in the stratosphere, and less intense for wavenumbers at which peaking occurs in the troposphere. According to Fig. 6.4.4, limb brightening actually occurs at wavenumbers between 600 and 900 cm^{-1}, whereas limb darkening occurs between 200 and 600 cm^{-1}. Thus emission orginates mainly from the stratosphere in the former case, and from the troposphere in the latter. The former condition is consistent with the plethora of emission features seen between 600 and 900 cm^{-1}, whereas the latter is consistent with a broad absorption feature at ~ 360 cm^{-1} instead of the broad haze emission features on either side that were originally suspected. Once this was understood, the identification of the S(0) and S(1) tropospheric absorption lines of molecular hydrogen quickly followed (Samuelson et al., 1981). The foregoing discussion illustrates how clarity can sometimes replace confusion upon a simple qualitative shift in perspective.

Much higher spectral resolution (~ 0.5 cm^{-1}) was obtained by the Infrared Space Observatory (ISO) in 1997 (see Coustenis et al., 1997; Coustenis & Taylor, 1999). A disk-average spectrum of Titan by the Short-Wavelength Spectrometer (SWS) of

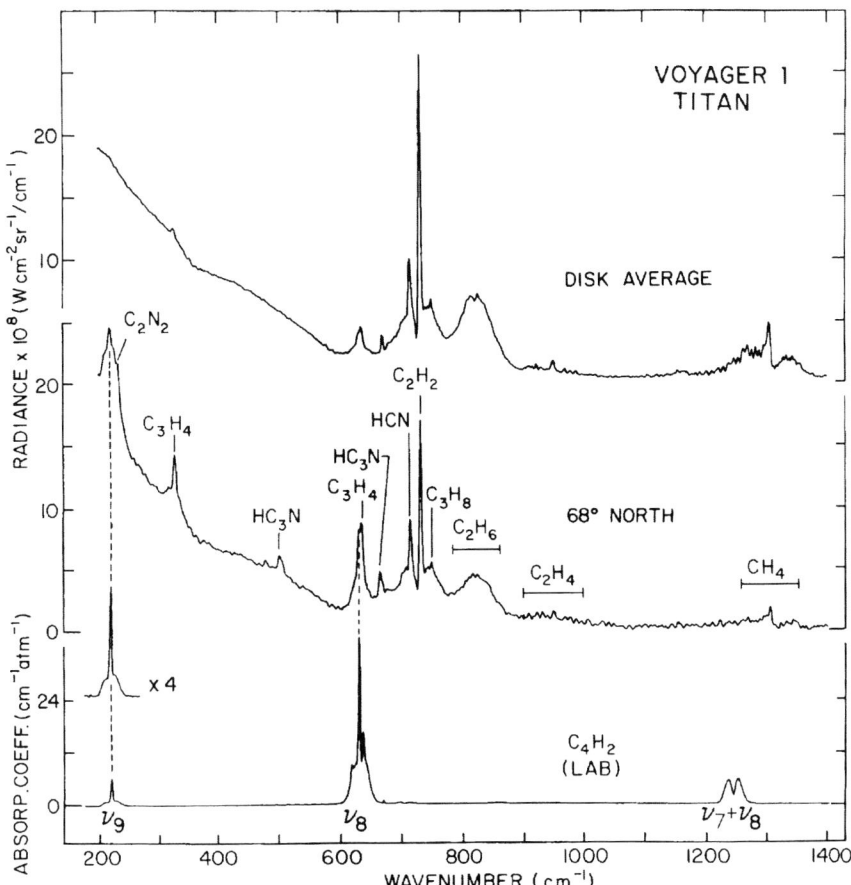

Fig. 6.4.3 Disk average and high latitude spectra of Titan recorded by Voyager 1. A laboratory spectrum of diacetylene (C_4H_2) is also shown (Kunde et al., 1981).

ISO is compared with an equatorial-region spectrum of Voyager IRIS in Fig. 6.4.5. Even though a much richer band structure is evident in the ISO spectrum, no new organic compound was identified from bands in the 600–1400 cm^{-1} spectral region. After diligent searching and averaging of spectra, however, two emission features at 226 and 254 cm^{-1} were identified as water vapor lines (Coustenis et al., 1998). This discovery added H_2O to the short list of oxygen-bearing compounds observed in Titan's stratosphere – CO and CO_2 are the only others observed to date.

A proper analysis of the troposphere requires accurate collision-induced absorption (CIA) coefficients for the various relevant gases. Once quantum-mechanical calculations for N_2–CH_4 CIA became available (Borysow & Tang, 1993), accurate atmospheric models became feasible. In order to suppress synthetic spectra sufficiently between 200 and 300 cm^{-1} to fit the data, it was necessary to fill the upper

Table 6.4.1 *Atmospheric composition of Titan*

Gas		Mole fraction			Comments-Ref.
Major components					
Nitrogen	N_2	0.757–0.99			Inferred indirectly (a)
Argon	Ar	0–0.07			Inferred indirectly (b, f)
Methane	CH_4	0.005–0.034			Stratosphere (a)
		0.049–0.071			Troposphere (b)
Hydrogen	H_2	0.0010–0.0013			(b, f)
		Equator	*North Pole*		
		~6 mbar	~0.1 mbar	~1.5 mbar	
Hydrocarbons					
Acetylene	C_2H_2	2.2×10^{-6}	4.7×10^{-6}	2.3×10^{-6}	(c)
Ethylene	C_2H_4	9.0×10^{-8}		3.0×10^{-6}	(c)
Ethane	C_2H_6	1.3×10^{-5}	1.5×10^{-5}	1.0×10^{-5}	(c)
Methylacetylene	C_3H_4	4.4×10^{-9}	6.2×10^{-8}	2.0×10^{-8}	(c)
Propane	C_3H_8	7.0×10^{-7}		5.0×10^{-7}	(c)
Diacetylene	C_4H_2	1.4×10^{-9}	4.2×10^{-8}	2.7×10^{-8}	(c)
Monodeuterated methane	CH_3D	1.1×10^{-5}			(c)
Nitriles					
Hydrogen cyanide	HCN	1.6×10^{-7}	2.3×10^{-6}	4.0×10^{-7}	(c)
Cyanoacetylene	HC_3N	$\leq 1.5 \times 10^{-9}$	2.5×10^{-7}	8.4×10^{-8}	(c)
Cyanogen	C_2N_2	$\leq 1.5 \times 10^{-9}$	1.6×10^{-8}	5.5×10^{-9}	(c)
Oxygen compounds					
Carbon dioxide	CO_2	1.4×10^{-8}		$\leq 7 \times 10^{-9}$	(c)
Carbon monoxide	CO			6×10^{-5}	Troposphere (d)
				4×10^{-6}	Stratosphere (e)
Water	H_2O			$(4–14) \times 10^{-9}$	~400 km (g)

(a) Lellouch *et al.* (1989)
(b) Samuelson *et al.* (1997)
(c) Coustenis *et al.* (1989a, 1989a, b, 1991)
(d) Lutz *et al.* (1983)
(e) Marten *et al.* (1988)
(f) Courtin *et al.* (1995)
(g) Coustenis *et al.* (1998)

troposphere with methane by factors of 1.5–2.0 times that of saturation (Courtain *et al.*, 1995). Methane clouds cannot persist under these conditions. Any cloud particles exposed to this degree of supersaturation would rapidly grow to centimeter-size and fall through the troposphere in less than two hours (Samuelson & Mayo, 1997; Samuelson *et al.*, 1997).

Therefore, it came as surprise when Griffith *et al.* (1998) found evidence of tropospheric clouds from ground-based near infrared reflection spectra. In particular,

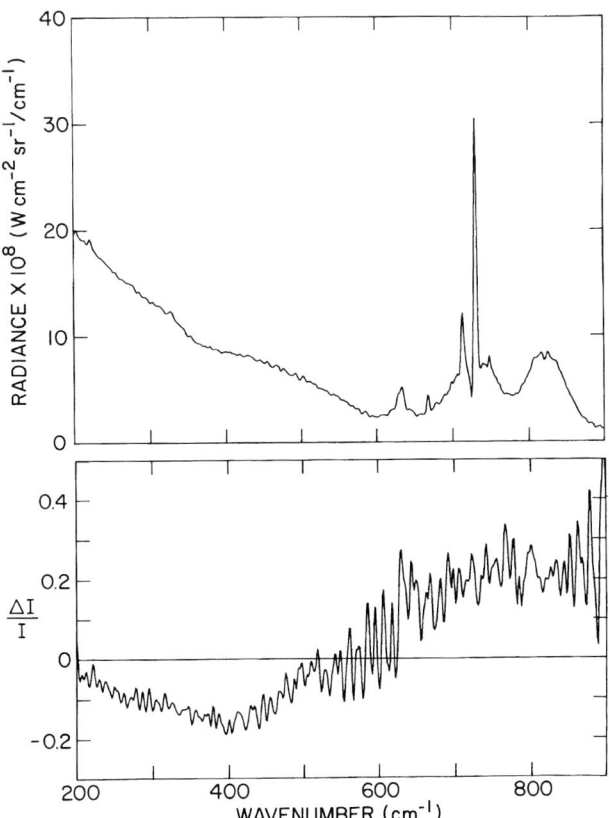

Fig. 6.4.4 Top panel: $I_\nu(57.2°)$, the average of 29 spectra near Titan's daytime east limb (average emission angle is 57.2°). Bottom panel: spectral limb function $\Delta I/I = [I_\nu(57.2°) - I_\nu(7.3°)]/I_\nu(52.7°)$, where $I_\nu(7.3°)$ is the average of 10 spectra near the center of the disk with an average emission angle of 7.3° (after Samuelson et al., 1981).

as shown in Fig. 6.4.6, data taken on September 5, 1995 show strong albedo enhancements compared with 16 nominal spectra taken on other dates. Deviations of the September 5 spectrum from the nominal spectra begin at a wavelengths of ~2.15 μm, where the weighting function peaks at about 15 km. Shorter wavelengths associated with greater atmospheric transparency show enhanced albedos compared with the nominal spectra, while longer wavelengths associated with lesser transparency yield albedos comparable with those of the nominal (not enhanced) spectra. This implies an effective reflecting level at an altitude of ~15 km in the September 5 data, consistent with a tropospheric methane cloud at this altitude. To date, no reconciliation of this interpretation with that of methane supersaturation exists, though both concepts appear well-supported by the data at hand. Hopefully, measurements from the Cassini spacecraft and Titan probe may resolve this issue.

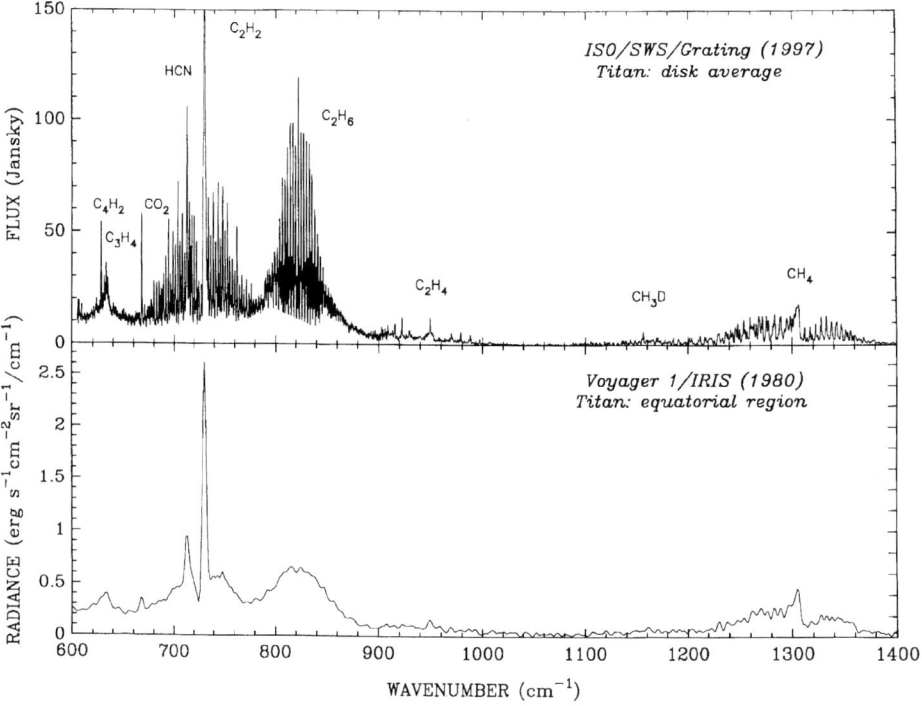

Fig. 6.4.5 Comparison of Infrared Space Observatory spectrum of Titan with that of Voyager 1 (Coustenis & Taylor, 1999).

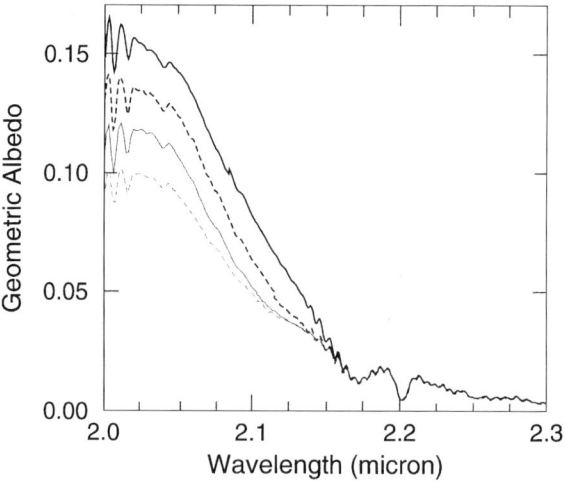

Fig. 6.4.6 Ground-based observed spectral albedos of Titan. The lower two spectra (faint solid and dashed) curves are nominal spectra taken at central longitudes of 76° and 236°, respectively. Differences at shorter wavelengths where the atmosphere is relatively transparent indicate differences in surface albedo at the two longitudes. The upper two (strong solid and dashed) curves are spectra taken respectively on September 4 and 5, 1995, and indicate that additional sources of reflectivity are present at these times (after Griffith et al., 1998).

6.5 Objects without substantial atmospheres

a. Tenuous atmospheres

In general, absorption cross sections due to electronic transitions of atoms and molecules are much larger than vibrational or rotational cross sections. As a result a tenuous atmosphere may be observed in the ultraviolet even though it cannot be detected in the thermal infrared. An example is the N_2 atmosphere of Neptune's largest satellite Triton, with a surface pressure of about 0.014 mbar. Even though no infrared signal was detected from Triton's atmosphere during the Voyager 2 fly-by, airglow and solar occultation measurements by the ultraviolet spectrometer showed that the major gas is N_2 with a trace of CH_4, and the temperature ranges between about 38 K near the surface to 95 K in the exosphere (Broadfoot *et al.*, 1989).

The planet Pluto as well as many satellites in the outer Solar System, such as Charon, Triton, and others belong to the group of relatively small objects with significant amounts of ices on the surface (see also Schmitt *et al.*, 1998). The atmospheric surface pressure is then controlled by the surface temperature, the vapor pressure of the frozen volatiles, and the escape rate of the gases. Even Mars, with its CO_2 polar caps, may be included in this group. Pluto and Charon are discussed together with comets and asteroids in Chapter 7.

Occasionally conditions are favorable for the detection of thermal signals from tenuous atmospheres. This usually requires the observation of an intrinsically strong spectral feature in a cold atmosphere against an unusually warm background (or vice versa), so that the temperature contrast is large, enhancing the feature. Such conditions prevailed at Io, the innermost Galilean satellite of Jupiter, during the 1979 Voyager 1 encounter. The 1361 cm^{-1} ν_3-band of sulfur dioxide (SO_2) was observed by the infrared spectrometer to be in absorption against a background having an effective brightness temperature between 200 and 210 K in this spectral region (Pearl *et al.*, 1979; see Fig. 6.5.1). The gas temperature was estimated to be about 130 K, providing a high thermal contrast and making detection possible. The surface hot spot Loki (see Fig. 6.5.2), contained in the field of view and associated with a region of local volcanic activity, was responsible for the elevated average surface temperature providing the background.

It is not clear if the observed sulfur dioxide gas was contained solely in the erupting plume from Loki, or was associated at least in part with an ambient SO_2 atmosphere concentrated on the illuminated side of Io (Ingersoll *et al.*, 1985). Although the Pele hot spot was much warmer than Loki at the time of the Voyager 1 encounter (650 K against 450 K; Pearl & Sinton, 1982), SO_2 absorption features were absent in the Pele spectrum. Possibly the plume of Pele was driven by a propellant different from that of Loki's; sulfur (S_2) has been suggested as the

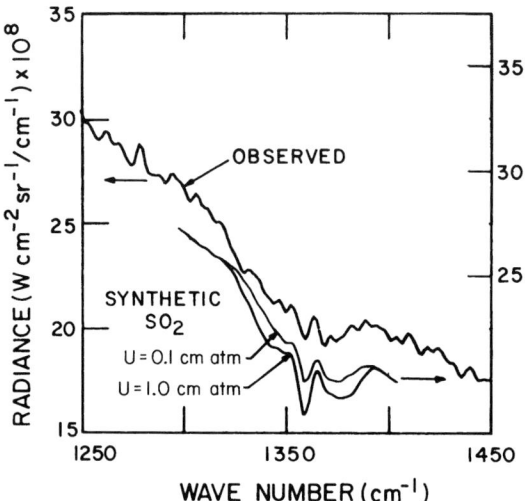

Fig. 6.5.1 Voyager 1 IRIS Io spectrum over the hot spot Loki Patera. The strong 1361 cm^{-1} ν_3-band of sulfur dioxide (SO_2) is seen in absorption (Pearl et al., 1979). Synthetic spectra for two SO_2 abundances are also shown for comparison.

driving agent for Pele by McEwen & Soderblom (1983) while SO_2 was probably the agent for Loki. Because S_2 is inactive in the infrared, no spectral signatures of this gas were available to be observed by IRIS, and no S_2 atmosphere, even if present in small amounts, could have been detected. Recently, S_2 was identified as the main propellant of the Pele plume in ultraviolet spectra recorded by the Hubble Space Telescope (Spencer et al., 2000a).

b. Surfaces

Though lacking the strong wavenumber variations that atmospheric spectra show as a result of gaseous vibration and rotation bands, spectra of solid and powdered surfaces can display their own characteristic signatures. More subdued than those of atmospheres (where thermal contrast is large), the spectra of solid bodies nonetheless provide significant information about the surface properties of these bodies.

The simplest form a surface spectrum can take in the thermal infrared is that of a single Planck function (blackbody). In such a case the temperature of the observed surface can be inferred directly. More often, however, a weighted sum of blackbody spectra is required to obtain an acceptable fit, especially if the observed spectrum covers an extended wavenumber range. This implies an inhomogeneous thermal structure across the field of view, which can arise from different causes.

Two common causes are lateral albedo variations, and variations in topography (surface roughness) across the field of view. The surface temperature is always

6.5 Objects without substantial atmospheres

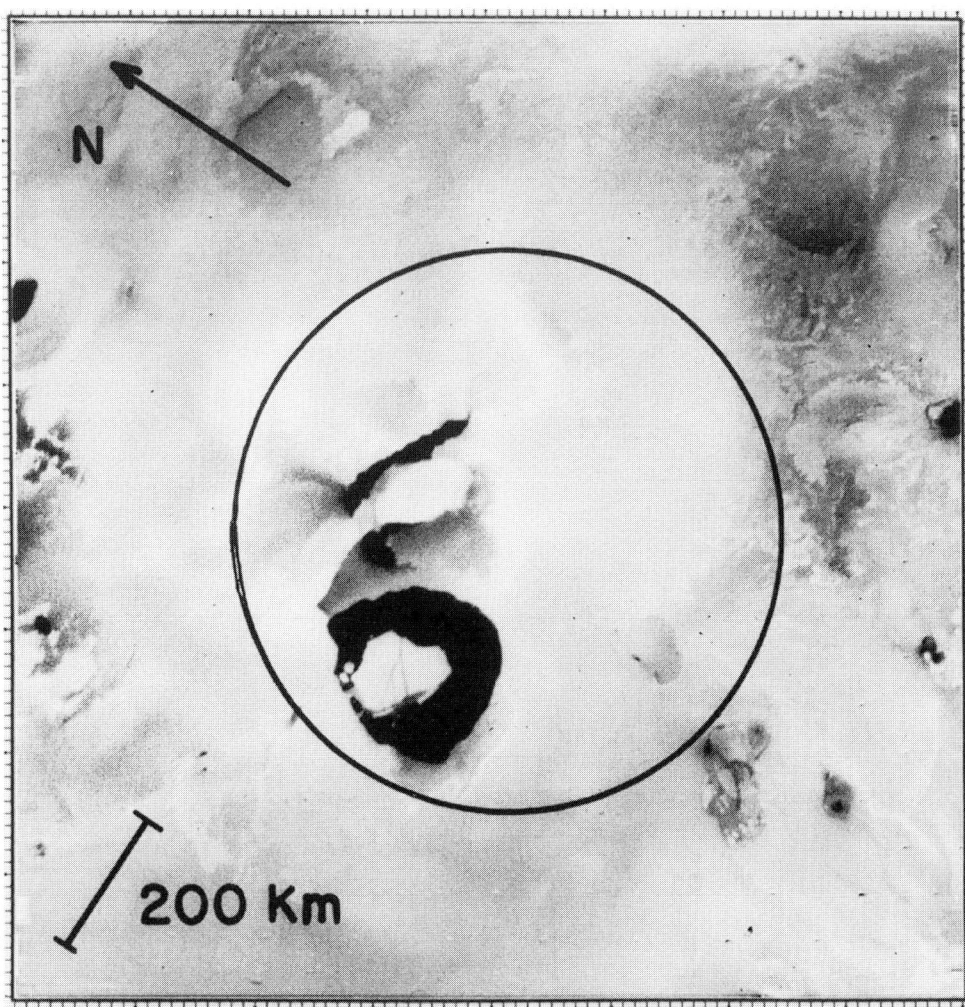

Fig. 6.5.2 Image of the Loki region of Io recorded by Voyager 1 (Smith *et al.*, 1979 *a*). The elongated dark feature above the dark area with the island was the source of a volcanic eruption at the time of the Voyager fly-by. The circle indicates the IRIS field of view corresponding to the spectrum of Fig. 6.5.1.

adjusting towards equilibrium, and regions of high albedo absorb less solar radiation, leading to lower equilibrium temperatures than surrounding darker areas. A comparable mechanism operates in areas of variable topography. Different surface slopes give rise to a diversity of solar incidence angles, leading to different rates of solar energy absorption within a given field of view. The thermal image of an arid area in New Mexico (Fig. 5.4.7) demonstrates such a case.

An example where both mechanisms operate is Callisto, Jupiter's outermost Galilean satellite, which is both heavily cratered and distinguished by a variegated

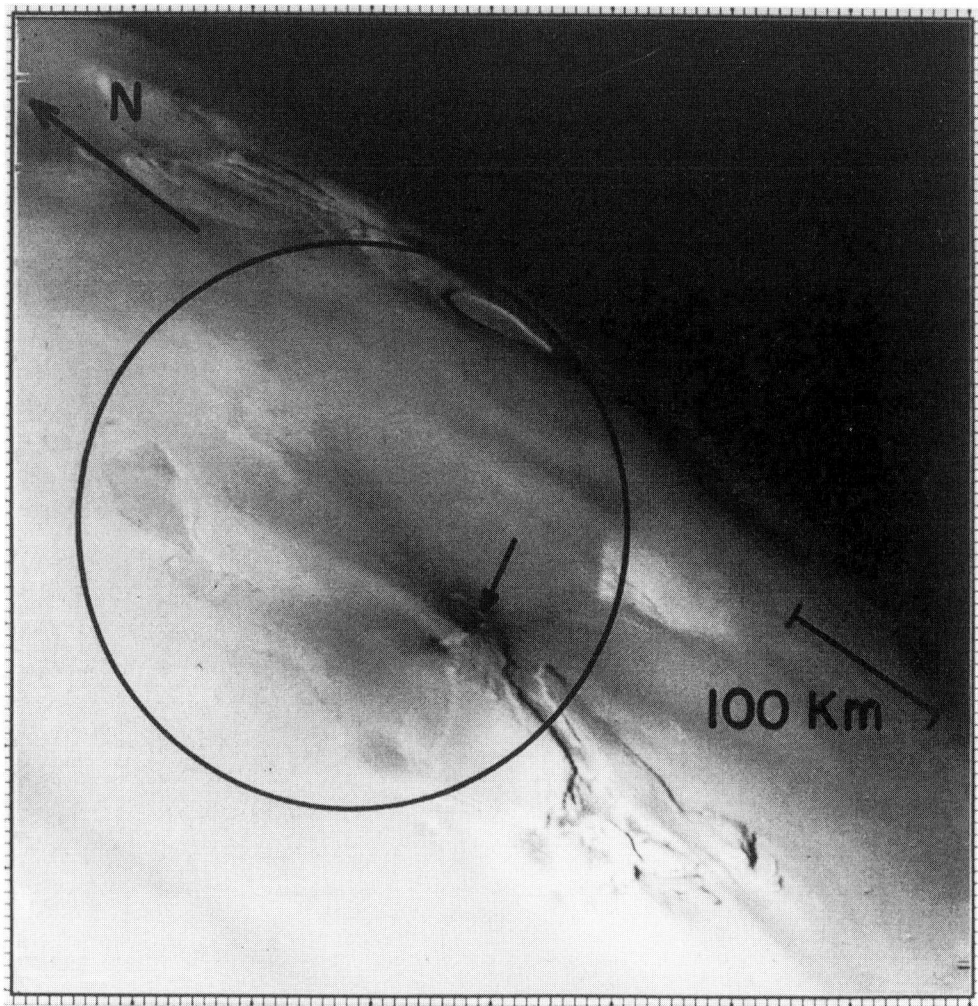

Fig. 6.5.3 Image of the Pele area of Io recorded by Voyager 1 (Smith *et al.*, 1979 *a*). The arrow points to the dark area, which is believed to be the warm spot in the IRIS field of view, indicated by the circle.

albedo. Dayside IRIS spectra always require more than one Planck function to obtain a satisfactory fit. On the other hand Mimas, the smallest of the six major icy satellites of Saturn, has a rather uniform albedo. Any variation of temperature across a limited region of its surface can be attributed to its extremely rugged terrain.

An extreme case of horizontal temperature variation is associated with Pele (Fig. 6.5.3), one of the hot spots on Io (Pearl & Sinton, 1982). Its spectrum is shown in Fig. 6.5.4. Even a casual inspection of the figure demonstrates the need for more than one blackbody to fit the data. At least three such functions are needed

6.5 *Objects without substantial atmospheres* 337

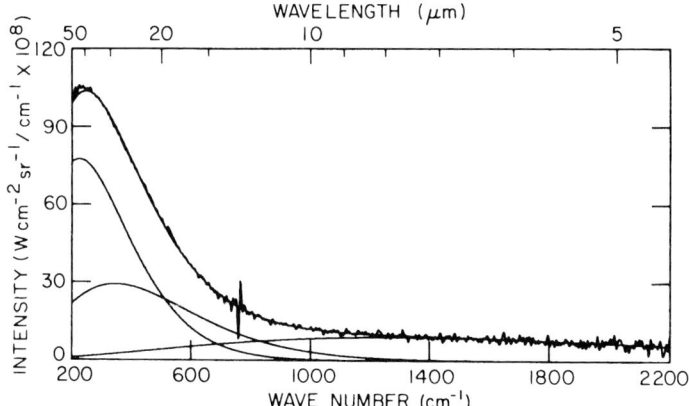

Fig. 6.5.4 Voyager 1 IRIS emission spectrum of the Pele region of Io and three blackbody spectra corresponding to different temperatures and filling different fractions of the field of view. The sum of the three blackbody spectra matches the measured spectrum well (Pearl & Sinton, 1982). The feature near 750 cm^{-1} is an artifact.

to match the observed IRIS spectrum satisfactorily. According to Fig. 6.5.4 the broad tail of the spectrum between about 1200 and 2200 cm^{-1} is due almost entirely to a very intense hot region ($T \sim 650$ K) covering less than 0.06% of the IRIS field of view. This hot spot corresponds to an area of 110 km^2 or a disk 6 km in radius. The same region also appears to be associated with an active volcanic plume (see, e.g., Smith *et al.*, 1979b; Morabito *et al.*, 1979), which accounts for the large range in temperature inferred for regions within the IRIS field of view.

Absorption and emission features can also appear in surface spectra. The two principal causes are emissivity differences and vertical temperature gradients in the material just beneath the surface. As a rule, water ice and minerals such as silicates are very opaque in the thermal infrared. Optical depth unity is located at levels too close to the surface to span a noticeable vertical variation in temperature. As a result, features in the spectra of the icy satellites of the major planets, and of the surfaces of such bodies as the Moon, Mars, and Mercury, are expected to arise almost solely from variations of emissivity with wavenumber.

Prabhakara & Dalu (1976) demonstrated from data acquired by the Nimbus 4 infrared spectrometer that the Earth's surface emissivity is significantly less than unity (within a narrow spectral interval centered about 1080 cm^{-1}) over arid and semiarid land forms. They attributed this low emissivity principally to quartz (SiO$_2$), and noted that clays tended to weaken the silicate absorption feature significantly.

Lunar infrared spectra between 750 and 1250 cm^{-1} were obtained by Potter & Morgan (1981) with a Michelson interferometer at the Wyoming Infrared Observatory. Using the Apollo 11 landing site as a standard, Potter and Morgan obtained

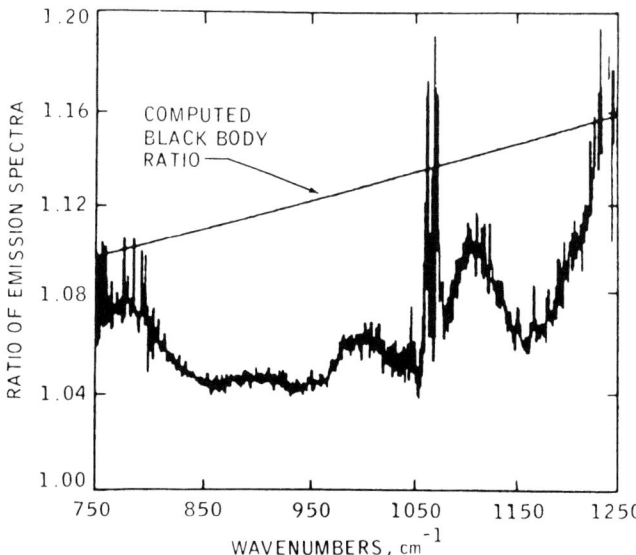

Fig. 6.5.5 Comparison of the observed lunar ratio spectrum for Tycho/Apollo 11 with the computed blackbody ratio spectrum (Potter & Morgan, 1981).

relative emissivity spectra of the Descartes Formation and the central peak of Tycho. Figure 6.5.5 shows their Tycho/Apollo 11 ratio spectrum, clearly demonstrating an emissivity difference between the two sites. Their Descartes/Apollo 11 ratio spectrum, shown in Fig. 6.5.6, also indicates emissivity effects, but is qualitatively different from the Tycho/Apollo 11 spectrum. An analysis of these spectra by Potter and Morgan suggested an excess of pyroxenes in the Descartes Formation and of plagioclases in the central peak of Tycho, relative to the Apollo 11 site. A higher than usual sodium content of the Tycho plagioclase was also inferred.

Unlike water ice and silicates, surface materials that are semi-transparent over most of the infrared spectrum can exhibit spectral features through vertical thermal contrast. If a non-negligible temperature difference exists over the optical depth range $0 \leq \tau < 1$, spectral features will be seen in emission or absorption, depending upon whether the temperature respectively decreases or increases inward immediately below the surface (see Figs. 4.2.1–4.2.2 and the related discussion). In this sense the transport of radiant energy exhibits the same qualitative behavior in solids and gases, and comparable analytic techniques apply.

The surface of Io appears to contain a large fraction of sulfur compounds (see, e.g., Sagan, 1979; Soderblom *et al.*, 1980; Pearl, 1988). Molecular sulfur and sulfur dioxide in particular are fairly transparent in the infrared, and surfaces composed predominantly of these or optically similar minerals should show infrared spectral features under favorable thermal conditions. Figure 6.5.7(a) is an IRIS ratio

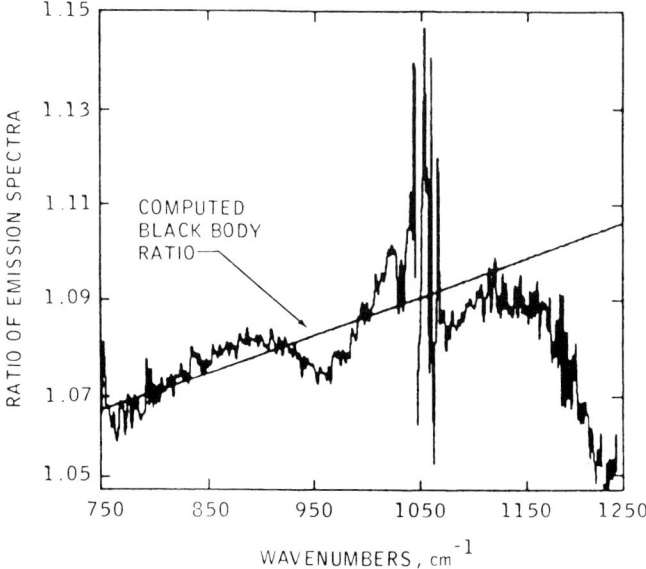

Fig. 6.5.6 Comparison of the observed ratio spectrum for Descartes/Apollo 11 with the computed blackbody ratio spectrum (Potter & Morgan, 1981).

spectrum of Io, while panel (b) is a synthetic ratio spectrum of a combination of S_8 and SO_2 [normalized somewhat differently from panel (a)].

The Io spectrum is an average of 78 daytime spectra, normalized to the broad thermal continuum. The ν_2-band of solid SO_2 is centered at 524 cm^{-1}, suggesting that the 525 cm^{-1} emission band in the Io spectrum be identified with this compound. It follows that the temperature decreases with depth, driven by the solar heating cycle, and other spectral features present should also be in emission. The 470 cm^{-1} emission feature is tentatively identified with S_8. A comparison of the two panels in Fig. 6.5.7 suggests that other compounds not yet identified contribute to the Io spectrum over the intervals 200–350 cm^{-1} and 550–650 cm^{-1}.

Nighttime spectra of Io tend to be featureless, implying the temperature gradient near the surface has been substantially reduced due to infrared cooling to space. A rate of cooling sufficient to cause a temperature gradient reversal would give rise to absorption features at positions identical to those of the daytime emission features. This has not been found on Io. Possibly conduction moderates the nighttime temperature gradient sufficiently to produce the bland spectra observed there. In any event there are real day/night differences in the ratio spectra, indicating that a vertical thermal gradient and not a wavenumber variation in emissivity is the basic cause of daytime spectral features.

Recently, the Galileo orbiter has observed Io from three relatively close approaches yielding data of very high spatial resolutions. For example, the Loki

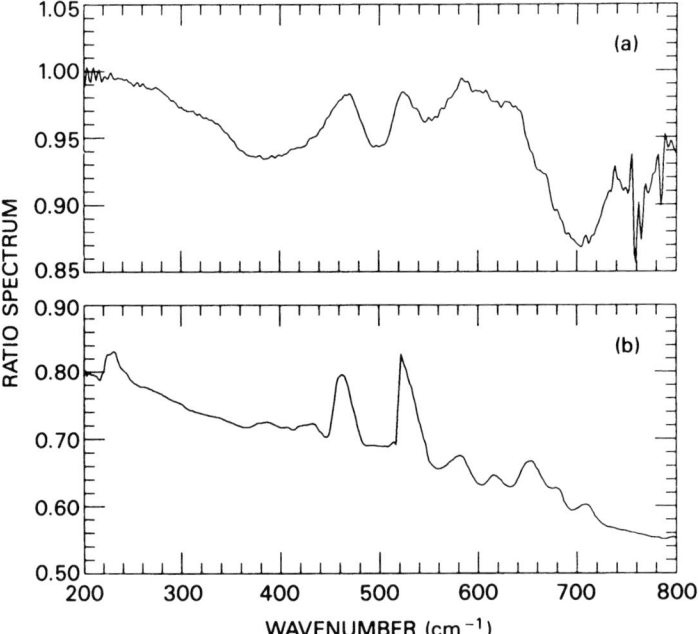

Fig. 6.5.7 (a) An average of 78 Voyager 1 IRIS spectra of Io, divided by an estimate of the thermal continuum, and (b) a synthetic emission spectrum of a solid sulfur–sulfur dioxide (S_8–SO_2) composite, assuming an exponential temperature profile with a temperature gradient $dT/dz = -25$ K cm^{-1} at the surface; the spectrum is normalized to the Planck intensity for a surface temperature of 130 K.

region shown in Fig. 6.5.2 was observed by several instruments. That region has been relatively constant in appearance while other volcanic regions on Io have changed substantially between the Voyager and Galileo encounters. The island in the dark 'lava lake' (caldera floor) and even the cracks in the island observed in 1979 still existed in 1999, although the plume region observed by Voyager just northeast of the lava lake has changed substantially. The Photopolarimeter-Radiometer on Galileo has obtained a contour map of the brightness temperature of the caldera using a radiometer channel centered at 16.8 μm (4.2 μm wide). The bulk of the caldera area showed temperatures between 220 and 240 K; however, a hotter region, with temperatures up to 320 K, was found in the south-west corner (Spencer et al., 2000b; see also McEwen et al., 2000). The surrounding lighter territory as well as the island showed much lower temperatures of about 120 to 200 K. The Loki lava lake was also observed by the NIMS instrument (Lopes-Gautier et al., 2000) with similar results. The lava on the caldera floor yielded brightness temperatures of 273 ± 6 K at about 5 μm. NIMS also measured other volcanic areas (Prometheus, for example), and obtained maps of frozen SO_2. Kieffer et al. (2000) found that the plume of Prometheus has wandered since the Voyager observation. The high temperatures of the Prometheus plume suggest the existence of a silicate magma ocean below the sulfur crust (Keszthelyi et al., 1999).

In addition to the data on Io, NIMS has recorded spectral maps of other Galilean satellites (McCord et al., 1997a, b). For a review of the Galilean satellites see also Showman & Malhotra (1999) and Johnson (2000). Current understanding of the interiors of the Galilean satellites has been reviewed by Guillot (1999). The NIMS maps of Europa show areas of high and low albedo. The spectral features of high albedo areas are characteristic of pure water ice with absorption bands at 1.04, 1.25, 1.5, and 2 μm. The spectra of the darker areas deviate from the water ice features somewhat, and show similarity to spectra of heavily hydrated salts of carbonates and sulfates. For example, natron ($Na_2CO_3*10H_2O$) and epsonite ($MgSO_4*6H_2O$) or ($MgSO_4*7H_2O$) are good candidates (McCord et al., 1998; Pappalardo et al., 1999). Since the darker areas often appear aligned with the numerous fractures in the surface ice, it was suggested that evaporation of brines emanating from the fractures cause the salt deposits.

The formation of the cycloidal surface features has been discussed by Hoppa et al. (1999). They estimate the thickness of the surface ice to be only about 1 km. The existence of an ocean below the layer of ice has been considered for some time (Carr et al., 1998; Kerr, 1999). Recent strong evidence of a global H_2O ocean beneath an ice layer has been obtained by the Galileo magnetometer (Kivelson et al., 2000, see also the perspective in the same *Science* issue, volume 289, page 1305, by D. Stevenson). It has even been speculated that such an ocean may harbor life. However, the probability of the existence of life in a dark ocean below a thick ice layer is rather low. Only bacteria within hydrothermally weathered rocks below the ocean, perculated with dissolved H_2, CH_4, or H_2S, could conceivably exist in such an environment (Gaidos et al., 1999).

Recently, solid hydrogen peroxide (H_2O_2) has been detected by NIMS on the leading, antiJovian quadrant of Europa (Carlson et al., 1999). A concentration of 0.13% was derived. It was suggested that this substance formed due to heavy bombardment of water ice by energetic particles from the Jovian radiation belt. Voyager results from the Galilean satellites can be found in the book by Morrison (1982).

7
Trans-Neptunian objects and asteroids

All planets from Mercury to Neptune and most of their satellites have been observed from Earth-based telescopes and at least once, some repeatedly, from spacecraft. Therefore, sufficient information was available to emphasize the physical principles in the discussions in Chapter 6. Trans-Neptunian objects and asteroids have been explored to a much lesser degree. Their small sizes, for many their large heliocentric distances, and their low surface temperatures prevented detailed exploration. Until recently, only a few samples of an enormous amount of objects have been investigated. Therefore, the treatment of these objects, grouped in this chapter, is primarily a summary of presently known properties. Section 7.1 discusses Pluto and its satellite Charon; Section 7.2 is devoted to comets; and Section 7.3 to asteroids.

7.1 Pluto and Charon

In 1930, Tombaugh discovered Pluto, the outermost known planet (Reaves, 1997; Marcialis, 1997). Several authors have derived the radius of Pluto with very small uncertainties; unfortunately, the derived values do not overlap. Consequently, only a broad range can be quoted (1145 to 1200 km) within which the true radius of Pluto may fall (Tholen & Buie, 1997). Pluto is by far the smallest planet of our Solar System; it is even smaller than many planetary satellites. Pluto's orbit is highly eccentric and inclined by more than $17°$ to the ecliptic plane (Malhotra & Williams, 1997). At perihelion (29.7 AU), Pluto is closer to the Sun than Neptune (30.1 AU), and at aphelion it reaches a heliocentric distance of almost 50 AU. Pluto's orbital period, 248.35 sidereal years, is locked in a $3:2$ ratio with that of Neptune (Cohen & Hubbard, 1965). The axis of rotation is nearly in the orbital plane; therefore, this small planet undergoes rather complex seasonal changes (Spencer *et al.*, 1997). Malhotra (1993, 1999) provides interesting discussions of the possible evolution of Pluto's orbit and that of other planets (see also Stern *et al.*, 1997).

In 1978, Christy & Harrington (1978) discovered Pluto's rather large, but close-by satellite Charon. The radius of Charon is between 600 and 650 km, which is more than half of that of Pluto (Tholen & Buie, 1997). In comparison, the lunar radius is 0.27 that of Earth. Charon orbits Pluto at a distance of 16.5 Pluto radii with an orbital period of 6.4 Earth days. It can safely be assumed that the bodies are tidally locked to each other, which means the rotation periods of Pluto and of Charon equal the orbital period of Charon (Dobrovolskis *et al.*, 1997). Charon must be an impressive sight observed from Pluto; hovering over the same equatorial area, it would appear nearly 7.5 times the diameter of the Moon as seen from Earth. Even more dramatic would be Pluto observed from the surface of Charon; its apparent diameter would be nearly 14 times the lunar diameter. In contrast to this, the diameter of the Sun subtends only 38 and 48 arcsec as seen from the aphelion and perihelion positions of Pluto, respectively. The maximum angular diameter of Jupiter seen from Earth is about 46 arcsec.

It is difficult to resolve Pluto and Charon using Earth-based telescopes, because their angular separation never exceeds 0.9 arcsec. An early picture showing both objects clearly resolved and separated was taken in 1990 with the 3.6 m Canada–France–Hawaii telescope on Mona Kea, Hawaii (see the note by Cruikshank *et al.*, *Science*, 27 August, 1999, page 1355). The complicated dynamics of the Pluto–Charon binary system is discussed by Dobrovolskis *et al.* (1997).

A 1987 stellar occultation by Charon did not reveal an atmosphere, but it did provide a fairly good estimate of the diameter of this satellite. Stellar occultations of small objects are rare. Fortunately, in 1988 Pluto occulted a star and the light curve was observed from a number of ground-based telescopes and from the air-borne Kuiper observatory (Millis *et al.*, 1993). Clear evidence of a tenuous atmosphere as well as fairly good estimates of the diameter were obtained. Another fortunate event was the alignment of Charon's orbital plane as seen from Earth. In 1988, that plane could be observed edge on and, consequently, Charon passed directly in front of Pluto and disappeared completely behind it. This orientation helped in the determination of the radii and in the separation of the spectra of both objects (Binzel & Hubbard, 1997).

Spectroscopic evidence collected from Pluto before the discovery of Charon is in general applicable to both bodies. For example, the first near infrared multispectral radiometry of Pluto (and Charon) showed the signature of methane. At this time, Pluto was close to perihelion (Cruikshank *et al.*, 1976). After the discovery of Charon, it became prudent to take advantage of the transit and occultation of Charon mentioned above to separate the spectra of both objects. Near infrared spectra were taken first with both bodies in the field of view and then at a time when Pluto occulted Charon completely. In the latter case, only Pluto contributed to the signal, while the difference spectrum (both objects minus Pluto only), was then due

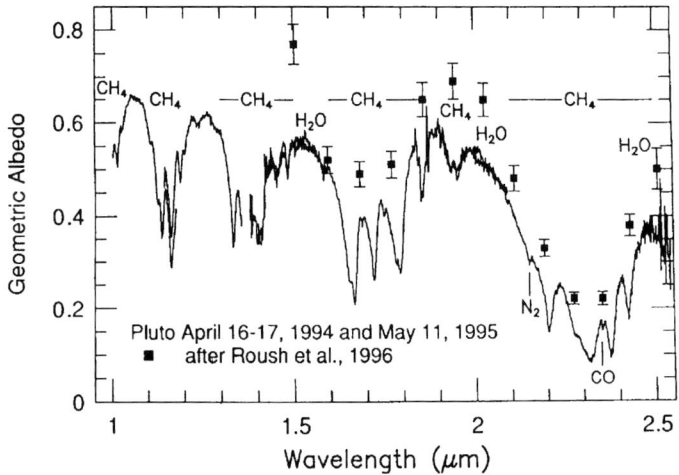

Fig. 7.1.1 Disk spectrum of Pluto, from Cruikshank *et al.* (1997).

to Charon alone. By this method it was found that the surface of Pluto contains frozen methane and that of Charon consists primarily of water ice. More recently Cruikshank *et al.* (1997) derived from near infrared spectra of Pluto that frozen N_2 was the dominant surface ice, and that frozen CH_4, CO, and H_2O were only minor constituents. The spectrum of Pluto shown by Cruikshank *et al.* (1997) is reproduced in Fig. 7.1.1.

From that spectrum and from the stellar occultation data, it can be assumed that N_2 is the dominant atmospheric constituent on Pluto (Yelle & Elliot, 1997; Summers *et al.*, 1997). Molecular nitrogen gas is not active in the infrared and, therefore, the near infrared spectrum of Pluto is dominated by CH_4 bands (gas and surface ice); however, CH_4 is only a minor atmospheric and surface constituent. Although predicted by theory, no ethane (C_2H_6) has been identified in the spectra. Because of losses to space, the atmospheric gases must be replenished continuously. Escape processes have been treated by Trafton *et al.* (1997). Review papers by Cruikshank *et al.* (1997, 1998) and by Yelle & Elliot (1997) discuss the surface and atmospheric composition of Pluto. More recently, good near infrared spectra between 1 and 2.5 μm of Pluto and of Charon have been obtained by Brown & Calvin (2000), using the 12 m Keck telescope on Mona Kea, Hawaii. The high angular resolution of the Keck telescope permitted the recording of spectra of Pluto and of Charon although both bodies were separated by only 0.9 arcsec. On Pluto, the presence of CH_4 ice on the surface has been confirmed. The signatures of N_2, CO, and H_2O ices are not apparent at this spectral resolution. On Charon, water ice in crystalline form and, in addition, a mixture of

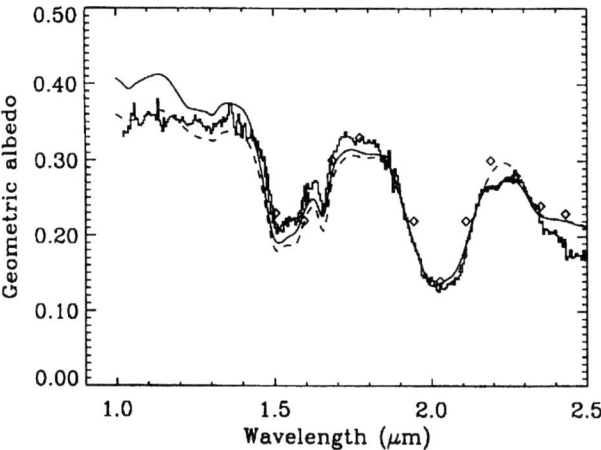

Fig. 7.1.2 Near infrared spectrum of Charon (Fig. 3 of Brown & Calvin, 2000). The histogram gives the data scaled to the albedo of Charon (Roush et al., 1996). The dashed line is a model consisting of only water ice and a dark neutral absorber. The solid line is a model in which ammonia and ammonia hydrate ices have been added to the water ice and dark absorber model.

ammonia and ammonia hydrate ices seem to fit the measured spectrum well, see Fig. 7.1.2.

Precise measurements of the orbital periods of Charon (6.387 223 ± 0.000 017 day) and the difficult measurement of the orbital wobble of both bodies with respect to their common center of mass lead to estimates of the masses of both (Tholen & Buie, 1997). Together with a measurement of the radii an estimate of the mean densities can be derived. The density of Pluto was found to be between 1.92 and 2.06 and that of Charon between 1.51 and 1.81 g cm^{-3}, indicating a substantial fraction of rocks in addition to ices for Pluto and a smaller amount of hydrated rocks below water ice for Charon. Consequently, Pluto is much more similar in composition to Neptune's retrograde satellite Triton (density = 2.043 ± 0.0121, radius = 1352.6 ± 2.4 km; see McKinnon et al., 1995), while Charon seems to be more similar to the smaller, icy satellites of Neptune and Uranus. For a discussion of Triton and the smaller satellites of Neptune see the book edited by Cruikshank (1995) and the paper by Quirico et al. (1999). The satellites of Uranus are discussed in the book edited by Bergstralh et al. (1991).

The history of the formation processes of the two now very close but distinctly different objects (Pluto and Charon) is discussed by Dobrovolskis et al. (1997) and by Stern et al. (1997). A collision scenario, similar to that invoked in the formation of the Earth–Moon system, is presently the preferred theory. Information on Pluto and Charon can also be found in articles by Lunine et al. (1989) and by Buie (1992).

7.2 Comets

Besides the planets and their satellites, the Solar System harbors a large number of smaller objects, ranging from hundreds of kilometers in size down to dust particles. If they consist of solid material and have at least the size of small boulders, they are called asteroids. If they enter the atmosphere of Earth and reach the surface, they are named meteorites. If they are very small and burn up on entry, they are referred to as meteors. If the bodies contain a substantial fraction of ices and develop tails as they come closer to the Sun, they qualify as comets. However, the classifications are not very consistent; for example, remnants of the dust tails of comets cause meteor showers, and older comets, once they have expended most of their volatile matter during many passes near the Sun, may not be distinguishable from asteroids. This section is devoted to comets and the next section (7.3) deals with asteroids.

Comets have been observed since antiquity. Commonly, they are defined as heavenly bodies with a luminous head and one, sometimes two, tails. Sir Edmund Halley recognized that the comet of 1682 was the same as that seen in 1607 and before in 1531. He predicted the next appearance of that comet, now called comet Halley, for the year 1759. Unfortunately, his death in 1742 prevented him from witnessing the realization of his prediction. By applying Kepler's laws and Newton's theory to the observations, it was clear to astronomers since Halley's time that many comets orbit the Sun in elongated, elliptical orbits, while others appear to have nearly parabolic ones.

Major advances in the understanding of comets came in 1950. At a comet conference in Liège, Belgium, Whipple (1950, 1951, 1963) suggested that the nuclei of comets are a conglomerate of ices, mostly that of water, and of silicate dust particles. The term dirty snowball was coined. Delsemme (1998) pointed out the importance of clathrates. As a comet approaches the inner Solar System, the temperature of the illuminated side of the nucleus rises and the ices start to evaporate. The emerging gases expand from the illuminated side in jets and drag along silicates and other particles. The particles leave the nucleus and the surrounding coma at high speed, but solar radiation pressure interacts with the particles and causes the dust tail to point in an arc away from the Sun. Some of the gas molecules fracture into daughter species, some form radicals. The ionized fractions interact with the magnetic field of the solar wind and form the plasma tail. This tail is often bluer, fainter, and does not coincide with the direction of the dust tail.

At the same Liège conference, Oort (1950) proposed the existence of a large reservoir of nuclei of comets at a distance of, according to most recent estimates, about 30 000 to 50 000 AU. This assemblage of nuclei is now known as the Oort cloud. Oort's prediction was based on a study of 19 highly elliptical orbits of 'new' comets – that is, comets that enter the inner Solar System presumably for the first time. In the cloud, occasionally, orbital perturbations from nearby stars, from tidal forces of the galactic plane, or from possible near collisions cause some of the nuclei in the Oort cloud to deviate from their orbits. Some move further away, others move towards the inner Solar System. They appear as 'new' comets, having rather large elliptical orbits and very long orbital periods. The gravitational interaction with the giant planets sometimes changes the orbital characteristics of comets. Again some may be ejected from the Solar System, while others are placed in orbits with smaller aphelion distances. The Oort cloud comets are not restricted to the ecliptic plane. The cloud is a spherical, not sharply defined region, containing possibly as many as 10^{12} or 10^{14} nuclei of comets (see Weissman, 1998).

Only one year after the conference mentioned above, Kuiper (1951) suggested that the mass distribution in the primordial cloud, which gave birth to the Solar System, did not end abruptly at the distance of Neptune. Although no major planet formed beyond Neptune (only the minor planet Pluto formed there), more distant remnants of the accretion disk still exist, many probably in the form of cometary nuclei. That assemblage of trans-Neptunian objects is now referred to as the Kuiper Belt. In recent discussions of these objects, Weissman & Levison (1997) as well as Gomes (1999) point out that Edgeworth (1949) had similar ideas and, therefore, the belt should more correctly be called the Edgeworth–Kuiper Belt. This belt may contain as many as 6.7×10^9 objects with a radius larger than 1 km within 50 AU (Weissman & Levison, 1997). It is believed to be the source of comets with periods of less than about 200 years and with orbits more or less in the ecliptic plane. Older comets, such as Halley and Encke, probably originated from that belt.

While the existence of the Kuiper or Edgeworth–Kuiper Belt appears plausible, that of the Oort cloud is not so obvious. At the 1965 Liège conference on comets, Öpik (1963, 1966) suggested that at the time of planetary formation, the accretion disk contained gases, solid bodies, and objects consisting of dust and ices. Mostly Jupiter and Saturn, but also Uranus and Neptune grew by the influx of much of that material. At the distance of Jupiter and beyond, the disk temperatures were low enough to preserve ices condensed on dust particles and grains. Ices of water and many organic substances existed originally in the interstellar cloud from which the Solar System formed (e.g., Irvine and Knacke, 1989). However, many of the particles coated with ice had coalesced into sizable objects, some as large as several tens of kilometers in diameter. They were in effect cometary nuclei. Some fell into the giant planets or came so close that the weak ice–dust structure broke up, as was vividly demonstrated by the Shoemaker–Levy 9 comet (see Moses, 1996; Nicholson *et al.*, 1995). Other nuclei passed relatively close to Jupiter, or one of the other giant planets, and were then diverted in many directions. Many traveled into the inner Solar System and ended up in the Sun, some were captured by the inner planets, for example by Earth, and brought with them large quantities of water. A few, called Centaurs, still circle the Sun well inside the orbit of Neptune (see Weissman & Levison, 1997). However, many others were ejected towards the limits of the gravitational influence of the Sun. There, the Oort cloud formed. A few years after the Second Liège Conference on Comets, Safranov (1972, 1977) expressed this scenario in a well-designed theory (see also the autobiography of Delsemme, 1998).

Since the 1950s, cometary science has advanced by more sophisticated computer models, but most dramatically by new data on the composition of comets. Spectra of the tails, the comas, and the nuclei of various bright comets, including those of Halley in 1986, have been obtained from ground-based and spacecraft observations. The image of the nucleus of comet Halley by the Giotto spacecraft showed the emission of jets from many localized areas. Spectrometry using the ultraviolet, visible, infrared, and radio frequencies, as well as mass spectrometry from spacecraft (Vega 2), provided new information on cometary composition. For summaries see Jessenberger *et al.* (1989), Delsemme (1992), Mumma (1992), Mumma *et al.* (1993), and Bockelee-Morvan *et al.* (1998).

Water (H_2O), formaldehyde (H_2CO), carbon monoxide (CO), carbon dioxide (CO_2), methane (CH_4), and hydrogen cyanide (HCN) have been identified as parent molecules, emanating from the nucleus of Halley. Grains of silicates, FeS, C, S, and particles consisting of complex organic substances, generally referred to as CHON particles, have been detected. It is expected that other comets would show a very similar composition. As the parent molecules enter the coma region, some of the molecules react chemically, some split into fragments (daughter species), and others become ionized. Some of these molecules recondense onto the particles as the gas cloud cools by adiabatic expansion.

Cometary tails remain in interplanetary space for some time, long after the comet has departed to the outer regions of the Solar System. The small particles scatter sunlight and contribute to the zodiacal light. A very large number of such particles must have existed during the early phase of planetary formation, at a time of direct bombardment of the Earth by many planetesimals and comets. Direct impact of comets may have provided water vapor, CO, and other elements, contributing, perhaps substantially, to our oceans and the atmosphere. However, the high velocities in impacts make it unlikely that complex organic molecules survived the resulting high temperatures and shock waves. In contrast, the small dust particles of cometary origin may very well have transported many molecules significant to the evolution of life on our planet (Delsemme, 1997, 1998). Even today, the Earth collects a substantial amount of interplanetary dust.

7.3 Asteroids

Asteroids are irregularly shaped bodies; most are stony, but a few have metallic composition. Sometimes layers of dust cover their surfaces. The majority of asteroids circle the Sun in the Asteroid Belt between 2.2 and 3.2 AU. The largest asteroid (Ceres) measures 950 km in diameter. About 4000 asteroids have been catalogued to be in excess of 1 km in diameter. However, the true number is estimated to be an order of magnitude higher. An enormous number of smaller bodies also exist within the belt. The heliocentric distribution of the asteroids within the belt is strongly controlled by gravitational resonances with Jupiter. The Asteroid Belt has been a threat to all spacecraft traveling to the outer planets. Fortunately, so far, no spacecraft has been lost due to collision with a sizable rock.

Numerous asteroids and smaller debris also exist outside the belt, that is at distances greater than 3.2 AU as well as smaller than 2.2 AU. In particular, the Apollo, Amor, and Atem groups of asteroids are of concern because their orbits cross, or come close to, that of Earth. Smaller objects, less than a few kilometers in diameter, are difficult to detect from Earth, but a collision with one of them could have catastrophic consequences, even if it had a diameter of only a few hundred meters. In 1995, an asteroid about 5 km in length passed between Earth and the Moon; it was not detected until days later. It is believed that a large asteroid impacted Earth 65 million years ago and was responsible for the extinction of the dinosaurs and many other species. It is quite possible that a similar event may have caused an even more devastating extinction of many forms of life nearly 250 million years ago (Ward *et al.*, 2000). The possibility that catastrophic impacts may occur again cannot be ruled out (Gehrels, 1994).

Using several thermal infrared channels in conjunction with the apparent visual magnitude allows an estimate of the temperature, diameter, and albedo of a small object, even when the telescope cannot fully resolve it. From several well-calibrated channels in the thermal infrared one may obtain the temperature by assuming a near blackbody emission of the object. Combining this information with the observed infrared intensity yields the apparent cross section. Visual magnitude and cross section lead then to the albedo. The broadband channels of the Infrared Astronomical Satellite (IRAS) especially have helped to generate catalogues of asteroids; see Matson (1986), Morrison & Lebofsky (1979), and Neugebauer *et al.* (1984).

The large number of asteroids in the main belt make it likely that many near encounters and collisions have occurred there over the lifetime of the belt. Orbits of some asteroids were redirected towards the inner as well as the outer regions of the Solar System and many were fragmented into smaller bodies. Smaller particles may have reattached themselves to larger objects, explaining the dust cover of many asteroids. However, very small dust particles are affected not only by gravitational forces, but also by radiation pressure from sunlight, and to a lesser degree by the solar wind and the Pointing-Robertson effect. An inventory of volatiles in asteroids is shown by Lebofsky *et al.* (1989).

Asteroids have been classified according to their orbit, albedo, color, and probable surface composition. Objects in the outer region of the Asteroid Belt (C-type bodies) tend to have generally a low albedo and sometimes exhibit the 3 μm water-of-hydration band. Their infrared spectra resemble that of carbonaceous chondrites. The inner region of the belt is more populated by objects (S-type) displaying a somewhat higher albedo and a more reddish tint, reminiscent of material with iron-bearing minerals, such as pyroxene and olivine. A few asteroids (M-type) show a high reflectivity at radar wavelengths, indicating a metallic composition. Iron–nickel meteorites originated from that class. The presence of distinctly metallic and stony asteroids in the belt suggests as their origin a small but geochemically differentiated parent body. Long ago, a small planet may have existed between the orbits of Mars and Jupiter. This body may have broken up, possibly by collisions with other large asteroids or from tidal forces, as Jupiter grew in mass by the influx of gases and remnants of the accretion disk.

In recent years several spacecraft came close to asteroids; the Galileo spacecraft passed near the small asteroid 951 Gaspra (29 October, 1991) and the S-type asteroid 243 Ida (28 August, 1993). In February 1996, the Near Earth Asteroid Rendezvous (NEAR) mission was launched with the objective to orbit the asteroid Eros. The first approach occurred on 23 December, 1998 at a distance of 3827 km; although the spacecraft motor failed near closest approach unique pictures and many other data have been obtained (Veverka *et al.*, 1999; Yeomans *et al.*, 1999). Eros travels in an elliptical orbit, not far beyond that of Earth (1.13 to 1.73 AU), and its orbital parameters are well known. Therefore, it was an easily reachable target for a rendezvous mission. The NEAR spacecraft visited Eros again in February 2000 and now orbits the asteroid at the remarkably close distance of 50 km (Yeomans *et al.*, 2000; Veverka *et al.*, 2000; Zuber *et al.*, 2000; Tromka *et al.*, 2000). Eros is a highly elongated body with dimensions $34 \times 13 \times 13$ km. Combining the gravitational disturbance on the spacecraft with the imaging data yielded a density of 2.67 ± 0.10 g cm^{-3}, which is similar to that obtained from the other S-type body, Ida. The C-type asteroid Mathilde, examined earlier by NEAR, showed a much lower density, ~ 1.3 g cm^{-3}, indicating a much lower packing density for the latter. The near infrared spectrometer of NEAR (0.8 to 2.5 μm) showed spectra with a general similarity to ground-based spectra of Eros. The 1.0 and 1.9 μm absorption bands are characteristic of pyroxene and olivine. Spectacular pictures of Eros and several other asteroids are shown by Asphaug (2000). In the same paper, the author argues that asteroids larger than 200 m are a loose conglomerate of boulders, gravel-size stones, sand, and dust. This conclusion is based on the observation that asteroids larger than 200 m never have rotation rates shorter than 2.2 hours. Faster rotating objects would be torn apart by centrifugal forces, unless they are of a monolithic nature. Objects smaller than 200 m have been observed with higher rotation rates. They are most likely solid bodies and not conglomerates. Further information on asteroids can be found in Lebofsky *et al.* (1989), Binzel *et al.* (1989), and Chapman (1992).

8
Retrieval of physical parameters from measurements

In Chapter 6 we examined planetary spectra using the knowledge that we gained in previous chapters, especially in Chapter 4, which was devoted to simple atmospheric models of radiative processes. By applying physical reasoning, we could extract a considerable amount of information on the conditions that gave rise to the measured spectra. This intuitive method is very important in analyzing data; however, if one desires more precise information in the form of numerical results, the strictly intuitive approach must be augmented by more sophisticated numerical methods. Such methods are the subject of this chapter.

In Section 8.1 we introduce numerical retrieval methods and apply them to atmospheric parameters in general. Section 8.2 is devoted to the retrieval of atmospheric temperature profiles. A large number of different numerical techniques is now available for this task. The retrieval of information on atmospheric composition is the subject of Section 8.3. Again, a wide range of methods must be considered. Cloud parameters and the properties of suspended particulates can also be deduced from infrared measurements. This topic is treated in Section 8.4. The determination of properties of solid surfaces is discussed in Section 8.5, while processes of finding the albedo and the total thermal emission of the Solar System objects are analyzed in Section 8.6.

8.1 Retrieval of atmospheric parameters

In spectral regions where the atmosphere is transparent and for objects without an atmosphere, the temperature, emissivity, and scattering properties of the solid surface determine the measured intensity, while, in opaque portions of the spectrum, atmospheric properties are dominant. In this section we treat the atmospheric retrieval problem and defer discussion of surface properties to Section 8.5.

To illustrate the principles of atmospheric parameter retrieval, let us consider an atmosphere in local thermodynamic equilibrium. The emerging radiance at a

8.1 Retrieval of atmospheric parameters

given wavenumber will contain contributions from a range of atmospheric levels; however, there will be a region of maximum contribution located near the level of unit optical depth. The relative contribution from each level is given by the weighting function as discussed in Chapter 4. If the vertical variation of the temperature and absorber mass is sufficiently smooth, then the brightness temperature corresponding to the radiance at the top of the atmosphere is representative of the kinetic temperature of the atmosphere in the vicinity of the unit optical depth level. If the distribution of the absorber with height is known, the unit optical depth level can be identified, and the temperature at that level can be estimated from a measurement of the radiance. Alternatively, if the atmospheric temperature profile is known, then the brightness temperature corresponding to the measurement can be used to infer the unit optical depth level, and the total absorber mass above that level can be estimated. If a set of measured radiances is available, rather than a single measurement, then information over a range of atmospheric levels can be inferred, provided the measurements span a spectral region with sufficient variation in atmospheric opacity. For example, measurements with high spectral resolution might be made within a single gaseous absorption line. Radiance measurements near the line center contain information from relatively high atmospheric levels while measurements in the less strongly absorbing line wing yield information from deeper regions. Similar considerations hold for measurements of lower spectral resolution within a molecular absorption band.

The extent to which various atmospheric parameters can be unambiguously retrieved from a set of measurements depends on the nature of the measurements, including spectral range and resolution, observational geometry, and signal-to-noise ratio. The fundamental problem associated with the quantitative interpretation of infrared planetary spectra is the formulation of effective methods for the extraction of the maximum information possible from a given set of measurements. A wide variety of approaches to this problem have been developed; here we will consider only a selected few.

The atmospheric parameters to be retrieved are physically related to the measurements through the radiative transfer process described in Chapter 2. The nature of this process imposes limitations on the information that can be extracted, even from error-free measurements, while measurement errors (noise) further reduce the information content. Both factors must be taken into consideration when choosing an appropriate method for a given retrieval problem.

For convenience the computational methods can be divided into two general classes, direct modeling and inversion. The two approaches are illustrated schematically in Fig. 8.1.1 along with the general radiative transfer process. The latter, depicted in part (A), can be regarded as a low-pass filter, permitting only information on the larger spatial scales in the vertical atmospheric structure to be preserved

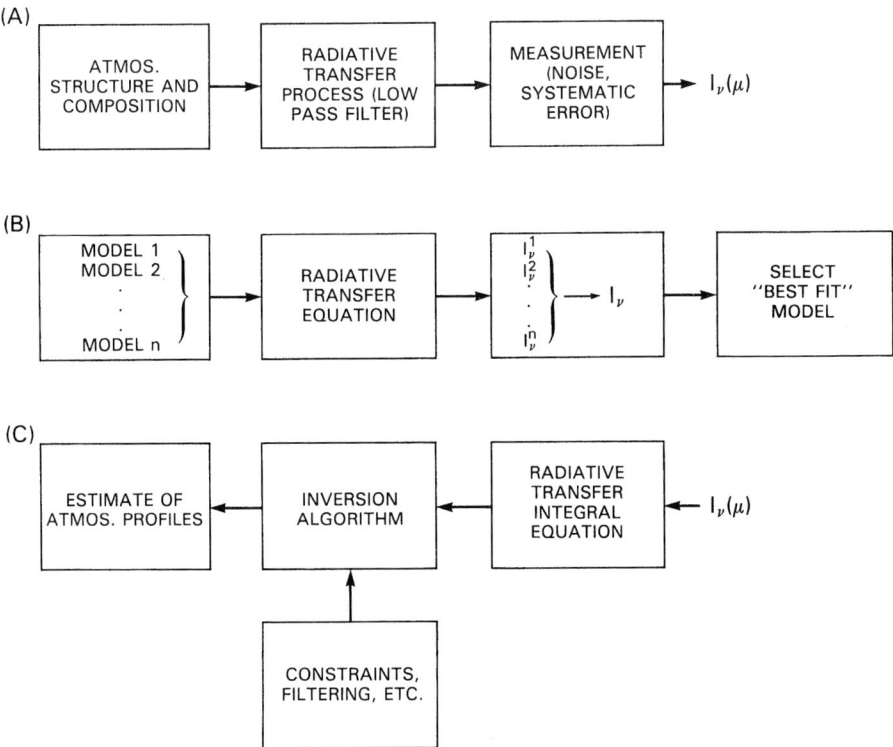

Fig. 8.1.1 Schematic of the atmospheric radiative transfer process and the retrieval of information on atmospheric parameters. (A) Relationship of the measured radiance, $I_\nu(\mu)$, to atmospheric parameters. The radiative transfer process acts as a low-pass filter, while the instrument used introduces random noise and possibly other effects. (B) Modeling approach to the extraction of atmospheric parameters from the measurements. (C) The inversion approach to the extraction of atmospheric parameters.

in the measured radiances at the top of the atmosphere. The information content is further reduced by the presence of instrumental noise. In the direct modeling approach to the retrieval problem (part (B)) a model atmosphere is used, along with the radiative transfer equation, to calculate spectral intensities that can be directly compared with measurements. The parameters of the model are then adjusted and the calculations are repeated until a reasonable fit to the data is obtained. The principal virtue of this approach is the conceptual simplicity of the direct use of the radiative transfer equation to calculate radiances even though the dependence on atmospheric parameters may be highly nonlinear. However, it can become computationally cumbersome to properly explore a sometimes complex parameter space, and the approach does not easily provide insight into the related questions of the actual information content of the measurements and the uniqueness of solutions.

We shall define the inversion approach, illustrated in Fig. 8.1.1(C), as the formulation of an algorithm that permits the atmospheric parameters to be retrieved directly from the measured radiances. This may involve the direct solution of an integral equation or may simply consist of the fitting of an analytic expression to calculated radiances, which can then be algebraically solved for the desired parameters. An inversion method usually provides an efficient, systematic means of obtaining a solution, and can also aid in assessing the information content of a measurement set and the uniqueness of the solutions. However, these methods can be difficult to implement when the dependence of the radiances on the atmospheric parameters sought is highly complex.

8.2 Temperature profile retrieval

a. General consideration

Knowledge of the temperature profile is essential for an understanding of the dynamical, thermodynamical, and chemical processes occurring in an atmosphere. In addition, the atmospheric temperature must first be obtained before information on gas or particle composition can be extracted from thermal emission measurements. More effort has been devoted to the retrieval of temperature profiles than any other atmospheric parameter and, as a consequence, this problem is better understood than other inverse radiative transfer problems. Study of this subject also provides a convenient means of illustrating some of the basic concepts of vertical profile reconstruction from remotely sensed data.

The thermal spectral radiance emerging from the top of the atmosphere as measured at the spacecraft can be written in the form

$$I(\mu, \nu) = \varepsilon_\nu B(\nu, T_s) \operatorname{Tr}(\mu, \nu, z_s) + \int_{z_s}^{z_t} B[\nu, T(z)] \frac{\partial \operatorname{Tr}(\mu, \nu, z)}{\partial z} \, dz, \qquad (8.2.1)$$

where μ is the cosine of the emission angle, ν is the wavenumber, $z = -\ln p$, p is the atmospheric pressure, and $B(\nu, T)$ is the Planck radiance at wavenumber ν and temperature T. The atmospheric transmittance from level z to the top of the atmosphere can be written

$$\operatorname{Tr}(\mu, \nu, z) = \int_{\delta\nu} \phi(\nu, \nu') e^{-\tau(\nu', z)/\mu} \, d\nu', \qquad (8.2.2)$$

where $\phi(\nu, \nu')$ is the instrument spectral response function extending over the spectral interval $\delta\nu$ with central wavenumber ν. The spectral interval over which ϕ is nonzero is assumed to be narrow compared to the scale over which the Planck function varies significantly. The first term in Eq. (8.2.1) represents emission from the planetary surface with emissivity $\varepsilon(\nu)$ and temperature T_s attenuated by the atmosphere. In the case of the giant planets with essentially infinitely deep atmospheres, this term is omitted. The second term is the atmospheric emission contribution, and the integral is taken from the surface z_s to the effective 'top' z_t above which contributions to the outgoing radiance are negligible. In writing Eq. (8.2.1), scattering has been neglected and the atmosphere is assumed to be in local thermodynamic equilibrium at all levels. We have also neglected a term representing reflection by the surface of downward thermal flux from the atmosphere, a term that will generally be small as long as $\varepsilon(\nu)$ does not differ greatly from unity.

While the forward problem, in which Eq. (8.2.1) is used to calculate the emergent spectral radiance from a model atmosphere is straightforward, the direct retrieval of temperature by inversion of measurements of $I(\mu, \nu)$ is 'ill-posed' in the sense that arbitrarily small changes in the measured radiances can produce finite changes in the retrieved temperature profile, resulting in a hypersensitivity to measurement errors. In addition, the problem will usually be underdetermined since it may be necessary to specify the temperature at more vertical numerical quadrature points then there are measurements. Nevertheless, it is possible to obtain physically meaningful solutions by imposing additional constraints that usually take the form of low-pass filtering, either explicitly or implicitly. This general type of inverse problem has been discussed extensively in the literature. An excellent summary of the basic philosophy of inverse problems encountered in astronomy is given by Craig & Brown (1986). There are a number of different approaches that have been applied to the atmospheric inverse radiative transfer problem; several of these are briefly discussed below.

b. Constrained linear inversion

We begin by using the numerical quadrature analog of Eq. (8.2.1) such that $T(z)$ is defined at n discrete atmospheric levels. Linearizing the resulting set of equations about a reference temperature profile $\mathbf{T}^\circ(z)$ yields

$$\Delta \mathbf{I} = \mathbf{K} \Delta \mathbf{T} \quad (8.2.3)$$

where the matrix element K_{ij} is the functional derivative at ν_i with respect to T at level z_j. The perturbation quantities are

$$\Delta \mathbf{T}_j = T(z_j) - T^\circ(z_j) \quad (8.2.4)$$
$$\Delta \mathbf{I}_i = I(\nu_i) - I^\circ(\nu_i) \quad (8.2.5)$$

where the radiance $I°(v_i)$ is calculated using $T°$. The standard approaches to the solution of Eq. (8.2.3), such as least squares, will usually result in unstable, non-physical results. Instead of solving Eq. (8.2.3) directly, a related, well-posed problem that incorporates physical constraints is considered. This general approach, sometimes called 'regularization', has been discussed extensively in the literature (see for example Twomey, 1963, 1977; Tikhanov, 1963; Craig & Brown, 1986; Press *et al.*, 1992). The version presented here follows Conrath *et al.* (1998).

Quite generally, $\Delta\mathbf{T}$ can be expanded in a set of basis vectors, i.e.,

$$\Delta\mathbf{T} = \mathbf{Fa} \tag{8.2.6}$$

where $\mathbf{F}$ is a matrix whose columns are the basis vectors, and the vector $\mathbf{a}$ contains the expansion coefficients. The solution we seek results from minimization with respect to $\mathbf{a}$ of the penalty function

$$Q = (\Delta\mathbf{I} - \mathbf{KFa})^T E^{-1} (\Delta\mathbf{I} - \mathbf{KFa}) + \gamma \mathbf{a}^T \mathbf{a}. \tag{8.2.7}$$

Here, E is the measurement error covariance matrix. The first term on the right hand side of Eq. (8.2.7) is the usual penalty function employed in least square fitting. In the second term, $\mathbf{a}^T\mathbf{a}$ is a measure of the departure of the solution from the reference or first guess profile, and the parameter γ determines the relative weight with which this constraint is applied. The minimization of Q is straightforward, and the details are given by Conrath *et al.* (1998). The resulting estimate for $\Delta\mathbf{T}$ is

$$\Delta\mathbf{T} = \mathbf{W}\Delta\mathbf{I}, \tag{8.2.8}$$

where

$$\mathbf{W} = \mathbf{SK}^T (\mathbf{KSK}^T + \gamma \mathbf{E})^{-1}. \tag{8.2.9}$$

The superscript T indicates matrix transposition. $\mathbf{S} = \mathbf{FF}^T$ is the two-point correlation matrix of the basis vectors; only this parameter appears in the solution and not the basis vectors themselves. The nonlinearity of the problem is taken into account through iterative application of Eq. (8.2.8). The error covariance matrix for the retrieval temperature profile due to instrument noise propagation is

$$\mathbf{V} = \mathbf{WEW}^T. \tag{8.2.10}$$

358   *Retrieval of physical parameters from measurements*

The square root of the diagonal elements of **V** are the rms errors at each level in the retrieved profile resulting from instrument noise propagation.

The functional derivatives with respect to temperature are calculated assuming that the temperature dependence of the atmospheric transmittance is weak relative to that of the Planck function so that **K** can be approximated as

$$K_{ij} = \frac{\partial B[\nu_i, T(z_j)]}{\partial T_i} \cdot \frac{\partial \operatorname{Tr}(\mu, \nu_i, z_j)}{\partial z_j}. \tag{8.2.11}$$

A set of functional derivatives (also referred to as contribution functions or kernels) for the Martian atmosphere, using selected wavenumbers in the 15 μm CO_2 absorption band, is shown in Fig. 8.2.1.

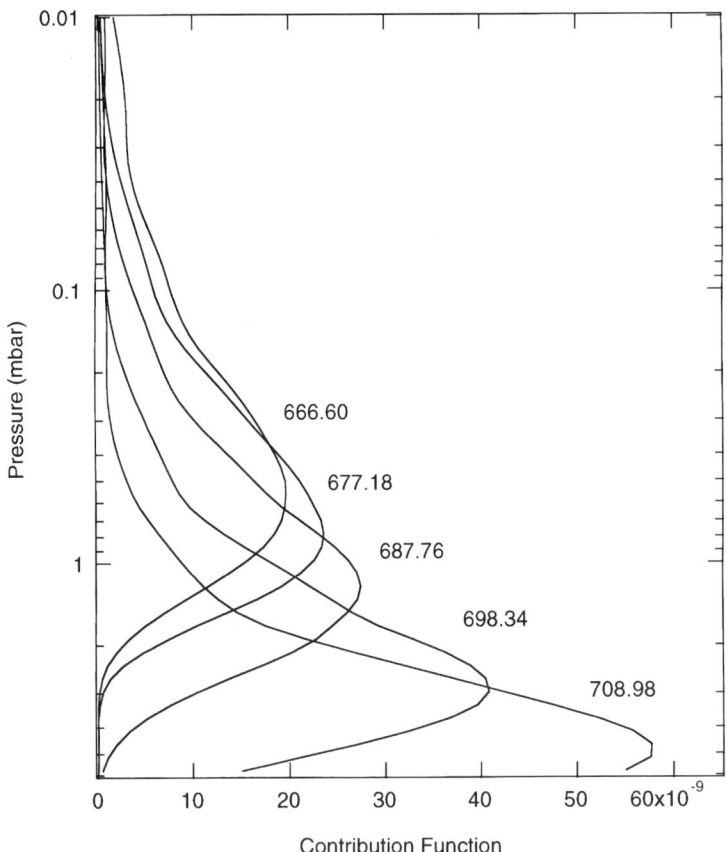

Fig. 8.2.1 Contribution functions (functional derivatives of radiance with respect to temperature) for the 15 μm CO_2 band in the atmosphere of Mars. These functions were calculated using Eq. (8.2.11) and are labeled by wavenumber (cm^{-1}). Units for the contribution functions are W cm^{-2} sr^{-1} cm K^{-1}. (After Conrath *et al.*, 2000.)

These are applicable to the 10 cm^{-1} spectral resolution measurements obtained by the Thermal Emission Spectrometer carried on the Mars Global Surveyor spacecraft (Christensen *et al.*, 1998), and points on both sides of the 15 μm CO_2 band are included. Rapid calculation of atmospheric transmittances is required for temperature retrieval from a large volume of spacecraft measurements, and the k-distribution method is used (Goody & Yung, 1989).

The correlation matrix **S** provides a convenient means of filtering the solution. Rather than specifying a specific set of basis functions, the form of **S** can be specified; for example, a convenient choice is

$$S_{ij} = \exp[-(z_i - z_j)^2/2c^2] \qquad (8.2.12)$$

where c is the correlation length in scale heights, and can be regarded as a measure of the width of the numerical filter imposed on the solution by the retrieval algorithm. The form of measurement error covariance matrix depends on the nature of the measurement errors. If the noise is independent throughout the interval used and is uncorrelated between any two spectral points, then

$$E_{ij} = N^2 \delta_{ij} \qquad (8.2.13)$$

where N is the Noise-Equivalent-Spectral-Radiance of the instrument.

This approach can be generalized to include multiple atmospheric parameters, providing there is adequate information content in the spectral radiance measurements. For example, it has been used to simultaneously infer both temperature profiles and molecular para-hydrogen profiles from Voyager IRIS measurements of Jupiter, Saturn, Uranus, and Neptune (Conrath *et al.*, 1998).

The retrieval method has been used extensively for temperature profile retrieval in both the terrestrial and other planetary atmospheres. Examples of profiles obtained by this technique for Earth, Mars, Jupiter, Saturn, Uranus, and Neptune are shown in Fig. 8.2.2. Also included is a Titan profile obtained from radio occultation data. The profiles for Earth and Mars were derived from measurements obtained with the Fourier transform spectrometers carried on Nimbus 3, 4, and Mariner 9, respectively. In both cases data from the 15 μm CO_2 absorption band were used. The profiles for the outer planets were obtained by inversion of measurements from the Voyager Fourier transform spectrometers. For Jupiter and Saturn, data from the S(0) and S(1) collision-induced H_2 lines between 200 and 600 cm^{-1} were used, along with measurements from the CH_4 ν_4-band centered near 1300 cm^{-1}. Because of the extremely low temperatures encountered on Uranus and Neptune, adequate signal-to-noise ratio for the retrieval of vertical thermal structures was obtained

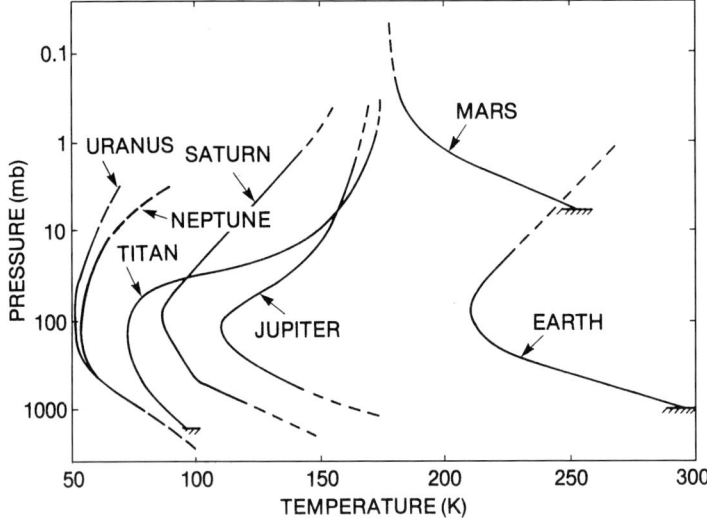

Fig. 8.2.2 Temperature profiles for various Solar System bodies. All profiles shown were obtained by inversion of infrared spectral measurements from Michelson interferometers with the exception of that of Titan, which was obtained from Voyager radio occultation data.

only within the S(0) line of hydrogen. Many data were acquired for each planet, providing extensive spatial coverage. Results of analyses of these data are discussed in greater detail in Chapter 9.

c. Relaxation algorithms

Another approach, widely used in remote sensing, is the relaxation method originally applied to the temperature retrieval problem by Chahine (1968). In the development of this algorithm it is assumed that measurements are available at a discrete set of wavenumbers, v_i ($i = 1, m$), which are associated with a set of weighting functions $W(v_i, z)$ whose maxima are well distributed over the atmospheric levels of interest. A first estimate of the temperature profile is obtained from the relation

$$B[v_i, T^0(z_i)] = I(v_i); \quad i = 1, m \qquad (8.2.14)$$

where z_i is the height of the maximum of the weighting function for frequency v_i. Equations (8.2.14) are solved for the set of temperatures $T^0(z_i)$ ($i=1, m$). Interpolation between the m levels is used to calculate the radiances, $I^0(n_i)$, corresponding

8.2 Temperature profile retrieval

to these temperatures. The solution is then iteratively improved using the relaxation algorithm

$$B[\nu_i, T^{n+1}(z_i)] = \frac{I(\nu_i)}{I^n(\nu_i)} B[\nu_i, T^n(z_i)]. \tag{8.2.15}$$

The procedure is continued until the differences between measured and calculated radiances reach the noise level.

Numerical experiments with this algorithm indicate that the low spatial frequency components of the retrieved profile converge more rapidly than the higher frequency components. Consequently, the basic strategy of this technique is to truncate the iteration before the high frequency terms can grow significantly. Several variations of the method have been formulated. In one approach, the linearly interpolated temperature profile is subjected to an explicit numerical filtering at each iterative step to suppress high frequency components in the solution (Conrath & Gautier, 1979). Another modification, proposed by Smith (1970), uses the weighting functions themselves to produce a smoothly interpolated profile. The method is especially useful for highly nonlinear problems, which cannot be easily cast into the form of a linear integral equation. Its application to the problem of constituent profile retrieval is discussed in Section 8.3.

d. Backus–Gilbert formulation

A retrieval formulation developed by Backus & Gilbert (1970) and used in geophysical inverse problems of the solid Earth has also found application in remote sensing of atmospheres. Although this formulation is developed from a somewhat different point of view, it is formally related to the constrained linear inversion as well as other techniques. We include a brief description here because of the additional physical insight it can provide into inversion theory.

For radiance measurements at m discrete frequencies, the linearized radiative transfer equation can be written

$$\Delta I_i = \int_0^\infty K_i(z) f(z) \, dz; \quad i = 1, m, \tag{8.2.16}$$

where $\Delta I_i = \Delta I(\nu_i)$, $f(z) = T(z) - T^0(z)$, and $K_i(z) = K(\nu_i, z)$. From Eq. (8.2.16) we cannot hope to retrieve $f(z)$ in complete detail, but only certain properties of the profile can be estimated. The measurements, ΔI_i, are proportional to certain averages of $f(z)$ with the averaging defined with respect to the radiative transfer kernels, $K_i(z)$. Linear combinations of the ΔI_is are sought that more nearly represent the properties of $f(z)$ than do the averages over $K_i(z)$ given by Eq. (8.2.16).

To address this problem let the estimated average property of $f(z)$ be written

$$\bar{f}(z) = \sum_{i=1}^{m} a_i(z) \Delta I_i, \qquad (8.2.17)$$

where the coefficients $a_i(z)$ depend on the specific retrieval algorithm. We can relate $\bar{f}(z)$ to $f(z)$ by substituting Eq. (8.2.16) into Eq. (8.2.17) to obtain

$$\bar{f}(z) = \int_0^\infty A(z, z') f(z') \, dz', \qquad (8.2.18)$$

where $A(z, z')$ is the so-called averaging kernel given by

$$A(z, z') = \sum_{i=1}^{m} a_i(z) K_i(z). \qquad (8.2.19)$$

The extent to which $\bar{f}(z)$ is useful depends on the properties of $A(z, z')$, such as its width relative to the individual $K_i(z)$s.

To pursue this line of reasoning further, it is necessary to define some measure of the width or shape of $A(z, z')$. Two quadratic measures of interest are

$$Q_1(z) = \int_0^\infty [A(z, z') - \delta(z - z')]^2 \, dz', \qquad (8.2.20)$$

and

$$Q_2(z) = \int_0^\infty (z - z')^2 A^2(z, z') \, dz'. \qquad (8.2.21)$$

By minimizing either of these quadratic forms, we can obtain a set of coefficients that provide the optimum vertical resolution in the sense of the particular form chosen. However, in the presence of measurement noise, the resulting retrievals would be unsatisfactory because of error propagation, and we must take the effects of noise explicitly into account.

The formal linear propagation of random measurement errors into $\bar{f}(z)$ can be written

$$\sigma_{\bar{f}}^2 = \sum_i a_i^2 \sigma_I^2, \qquad (8.2.22)$$

8.2 Temperature profile retrieval

where the errors are assumed to be uncorrelated between frequencies with a variance independent of frequency given by σ_I^2. Instead of minimizing one of the quadratic forms, $Q(z)$, we must minimize a linear combination of the form

$$q(z) = wQ(z) + (1 - w)r\sigma_I^2(z), \qquad (8.2.23)$$

where w is the relative weight placed on the narrowness of the averaging kernel versus the error propagation, and r is an arbitrary scaling factor.

If we choose $Q(z) = Q_1(z)$ and minimize $q(z)$, then $\bar{f}(z)$ takes the form

$$\bar{f}(z) = \mathbf{K}^T(z) \left[\int_0^\infty \mathbf{K}(z') \mathbf{K}^T(z') \, dz' + \frac{1-w}{w} r\sigma_I^2 \mathbf{1} \right]^{-1} \Delta \mathbf{I}, \qquad (8.2.24)$$

where $\mathbf{K}(z)$ and $\Delta \mathbf{I}$ are column matrices whose elements are given by $K_i(z)$ and ΔI_i respectively, and $\mathbf{1}$ is the unit matrix.

Thus, from the Backus–Gilbert point of view Eq. (8.2.24) is that retrieval for which the averaging kernel lies closest to a Dirac-delta function in the least squares sense, subject to the constraint imposed by a limited set of measurements in the presence of measurement error.

Extensive use has been made of the form $Q(z) = Q_2(z)$ in Eq. (8.2.23) with a measure of the width of $A(z, z')$ defined as $s(z) = 12 Q_2(z)$. The factor 12 is included so that $s(z)$ gives the correct width for a rectangular function of unit area. Minimizing $q(z)$, Eq. (8.2.23), subject to the additional constraint that $A(z, z')$ be 'unimodular', i.e.,

$$\int_0^\infty A(z, z') \, dz' = 1, \qquad (8.2.25)$$

yields

$$\bar{f}(z) = \frac{\mathbf{u}^T \mathbf{W}^{-1}(z) \Delta \mathbf{I}}{\mathbf{u}^T \mathbf{W}^{-1}(z) \mathbf{u}}, \qquad (8.2.26)$$

where

$$W_{ij} = w \int_0^\infty (z' - z)^2 K_i(z') K_j(z') \, dz' + (1 - w) r\sigma_I^2 \delta_{ij} \qquad (8.2.27)$$

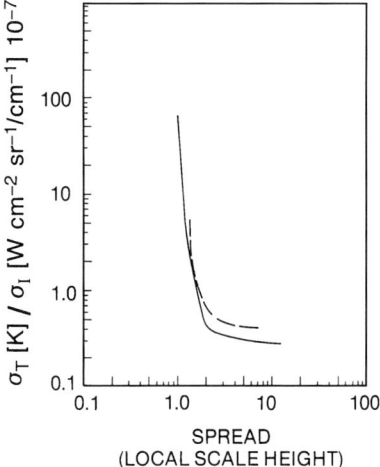

Fig. 8.2.3 Examples of trade-off curves for measurements within the 15 μm CO_2 band in the Earth's atmosphere as defined in the Backus–Gilbert theory. The spread is a measure of the vertical resolution for the retrieved profiles, while σ_T/σ_I is the ratio of the formal statistical error in the retrieved profile to the rms noise level in the measurements. The broken and solid curves are for sets of measurements containing six and twelve spectral intervals, respectively.

and

$$u_i = \int_0^\infty K_i(z')\,dz'. \qquad (8.2.28)$$

The selection of w in Eq. (8.2.27) is arbitrary and its choice involves a trade-off between maximization of vertical resolution and minimization of measurement noise. In fact at each level for which $f(z)$ is to be calculated, a trade-off curve can be constructed expressing the relationship between σ_I^2 and the vertical resolution as measured by $s(z)$. As an example, Fig. 8.2.3 shows trade-off curves for retrieval of the temperature at the 49 mbar level in the terrestrial stratosphere, using measurements in the 15 μm CO_2 band.

The 'elbow' shape observed in Fig. 8.2.3 is a general characteristic of trade-off curves. As the vertical resolution is degraded the noise propagation decreases sharply until a point is reached beyond which little improvement is realized. While the Backus–Gilbert formulation explicitly considers the trade-off between resolution and error propagation, this general property of profile retrieval is apparent in the filtering approach discussed earlier. As vertical resolution is increased by the addition of higher spatial frequency components to the solution, these components amplify the noise.

The inversion algorithm Eq. (8.2.26) resulting from this formulation tends to be computationally cumbersome since a matrix inversion must be performed for each atmospheric level considered. While it may be feasible to use the algorithm when a limited number of cases is to be treated, it is not practical when retrievals from many sets of measurements are required. However, the formulation does provide a convenient framework for investigation of the potential information content of a planned experiment from a simple physical point of view.

e. Statistical estimation

Statistical estimation theory provides a powerful framework for the development of retrieval methods. Rodgers (2000) gives an excellent overview of this subject. Algorithms which yield 'optimum' solutions in the statistical sense have found extensive applications in the remote temperature sounding of the terrestrial atmosphere where much *a priori* information is available.

As an example of the statistical approach we consider a formulation that follows the works of Foster (1961), Strand & Westwater (1968), and Rodgers (1970). Consider a representative ensemble of atmospheric temperature profiles defined by a mean profile $\langle T \rangle$ and a covariance matrix

$$S_{ij} = \langle (T_i - \langle T_i \rangle)(T_j - \langle T_j \rangle) \rangle. \qquad (8.2.29)$$

The ith element of the vector $\mathbf{T}$ is the temperature at the ith atmospheric level, and the angular brackets denote an ensemble mean. Expansion of $B[\nu, T(z)]$ about the mean temperature profile, and substitution into Eq. (8.2.1), permits us to write an expression in discrete form analogous to Eq. (8.2.3),

$$\Delta \mathbf{I} = K\mathbf{f} + \varepsilon, \qquad (8.2.30)$$

where

$$\mathbf{f} = \mathbf{T} - \langle \mathbf{T} \rangle \qquad (8.2.31)$$

and $\Delta \mathbf{I}$ is the vector of measured radiances minus the radiances calculated using the ensemble mean temperature profile. The measurement error has been included explicitly as the vector ε, which is assumed to have zero mean and a known covariance matrix

$$E_{ij} = \langle \varepsilon_i \, \varepsilon_j \rangle. \qquad (8.2.32)$$

We seek an estimated solution $\tilde{\mathbf{f}}$ of the form

$$\tilde{\mathbf{f}} = \mathbf{L}^{-1}\Delta\mathbf{I}. \tag{8.2.33}$$

The matrix L^{-1} is chosen such that the mean square deviation of the solution from the temperature profiles of the ensemble is a minimum, i.e.,

$$\sum_i \langle (f_i - \tilde{f}_i)^2 \rangle = \text{minimum}. \tag{8.2.34}$$

This leads to a solution in the form

$$\tilde{\mathbf{f}} = \mathbf{SK}^T(\mathbf{KSK}^T + E)^{-1}\Delta\mathbf{I}. \tag{8.2.35}$$

This approach works best when the solution does not deviate greatly from the mean properties of the statistical ensemble. By exploiting correlations between large- and small-scale vertical structure in the profiles, the method can yield solutions with higher vertical resolution than that intrinsic to the radiance measurements alone. Note that Eq. (8.2.35) is formally equivalent to the constrained solution defined by Eqs. (8.2.8) and (8.2.9) with $\gamma = 1$. However, the matrix $\mathbf{S}$ is given a different conceptual interpretation.

Because of the lack of statistical information, this approach in its most complete form has not been used in the analysis of infrared data from planetary missions. However, one limiting case is of particular interest in the planetary context. If the profile and noise covariance matrices are approximated by

$$S_{ij} = \sigma_T^2 \delta_{ij} \tag{8.2.36}$$

and

$$E_{ij} = \sigma_N^2 \delta_{ij}, \tag{8.2.37}$$

we obtain

$$\tilde{f} = K^T \left(KK^T + \frac{\sigma_N^2}{\sigma_T^2} 1 \right)^{-1} \Delta\mathbf{I}. \tag{8.2.38}$$

This is known as the 'minimum information' or 'maximum entropy' estimation.

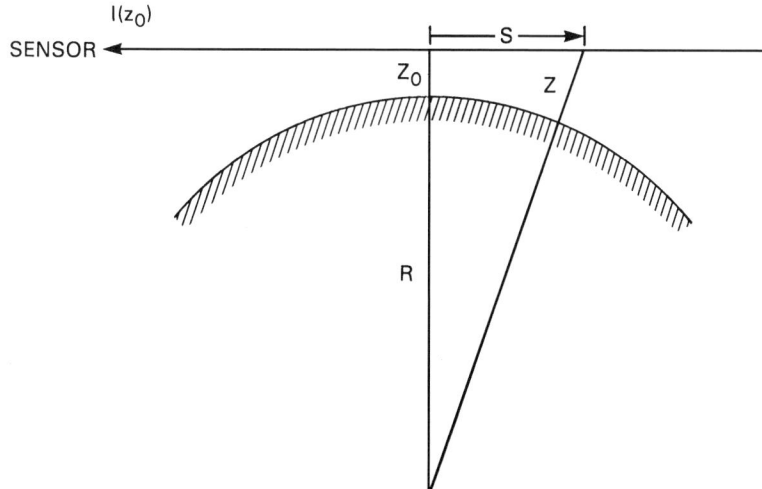

Fig. 8.2.4 Geometry for limb-tangent measurements. A ray is depicted with tangent height z_0 for a planet with radius R. Distance from the tangent point to an arbitrary point along the ray path is denoted by s, and z is the height of the point above the planetary surface.

f. Limb-tangent geometry

Providing the instrumental field of view can be made sufficiently small, measurements tangential to the planetary limb can produce atmospheric profiles with a vertical resolution higher than that achievable from nadir viewing measurements. In addition, the very long path length with a cold space background permits retrieval of trace gas abundances that could not otherwise be accomplished. This approach has been used extensively to determine thermal structure and trace gas distributions in the Earth's stratosphere and mesosphere.

The basic geometry of the observations is shown in Fig. 8.2.4. A ray path traverses the atmosphere with tangent altitude z_0 and reaches the sensor at $s = -\infty$. Here, z denotes geometric height, s the distance along the path, and R the planetary radius. The radiance measured by an idealized sensor of infinite spatial resolution can be written

$$I(z_0) = \int_{\delta \nu} \int_{-\infty}^{+\infty} \phi(\nu)\, B[\nu, T(s)] \frac{\partial}{\partial s}\left[e^{-\tau(\nu,s)}\right] ds\, d\nu, \quad (8.2.39)$$

where $\tau(\nu, s)$ is the optical thickness along the ray path from $-\infty$ to s, and $\phi(\nu)$ is the normalized spectral response of the sensor. In order to compare observations in the limb-tangent geometry with nadir viewing measurements, we write Eq. (8.2.39) in the form

$$I(z_0) = \int_{z_0}^{\infty} B[T(z)] W(z_0, z) \, dz, \qquad (8.2.40)$$

where the weighting function in this case is defined as

$$W(z_0, z) = \int_{\delta\nu} \phi(\nu) \left\{ \frac{\partial}{\partial s} \left[e^{-\tau(\nu, s)} \right]_- - \frac{\partial}{\partial s} \left[e^{-\tau(\nu, s)} \right]_+ \right\} \frac{ds}{dz} \, d\nu. \qquad (8.2.41)$$

The subscripts (−) and (+) refer to that portion of the ray path between the sensor and the tangent point and that beyond the tangent point, respectively. We have again assumed that the variation in the Planck radiance over the spectral band width of the instrument, as defined by $\phi(\nu)$, can be neglected. The weighting function defined by Eq. (8.2.41) is for infinite resolution and must be convolved with the spatial response of the instrument. If a series of measurements is obtained over a range of tangent heights z_0, then Eq. (8.2.40) can be inverted to obtain an estimate of $T(z)$ using any of the algorithms discussed above.

The detailed behavior of the limb-tangent weighting functions depends on the opacity of the spectral region chosen. In the optically thin case, the weighting functions tend to be quite narrow, but because of the relatively small atmospheric emissivity, it is often difficult to achieve a good signal-to-noise ratio. On the other hand, as the atmospheric optical thickness increases, the weighting functions become broader, approaching widths similar to those for the nadir viewing case. A reasonable trade-off is usually obtained for an optical thickness at the tangent point of order unity. In practice, it is advantageous to obtain measurements in at least two spectral intervals of different opacities. Determination of the limb-tangent height with the necessary precision requires a knowledge of the spacecraft position and orientation, which can exceed that obtainable from most spacecraft. However, if the relative pointing of the sensor axis from one measurement to the next is known, or several sensors are used in simultaneous measurements, then $T(p)$ can be determined by requiring that the solution obtained from a relatively transparent spectral interval coincides with that from a more opaque interval. This general principle has been used in most of the limb-tangent stratospheric and mesospheric sounding investigations carried out thus far (see, for example, Gille et al., 1984).

8.3 Atmospheric composition

The thermal infrared portion of the spectrum is rich in absorption bands of polyatomic gases and can provide extensive information on the composition of planetary atmospheres. The type of information available ranges from the identification of constituents by the detection of their spectral features to the retrieval of detailed vertical distributions of gaseous abundances. Factors that govern the information available from a given set of data include the spectral resolution and signal-to-noise ratio of the measurements, the degree of overlap of various atmospheric spectral features, the contrast of the spectral features as determined by the atmospheric temperature lapse rate, and the presence of other potentially interfering opacity sources such as clouds and particulates.

a. Principles

First, we briefly review the principles that permit identification of atmospheric gases and the determination of their abundances. In the thermal infrared the maximum contribution to the radiance at a given wavenumber comes from an atmospheric level determined by the magnitude of the absorption coefficient and the abundance of the absorbing gas. Consider a hypothetical absorption or emission feature, such as an individual line or the contour of an unresolved band. At the limb the optical path length $\tau(\nu, s)$ through the atmosphere above a certain level, z_0, is finite (see Fig. 8.2.4), and the variation of $\tau(\nu, s)$ with ν describes an emission feature against the cold background of space. For nadir viewing measurements, however, a contrast in radiance will be apparent between the center of the feature and its wings only if the atmospheric temperatures corresponding to the respective weighting functions are different. Thus, the apparent strength of the observed feature depends on the atmospheric temperature lapse rate as discussed in Chapter 4. Best quantitative results are obtained when the lapse rate is large; when the lapse rate approaches zero, the spectral feature vanishes. If the spectral resolution of the measurements is sufficiently high to define the detailed contour of the absorption feature, if the lapse rate is nonzero, and if the noise level is small relative to the total radiance contrast across the feature, then it may be possible to extract information on the vertical distribution of the absorbing gas over some range of heights. If the feature cannot be resolved in detail, it may still be possible to estimate the total abundance of the gas within an atmospheric column.

b. Feature identification

If the measurements are of an exploratory nature, the initial goal is the identification of the atmospheric gases that produce observable spectral features. An example of this task is the analysis of the Voyager infrared spectra of Saturn's satellite Titan (Kunde *et al.*, 1981; Maguire *et al.*, 1981; Samuelson *et al.*, 1981). A Titan spectrum is shown in Fig. 6.4.2 along with laboratory spectra of HC_3N and C_2N_2.

The 4.3 cm^{-1} resolution of the Voyager infrared spectrometer does not permit the detection of individual spectral lines, but the Q-branches of numerous absorption bands stand out as sharp features, and for some bands the P- and R-branch contours are also evident. The features appear in emission because the bands are opaque, and all contributions to the radiances originate in the stratosphere where the temperature is increasing with height. In most cases atmospheric constituents can be identified by comparison of the locations of the observed spectral features with those of candidate gases obtained from laboratory measurements as shown in Fig. 6.4.2. However, for weak features more sophisticated approaches must sometimes be used.

c. Correlation analysis

Statistical techniques are powerful tools, which can be used in the search for minor constituents in planetary spectra. Correlation analysis is particularly applicable to the detection of gases with many, generally weak spectral features. Correlation analysis is best applied when the signatures of a suspected constituent are of the same magnitude or possibly even less than the noise level of the instrument. Under such conditions visual inspection of a spectral region where known lines of a particular gas should appear may not be conclusive. The advantage of correlation analysis is that many spectral positions can be searched simultaneously.

Correlation expresses the degree of linear coherence between two functions, F and G. We define the covariance function, $C(\delta)$, by

$$C(\delta) = \lim_{x \to \infty} \frac{1}{2x} \int_{-x}^{+x} F(y)\, G(y + \delta)\, dy. \tag{8.3.1}$$

If F equals G we obtain the autocovariance, also called the autocorrelation function, although the latter name is sometimes reserved for the normalized version, $C(\delta)/C(0)$. Correlation analyses play important roles in communication theory, where F and G normally are functions of time ($y = t$) and the lag δ is a time delay; in our case y represents the wavenumber and δ a wavenumber shift. The mathematical formulation of correlation theory can be found in books on information and communication theory, such as Goldman (1953).

In all cases of concern, F and G are given with finite spectral resolution, in other words, at discrete sample points. The integration of Eq. (8.3.1) must then be replaced by a summation. Furthermore, the functions can only be specified over a finite spectral range and not to infinity; consequently, we can find only an approximation to the true covariance function. Care must be exercised to take the spectral interval sufficiently wide and select proper boundaries beyond which

8.3 Atmospheric composition

spectral information on F and G is not available. As a rule of thumb, the range should be at least 5 to 10 times the maximum wavenumber shift. In discrete form the covariance function is

$$C(\delta_m) = \frac{1}{2N+1} \sum_{n=-N}^{+N} F(\nu_n) G(\nu_n + \delta_m), \qquad (8.3.2)$$

m and n are summation indices and N defines the limits over which the correlation is performed. Spectral correlation analysis has recently been applied to the ATMOS data by Beer & Norton (1988).

d. Abundance determination

The next level of compositional interpretation is the quantitative estimation of the abundances of the gases present. If the constituent is believed to be uniformly mixed in the atmosphere, then an attempt can be made to determine the single parameter describing the amount of gas present, that is, the constant mole fraction or the column abundance above some level. If the gas is not uniformly mixed, but the spectral information is minimal, then a one-parameter description, such as the vertical mean mole fraction, may still be used. If a significant amount of spectral information is available, it may be possible to obtain a detailed description of the vertical distribution of the gas. Many different approaches have been devised to obtain abundance information from measured spectra. As discussed earlier in this chapter it is convenient to divide these into two groups: direct comparison methods and inversion methods.

In the direct comparison approach, an atmospheric model is assumed, and the radiative transfer equation is used to calculate a spectrum, which is then compared with the measurements along the lines discussed in Chapters 4 and 6. The parameters of the model are varied and calculations are repeated until the theoretical spectrum agrees with the measured spectrum to within prescribed limits. This approach requires an independent knowledge of the temperature as a function of pressure in the atmosphere, which may be obtained either from another portion of the spectrum or from auxiliary measurements. A knowledge of the absorption coefficients for the gases that are optically active in the portion of the spectrum being analyzed is also necessary. The coefficients can be obtained from a combination of theory and laboratory measurements, as discussed in Chapter 3.

With this approach, the unknown vertical distribution of the gas to be studied is usually parameterized in some way so that only a small number of quantities need be considered. These parameters can then be varied in a systematic way until a satisfactory fit to the measured spectrum is achieved. Some measure of the quality of fit, such as the rms residuals between calculated and measured spectra, is usually adopted, and an attempt is made to minimize this quantity. Even for a small number of parameters, the search of the parameter space can require an extensive computational effort, and establishment of the uniqueness of the resulting solution may not be straightforward.

When the gaseous constituent is believed to be uniformly mixed, and the temperature profile is known by some independent means, then the direct comparison of measured and calculated spectra is probably the most efficient means of determining the abundance. The determination of the carbon isotopic abundance ratio $^{12}C/^{13}C$ in the Jovian atmosphere provides one example of the application of this method. Courtin et al. (1983) obtained the abundance of the isotopic form of methane, $^{13}CH_4$, from Voyager infrared measurements in the methane ν_4-band complex near 1300 cm^{-1}. A mean of 1260 spectra within a portion of this band is displayed in Fig. 8.3.1. The features are seen in emission since the principal contributions to the spectral radiance originate in the stratosphere where the temperature is increasing with height. The opacity in this spectral region is dominated by the more common isotopic form of methane, $^{12}CH_4$; however, some regions, such as the Q-branch of $^{13}CH_4$ near 1295 cm^{-1}, are sensitive to the less abundant form.

It was first necessary to retrieve the upper stratospheric thermal structure. This was accomplished using measurements at 1305 and 1358 cm^{-1}, which are insensitive to $^{13}CH_4$. The mole fraction of $^{12}CH_4$ was assumed to be known. Spectra were then calculated for various values of $^{13}CH_4$ and compared with the measured spectra as shown for three cases in Fig. 8.3.1. The theoretical curves in the figure are labeled according to the $^{12}C/^{13}C$ ratio. By minimizing the mean residual between measured and calculated spectra between 1292 and 1300 cm^{-1}, Courtin et al. obtained a $^{12}C/^{13}C$ ratio of 178 which is approximately 1.8 times the terrestrial value.

e. Profile retrieval

Although the basic principles of the retrieval of vertical composition profiles from infrared measurements by inversion of the radiative transfer equation are the same as the retrieval of temperature profiles discussed in Section 8.2, the composition problem is usually more difficult to deal with in practice. The optical depth at a given level in the atmosphere is determined by an integration over the optically active gas profile from that level to the effective top of the atmosphere. Calculation of the radiance at the top of the atmosphere then requires an integration of the source function over all optical depths from the lower boundary to the top of the atmosphere. Thus the desired abundance profile is embedded within a double integration.

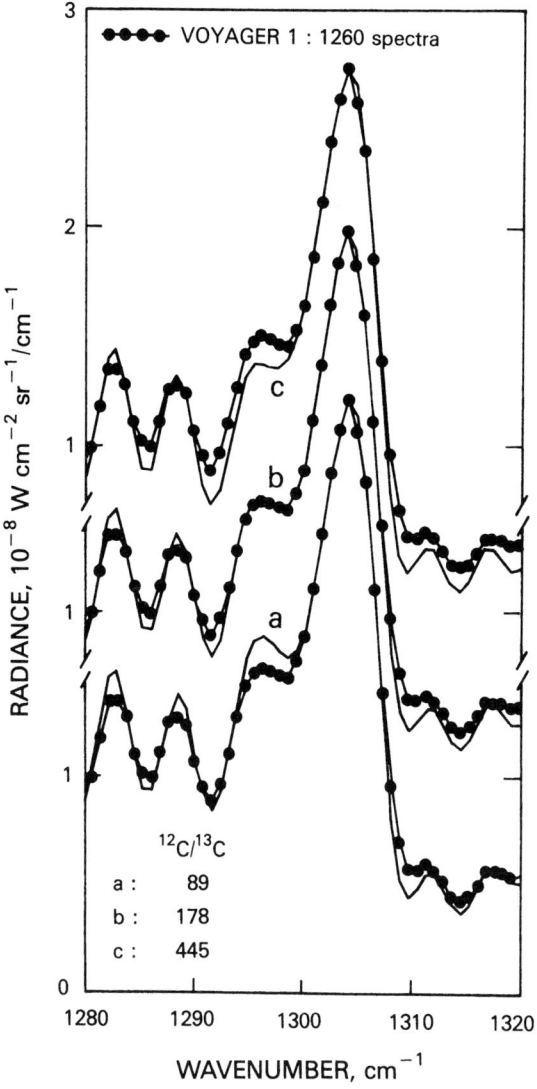

Fig. 8.3.1 Determination of the isotopic ratio $^{12}C/^{13}C$ in the atmosphere of Jupiter using Voyager IRIS measurements. The curves with the superposed dots represent the average of 1260 Voyager IRIS spectra of Jupiter, and the curves without dots are spectra calculated from model atmospheres. The major spectral feature centered near 1304 cm^{-1} is the Q-branch of the ν_4-band of $^{12}CH_4$ while the smaller peak on the low wavenumber shoulder of the major feature is the Q-branch of $^{13}CH_4$. The best fit for both features is obtained for comparison 'b', corresponding to an isotopic carbon ratio of 178 (Courtin et al., 1983).

One approach to the problem of retrieval of gas profiles by inversion is linearization and iteration. The unknown profile is expressed in terms of a set of discrete parameters; these may consist of the values of the gas mole fraction at the quadrature points used in the numerical integrations required to calculate the radiance or they may be parameters in some analytical representation of the profile. Let us designate the set of parameters as $q_1, q_2, \ldots, q_n$. Then the theoretical radiance at the ith wavenumber ν_i can be expressed as

$$I_i = f_i(q_1, q_2, \ldots, q_n). \tag{8.3.3}$$

We can then expand about a first guess for the qs and obtain to first order

$$\Delta I_i = \sum_j \frac{\partial f_i}{\partial q_j} \Delta q_j \quad (i = 1, m) \tag{8.3.4}$$

where ΔI_i is taken to be the difference between the measured radiance and the radiance calculated using the first guess, and the Δqs are the linear corrections to the first guess. In principle, this set of equations (8.3.4) can be solved by using any of the linear methods described above in the discussion of temperature retrievals. From the resulting solution, an improved set of qs is obtained, which is then used as a new guess, and the procedure is iterated until some convergence criterion is met. This approach has the advantage that it permits application of the well-understood linear inversion techniques. However, it can prove difficult to apply in practice. The matrix of derivatives that is required at each iterative step must usually be estimated numerically, and this can be a time consuming computational process. In strongly nonlinear cases, it may be difficult to insure that the first guess lies within the region of convergence in the parameter space.

An alternate approach is to apply a relaxation technique similar to the method of temperature inversion discussed in Subsection 8.2.c. In this case the number of parameters describing the gas profile is chosen equal to the number of wavenumbers for which we have measurements. The radiance at each wavenumber ν_i is associated with a gas mole fraction q_i at the atmospheric level to which the radiance is most sensitive, i.e., near the peak of the contribution function. A first guess, q_i^0 $(i = 1, m)$, is introduced and used in the radiative transfer equation to calculate a set of radiances, $I^0(\nu_i)$. In order to carry out the radiance calculation it is necessary to adopt some form of interpolation between the levels for which q is initially specified. An improved solution q_i^1 is then obtained using the relaxation relation

$$q_i^1 = \frac{I(\nu_i)}{I^0(\nu_i)} q_i^0 \quad (i = 1, m) \tag{8.3.5}$$

where the $I(\nu_i)$ are measured radiances. The procedure can be iterated until either the residuals between measured and calculated radiances reach some prescribed level or the change in the parameters from one iteration to the next is less than some

8.3 Atmospheric composition

specified amount. Convergence is usually found to be rather slow; however, for many applications, it may be computationally more efficient than the linear method because it eliminates the need for evaluation of the elements of the derivative matrix. This approach, or variations on this approach, has been used extensively for determining gas composition profiles in both the Earth's atmosphere and the atmospheres of other planets. As an example, we consider the retrieval of an ammonia profile on Jupiter from measurements made by the Voyager infrared spectrometer (Kunde et al., 1982). The portion of the spectrum between 850 and 1100 cm^{-1} was used, as shown in Fig. 8.3.2. A modification of the relaxation method as proposed by Smith (1970) was applied to the data. The retrieved NH$_3$ profile is shown in Fig. 8.3.3 where it is compared with the profile that would exist if ammonia were saturated at the local atmospheric temperature. The atmospheric temperature profile employed in the analysis was obtained by inversion of measurements within the collision-induced S(0) and S(1) hydrogen lines centered at 354 cm^{-1} and

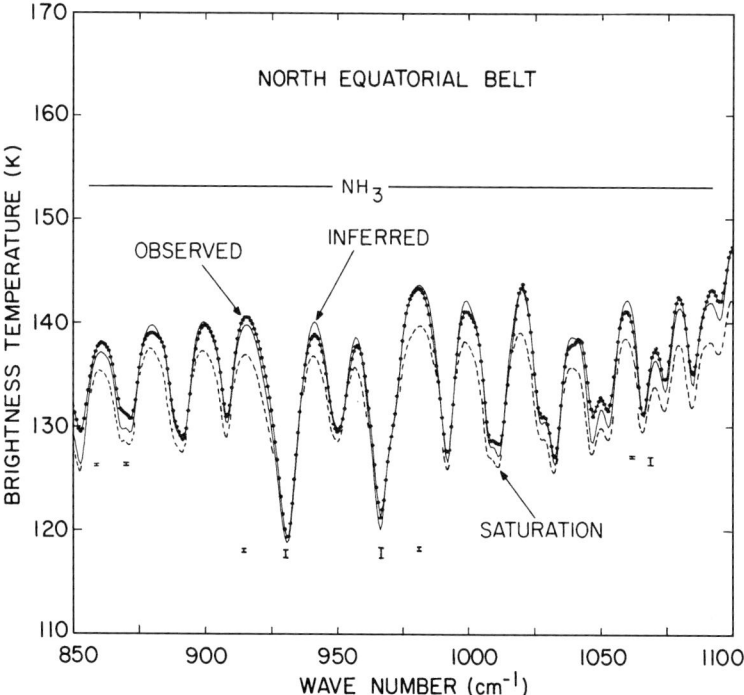

Fig. 8.3.2 Determination of the ammonia (NH$_3$) abundance in the atmosphere of Jupiter by inversion of a portion of a spectrum obtained by the Voyager infrared spectroscopy experiment. The measured spectrum is denoted by the curve with dots. The solid curve is a spectrum calculated using the vertical distribution of ammonia shown in Fig. 8.3.3. For comparison, a spectrum calculated assuming a saturated ammonia distribution is indicated by the broken curve. Sample error bars are shown below the curves (Kunde et al., 1982).

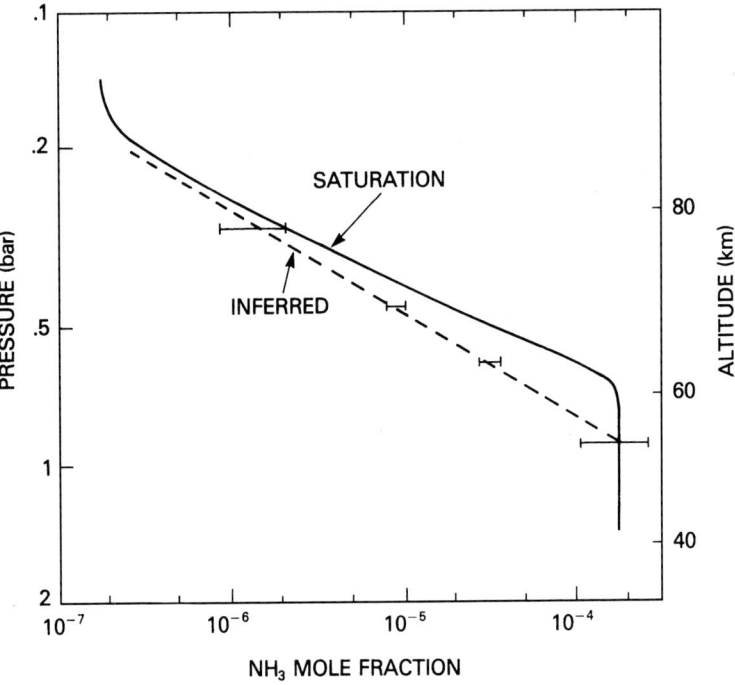

Fig. 8.3.3 Vertical ammonia distribution in Jupiter's North Equatorial Belt obtained by inversion of the Voyager IRIS measurements shown in Fig. 8.3.2. The inferred mole fraction versus pressure is shown as the broken curve with estimated error bars superposed. The solid curve represents the saturation mole fraction for comparison (Kunde *et al.*, 1982).

602 cm^{-1}, respectively. The synthetic spectrum calculated using the retrieved NH$_3$ and temperature profiles is compared with the measured spectrum in Fig. 8.3.2.

f. Simultaneous retrieval of temperature and gas abundance

In the absence of independent, direct determinations of planetary profiles such as *in situ* measurements from an entry probe, it becomes necessary to infer both atmospheric temperature and composition from remotely sensed measurements alone. An example is the determination of the helium abundance in the atmospheres of the giant planets. One approach uses a combination of the thermal emission spectrum and the radio signal from the spacecraft occulted by the atmosphere. The radio occultation measurements can be inverted to obtain an atmospheric refractivity profile from which a profile of temperature versus pressure can be calculated if the atmospheric composition is known. An initial atmospheric composition is assumed, and the resulting temperature profile is used to calculate a theoretical spectrum that is compared with a measured spectrum acquired near the occultation point.

The composition is adjusted until the calculated and measured spectra are brought into agreement. Since the atmospheres of the giant planets are composed primarily of molecular hydrogen and helium with small admixtures of heavier components, this method can be used to determine the atmospheric ratio of helium to hydrogen. This approach was applied to a combination of Voyager radio occultation (RSS) data and Voyager infrared spectrometer (IRIS) measurements to obtain the helium abundance in the atmospheres of Jupiter, Saturn, Uranus, and Neptune (Gautier *et al.*, 1981; Conrath *et al.*, 1984, 1987, 1991a, 1993).

More recently, *in situ* measurements of helium have been made from the Galileo descent probe into Jupiter using two separate instruments, a dedicated helium abundance detector (von Zahn & Hunten, 1996; von Zahn *et al.*, 1998) and a mass spectrometer (Niemann *et al.*, 1996; 1998). A volume mixing ratio of $He/H_2 = 0.157 \pm 0.003$ was obtained, compared to the Voyager value of 0.11 ± 0.032. Although the difference between the Voyager and Galileo results is not extremely large when the combined error bars are taken into account, results suggest the likely presence of a systematic error. Since the two independent Galileo determinations are in agreement, it is assumed that the errors are in the Voyager measurements. Possible error sources in the IRIS data have been examined, but none has been identified that could produce the observed differences; a complete analysis of error sources in the radio occultation has yet to be carried out.

The disagreement between the Voyager and Galileo results has raised the question of the validity of the Voyager helium determination for the other three giant planets, prompting Conrath & Gautier (2000) to reexamine the problem. Based on an earlier suggestion of Gautier & Grossman (1972), they abandon the constraint of the ratio occultation data, using a direct inversion of IRIS spectra alone. They used the spectral region between 230 and 600 cm^{-1}, which is dominated by the S(0) and S(1) lines and a portion of the translational continuum of the collision-induced absorption spectrum of molecular hydrogen. The detailed shape of this portion of the spectrum is sensitive to the He abundance through the contributions of H_2–He interactions to the collision-induced H_2 spectrum. The spectral shape depends also on the ortho–para ratio of molecular hydrogen, which is known to depart from its thermal equilibrium value in the atmospheres of all four giant planets (see for example Conrath *et al.*, 1998). Consequently, it is necessary to attempt to simultaneously infer the temperature profile, the para-hydrogen profile, and the He/H_2 ratio from the spectrum. This is possible only if there are multiple spectral regions that sound essentially the same atmospheric layers, but have differing sensitivities to the parameters being inferred. Unfortunately, this condition is not met in the IRIS Jupiter spectra because of strong gaseous NH_3 and cloud opacity near the low wavenumber end of the spectral range, but this restriction is less severe at the lower temperatures of Saturn. The constrained algorithm discussed in Section 8.2.b was extended to include all three parameters, and was applied to IRIS Saturn spectra, yielding values of He/H$_2$ between 0.11 and 0.16. This range of values is significantly larger than the value of 0.034 ± 0.024 previously obtained by Conrath *et al.* (1984) using the RSS–IRIS method. This again suggests the possibility of systematic errors in the original analysis.

While the determination of helium in the atmospheres of the giant planets has been used as an example here, the approach for simultaneously retrieving temperature in combination with other atmospheric parameters by direct inversion of thermal emission spectra can be applied more generally. The basic requirement is that there be multiple points within the spectral range that are 'redundant' in the sense that the atmospheric pressure level of unit optical depth is essentially the same for all points while the sensitivity of the spectrum to various parameters differs from point to point. Each particular application must be studied in order to determine the actual information content of the measurements and the precision of the retrievals.

g. Limb-tangent observations

Information on the vertical distribution of gaseous constituents can also be obtained from limb-tangent observations, that is, observations for which the ray paths traverse the atmosphere at the limb. As in the case of temperature profile retrieval, higher vertical resolution can be obtained with this observational geometry than from nadir viewing measurements. In addition, the long atmospheric path obtained in this way is advantageous when attempting to extract information on weakly absorbing trace constituents. This geometry has been employed both for thermal emission and solar absorption measurements.

We first consider the analysis of solar absorption limb measurements. The geometry of the situation is similar to that illustrated in Fig. 8.2.4 except in this case the Sun serves as a source located at $s = +\infty$. Solar radiation transmitted through the atmosphere is measured by a sensor at the spacecraft. As the spacecraft moves, the ray path between the Sun and the sensor passes through different parts of the atmosphere. If it is assumed that thermal emission from the atmosphere is negligible compared with the transmitted solar radiation, the transmittance, $\tilde{\tau}$, of the atmosphere along the ray path is given by

$$\tilde{\tau} = \frac{I(\nu, h)}{I_0(\nu)}. \tag{8.3.6}$$

Here $I(\nu, h)$ is the measured radiance at wavenumber ν for the ray path whose closest approach to the planet at the ray tangent point is h, and $I_0(\nu)$ is the solar radiance as observed by the sensor in the absence of the intervening atmosphere.

The transmittance can be related to the abundance of the absorbing gas by

$$\tilde{\tau}(\nu, h) = \int_{\Delta\nu} d\nu' \phi(\nu, \nu') \exp\left[-2 \int_0^\infty k(\nu') q n \, ds\right] \quad (8.3.7)$$

where $\phi(\nu, \nu')$ is the instrument spectral response function centered at wavenumber ν with width $\Delta\nu$, $k(\nu)$ is the monochromatic absorption coefficient, n is the total atmospheric number density, q is the mole fraction of the gaseous absorber, and s is the distance along the ray path measured from the tangent point. The integration along the ray path must take atmospheric refraction into account.

Usually, measurements are available for a series of ray tangent point heights $z_0 = h_i, i = 1, \ldots, m$. It is convenient to divide the atmosphere into m discrete layers with the altitude of the bottom of the ith layer equal to h_i. If the gas mole fraction, q_i, within each layer is treated as a constant, then the transmittance of the atmosphere along the ray path with tangent point altitude h_i is a function of the gas mole fraction in each of the layers from the top down to and including the ith layer, but is not dependent on the mole fraction in the deeper layers. We can then express the transmittances in the functional forms

$$\tilde{\tau}(\nu, h_1) = f_1(\nu; q_1)$$
$$\tilde{\tau}(\nu, h_2) = f_2(\nu; q_1, q_2)$$
$$\vdots$$
$$\tilde{\tau}(\nu, h_m) = f_m(\nu; q_1, q_2, \ldots, q_m). \quad (8.3.8)$$

The first of these equations can be solved to obtain q_1 from measurements of $\tilde{\tau}(\nu, h_1)$. This value is substituted into the second equation, which is then used along with measurements of $\tilde{\tau}(\nu, h_2)$ to determine q_2. The procedure is continued until q_m is determined from the last of the equations. If $\tilde{\tau}(\nu, h_i)$ is obtained for a number of values of ν within an appropriate spectral range, the qs can be determined using nonlinear least squares solutions of Eq. (8.3.8). The retrieval method outlined here is usually called 'onion peeling' in the remote sensing literature.

The limb-tangent solar absorption method has been used to obtain profiles of trace gases in the Earth's stratosphere. An example of an investigation using this approach is the Atmospheric Trace Molecule Spectroscopy (ATMOS) Michelson interferometer carried on the Shuttle-borne Spacelab 3 in 1985. A large number of atmospheric transmission spectra were acquired in the 600 to 4700 cm^{-1} spectral region (Farmer & Raper, 1986; Farmer & Norton, 1989). Vertical profiles of various trace gases have been obtained from these data (Park et al., 1986).

The principal advantage of the solar occultation approach is the relatively strong signal obtained by using the Sun as a source. Its principal disadvantage is that the measurements can be acquired only at local sunrise or sunset.

Thermal emission data acquired in the limb viewing mode have also been used to obtain composition measurements in the terrestrial atmosphere. The geometry is the same as that given in Fig. 8.2.4, only in this case thermal emission from the atmosphere itself is observed rather than transmitted solar radiation. The radiance measured by the spacecraft sensor along a ray path with tangent height h can be written

$$I(h) = \int_{\Delta \nu} d\nu' \phi(\nu, \nu') \int_{-\infty}^{\infty} B(\nu, T) \frac{\partial}{\partial s} \left[\exp\left(- \int_{s}^{\infty} kqn \, ds' \right) \right] ds. \quad (8.3.9)$$

To retrieve profiles of gas mole fraction, $q(z)$, from measurements of $I(h)$, one can again resort to the onion peeling approach. In this case, there are two additional complications. Knowledge of the temperature profile $T(z)$ is required in order to specify the Planck radiance $B(\nu, T)$ at all points along the ray path, and a double integration along the ray path is required. However, the radiance $I(h)$ depends only on the values of q for atmospheric levels above $z = h$. The atmosphere can again be divided into discrete layers and the values of q can be determined sequentially, starting at the top layer and working downward. This method can be employed without restrictions in local time, but tends to be less sensitive to trace species than the solar transmission method. A number of instruments have been flown in Earth orbit which have made limb emission measurements for the purpose of obtaining information on trace gases in the terrestrial middle atmosphere (see, for example, Remsberg *et al.*, 1984).

8.4 Clouds and aerosols

The inference of cloud characteristics is based on much less sophisticated approaches than those for determining thermal structure and gas abundances. Clouds tend to be quite inhomogeneous compared with gaseous mixtures and require more parameters for adequate definition. Also, the appropriate equation of transfer [Eq. (2.1.40)] is considerably more complex than Eq. (8.2.1), and not nearly as amenable to inversion techniques. Even so, direct techniques are sometimes capable of leading to rather definitive conclusions about cloud and aerosol systems. We illustrate with an example concerning the abundance of the photochemical aerosol in Titan's stratosphere.

8.4 Clouds and aerosols

a. Small absorbing particles

Dielectrics tend not to be completely transparent in the thermal infrared, leading to some absorption in particles of all sizes. According to Figs. 3.8.3 and 3.8.4, extinction cross sections of very small particles ($a \ll \lambda$) are due almost exclusively to absorption ($Q_E \sim Q_A$) as long as $Q_A > 0$. This is verified by expanding the ψ- and ζ-functions and their derivatives in Eq. (3.8.48) in power series of the argument, retaining only the lowest order terms. These terms are then used to evaluate a_n and b_n in Eqs. (3.8.50) and (3.8.51), leading to expressions for S(0) in Eq. (3.8.67) and Q_E in Eq. (3.8.75). The lowest order absorption term is found to be

$$Q_A = 4\left(\frac{\mu_1}{\mu_2}\right)\left(\frac{2\pi a}{\lambda}\right) \frac{6 n_r n_i}{\left(n_r^2 - n_i^2 + 2\frac{\mu_2}{\mu_1}\right)^2 + 4 n_r^2 n_i^2}, \qquad (8.4.1)$$

whereas the lowest order scattering term is proportional to the fourth power of $2\pi a/\lambda$. Hence, if $a \ll \lambda$, and n_i is not negligible, scattering is unimportant and Eq. (8.4.1) suffices to describe the extinction properties of a single particle. Setting $\mu_1/\mu_2 = 1$, the extinction coefficient for a volume element becomes

$$N \chi_E = \frac{48 \pi^2 a^3}{\lambda} N \frac{n_r n_i}{\left(n_r^2 - n_i^2 + 2\right)^2 + 4 n_r^2 n_i^2}. \qquad (8.4.2)$$

As usual, N is the particle number density and χ_E is the extinction cross section per particle. It is assumed that both the particles and the atmosphere in which they are imbedded are nonmagnetic, which requires that $\mu_1 = \mu_2 = \mu_0$, the magnetic permeability of free space.

Equation (8.4.2) leads to a great simplification because no scattering terms are involved in the equation of transfer. On the other hand, outside the range of infrequent strong absorption bands, both n_r and n_i tend to vary slowly with wavenumber, implying that vertical structure is difficult to infer from spectral information. This is in contrast to gaseous absorption, where vibration–rotation bands give rise to large opacity variations over relatively small spectral intervals.

Limb-tangent observations are much better suited for inferring the vertical structure of an absorbing aerosol. Density rather than spectral variations are responsible for providing vertical contrast. As long as the particles are sufficiently small and absorbing, Eq. (8.3.9) can be used to infer their vertical distribution and abundance.

b. Titan's stratospheric aerosol

The photochemical aerosol in Titan's stratosphere provides a good example. Several lines of evidence (Hunten *et al.*, 1984) indicate that submicron size particles are responsible for the observed photometric and polarimetric properties of this aerosol. Refractive indices of a presumably comparable material were derived by Khare *et al.* (1984) by analyzing laboratory spectra of a polymeric residue obtained from electric discharge in an N_2–CH_4 gas mixture. Values of $n_r = 1.8$ and $n_i = 0.17$ appear to be reasonable averages over the 250–600 cm^{-1} spectral range. This combination of large n_i and small $(a/\lambda)(n_r - 1)$ ensures that scattering is unimportant, and Eq. (8.3.9), modified to account for a finite field of view, is appropriate.

Spectra of Titan's north polar limb, shown in Fig. 8.4.1, were obtained by the Voyager 1 spectrometer (IRIS). The continuum between gaseous emission features is due mainly to a stratospheric aerosol, although tropospheric emission also contributes to the spectrum shown in the lower panel. Because each field of view extends over several scale heights, it is necessary to convolve Eq. (8.3.9) with the field of view; the onion peeling approach is not practical. Instead, the equation is solved directly, varying the aerosol parameters until the spectral continua are fit simultaneously for a given model.

In practice, because of the large fields of view, vertical resolution in the stratosphere is limited to two levels. Continuity of particle number density N across the boundary $z = z_0$ between the two levels is required in the modeling, but dN/dz is allowed to be discontinuous. We replace Eq. (4.2.5) with

$$N = N_0 \left(\frac{T}{T_0}\right)^{-(1-mg/ck\Gamma)}, \tag{8.4.3}$$

where the reference level is the common boundary, z_0, and c is defined below.

As $\Gamma \to 0$ the atmosphere becomes isothermal, and Eq. (8.4.3) transforms into

$$\lim_{\Gamma \to 0} N = N_0\, e^{-(z-z_0)/cH}, \tag{8.4.4}$$

where

$$H = \frac{kT}{mg} \tag{8.4.5}$$

is the pressure scale height. Hence, the parameter c in Eq. (8.4.4) is a scale factor for adjusting the aerosol scale height in units of H. Its meaning becomes somewhat more general in Eq. (8.4.3) where $\Gamma \neq 0$, but may still be used as a scale height factor.

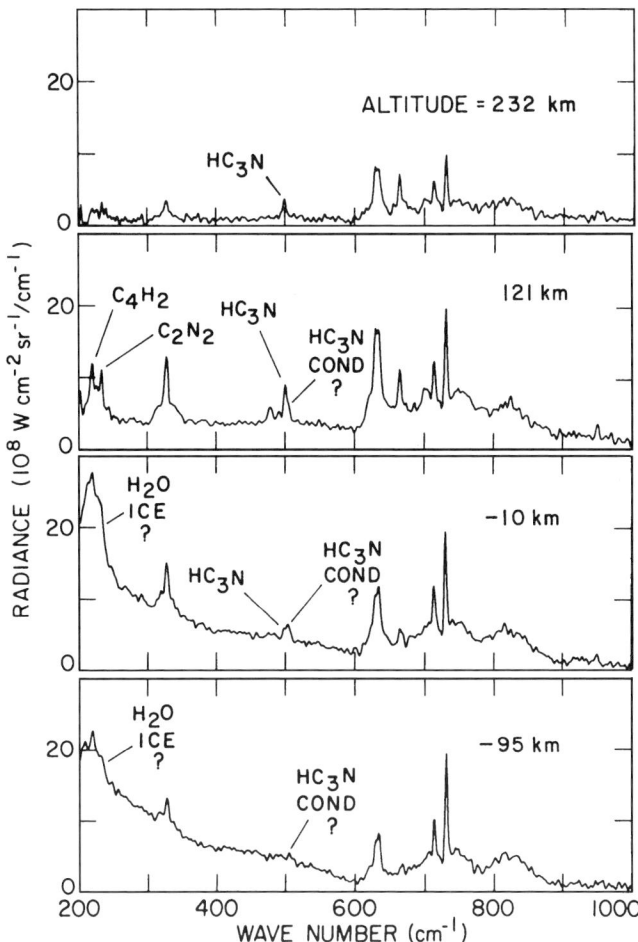

Fig. 8.4.1 Voyager 1 IRIS spectra viewed along the tangent path over the north polar hood of Titan. Numerical values refer to the vertical distance at which the ray defining the center of the IRIS field of view passes above the apparent horizon as seen from the spacecraft (Samuelson, 1985).

In fitting the continua shown in Fig. 8.4.1, it is found that, above z_0, $c \sim 1.5$ independently of wavenumber. Below z_0, c is in the range $0.6 < c < 1.5$, a function of both ν and z_0. The exact location of z_0 cannot be inferred from the data. The variation of c with ν below z_0 implies either a rapid increase in particle size, so that Eq. (8.4.2) is no longer a suitable approximation, or a change in composition. The latter is quite possible, as the lower stratosphere is the region where organic gases are expected to condense (Maguire et al., 1981; Sagan & Thompson, 1984). A value of $c > 1.0$ in the middle and upper stratosphere is consistent with the source of the aerosol (presumably photochemical in origin) being located at higher altitudes.

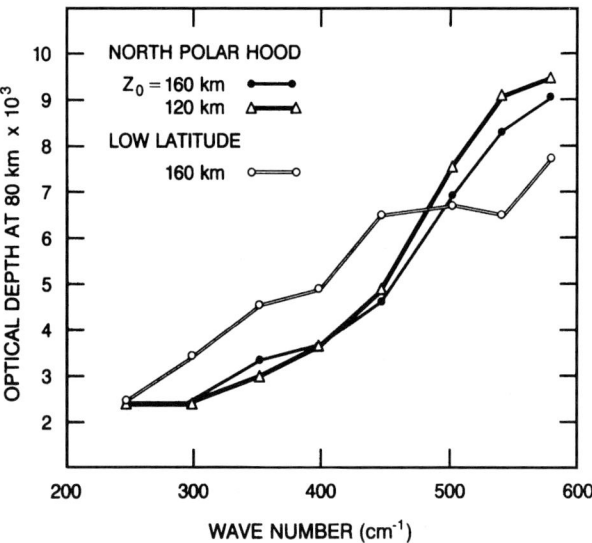

Fig. 8.4.2 Normal optical depths of Titan's continuum at 80 km of both the north polar hood and the region around the equator. The parameter z_0 is an altitude at which the opacity gradient is permitted to be discontinuous in the models (after Samuelson & Mayo, 1991).

The integrated optical depth down to $z = 80$ km,

$$\tau = \int_{80 \text{ km}}^{\infty} N \chi_E \, dz, \tag{8.4.6}$$

is shown in Fig. 8.4.2 for two values of z_0. It is gratifying to see the results are essentially independent of z_0, since the parameter itself is so ill-defined by the data. Also shown are results for the equatorial region of Titan, although the effective field of view of the instrument is much larger for these data, and no vertical resolution is possible. Both data sets show a wavenumber dependence for the opacity consistent with small particles [see Eq. (8.4.2) and Figs. 3.8.3 and 3.8.4]. Little information is available below $z = 80$ km, principally because the temperature is decreasing rapidly with decreasing altitude in the lower stratosphere, and emission is very weak there.

The volume q_v occupied by the aerosol particles per unit volume of atmosphere (volume mixing ratio) is

$$q_v = \tfrac{4}{3}\pi a^3 N, \tag{8.4.7}$$

while the volume extinction coefficient $N\chi_E$ is determined by fitting the data [see

Eq. (8.4.6) and Fig. 8.4.2]. Hence, q_v can be inferred from Eq. (8.4.2) if n_r and n_i are known. The values suggested by Khare *et al.* (1984) lead to

$$q_v \sim 2.5 \times 10^{-13} \tag{8.4.8}$$

at $z = 160$ km. Assuming a bulk density of unity and adopting an atmospheric density of 4.1×10^{-6} g cm^{-3} at 160 km yields a mass mixing ratio

$$q_m \sim 6 \times 10^{-8} \tag{8.4.9}$$

at this level. As long as the particles are small their cross sections and masses are both proportional to the same power of the radius, and mass mixing ratios may be obtained by the above method.

8.5 Solid surface parameters

a. Surface temperature

In the thermal infrared the spectral intensity, I_ν, measured by a radiometric instrument facing the opaque surface of an astronomical object can be written

$$I_\nu(\theta) = \varepsilon_\nu(\theta) B_\nu(T), \tag{8.5.1}$$

[see Eq. (2.4.5)]. The surface emissivity, ε_ν, depends on wavenumber, ν, and emission angle, θ, and may also be a function of the plane of polarization (see Fig. 1.6.1). An azimuthal dependence of ε can generally be neglected. However, topography or the presence of boulders illuminated by the evening Sun may introduce an azimuthal dependence in the thermal emission, even after sunset. If the surface material is partially transparent at certain wavenumbers, the temperatures of the subsurface layers as well as of the surface itself may have to be considered; the emission must be treated as a radiative transfer problem, quite analogous to the case of an atmosphere with a spectrally variable absorption. As discussed in Subsection 6.5.b this situation apparently exists on Io. In this section we assume a surface layer isothermal with depth where the measured intensity is specified by Eq. (8.5.1).

The simplest interpretation of the radiance measured by a single-channel radiometer is through the brightness temperature, which is the temperature of a blackbody that emits the same intensity, I_ν, at the wavenumber ν as does the object of interest. To find the kinetic temperature of the surface from the brightness temperature one must know the emissivity of the material. Rarely are the composition and texture of the surface known well enough to permit computations of ε. In reality one must choose the spectral range and polarization of the radiometer to assure measurement in a region of high emissivity. The measured brightness temperature is then close to, but always less than, the kinetic temperature of the surface.

We consider now measurements with a nonpolarizing radiometer, that is, with an instrument that averages over both planes of polarization. All infrared radiometers flown on planetary spacecraft to date are of this type. If we also exclude measurements at large emission angles, say larger than 45°, then the thermal emissivity of a smooth, polished surface can be approximated from Eqs. (1.6.15) and (1.6.16) as

$$\varepsilon_\nu \approx 4n(1+n)^{-2}. \tag{8.5.2}$$

Materials of remote sensing interest have refractive indices ranging from approximately 1.3 (water ice) to 3.2 (hematite, Fe_2O_3). Most silicate minerals also fall within these limits. Consequently, the emissivities would be between 0.73 and 0.98. But planetary and satellite surfaces are far from polished. An exception to this rule, the lava lakes on Io, are discussed in Subsection 8.5.c. Most real surfaces are rough on the scale of infrared wavelengths. Even relatively smooth, sandy or powdery surfaces contain gaps and holes between individual grains that trap radiation and decrease the reflectivity. The radiometer averages over the exposed solid surfaces, and the small voids act as little, almost black cavities. The emissivity of such a surface is generally higher than the values just mentioned. However, scattering also becomes important. The particles are generally in contact with one another, and the simple scattering model introduced in Chapter 2 for individual, well-separated particles is no longer valid. At present, a completely satisfactory theory (comparable to the Mie theory of spherical, homogeneous, well-separated particles described in Section 3.8) does not exist for sands and powders. Due to near-field and shape effects, matters are very complicated and one has to fall back on semi-empirical theories to explain the results of laboratory measurements of the emissivity or reflectivity of mineral powders of different particle sizes. Emslie & Aronson (1973) have attempted to understand the infrared properties of such surfaces. For visible wavelengths Hapke & Wells (1981), and Hapke (1981, 1984, 1986) have developed a scattering theory for powdery surfaces. Many aspects of the measured reflectivities and emissivities can be explained by their theories. Hovis & Callahan (1966), Logan & Hunt (1970), Hunt & Logan (1972), Aronson & Emslie (1973), and others have measured the reflectivity of sands of different materials and grain sizes. In general, they found the infrared reflectivity to be lower for powdery surfaces than for coarser sand of the same mineral, but exceptions to this rule can be found. In the visible and near infrared the reflectivity is generally high. Most powders appear white to the eye. The reflectivity varies somewhat with wavelength, but for most minerals it is relatively low in the infrared; that is, the emissivity often

8.5 Solid surface parameters

exceeds 0.8. Not knowing the precise value, many investigators have assumed ε to be 0.9, and independent of wavenumber and emission angle.

Several precautions can be taken to assure a good estimate of the true surface temperature. As indicated in Eq. (8.5.1), the emissivity normally depends on the wavenumber. With a spectrometer or multichannel radiometer one may search for a dispersion region of the surface material where the refractive index varies strongly with wavenumber (see Subsection 3.7.b). Near the index minimum the emissivity has a maximum. In addition to the composition the emissivity strongly depends on particle size and surface texture. A spectral search for an emissivity maximum is an improvement over the use of an arbitrarily chosen spectral interval. The maximum in the Martian brightness temperature near $1280 \, \text{cm}^{-1}$, shown in the upper spectrum of Fig. 6.2.8, may be an example of such a case.

At wavenumbers where the refractive index of the material equals that of the substance it is imbedded in (CO_2 gas in the Martian case), the suspended atmospheric dust is least scattering, but still absorbing. The frequency where the index of the particles equals that of the environment is called the Christiansen frequency. If atmospheric gas and dust absorption are not excessive, the spectral region surrounding this frequency may provide a suitable window for a measurement of the surface temperature.

Another approach to find conditions for $\varepsilon \sim 1$ suggests itself from Fig. 1.6.1. The emissivity corresponding to a polarization parallel to the plane of incidence (the plane through the surface normal and the view direction) is consistently higher than that of the orthogonal component. For silicates ($n \sim 1.4$ to 2) $\varepsilon_\parallel$ approaches unity for emission angles between $40°$ and $60°$. On a rough surface the local surface normals are more or less randomly grouped around the average, large-scale normal. Therefore, the surface emissivity is an average over θ, concentrated around the large-scale value of θ. In either case $\varepsilon_\parallel$ is on the average larger than $\varepsilon_\perp$. To date, only nonpolarimetric measurements have been carried out in the thermal infrared from spacecraft. However, 5 GHz microwave measurements of snow surfaces on Earth show the polarization effect very clearly (Fung & Eom, 1981; Ulaby *et al.*, 1986).

Multichannel infrared radiometers have observed Mars from Mariner 6, 7, and 9, as well as from the Viking orbiters (Neugebauer *et al.*, 1971; Chase *et al.*, 1972; Kieffer *et al.*, 1973). Maps of the diurnal temperature variations and thermal inertia of Mars have been published. In the Viking observations the brightness temperature of the surface ranged from 130 K in the south polar region to 290 K shortly after local noon at low latitudes. Similar measurements on Mercury have been carried out from Mariner 10 at two wavelength intervals: 8.5–14 μm and 34–55 μm (Chase *et al.*, 1976). Because of the proximity of Mercury to the Sun (mean distance 0.387 AU) the subsolar surface temperature is expected to be quite high. Assuming a planetary albedo of 0.1 and a surface emissivity of 0.9 results in an equilibrium temperature of about 635 K; for comparison lead melts at 600 K. The trajectory of Mariner 10 prevented observation of the subsolar point, but two swaths across the evening and morning terminators and across the dark side have been obtained (see Subsection 8.5.b). A minimum temperature of about 100 K was found near the antisolar longitude of Mercury.

Early lunar infrared measurements from the Earth's surface have been discussed by Sinton (1962). Very detailed measurements of the Moon have been carried out by Saari & Shorthill (1967, 1972) and Saari et al. (1972) with the 152 cm telescope on Mt Wilson and the 188 cm telescope of the Helwan observatory in Egypt. Spatial resolutions were 8 and 10 seconds of arc, respectively. The infrared spectral interval was from 10 μm to 12 μm. Simultaneous observations have been made at visible wavelengths (0.35–0.55 μm) with identical spatial resolutions. Altogether the Moon was observed at 23 phase angles, more or less evenly distributed over a full lunar cycle.

If the spectral coverage of the spectrometer or multichannel radiometer is wide enough, detailed information on surface temperatures can be obtained by analyzing the shape of the spectrum. An example provided by the Voyager infrared measurements of Io is discussed in Subsection 6.5.b.

b. Thermal inertia

On quiescent planets or satellites without substantial atmospheres, the surface temperature is determined by a balance between incident solar flux, thermally emitted radiation, and conductive heat transport into or out of the opaque surface. By measuring the surface temperature and the bolometric albedo the absorbed and emitted radiation can be found and the conductive flux into the solid body derived. After sunset or during a solar eclipse the cooling rate of the surface depends on the 'thermal inertia' of the subsurface layers. A study of such cooling rates provides a sensitive means of discriminating between powdery, sandy, or solid rock surfaces. We now review the theory behind such an analysis, and discuss examples of thermal inertia measurements.

The theory of thermal conduction in solids was first applied to lunar studies by Wesselink (1946). The basic equation of heat conduction is

$$\nabla^2 T - \frac{\rho c}{k}\frac{\partial T}{\partial t} = 0, \qquad (8.5.3)$$

where T [K] is the temperature, ρ [kg m^{-3}] the density, c [J kg^{-1} K^{-1}] the specific heat per unit mass, and k [J s^{-1} m^{-1} K^{-1}] the thermal conductivity. The term

8.5 Solid surface parameters

$k\rho^{-1}c^{-1}$ is called the temperature conductivity, or thermal diffusivity, a^2. If the surface is flat and horizontally homogeneous only the depth below the surface, x, enters as a space coordinate, and Eq. (8.5.3) reduces to

$$\frac{\partial^2 T}{\partial x^2} - \frac{1}{a^2}\frac{\partial T}{\partial t} = 0. \tag{8.5.4}$$

To separate the variables we take the usual Bernoulli product approach,

$$T(x,t) = X(x)\,e^{i\omega t}. \tag{8.5.5}$$

In anticipation of the periodic nature of solar illumination we use a periodic function for the time factor. In general, solar radiation is not a simple sinusoidal term, but must be expressed by a Fourier series. To simplify notation we treat only the lowest diurnal frequency of the series; higher order terms can be considered later by superposition of solutions. Substituting Eq. (8.5.5) into Eq. (8.5.4) leads to

$$X'' - \frac{i\omega}{a^2}X = 0. \tag{8.5.6}$$

Expressing X by an exponential function of depth,

$$X(x) = \alpha\,e^{\lambda x}, \tag{8.5.7}$$

yields a characteristic equation for λ,

$$\lambda^2 - \frac{i\omega}{a^2} = 0;\quad \lambda = \pm(1+i)\frac{1}{a}\left(\frac{\omega}{2}\right)^{\frac{1}{2}}. \tag{8.5.8}$$

Since the exponential term must approach zero at large depths, only the negative root is acceptable, so that

$$T(x,t) = \alpha\exp\left[-\frac{1}{a}\left(\frac{\omega}{2}\right)^{\frac{1}{2}}x\right]\exp\left[i\omega\left(t - \frac{x}{a(2\omega)^{\frac{1}{2}}}\right)\right]. \tag{8.5.9}$$

The temperature is periodic in time and space with an amplitude exponentially decreasing with depth. The velocity of the thermal wave shows dispersion, that is, it depends on the frequency,

$$v = a(2\omega)^{\frac{1}{2}}. \tag{8.5.10}$$

The scale depth, x_1, where the amplitude decreases by $1/e$, is

$$x_1 = a\left(\frac{2}{\omega}\right)^{\frac{1}{2}} = a\left(\frac{\tau}{\pi}\right)^{\frac{1}{2}}, \qquad (8.5.11)$$

where τ is the period of the illumination function. Typically, x_1 is in the range from a few mm to dm. The scale depth probed by eclipses (short τ) and by diurnal heating (long τ) can be significantly different, which provides a crude probe of homogeneity with depth.

The conductive thermal energy flux into the surface is

$$\pi F = -k\frac{\partial T}{\partial x}. \qquad (8.5.12)$$

Differentiating T in Eq. (8.5.9) with respect to x, multiplying with k, and setting x equal to zero yields the flux, πF_0, at the surface boundary,

$$\pi F_0 = (1 + \mathrm{i})(k\rho c)^{\frac{1}{2}}\alpha\left(\frac{\omega}{2}\right)^{\frac{1}{2}} \mathrm{e}^{\mathrm{i}\omega t}. \qquad (8.5.13)$$

The quantity $(k\rho c)^{\frac{1}{2}}$ has the dimension of $[\mathrm{J\ s^{-1}\ m^{-2}\ K^{-1}}]$; it is called the thermal inertia of the material. The complex factor indicates a phase shift between the driving function and the flux.

The conductive thermal flux at the surface, πF_0, must be balanced by the radiative fluxes from the Sun and the thermal emission from the surface,

$$\pi F_0(t) = \frac{\pi S}{D^2}(1-A)G(t) - \varepsilon\sigma T_0^4(t), \qquad (8.5.14)$$

where πS is the solar constant, D the heliocentric distance of the object in Astronomical Units, A is the Bond albedo, ε the emissivity, and $G(t)$ the time-dependent illumination of the surface element under consideration. Since we have assumed a solution consisting of a linear superposition of Fourier components in time, the thermal emission term in Eq. (8.5.14) must be linearized about a mean temperature. This procedure is adequate for small changes in T_0, but in most practical applications the amplitude of the temperature fluctuation is large, and numerical methods for solving the highly nonlinear problem must be applied. For the illuminated side $G(t)$ is the cosine of the illumination angle; on the dark side $G(t)$ is zero.

To determine the thermal inertia, one fits a series of temperature measurements taken over an illumination cycle or during an eclipse to numerical solutions for the

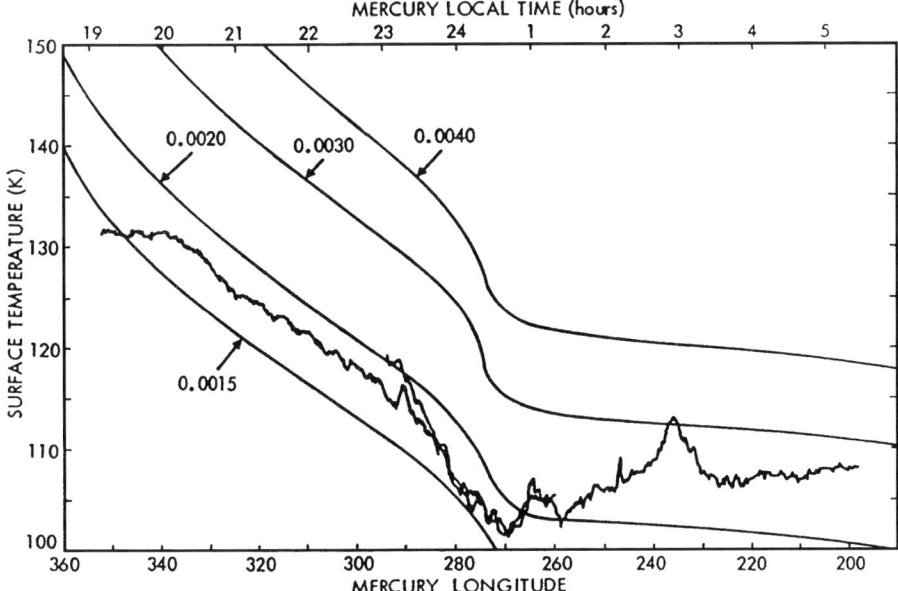

Fig. 8.5.1 Nighttime surface temperatures of Mercury compared with model curves for homogeneous surfaces at the given thermal inertias (in cal cm^{-2} s$^{-\frac{1}{2}}$ K^{-1}) (Chase *et al.*, 1976).

surface temperature from Eqs. (8.5.13) and (8.5.14). An example of such a procedure is shown in Fig. 8.5.1 for Mercury (Chase *et al.*, 1976). Since Mercury's rate of revolution about the Sun of 88 Earth days is relatively slow ($\omega = 2\pi/88$ days $= 0.826 \times 10^{-6}$ s^{-1}), the scale depth is relatively large. Also, because of Mercury's high orbital eccentricity ($\varepsilon = 0.206$) and the 2:3 spin–orbit resonance, the solar illumination, $G(t)$, and the time-dependent factor in $F_0(t)$ are complicated functions of time and surface location (Chase *et al.*, 1976). As shown in Fig. 8.5.1 the procedure gives reasonable results for local times between sunset and midnight, but deviations from expectation are seen after that time. The thermal inertia derived from the early period is characteristic of powdery surfaces, quite similar to that found on the Moon. The data after midnight suggest the presence of a horizontally inhomogeneous surface; outcrops of solid rocks or boulders covering only a few percent of the area can account for the measured higher inertias near 3 o'clock local time.

The technique is surprisingly successful, considering the approximations made in the derivations. The terms ρ, c, and k are in reality far from being constant. For example, measurements on lunar samples returned by the Apollo 11 astronauts indicate changes in the heat capacity over a factor of three between 100 K and 300 K (Robie *et al.*, 1970; see also Keihm *et al.*, 1973). The density is bound to increase with depth, and the heat conduction also depends on temperature. More complex models, taking these dependences into account, and using several layers of distinctly different properties have been employed to provide better agreement between measurements and theory (Linsky, 1966; Brown & Matson, 1987). However, with more degrees of freedom introduced by vertical inhomogeneity, the solutions become less unique.

c. Refractive index and texture

Up to the present, most of the compositional information of solid body surfaces has been derived from conventional photometric and spectrometric measurements, ranging from the ultraviolet through the visible and into the near infrared. A recent summary of such observations and their interpretation for the Galilean satellites has been provided by Sill & Clark (1982). We do not analyze these methods in this section, but discuss the application of polarization measurements in the near infrared. This technique has recently been applied to the volcanic features of Io by Goguen & Sinton (1985) and Sinton *et al.* (1988). From ground-based observations, which do not resolve locations on Io, Sinton and his collaborators derived the location, approximate size, and lower limits of the refractive index of the surface material of certain hot spots. This information could be obtained because at these wavelengths (4.8 μm and 3.8 μm) the hot spots contribute a significant fraction to the total infrared flux from the full disk, and the apparent smoothness of the hot spot surface at a scale of ~ 5 μm causes strong polarization at large emission angles. At a phase angle near zero degrees, reflected sunlight is expected to be nearly unpolarized.

To take full advantage of this method the 'effective' geometric albedo, the degree of polarization, and the azimuth angle of the polarization were measured over a wide range of orbital positions of Io. Panel (a) of Fig. 8.5.2 (Fig. 1 of Goguen & Sinton) shows the effective geometric albedo of the full disk of Io computed for a hypothetical hot spot area of 2000 km^2 at a brightness temperature of 450 K located at a meridian of 180° and at a latitude of 15°N. The calculations assume a refractive index of 1.5 for the hot spot material and a geometric albedo of 0.8 for the rest of Io's disk. The increase of the geometric albedo to a value higher than unity is a result of the additional thermal emission from the spot.

The degree of linear polarization of the whole disk is shown in panel (b). The degree of polarization is defined by $V = (F_v/F_t)(\varepsilon_\perp - \varepsilon_\parallel)/(\varepsilon_\perp + \varepsilon_\parallel)$. The flux from the volcano is F_v and the total flux (background plus hot spot) is F_t. The emissivities parallel and perpendicular to the plane of view have the same meaning as in Eqs. (1.6.15) and (1.6.16). The degree of polarization falls to zero at the limb, that is, when the spot disappears from view, and has a minimum when the spot is at

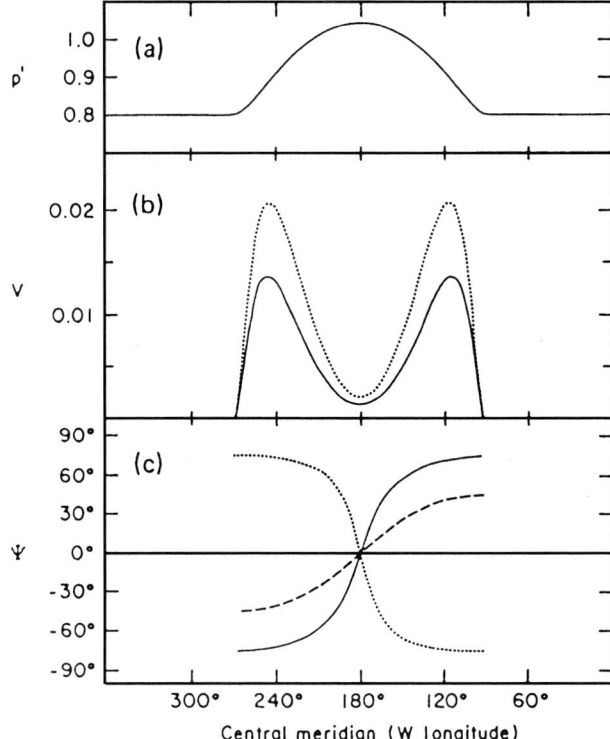

Fig. 8.5.2 Variation of flux and linear polarization with orbital longitude due to a single standard hot spot of area 2000 km^2 ($T = 450$ K) located at 180°W, 15°N on a model planet with $n = 1.5$ and a true geometric albedo of 0.8. In each frame of the figure, the solid line is the signature of this standard spot and the broken lines show the changes in this signature if one of the parameters is varied from the standard configuration. (a) The quasi-geometric albedo p'; (b) the degree of linear polarization V (dotted line, $n = 1.8$); (c) azimuth of linear polarization ψ (dashed line, latitude 45°N; dotted line, latitude 15°S) (Goguen & Sinton, 1985).

the meridian passing through the center of the disk. The active area is then viewed at a minimum emission angle equal to the latitude of the spot. The solid line in the figure is for a refractive index of 1.5 and the dotted line for an index of 1.8.

Panel (c) shows the position angle (angle between the hot spot–sub Earth point line and the north direction) as a function of the central meridian. The solid curve is for a spot at 15°N, the dashed line for one at 45°N, and the dotted line for one at 15°S. If more than one hot spot contributes to the flux then the signal is the superposition of the contributions from all spots.

By applying this model to polarization measurements at 4.76 μm from the 3 m infrared telescope at the Mauna Kea observatory in Hawaii, Goguen & Sinton (1985) detected three areas of strong emission and polarization. A least squares fit of a three source model to the observations identifies the strongest source of ~3057 ± 118 km² at a longitude of ~288°W and a latitude of ~ +20°N. This source is very close to Loki which was an active hot spot and plume area during the Voyager encounter. The second spot identified by Goguen & Sinton is somewhat smaller (1474 ± 212 km²) at longitude ~72°W and latitude +6.44°N. This location is near Ra Patera, also identified as a volcanic area by the Voyager cameras. The third spot of 1723 ± 116 km² is at longitude ~322°W and latitude ~ 11.6°N. It does not correspond to any plume or hot spot discovered by Voyager. Reanalysis of the same data by Sinton *et al.* (1988) takes into account a small amount of residual disk polarization. The spots shift somewhat and the association of the first spot with the Loki area becomes doubtful. In the same paper Sinton *et al.* report on more recent (1986) measurements at 3.8 μm and 4.8 μm. A general decrease in activity seems to have taken place between 1984 and 1986; Loki was no longer active. Changes in the position and strength of volcanic activity were also observed between the two Voyager encounters (Smith *et al.*, 1979a, b). Apparently, the volcanic activities on Io change on time scales of a few years.

The best fit to the data yields a refractive index of 1.51 (+0.06/−0.08) for 4.8 μm and 1.11 ± 0.03 for the 3.8 μm measurements of 1986. These values are lower limits that would apply to perfectly smooth surfaces; however, if a few fractures exist the data are consistent with larger indices of the surface material. Therefore, the present measurements are not restrictive enough to discriminate between basalts and sulfur lakes, for example. The data have been interpreted using a flat surface model. However, this is not required. The natural curvature on the surface of Io (radius 1815 km) results in a variation of emission angle of about 6° for the 'lava lake' Loki, measured vertically, and somewhat less, measured under lower viewing angles. As shown in Fig. 1.6.1 a change in the emission angle of a few degrees does not affect the conclusions much at all. Even a model of small, adjacent, individual areas (frozen puddles, possibly overlapping) would fit the data equally well. In either case, the hot spot area must be remarkably smooth (glossy) at a scale of a few μm to show such a strong polarization.

8.6 Photometric investigations

a. Introduction

Photometric measurements of planets and satellites often serve one of two distinct purposes. In the first case the local properties of surfaces or of clouds are sought. If this is accomplished for many places on a planet, then statements on the global characteristics can also be made. An instrument suitable for determination of local reflection properties requires a narrow field of view with many narrow-band

8.6 Photometric investigations

channels at different wavelengths and a polarization measurement capability. Ideally, many characteristic areas of a planet, such as densely clouded or clear areas, should be measured over a wide range of Sun and emission angles. The photopolarimeters on Pioneer (Gehrels *et al.*, 1974; Gehrels, 1976) and on Voyager (Lillie *et al.*, 1977; Lane *et al.*, 1982) are examples of this type of instrumentation. However, we will not follow this particular aspect of research; these investigations have been presented by Tomasko *et al.* (1978, 1980), West *et al.* (1986), Nelson *et al.* (1987), and references therein.

In the second case, the objective is to determine the energy balance parameters, that is, the bolometric albedo and the spectrally integrated thermal emission. The bolometric albedo is the fraction of incident solar flux that is scattered by a particular planetary area in all directions. The bolometric albedo of the whole planet is the Bond albedo. For some investigations, such as climatological studies on Earth, the energy balance on a regional or latitudinal scale is of interest. For other investigations, such as studies of the internal structure and evolution of the giant planets, the global heart balance is of prime interest.

To be well-suited for albedo measurements a radiometer should have a wide spectral range with a flat spectral response to register as much of the reflected solar energy as possible. Polarization information is not required; on the contrary, the instrument should be insensitive to polarization. Measurements over a full range of phase and azimuth angles are needed to derive either the local or the global albedo. Examples of such photometers are the instruments of the Earth Radiation Budget Experiment (ERBE) and the IRIS radiometer on Voyager. Both the objective of obtaining the local scattering properties and that of obtaining the Bond albedo require instruments with good radiometric calibration. In each case a well-calibrated spectrometer would be preferable to a radiometer, but only radiometers have been used in space for these purposes so far.

To find the thermal emission from a specified planetary region, the infrared intensities must be measured over all emission angles, but an azimuthal dependence is generally absent. Thermal emission measurements of the Earth on a local scale have been obtained by the long wavelength instruments on ERBE as well as on other investigations. Historical aspects, instrumentation, and results are discussed by Hunt *et al.* (1986), House *et al.* (1986), Kopia (1986), Barkstrom & Smith (1986), Luther *et al.* (1986), and several other papers in the same issue of *Reviews of Geophysics*, Vol. 24, No. 2 (1986). For Jupiter the meridional energy balance was investigated by Pirraglia (1984).

To find the total energy emitted, full disk measurements must be performed from all directions, to permit integration of the emission over 4π steradians. In addition the spectrum must be integrated from zero wavenumbers to an upper limit that includes most of the thermal emission, but excludes reflected sunlight. This limit is about 2500 cm^{-1} for Jupiter. For Saturn it is sufficient to integrate up to about 1500 cm^{-1} and for Uranus and Neptune to about 500 cm^{-1}. The Voyager infrared spectrometer measured infrared emission spectra from about 200 cm^{-1} to these upper limits. With extrapolation to lower wavenumbers, based on simultaneously determined temperature and composition information, the total thermal emission of Jupiter, Saturn, Uranus, and Neptune have been found (Hanel *et al.*, 1981*b*, 1983; Pearl *et al.*, 1990; Conrath *et al.*, 1989*b*). As an example we treat measurements of the total emission of the giant planets in Subsection 8.6.c.

b. The Bond albedo

A determination of the planetary albedo requires a large set of individual photometric measurements. Two approaches may be followed in collecting the data. In one approach the flux from a surface element is determined by recording the intensities emanating from that element in all directions and calculating the local flux according to the definition (Chandrasekhar, 1950),

$$\pi F = \int_0^{2\pi} \int_0^{\frac{\pi}{2}} I(\delta, \phi) \sin \delta \cos \delta \, d\delta \, d\phi, \tag{8.6.1}$$

where δ is the zenith and ϕ the azimuth angle. This, in essence, is the approach needed to determine the energy balance terms on a local or regional scale. To find the global values many individual flux measurements must then be integrated over the whole globe. While this approach is conceptually straightforward, in reality it is difficult to implement. Alternatively, one may measure the flux from the whole planetary disk as it appears to a distant observer and then integrate the individual disk measurements over all directions, that is, over 4π steradians. Although this method is equivalent to the first approach, it requires fewer individual measurements than the first method, leading to a more feasible measurement strategy.

To gain an understanding of the geometry and the radiometric quantities involved, consider a spherical planet, as shown in Fig. 8.6.1, with the Sun illuminating the whole upper hemisphere. The direction towards the observer is indicated by an arrow in the figure. Also shown is the surface element, da, which scatters part of the incident solar radiation in that direction. The coordinate system of Fig. 8.6.1 reflects the relative positions of the planetary surface element with respect to the Sun and the observer and does not correspond to the conventional latitude and longitude system. The angle between the direction to the observer and the normal at the surface element, da, is the emission angle, ε (observer–center of planet–surface element). The angle between the directions to the observer and the Sun is the phase angle, θ (observer–center of planet–Sun). The angle between the Sun direction

8.6 Photometric investigations

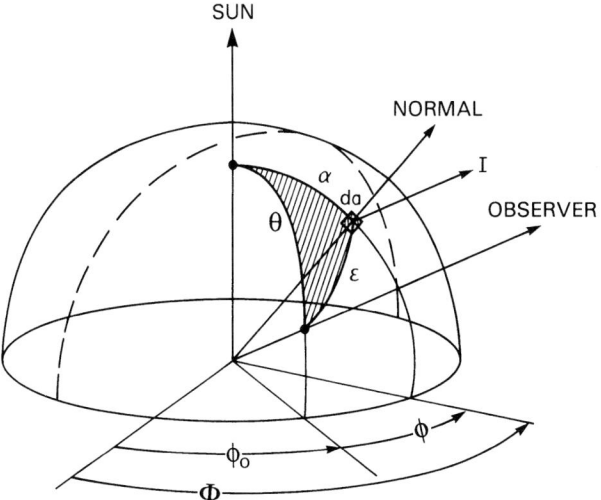

Fig. 8.6.1 Spherical coordinate system. The polar axis points towards the Sun. Directions normal to the surface element da and towards the distant observer are also indicated. The emission angle is ε, the phase angle θ, and the Sun angle α.

and the surface normal at da is the illumination or Sun angle, α (Sun–center of planet–surface element). Emission and illumination cosines are often expressed by $\mu = \cos\varepsilon$ and $\mu_0 = \cos\alpha$.

The bidirectional reflectivity, ρ, of the area element da is a function of the location and the inclination of the element with respect to the observer and the Sun. The reflectivity depends, therefore, on the angles, α, θ, and ε, and is defined by

$$\rho(\alpha, \theta, \varepsilon) = \frac{\pi I(\alpha, \theta, \varepsilon)}{\mu_0 \pi F_s/D^2}, \tag{8.6.2}$$

where $I(\alpha, \theta, \varepsilon)$ is the intensity in the direction of the observer, μ_0 is the cosine of the Sun angle, and $\pi F_s/D^2$ is the solar flux at the heliocentric distance, D, measured in Astronomical Units.

The specific intensity depends on four parameters, $(\mu, \phi; \mu_0, \phi_0)$. The azimuth angles, ϕ and ϕ_0, have an arbitrary origin, and $\Phi = \phi_0 + \phi$ (see Fig. 8.6.1). The spherical cosine law, applied to the shaded triangle displayed in Fig. 8.6.1, establishes a relation between ε, α, θ, and ϕ,

$$\cos\varepsilon = \cos\alpha \cos\theta + \sin\alpha \sin\theta \cos\phi. \tag{8.6.3}$$

Therefore, the reflectivity in Eq. (8.6.2) can be expressed by a function of (α, θ, ϕ) as well. From Eq. (8.6.2) the intensity I, scattered by the surface element, da, towards the observer can be written

$$I(\alpha, \theta, \phi) = \frac{F_s}{D^2} \rho(\alpha, \theta, \phi) \cos \alpha. \tag{8.6.4}$$

Integration of Eq. (8.6.4) over the apparent planetary disk yields the observable mean disk intensity at phase angle, θ, and azimuth angle, ϕ_0,

$$\bar{I}(\theta, \phi_0) = \frac{1}{\pi R^2} \frac{F_s}{D^2} \int \rho(\theta, \alpha, \phi) \cos \alpha \cos \varepsilon \, da, \tag{8.6.5}$$

where R is the planetary radius. The cosine of the emission angle, ε, is a function of α, θ, and ϕ, as given in Eq. (8.6.3). Since

$$da = R^2 \sin \alpha \, d\alpha \, d\phi, \tag{8.6.6}$$

the integration of Eq. (8.6.5) must be performed with respect to α and ϕ. In this calculation of the mean disk intensity, the integration over the planet must be restricted to the illuminated hemisphere, $\alpha \leq \pi/2$, as well as to the part visible to the distant observer, $\varepsilon \leq \pi/2$. The horizon seen by the observer corresponds to $\varepsilon = \pi/2$, that is, $\cos \varepsilon_{\text{horizon}} = 0$. From Eq. (8.6.3) we obtain

$$\cos \varepsilon_{\text{horizon}} = \cos \alpha^* \cos \theta + \sin \alpha^* \sin \theta \cos \phi = 0, \tag{8.6.7}$$

where α^* is the illumination angle at the observer's horizon. Solving Eq. (8.6.7) for α^* yields

$$\alpha^* = \arctan\left(\frac{-1}{\tan \theta \cos \phi}\right). \tag{8.6.8}$$

If θ is between zero and $\pi/2$ the disk measurement, $\bar{I}(\theta, \phi_0)$, can be expressed by the sum of two integrals

$$\bar{I}(\theta, \phi_0) = \frac{F_s}{\pi D^2} \left[\int_{-\pi/2}^{+\pi/2} \int_0^{\pi/2} G(\alpha, \theta, \phi) \, d\alpha \, d\phi + \int_{\pi/2}^{3\pi/2} \int_0^{\alpha^*} G(\alpha, \theta, \phi) \, d\alpha \, d\phi \right], \tag{8.6.9}$$

where

$$G(\alpha, \theta, \phi) = \rho(\alpha, \theta, \phi) \sin \alpha \cos \alpha (\cos \alpha \cos \theta + \sin \alpha \sin \theta \cos \phi). \tag{8.6.10}$$

The first integral in Eq. (8.6.9) is over the fully visible and illuminated segment, while the second integral covers the segment containing the illuminated horizon, that is, the apparent planetary limb. If θ is between $\pi/2$ and π, the planet appears to the observer as a crescent and only one integral needs to be evaluated,

$$\bar{I}(\theta, \phi_0) = \frac{F_s}{\pi D^2} \int_{-\pi/2}^{+\pi/2} \int_{\alpha^*}^{\pi/2} G(\alpha, \theta, \phi) \, d\alpha \, d\phi. \qquad (8.6.11)$$

Ideally, one attempts to derive an analytical expression for the disk measurements and their dependence on the phase angle. Conceivably, a family of such expressions could serve as interpolation functions to cover gaps between sparsely available data points. In reality integration of Eq. (8.6.9) is far from simple. In most cases $\rho(\alpha, \theta, \phi)$ is not well known. Alternatively, disk measurements must be made for many values of phase and azimuth angles, θ and ϕ_0, respectively. Then one can integrate the individual disk measurements over 4π steradians, which yields the total power reflected by the planet,

$$P_{\text{reflected}} = \pi R^2 \int_0^{2\pi} \int_0^{\pi} \bar{I}(\theta, \phi_0) \sin \theta \, d\theta \, d\phi_0. \qquad (8.6.12)$$

Often the simplifying assumption is made that the mean disk intensity depends only on θ and not on ϕ_0. This assumption is reasonable for relatively uniform bodies, such as Uranus, and to a lesser degree also for the other outer planets and Titan. It is only a crude approximation for Earth, which has bright polar caps, clouds, and a dark ocean surface. With the assumption of $\bar{I}$ being independent of ϕ_0, the integration over ϕ_0 is trivial,

$$P_{\text{reflected}} = 2\pi^2 R^2 \int_0^{\pi} \bar{I}(\theta) \sin \theta \, d\theta. \qquad (8.6.13)$$

The reflected power divided by the incident power intercepted by the object gives the albedo. If the albedo is measured over narrow spectral intervals, then the spectral albedo is determined. If the spectral response of the instrument is flat over all wavelengths pertinent to solar radiation (from 0.2 μm to 4 μm, for example), the global bolometric or Bond albedo is obtained.

The incident power is the planetary cross section times the solar flux at the planetary distance from the Sun,

$$P_{\text{incident}} = \pi R^2 \frac{\pi F_s}{D^2}. \qquad (8.6.14)$$

The albedo becomes

$$A = \frac{P_{\text{reflected}}}{P_{\text{incident}}} = \frac{2}{F_s/D^2} \int_0^\pi \bar{I}(\theta) \sin\theta \, d\theta. \qquad (8.6.15)$$

Direct evaluation of this equation is most appropriate for space-borne measurements where the mean disk intensities can be found by direct measurements or derived from several local measurements for a particular phase angle with the help of an auxiliary empirical function, such as that of Minnaert (1941) (see Pearl *et al.*, 1990).

For ground-based observations it has become customary to express the albedo as the product of two functions, p and q,

$$A = pq = \frac{\bar{I}(0)}{F_s/D^2} \cdot 2 \int_0^\pi \frac{\bar{I}(\theta)}{\bar{I}(0)} \sin\theta \, d\theta. \qquad (8.6.16)$$

The first factor,

$$p = \frac{\bar{I}(0)}{F_s/D^2}, \qquad (8.6.17)$$

is the geometric albedo; it can be determined for the outer planets from Earth-based observations. The second factor,

$$q = 2 \int_0^\pi \frac{\bar{I}(\theta)}{\bar{I}(0)} \sin\theta \, d\theta, \qquad (8.6.18)$$

is the phase integral. By definition, the phase function, $\bar{I}(\theta)/\bar{I}(0)$, is unity at zero phase angle. For the outer planets a measurement of the phase integral cannot be performed from Earth, and data from a space platform are required. Considering Eq. (8.6.9), (8.6.10), and (8.6.17), the geometric albedo is

$$p = \frac{1}{\pi} \int_0^{2\pi} \int_0^{\pi/2} \rho(\alpha, \phi) \sin\alpha \cos^2\alpha \, d\alpha \, d\phi. \qquad (8.6.19)$$

For the special case of a Lambert surface, where ρ is a constant, p can be evaluated analytically,

$$p_{\text{Lambert}} = \tfrac{2}{3}\rho. \qquad (8.6.20)$$

However, real surfaces and atmospheres deviate from the simplifying assumption that ρ is constant, and one may try to find an expression for ρ from radiative transfer calculations in a scattering and absorbing atmosphere and integrate Eq. (8.6.19), or, best of all, the full disk intensity at zero phase angle can be measured.

The evaluation of the phase integral, Eq. (8.6.18), is more complicated. As mentioned before, a measurement of the phase function of the outer planets, $\bar{I}(\theta)/\bar{I}(0)$, requires observations from a space platform. Actual measurements obtained so far have produced only a few values of $\bar{I}(\theta)$. Alternatively, one may try to find $\rho(\phi, \theta, \alpha)$ from radiative transfer calculations and evaluate Eqs. (8.6.9) and (8.6.11) to obtain estimates for $\bar{I}(\theta)$, as has been done by Pollack et al. (1986). The phase function for a Lambert sphere can be determined analytically, and the corresponding phase integral is $q = \frac{3}{2}$. Therefore, a Lambert sphere has the Bond albedo ρ.

To obtain the Bond albedo, integration of the spectral albedo, $A(\lambda)$, must also be performed over all wavelengths. This integration can be accomplished either by measuring the spectral albedo first and then integrating the data, or by designing the radiometer with a sufficiently flat response function, so that the integration over wavelength takes place at the detector.

Recent determinations of the geometric albedos, the phase integrals, and the Bond albedos for the outer planets are summarized in Table 8.6.1. References indicating the sources of the quoted values are also shown. The measurement of the Saturn albedo is complicated by the existence of the ring system, which not only casts a shadow on Saturn but also scatters additional sunlight towards that planet. Both effects vary over the Saturnian year due to the relatively large inclination of the equator towards the orbital plane ($\sim 27°$).

Finally, one may ask which type of spacecraft orbit yields a sufficiently dense set of data, that is, which orbit provides sufficient coverage in phase and azimuth

Table 8.6.1 *Albedo of the outer planets*

Object	Jupiter (a)	Saturn (b)	Uranus (c)	Neptune
Geometric albedo	0.274 ± 0.012	0.242 ± 0.012	0.215 ± 0.046	0.215 ± 0.050 (d)
Phase integral	1.25 ± 0.010	1.42 ± 0.10	1.40 ± 0.14	1.35 ± 0.16 (e)
Bond albedo	0.343 ± 0.032	0.342 ± 0.030	0.300 ± 0.049	0.29 ± 0.067
Absorbed solar power (10^{16} W)	50.14 ± 2.48	11.14 ± 0.50	0.526 ± 0.037	0.204 ± 0.019

(a) Hanel *et al.* (1981b)
(b) Hanel *et al.* (1983)
(c) Pearl *et al.* (1990)
(d) Pearl & Conrath (1991). Uncertainties in the geometric albedo (and therefore in the Bond albedo) may be larger due to uncertainties in the radiometer calibration (see Pearl & Conrath, 1991).
(e) Pollack *et al.* (1986)

angles? Clearly, a polar orbit slowly precessing with respect to the Sun–planet line permits disk measurements from all possible directions. With a perfect polar orbit, which does not precess in inertial space, it will take half a planetary year to acquire the needed disk measurements. This corresponds to 42 and 82 Earth years for Uranus and Neptune, respectively, not very attractive propositions. However, the precession rate of a spacecraft orbit may be modified by taking advantage of gravitational forces resulting from the oblateness of a rotating planet. For example, many spacecraft have been placed in Sun-synchronous, near polar orbits around Earth. In that case the orbital precession is chosen to equal the apparent motion of the Earth–Sun vector. Orbit after orbit, the spacecraft passes the equator near local noon and midnight, or at any other local time chosen. Meteorological observations can then be made quasi-synoptically, that is, at the same local time every day. Such orbits belong to the retrograde class. For measurements designed to derive the albedo just the opposite of a synoptic observation is desired. A prograde orbit would be chosen so that the precession advances with respect to the planet–Sun vector, and the desired disk measurements could be accomplished in somewhat shorter time. Such a measurement program is feasible, although it probably would still take considerable time with a single spacecraft. It is unlikely, however, that a spacecraft orbit will ever be optimized just for the benefit of an albedo investigation; spacecraft orbits are necessarily compromises among many, often conflicting requirements. In reality, matters are much worse. Albedo measurements of the outer planets have been made from Pioneer and Voyager. These spacecraft followed fly-by trajectories that limited the sampling of disk measurements to rather small sets, inadequately representing the required phase-angle, azimuth-angle space.

c. Thermal emission

As mentioned in Subsection 8.6.a, a determination of the emitted planetary power requires measurements of the spectrally integrated disk intensity over all directions, that is, over 4π steradians. Since thermal emission does not directly depend on the solar flux, calculations of the total thermal emission are simpler than those required to find the Bond albedo. It is sufficient to use the conventional latitude–longitude system. The coordinates of the direction towards the observer are θ and ϕ_0, and those of the area element, da, are α and ϕ. However, to save ourselves cumbersome trigonometric transformations, we use the emission angle ε and the azimuth angle ψ with reference to the subobserver point as coordinates of da. The full disk intensity at wavenumber ν, which can be measured by a distant observer, is then

$$\bar{I}_\nu(\theta, \phi_0) = \frac{1}{\pi R^2} \int_a I_\nu \cos \varepsilon \, da, \qquad (8.6.21)$$

where the integration is over the exposed hemisphere. The surface element, da, is

$$da = R^2 \sin\varepsilon\, d\psi\, d\varepsilon, \tag{8.6.22}$$

and R is the planetary radius. If the atmosphere is in thermodynamic equilibrium, I_ν is given by Eq. (2.4.1). For the giant planets $\tau_s \gg 1$, and Eq. (8.6.21) can be replaced with

$$\bar{I}_\nu(\theta, \phi_0) = \frac{1}{\pi} \int_0^{2\pi} \int_0^{\pi/2} \int_0^\infty e^{-\tau(\nu)/\mu} B_\nu(\tau) \sin\varepsilon\, d\tau\, d\varepsilon\, d\psi. \tag{8.6.23}$$

In the most general case the atmospheric temperature at a certain optical depth, $T(\tau)$, will vary somewhat from place to place on a planet, consequently the Planck function $B[T(\tau)]$ will depend on ε and ψ as well. However, actual measurements on the giant planets have shown a remarkable uniformity of the individual temperature fields. Even between polar and equatorial regions, where the contrast in solar energy deposition is extreme, the temperature differences on a constant pressure surface do not exceed a few degrees kelvin. Therefore, in first order it is justified to assume $B(\tau)$ to be independent of location. Although $B(\tau)$ is a strongly varying function of τ, horizontal uniformity allows Eq. (8.6.23) to be reduced to

$$\bar{I}_\nu(\theta, \phi_0) = 2 \int_0^{\pi/2} \int_0^\infty e^{-\tau(\nu)/\mu} B_\nu(\tau) \sin\varepsilon\, d\tau\, d\varepsilon. \tag{8.6.24}$$

Integration with respect to ε ($\mu = \cos\varepsilon$) is not so straightforward. First, we substitute $\cos\varepsilon = x^{-1}$, yielding $\sin\varepsilon\, d\varepsilon = x^{-2}\, dx$, and

$$\bar{I}_\nu(\theta, \phi_0) = 2 \int_1^\infty \int_0^\infty \frac{e^{-\tau x}}{x^2} B_\nu(\tau)\, d\tau\, dx. \tag{8.6.25}$$

The exponential integral of nth order is defined by

$$\int_1^\infty \frac{e^{-\tau x}}{x^n} dx = E_n(\tau). \tag{8.6.26}$$

The exponential integrals are tabulated by Kourganoff (1952), for example. Then Eq. (8.6.25) may be expressed

$$\bar{I}_\nu(\theta, \phi_0) = 2 \int_0^\infty E_2(\tau) B_\nu(\tau)\, d\tau. \tag{8.6.27}$$

Table 8.6.2 *Thermal emission of outer planets*

Object	Jupiter (a)	Saturn (b)	Uranus (c)	Neptune (d)
Effective temperature (K)	124.4 ± 0.3	95.0 ± 0.4	59.1 ± 0.3	59.3 ± 0.8
Total emitted power (10^{16} W)	83.65 ± 0.84	19.77 ± 0.32	0.560 ± 0.011	0.534 ± 0.036

(a) Hanel *et al.* (1981*b*)
(b) Hanel *et al.* (1983)
(c) Pearl *et al.* (1990)
(d) Pearl & Conrath (1991)

The function $E_2(\tau)$ may be considered a weighting function of $B(\tau)$ to yield the planetary flux in the direction (θ, ϕ_0), and it also is a function of ν. To find the total power emitted by a planet at wavenumber ν one has to multiply the disk intensity by the apparent cross section, πR^2, and integrate over all directions,

$$P_{\text{emiss.}}(\nu) = \frac{\pi R^2}{4\pi} \int_0^{2\pi} \int_0^{\pi} \bar{I}_\nu(\theta, \phi_0) \sin\theta \, d\theta \, d\phi_0. \qquad (8.6.28)$$

With the previously made assumption of horizontal uniformity, Eq. (8.6.28) can easily be integrated with respect to θ and ϕ_0,

$$P_{\text{emiss.}}(\nu) = 2\pi R^2 \int_0^\infty E_2(\tau) B_\nu(\tau) \, d\tau. \qquad (8.6.29)$$

Integration in the spectral domain also needs to be performed. Except for the factor πR^2, Eq. (8.6.29) is identical to Eq. (8.6.27). Of course, on a horizontally uniform object it is sufficient to measure the flux in one direction to obtain the emitted power in all directions. Indeed, this was the procedure applied to find the total emission of Jupiter and Saturn (Hanel *et al.*, 1981*b*, 1983). Complications due to the planetary oblateness and the presence of the rings had to be dealt with.

On Uranus, Voyager IRIS measurements indicated that mid-latitude regions in both hemispheres are 1–2 K cooler than the rest of the planet. Since these regions represent a substantial fraction of the total planetary area, the variation was taken into account by Pearl *et al.* (1990) in deriving the total thermal emission. The planet was divided into latitude bands 10 degrees wide, and a temperature profile was inferred for each band from IRIS data. The thermal flux was then calculated for each band, and the results were integrated over the planetary surface area to obtain the total thermal emission. The effective temperature of Neptune, 59.3 ± 0.8 K (Pearl & Conrath, 1991), is similar to that of Uranus. The results of several thermal emission measurements of the outer planets are summarized in Table 8.6.2.

9
Interpretation of results

The preceding chapter demonstrates how the basic thermal, compositional, and cloud structures of planetary atmospheres can be inferred from infrared measurements. Some information on surface properties is also available. So far, however, there has been no discussion of how underlying physical processes cause these structures to develop and evolve. That is the purpose of this chapter.

We divide the discussion into four topics. In Section 9.1 we are concerned with the one-dimensional thermal equilibrium configuration of an atmosphere in the absence of internal motion. In Section 9.2 we expand the temperature field to three dimensions and investigate the dynamical properties of atmospheres. In Section 9.3 we address the question of how determinations of chemical composition imply the evolution of planets and the Solar System as a whole. Finally, in Section 9.4 we review measurements of the excess heat emitted by the planets, and discuss the importance of these measurements for determining the status of planetary evolution in the present epoch.

9.1 Radiative equilibrium

The absorption of solar radiation leads to heating within the atmosphere, while cooling is achieved by the emission of infrared radiation. Thermal gradients are established, and the magnitudes and directions of these gradients, coupled with the forces of gravity and planetary rotation, give rise to imbalances in local pressure fields that lead to atmospheric motions. These internal motions are responsible for additional energy transport, and it is the balance of the dynamical and radiative heating and cooling rates that determines the ultimate thermal structure of the atmosphere.

The concept of radiative equilibrium provides a starting point for understanding the various physical processes involved. It leads to a useful (sometimes very good) approximation for the vertical thermal structure of the atmosphere. Comparison of measured atmospheric temperatures with the radiative equilibrium solution can serve as a useful point of departure for examining atmospheric motions. In this section fundamental principles are of prime concern and we approach the subject from an analytic rather than a numerical perspective. As a result we sacrifice some precision in determining the thermal structures of real atmospheres, but gain in a physical understanding of their basic differences and the reasons why these differences arise. The discussion that follows is based on a detailed treatment by Samuelson (1983).

a. Governing principles

An atmosphere is in a state of radiative equilibrium when radiative processes are the exclusive means of energy transport, and there is no net heating or cooling at any point in the atmosphere. All energy sources and sinks are exterior to the atmosphere. Examples of such sources are the Sun and radioactive heating in the planetary interior. Outer space is the usual sink.

The criterion for radiative equilibrium in a plane-parallel atmosphere is that

$$\frac{d}{dz}\left(\int_\nu F_\nu \, d\nu\right) = 0 \qquad (9.1.1)$$

at every level, where πF_ν is the monochromatic net flux at level z. The integration is carried out over the entire electromagnetic spectrum.

All substantial planetary atmospheres in the Solar System have well-developed tropospheres, and, except for Venus and Mars, have stratospheric temperature inversions as well. The former implies heating from below, whereas the latter requires heating from above. Two external heating sources are therefore required. Although the opacity in real atmospheres is extremely variable with wavenumber, the essential physics can be distilled by treating the opacity as independent of wavenumber over each of three large spectral intervals, or channels.

One channel consists of solar 'visible' radiation (superscript v) that is absorbed in the stratosphere, causing a temperature inversion. A second solar channel, consisting of conservatively scattered radiation (superscript c), diffuses or is directly transmitted through the atmosphere and is partially absorbed by the planetary surface. Because no radiation is absorbed before reaching the surface this channel does not contribute directly to atmospheric heating. Instead, heating at the surface gives rise to a source of thermal radiation, which in turn is responsible for the formation of a troposphere. Such mechanisms as radioactive heating in the interior and the conversion of gravitational potential energy to heat deep in the atmosphere can also contribute to this source, as they are all formally indistinguishable. Finally, a single thermal infrared channel (superscript ir) is responsible for the redistribution of thermal energy in the atmosphere and cooling to space. The requirement that the algebraic sum of flux divergences of the individual radiation fields equals zero at every atmospheric level is the criterion that defines the thermal structure under the condition of radiative equilibrium.

Of the three channels in our simplified treatment of the radiation field, only the visible and infrared fields contribute to heating and cooling within the atmosphere itself. Heating by the conservative field is restricted to the planetary surface. It follows from Eq. (9.1.1) that

$$\frac{dF(\tau^{ir})}{dz} = -\frac{dF(\tau^{v})}{dz}, \quad (9.1.2)$$

where $\pi F(\tau^{v})$ and $\pi F(\tau^{ir})$ are the visible and infrared fluxes at the level z.

A slightly more restrictive (but otherwise equivalent) criterion for radiative equilibrium is the requirement that all radiative net fluxes must equal the flux πF_n from the deep interior. This latter flux is a positive constant, and may arise from radioactive decay or from the conversion of gravitational potential energy to internal (heat) energy. We have

$$F(\tau^c) + F(\tau^v) + F(\tau^{ir}) = F_n, \quad (9.1.3)$$

where the two solar components contain contributions from the reduced incident solar flux at the level z as well as contributions from the diffuse intensity fields at this level.

b. The solar radiation field

The solar flux crossing a horizontal plane depends on the Sun's elevation angle but is independent of azimuth. The infrared radiation field is always azimuthally symmetric. Therefore, because only fluxes contribute to heating rates, we are interested solely in azimuth-independent radiation fields in our analysis. By analogy with Eq. (2.5.10), the appropriate equation for describing the transfer of radiation in the visible channel is

$$\mu \frac{dI(\tau^v, \mu)}{d\tau^v} = I(\tau^v, \mu) - \frac{1}{2} \sum_{\lambda=0}^{N} \tilde{\omega}_\lambda^v P_\lambda(\mu) \int_{-1}^{+1} P_\lambda(\mu') I(\tau^v, \mu') d\mu'$$

$$- \frac{1}{4} F^v e^{-\tau^v/\mu_0} \sum_{\lambda=0}^{N} \tilde{\omega}_\lambda^v P_\lambda(\mu) P_\lambda(-\mu_0), \quad (9.1.4)$$

where πF^v is the solar flux in this channel. A comparable equation governs the transport of radiation in the conservative channel. The principal difference is that

$$\tilde{\omega}_0^c \equiv 1 \qquad (9.1.5)$$

in the conservative channel, whereas $0 \leq \tilde{\omega}_0^v < 1$ in the partially absorbing visible channel, leading to certain fundamental differences in the formal solutions.

Because only fluxes are physically significant in our problem, the two-stream approximation is deemed adequate. Hence the radiation fields are restricted to the directions $\mu_0 = \mu_1 = -\mu_{-1} = 1/\sqrt{3}$. It can be shown (Samuelson, 1983) that a solution of Eq. (9.1.4) leads to a flux divergence

$$\frac{dF(\tau^v)}{d\tau^v} = \left(1 - \tilde{\omega}_0^v\right) F^v \left(\sum_\alpha L_\alpha e^{-k_\alpha \tau^v} + \gamma_0' e^{-\tau^v/\mu_0} \right) \quad (\alpha = \pm 1), \qquad (9.1.6)$$

where $\pi F(\tau^v)$ is the total net flux (diffuse plus direct) in the visible channel. The quantity

$$\gamma_0 = \tilde{\omega}_0^v \gamma_0' \qquad (9.1.7)$$

is given by Eq. (2.5.32). The roots k_1 and k_{-1} are defined by Eq. (2.5.24), and the integration constants L_1 and L_{-1} can be found from the two boundary conditions

$$I(0, \mu_{-1}) = 0 \qquad (9.1.8)$$

and

$$F_+ = -AF_-, \qquad (9.1.9)$$

where F_+ and F_- are the upward and downward components of $F(\tau_1^v)$, respectively, and A is the surface albedo. Comparable conditions govern the conservative channel.

A further condition,

$$F_D(0) = -a^v \mu_0 F^v, \qquad (9.1.10)$$

relates the diffuse net flux $\pi F_D(0)$ at $\tau^v = 0$ to the albedo a^v at the top of the atmosphere. Again, a corresponding expression is valid for the conservative channel. Equation (9.1.10), coupled with the boundary conditions, eventually leads to the auxiliary relation

$$[(1+a^{\text{v}})(1+A)f^2 - (1-a^{\text{v}})(1-A)](1 - e^{-2k_1\tau_1^{\text{v}}})$$
$$+ 2f(a^{\text{v}} - A)(1 + e^{-2k_1\tau_1^{\text{v}}}) = 0, \qquad (9.1.11)$$

where f is given by [see Eq. (4.1.10)]

$$f = \left(\frac{1 - \tilde{\omega}_0^{\text{v}}}{1 - \frac{1}{3}\tilde{\omega}_1^{\text{v}}}\right)^{1/2}. \qquad (9.1.12)$$

Equation (9.1.11) is a relation connecting the macroscopic quantities a^{v}, A, and τ_1^{v} with the single scattering parameters $\tilde{\omega}_0^{\text{v}}$ and $\tilde{\omega}_1^{\text{v}}$. If $\tau_1^{\text{v}} = \infty$, the physically meaningful solution to Eq. (9.1.11) is

$$f_\infty = \frac{1 - a^{\text{v}}}{1 + a^{\text{v}}}, \qquad (9.1.13)$$

in agreement with Eq. (4.1.30). The conservative channel counterpart to Eq. (9.1.11) can be shown to be

$$(1 - \langle \cos\theta \rangle^{\text{c}})\sqrt{3}\tau_1^{\text{c}} = \frac{2(a^{\text{c}} - A)}{(1 - a^{\text{c}})(1 - A)}, \qquad (9.1.14)$$

where the asymmetry factor $\langle \cos\theta \rangle^{\text{c}}$ is defined by Eq. (4.1.17). Equation (9.1.14) is an expression relating the optical thickness of the atmosphere to the planetary and surface albedos. It is clear that for positive τ_1^{c}, a^{c} must always exceed A; and as $\tau_1^{\text{c}} \to \infty$, $a^{\text{c}} \to 1$, as it must in a nonabsorbing atmosphere. Expressions (9.1.11) through (9.1.14) will be found useful in discussing asymptotic solutions to the radiative equilibrium temperature profile.

Finally, in order to assign the distribution of solar heating rates correctly, it is necessary to quantify the relative fractions of visible and conservative radiation incident at the top of the atmosphere. If q is the fraction associated with the visible channel, we have

$$F^{\text{v}} = qF_0; \quad F^{\text{c}} = (1-q)F_0, \qquad (9.1.15)$$

where πF_0 is the total flux crossing a plane perpendicular to the direction of propagation at the distance of the planet in question.

c. Thermal radiation and the temperature profile

The Sun and energy conversion in the interior provide the sources of heating for the atmosphere, and the distribution of opacity sources determines the local heating rates. Thermal equilibrium is achieved when the temperature is adjusted so that the heating and cooling rates are equal in magnitude at every level.

The infrared radiation field serves as both the thermostat and distributor of thermal energy in this adjustment. It regulates the overall magnitude of the temperature field by emitting to space, and it smoothes out thermal gradients through a continuing process of absorption and reemission throughout the atmosphere. We now examine explicitly the means by which this is achieved.

The equation of transfer governing the infrared radiation field is found from Eq. (2.5.10) to be

$$\mu \frac{dI(\tau^{ir}, \mu)}{d\tau^{ir}} = I(\tau^{ir}, \mu) - \frac{1}{2} \sum_{\lambda=0}^{N} \tilde{\omega}_\lambda^{ir} P_\lambda(\mu) \int_{-1}^{+1} P_\lambda(\mu') I(\tau^{ir}, \mu') d\mu'$$
$$- \left(1 - \tilde{\omega}_0^{ir}\right) B(\tau^{ir}). \tag{9.1.16}$$

Multiplying both sides by $d\mu$ and integrating over the interval $(-1, +1)$ yields

$$B(\tau^{ir}) = \frac{1}{2} \int_{-1}^{+1} I(\tau^{ir}, \mu) d\mu - \frac{1}{4(1 - \tilde{\omega}_0^{ir})} \frac{dF(\tau^{ir})}{d\tau^{ir}}, \tag{9.1.17}$$

where [see Eq. (1.8.4)]

$$F(\tau^{ir}) = 2 \int_{-1}^{+1} \mu I(\tau^{ir}, \mu) d\mu. \tag{9.1.18}$$

Hence Eq. (9.1.16) becomes

$$\mu \frac{dI(\tau^{ir}, \mu)}{d\tau^{ir}} = I(\tau^{ir}, \mu) - \frac{1}{2} \sum_{\lambda=0}^{N} \tilde{\omega}_\lambda P_\lambda(\mu) \int_{-1}^{+1} P_\lambda(\mu') I(\tau^{ir}, \mu') d\mu' + \frac{1}{4} \frac{dF(\tau^{ir})}{d\tau^{ir}}, \tag{9.1.19}$$

where

$$\tilde{\omega}_0 = 1; \quad \tilde{\omega}_\lambda = \tilde{\omega}_\lambda^{ir} \quad (\lambda > 0). \tag{9.1.20}$$

Because the atmosphere is in a state of radiative equilibrium, the rate of cooling must equal the rate of heating at every level in the atmosphere. This requirement is

9.1 Radiative equilibrium

expressed quantitatively by Eq. (9.1.2), which, when substituted into Eq. (9.1.19), allows us to replace the inhomogeneous term in the latter equation with a known function. The necessary transformation of independent variables is given through the relations

$$\left. \begin{array}{l} d\tau^{ir} = -N(z)\chi_E^{ir} dz \\ d\tau^{v} = -N(z)\chi_E^{v} dz \end{array} \right\}, \quad (9.1.21)$$

where $N(z)$ is the particle number density and χ_E^{ir} and χ_E^{v} are particle extinction cross sections in the infrared and visible, respectively. If the atmosphere is vertically homogeneous, the extinction cross section ratio

$$\beta = \frac{\chi_E^{v}}{\chi_E^{ir}} \quad (9.1.22)$$

is independent of height, and Eq. (9.1.21) yields

$$\tau^{v} = \beta \tau^{ir}. \quad (9.1.23)$$

This condition can be combined with Eq. (9.1.6) to write the inhomogeneous term in Eq. (9.1.19) as a function of the independent variable τ^{ir}, and the complete equation, with the help of Eq. (9.1.15), becomes

$$\mu \frac{dI(\tau^{ir}, \mu)}{d\tau^{ir}} = I(\tau^{ir}, \mu) - \sum_{\lambda=0}^{N} \tilde{\omega}_\lambda P_\lambda(\mu) \int_{-1}^{+1} P_\lambda(\mu') I(\tau^{ir}, \mu') d\mu'$$
$$- \frac{1}{4}(1 - \tilde{\omega}_0^{v}) \beta q F_0 \left(\sum_\alpha L_\alpha e^{-k_\alpha \beta \tau^{ir}} + \gamma_0' e^{-\beta \tau^{ir}/\mu_0} \right)$$
$$(\alpha = \pm 1). \quad (9.1.24)$$

An analytic solution to Eq. (9.1.24) has been obtained by the method of discrete ordinates (Samuelson, 1983), which is a generalization of the two-stream approximation considered in this chapter. The solution is then used to obtain expressions for the flux from Eq. (9.1.18) and the Planck intensity from Eq. (9.1.17). The upper boundary condition

$$I(0, \mu_{-1}) = 0, \quad (9.1.25)$$

and the condition for radiative equilibrium, given by Eq. (9.1.3), are used to evaluate the two integration constants resulting from the two-stream approximation.

No further unknowns remain to be evaluated. As a result conditions at the surface cannot be freely chosen but must be calculated from derived quantities. A straightforward evaluation demonstrates that, in the limit of large optical thickness,

$$I(\tau_1^{ir}, \mu_1) - B(\tau_1^{ir}) \sim \tfrac{1}{4}[(1-q)(1-a^c)F_0 + \sqrt{3}F_n], \quad (\beta\tau_1^{ir} \gg 1), \quad (9.1.26)$$

where τ_1^{ir} is the infrared optical thickness of the atmosphere.

Equation (9.1.26) demonstrates the existence of a thermal discontinuity at the surface even in the absence of an internal heat source. The surface is warmer than the atmosphere immediately above it, a condition that is unstable against convection. Thus, a planetary boundary layer is formed, the thickness of which depends upon the magnitude of the net flux crossing the surface. If the internal heat source is negligible, the temperature discontinuity is inversely proportional to the conservative optical thickness τ_1^c of the atmosphere [see Eq. (9.1.14)]. As $\tau_1^c \to \infty$ the thermal discontinuity goes to zero. On the other hand, if the internal heat source dominates and the atmosphere is very deep, turbulent convection is likely throughout most of the troposphere. This depends critically upon the infrared opacity, and whether the magnitude of the thermal gradient under conditions of radiative equilibrium exceeds that for convective stability.

We are now ready to express the radiative equilibrium thermal profile in terms of known parameters. We define the quantities

$$h = \left(\frac{1 - \tilde{\omega}_0^{ir}}{1 - \tfrac{1}{3}\tilde{\omega}_1^{ir}}\right)^{1/2}, \quad (9.1.27)$$

$$f' = \frac{1 - a^v}{1 + a^v}, \quad (9.1.28)$$

and

$$\beta' = \beta \frac{1 - \tilde{\omega}_0^v}{1 - \tilde{\omega}_0^{ir}}. \quad (9.1.29)$$

The expression for the Planck intensity can then be written

$$B(\tau^{ir}) = \tfrac{1}{4}F_0\{g_0 + g_1\tau^{ir} + g_2 e^{-k_1\beta\tau^{ir}}[1 + g_3 e^{-2k_1\beta(\tau_1^{ir} - \tau^{ir})}]\}, \quad (9.1.30)$$

where

$$g_0 = (1-q)(1-a^c) + \sqrt{3}\frac{F_n}{F_0} + q(1+a^v)\left[\frac{1}{\beta'}\left(\frac{f}{h}\right)^2 + f'\right], \quad (9.1.31)$$

$$g_1 = \sqrt{3}\left(1 - \frac{1}{3}\tilde{\omega}_1^{ir}\right)\left[(1-q)(1-a^c) + \sqrt{3}\frac{F_n}{F_0}\right], \quad (9.1.32)$$

$$g_2 = \frac{q}{2}(1+a^v)\left(1+\frac{f'}{f}\right)\left[\beta' - \frac{1}{\beta'}\left(\frac{f}{h}\right)^2\right], \quad (9.1.33)$$

and

$$g_3 = \frac{f - \frac{1-A}{1+A}}{f + \frac{1-A}{1+A}}. \quad (9.1.34)$$

Its relationship to the temperature is given by

$$\pi B(\tau^{ir}) = \sigma T^4(\tau^{ir}), \quad (9.1.35)$$

where σ is the Stefan–Boltzmann constant.

In the foregoing the flux of incident solar radiation relative to the local vertical is $-\mu_0 \pi F_0$, whereas the incident solar flux averaged over a spherical planet is $-\pi F_0/4$. This implies a reduced flux $\pi F_0'$ should replace πF_0 under conditions of global radiative equilibrium. If $\mu_0 = 1/\sqrt{3}$,

$$F_0' = \frac{\sqrt{3}}{4} F_0. \quad (9.1.36)$$

The correspondingly reduced Planck intensity should be representative of global conditions averaged over all latitudes and hour angles.

d. General atmospheric properties

The emitted thermal flux $\pi F(0)$ at $\tau^{ir} = 0$ is a quantity of considerable importance. If there is a relatively negligible internal heat flux, the emitted flux is in global balance with the absorbed solar flux. This situation prevails at Venus, Earth, Mars, Titan, and possibly Uranus. As discussed in detail in Section 9.4, a non-negligible internal heat flux exists for Jupiter, Saturn, and Neptune. In these cases the emitted flux is in substantial excess of that required to balance the incoming solar flux absorbed by the atmosphere and surface.

Except for Earth and Mars, the integrated thermal optical thickness tends to be large, and $k_1 \beta \tau_1^{ir} \gg 1$ for the other planets cited above. In these cases, Eq. (9.1.13) is an acceptable approximation, and the outgoing flux $\pi F(0)$ can be shown to reduce to

$$F(0) = F_0 \left[\frac{1}{\sqrt{3}} (1-q)(1-a^c) + \frac{F_n}{F_0} + \frac{1}{\sqrt{3}} q(1-a^v) \right]. \tag{9.1.37}$$

We now consider how the thermal profile of an atmosphere in radiative equilibrium is related to the emitted thermal flux. The Planck intensity at $\tau^{ir} = 0$, according to Eq. (9.1.30), is given by

$$B(0) = \tfrac{1}{4} F_0 (g_0 + g_2). \tag{9.1.38}$$

We still assume $k_1 \beta \tau_1^{ir} \gg 1$, which, from Eqs. (9.1.13) and (9.1.28), implies $f' \sim f$. With the aid of Eqs. (9.1.28), (9.1.29), (9.1.31), and (9.1.33), Eq. (9.1.38) becomes

$$B(0) = \frac{\sqrt{3}}{4} F_0 \left\{ \frac{1}{\sqrt{3}} (1-q)(1-a^c) + \frac{F_n}{F_0} + \frac{1}{\sqrt{3}} q(1-a^v) \right.$$
$$\left. \times \left[1 + \beta \left(\frac{1+a^v}{1-a^v} \right) \frac{1-\tilde{\omega}_0^v}{1-\tilde{\omega}_0^{ir}} \right] \right\}. \tag{9.1.39}$$

A comparison of Eqs. (9.1.37) and (9.1.39) yields a minimum value $B(0) \geq \sqrt{3} F(0)/4$. The exact value depends on β, the ratio of visible-to-infrared extinction coefficients. As β increases from zero to larger values, $B(0)$ increases accordingly, and, for sufficiently large β, a temperature inversion is introduced in the stratosphere.

At the other extreme, in the absence of an internal heat source, $B(\tau_1^{ir})$ approaches a finite limit as $\tau_1^{ir} \to \infty$. According to Eq. (9.1.14),

$$\lim_{\tau_1^{ir} \to \infty} (1 - a^c) \tau_1^{ir} = \frac{2}{\sqrt{3(1 - \langle \cos \theta \rangle^c) \beta^c}}, \tag{9.1.40}$$

where

$$\beta^c = \frac{\tau^c}{\tau^{ir}} = \frac{\tau_1^c}{\tau_1^{ir}} \tag{9.1.41}$$

9.1 Radiative equilibrium

is the ratio of conservative-to-infrared extinction coefficients. From Eq. (9.1.30) we obtain

$$B(\infty) = \lim_{\tau_1^{ir} \to \infty} \tfrac{1}{4} F_0 \left(g_0 + g_1 \tau_1^{ir} \right), \quad (9.1.42)$$

which, from Eqs. (9.1.13), (9.1.27)–(9.1.29), (9.1.31), (9.1.32), and (9.1.40), becomes

$$B(\infty) = \frac{1}{4} F_0 \left\{ \frac{2(1-q)}{\beta^c} \frac{1 - \tfrac{1}{3}\tilde{\omega}_1^{ir}}{1 - \langle \cos\theta \rangle^c} + q(1-a^v)\left[1 + \frac{1 - \tfrac{1}{3}\tilde{\omega}_1^{ir}}{1 - \tfrac{1}{3}\tilde{\omega}_1^v} \left(\beta \frac{1-a^v}{1+a^v} \right)^{-1} \right] \right\}. \quad (9.1.43)$$

By postulate $F_n = 0$. Of course, if there is an internal heat source and $F_n > 0$, $B(\tau_1^{ir})$ increases without bound as $\tau_1^{ir} \to \infty$.

It is clear that $B(\infty)$ increases as either β^c or β decreases. In either case the atmosphere becomes more transparent to one of the solar channels compared with the thermal channel. A given amount of sunlight penetrates to relatively greater optical depths, and heating at depth (the greenhouse effect) is enhanced.

A somewhat unexpected result is that, as $\tau_1^c \to \infty$, the magnitude of the greenhouse effect becomes independent of the surface albedo A ($A = 1$ leads to ambiguity, however, as no absorption of solar energy occurs at the surface in this case). Apparently, as $\tau_1^c \to \infty$, the radiation reflected from the surface is backscattered by the atmosphere onto the surface again, compensating for the original lack of absorption by the surface. This compensation is complete only because $\tilde{\omega}_0^c = 1$, resulting in no loss due to atmospheric absorption, and also because $\tau_1^c = \infty$, resulting in no direct loss to space. In effect the radiation in the conservative channel is trapped between the atmosphere and surface until it is finally absorbed by the surface and converted into infrared radiation.

This mechanism contributes to the elevated surface temperature of $\sim$750 K observed on Venus (see Fig. 9.1.1) in spite of the high planetary albedo. A combination of large $(1 - q)$ (i.e., fraction of incident solar radiation in the conservative channel), small β^c, and large $\langle \cos\theta \rangle^c$ (forward scattering, resulting in deep penetration) leads to a maximum greenhouse effect. In the case of Venus, the sulfuric acid clouds contain particles that are both highly reflecting and forward scattering. Considerable radiation penetrates to levels where collision-induced absorption of carbon dioxide greatly increases the infrared opacity. Although the atmosphere of Venus is far from being vertically homogeneous, the effective extinction coefficient ratio β^c is very low, giving rise to the elevated temperatures observed near the surface.

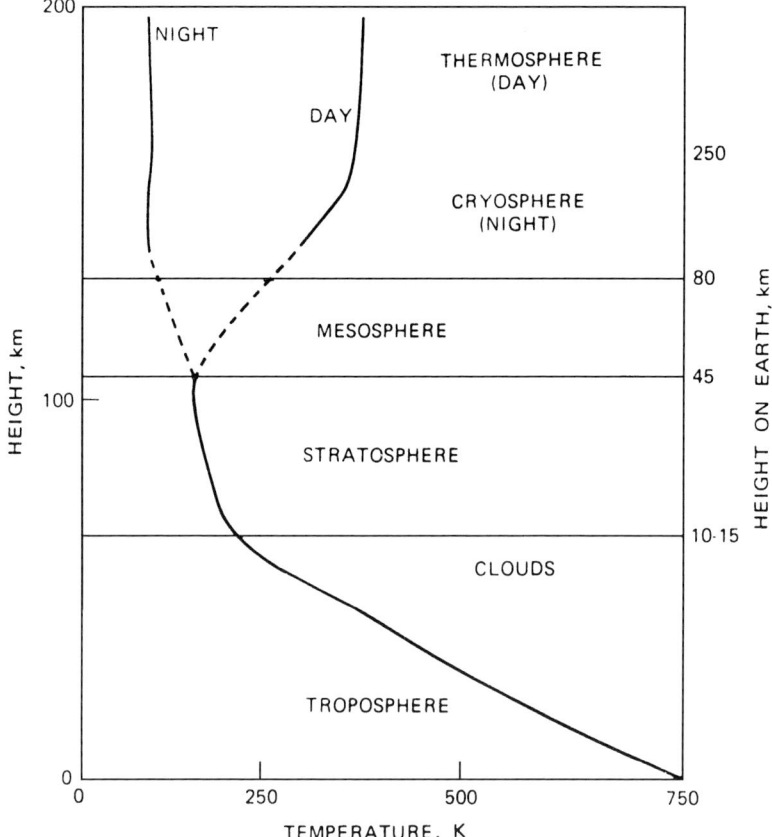

Fig. 9.1.1 Typical temperatures for the Venus atmosphere. The discussion in this section is relevant only below 100 km. Heights for Earth are shown for comparison (Fimmel et al., 1983).

According to Figs. 8.2.2 and 9.1.1, a greenhouse effect appears to exist for all planets with substantial atmospheres, although in the cases of Jupiter, Saturn, and Neptune the effect is overwhelmed at very large depths by an internal heat source. This heat source leads to dynamically active lower tropospheres in these major planets, a subject discussed in the next section.

A further simplification can be accomplished by eliminating the conservative channel along with the internal heat source. This leads to greater insight on how atmospheric heating from above contributes to the thermal profile. Equations (9.1.37), (9.1.39), and (9.1.43) respectively reduce to

$$F(0) = \frac{1}{\sqrt{3}} F_0 q (1 - a^v), \tag{9.1.44}$$

$$B(0) = \frac{1}{4} F_0 q (1 - a^v) \left[1 + \beta \left(\frac{1 + a^v}{1 - a^v} \right) \frac{1 - \tilde{\omega}_0^v}{1 - \tilde{\omega}_0^{ir}} \right], \tag{9.1.45}$$

and

$$B(\infty) = \frac{1}{4}F_0 q(1-a^{\rm v})\left[1 + \frac{1}{\beta}\left(\frac{1+a^{\rm v}}{1-a^{\rm v}}\right)\frac{1-\tilde{\omega}_1^{\rm ir}}{1-\frac{1}{3}\tilde{\omega}_1^{\rm ir}}\right]. \tag{9.1.46}$$

With the aid of Eqs. (9.1.12) and (9.1.13), Eqs. (9.1.45) and (9.1.46) lead to

$$1 - \frac{B(0)}{B(\infty)} = \frac{1 - \left(\beta\frac{\kappa_1^{\rm v}}{\kappa_1^{\rm ir}}\right)^2}{1 + \beta\frac{\kappa_1^{\rm v}}{\kappa_1^{\rm ir}}\left(\frac{1-\tilde{\omega}_0^{\rm ir}}{1-\frac{1}{3}\tilde{\omega}_1^{\rm ir}}\right)^{1/2}}, \tag{9.1.47}$$

where [compare with Eq. (2.5.24)]

$$\left.\begin{aligned}\kappa_1^{\rm v} &= \left[(1-\tilde{\omega}_0^{\rm v})(1-\tfrac{1}{3}\tilde{\omega}_1^{\rm v})\right]^{1/2}\\ \kappa_1^{\rm ir} &= \left[(1-\tilde{\omega}_0^{\rm ir})(1-\tfrac{1}{3}\tilde{\omega}_1^{\rm ir})\right]^{1/2}\end{aligned}\right\}. \tag{9.1.48}$$

If scattering in the thermal infrared is negligible (as is often the case), a particularly simple expression results:

$$B(0) = \beta \kappa_1^{\rm v} B(\infty). \tag{9.1.49}$$

The atmosphere is warmer at the top or bottom depending on whether $\beta\kappa_1^{\rm v}$ is, respectively, greater than or less than unity.

In the absence of scattering in the infrared, and with the aid of Eqs. (9.1.12), (9.1.13), and (9.1.48), Eq. (9.1.45) reduces to

$$B(0) = \tfrac{1}{4}F_0 q(1-a^{\rm v})(1+\beta\kappa_1^{\rm v}). \tag{9.1.50}$$

Combining Eqs. (9.1.44), (9.1.49), and (9.1.50) yields

$$B(0) = \frac{\sqrt{3}}{4}F(0)\left[1 + \frac{B(0)}{B(\infty)}\right], \tag{9.1.51}$$

from which it follows that

$$\frac{1}{\frac{\sqrt{3}}{4}F(0)} = \frac{1}{B(0)} + \frac{1}{B(\infty)}, \tag{9.1.52}$$

a linear relation connecting the reciprocals of the emitted thermal flux and the

Planck intensity extremes. A remarkable feature of this relation is its generality; it is quantitatively independent of particular model details within rather wide constraints.

The Planck intensity extremes can also be expressed as

$$\left.\begin{array}{l} B(0) = \dfrac{\sqrt{3}}{4} F(0)\left(1 + \beta \kappa_1^{\rm v}\right) \\[6pt] B(\infty) = \dfrac{\sqrt{3}}{4} F(0)\left(1 + \dfrac{1}{\beta \kappa_1^{\rm v}}\right) \end{array}\right\} \quad (9.1.53)$$

in this simple model. Either $B(0)$ or $B(\infty)$ can increase without bound as β takes on its limiting values in the range $0 \le \beta \le \infty$, while the other reduces to a lower limit of $\sqrt{3}F(0)/4$. Thus heating from above can result in either a temperature inversion or a greenhouse effect, but not both at the same time.

In order to have a temperature inversion and a greenhouse effect simultaneously, it is necessary to include heating from both above and below. Differentiation of Eq. (9.1.30) leads to

$$\frac{\partial B(\tau^{\rm ir})}{\partial \tau^{\rm ir}} = \tfrac{1}{4} F_0 \left\{ g_1 - k_1 \beta g_2 {\rm e}^{-k_1 \beta \tau^{\rm ir}} \left[1 - g_3 {\rm e}^{-2k_1 \beta(\tau_1^{\rm ir} - \tau^{\rm ir})}\right]\right\}. \quad (9.1.54)$$

According to Eq. (9.1.34), $|g_3| < 1$, and the quantity in square brackets in Eq. (9.1.54) cannot be negative. Thus it is possible for a temperature minimum to occur $[\partial B(\tau^{\rm ir})/\partial \tau^{\rm ir} = 0$ at $\tau^{\rm ir} = \tau_{\rm m}^{\rm ir}$, say] if g_2 is sufficiently large. In practice $\tau_{\rm m}^{\rm ir}$ tends to be much smaller than $\tau_1^{\rm ir}$. Also, as a general rule, $2k_1 \beta \tau_1^{\rm ir} \gg 1$ (Mars and possibly Earth are exceptions), implying that

$$g_1 = k_1 \beta g_2 {\rm e}^{-k_1 \beta \tau_{\rm m}^{\rm ir}}, \quad (9.1.55)$$

approximately. This in turn requires that

$$g_1 < k_1 \beta g_2 \quad (9.1.56)$$

if a temperature minimum is to exist.

It is of considerable interest, then, to evaluate the ratio $k_1 \beta g_2 / g_1$. If the ratio exceeds unity a temperature minimum is possible – otherwise it is not. Upon substituting Eqs. (9.1.12), (9.1.13), and (9.1.27) through (9.1.29) into Eqs. (9.1.32) and

(9.1.33), we find after some reduction that, for $2k_1\beta\tau_1^{ir} \gg 1$,

$$k_1\beta\frac{g_2}{g_1} = \frac{q(1-a^v)}{(1-q)(1-a^c)+\sqrt{3F_n/F_0}}\left[\left(\beta\frac{\kappa_1^v}{\kappa_1^{ir}}\right)^2 - 1\right], \quad (9.1.57)$$

where k_1 and the κs are given by Eqs. (2.5.24) and (9.1.48), respectively.

Although detailed line by line calculations of molecular opacities are required to obtain rigorous solutions, the simple models considered in this section are adequate for discussing certain general characteristics of the various atmospheres. The most important factor in Eq. (9.1.57) is the one in brackets. The carbon dioxide atmospheres of Venus and Mars do not absorb strongly in the visible and near infrared. Thermal emission is controlled principally by the ν_2-band at 667 cm^{-1}. It appears that $\beta\kappa_1^v < \kappa_1^{ir}$ for the stratospheres of these two planets, yielding a negative value for the quantity in brackets. Thus, a temperature minimum cannot form near the tropopause regardless of the exact values of the other quantities in the equation. This is consistent with the vertical thermal structures of these two planets, shown in Figs. 8.2.2 and 9.1.1.

On Earth ozone absorbs strongly in the ultraviolet. Ozone emission centered at 1040 cm^{-1} cannot compensate completely, and β is sufficiently large to form a stratospheric inversion. On Jupiter and Saturn methane plays the same role as ozone on Earth. Many strong absorption bands occur in the near infrared, but only the methane ν_4-band at 1304 cm^{-1} is available for stratospheric emission. Thermal infrared bands of ethane and acetylene also contribute to stratospheric cooling, but are not sufficient to overcome heating provided by solar absorption in the near infrared CH$_4$ bands. In addition, photochemistry of the various hydrocarbons can lead to the formation of small, dark aerosol particles. According to the discussion in Section 8.4, this aerosol can augment the large values of β.

Because they are so alike in other ways, Uranus and Neptune might also be expected to have similar stratospheric thermal structures, but as Fig. 8.2.2 shows, Neptune's thermal gradient is somewhat larger. Two reasons can be advanced for this difference, both related to Neptune's large internal heat source (see Section 9.4). Possibly the high degree of turbulent convection required in the troposphere to transport the excess heat flux results in convective penetration of the tropopause. If the velocity of penetration is larger than the Stokes terminal velocity for condensed methane particles (this depends on particle size), it may be possible to pump condensed methane through the tropopause, injecting larger amounts of methane into Neptune's stratosphere (where it again vaporizes) than can be accounted for from simple cold trap theory. A second possibility is that dynamical rather than radiative heating is responsible for Neptune's temperature inversion. Waves generated by dynamical activity in the troposphere may be dissipated in the stratosphere, for example, releasing their energy as heat.

Titan has the most extreme temperature inversion of all (see Fig. 8.2.2). The quantity in brackets in Eq. (9.1.57) has a value $\sim 10^3$, while the factor multiplying it is ~ 2. This leads to an unusually large thermal gradient in the lower stratosphere. The principal contributor to the large value of β is a thick stratospheric aerosol, the product of photochemistry and charged particle bombardment in an atmosphere rich in organic gases.

In spite of this dense aerosol, enough near infrared radiation penetrates Titan's atmosphere to heat the surface slightly and create a small greenhouse effect. The principal source of thermal infrared opacity is collision-induced absorption due to N_2–CH_4–H_2 combinations. There is a fairly transparent window between 400 and 600 cm^{-1}, however, due to the small amount of hydrogen in Titan's atmosphere, and this severely limits the magnitude of the effect. The atmospheres of the major planets do not suffer from the same difficulty, since they are composed principally of hydrogen, and the window between 400 and 600 cm^{-1} is effectively closed.

9.2 Atmospheric motion

Infrared remote sensing has contributed significantly to our understanding of the dynamics of planetary atmospheres. In the case of the Earth emphasis has been placed on the global acquisition of thermal structure for use in numerical weather forecasting. Retrieved temperatures and trace constituents have also contributed to basic research on the coupling of radiation, chemistry, and dynamics in the terrestrial middle atmosphere. A large and rapidly expanding literature exists in these areas (see, for example, Andrews *et al.*, 1987), but it is not reviewed here. We confine our attention to the atmospheres of several of the other Solar System bodies where the principal concerns are the identification of basic dynamic regimes and the study of fundamental heat and momentum transfer processes rather than detailed prediction of atmospheric behavior.

We concentrate on the information obtained from infrared spectroscopy and radiometry, both directly and in conjunction with other data sets, such as those from visible imaging. To provide the necessary background for the subjects of this section, we first review the equations of fluid motion and the succession of approximations leading to a tractable set of equations that can be used to describe the motion of a planetary atmosphere. For most of the cases considered, geostrophic balance and the associated thermal wind equations play major diagnostic roles in the inference of atmospheric motions from remotely sensed temperatures. For this reason, the derivation of these relations will be discussed in some detail. Other more specialized approximations will be introduced as required in the discussions of phenomena observed in specific planetary atmospheres.

Remotely sensed data suitable for studies of atmospheric dynamics now exist for a variety of atmospheres. Mars provides an example of a rapidly rotating, shallow atmosphere with strong radiative forcing. Jupiter, Saturn, Uranus, and Neptune are rapidly rotating planets with massive atmospheres. Finally, Venus and Titan are both examples of slowly rotating bodies with deep atmospheres. Each of these cases is considered.

a. Governing equations

The equations required to describe the motions of a planetary atmosphere include Newton's second law, the mass continuity equation, the first law of thermodynamics, and an equation of state for the atmospheric gas. These relations are briefly reviewed from a general point of view. More detailed discussions of the governing equations and their applications can be found in texts on dynamical meteorology and geophysical fluid dynamics, such as Pedloskey (1979), Haltiner & Williams (1980), Holton (1992), and Salby (1996).

In an inertial or nonaccelerating reference frame, Newton's second law can be written

$$\left(\frac{d}{dt}\mathbf{V}_I\right)_I = \mathbf{G}, \qquad (9.2.1)$$

where the subscript I indicates that the velocity and acceleration are with respect to the inertial frame. The resultant total force per unit mass is denoted by $\mathbf{G}$. For applications to planetary atmospheres, we are generally interested in the flow velocity, $\mathbf{V}$, relative to a coordinate system fixed in the rotating planet rather than the velocity, $\mathbf{V}_I$, in an inertial system. If the planetary angular velocity is $\mathbf{\Omega}$, then the relationship between $\mathbf{V}$ and $\mathbf{V}_I$ is

$$\mathbf{V}_I = \mathbf{V} + \mathbf{\Omega} \times \mathbf{r}, \qquad (9.2.2)$$

where $\mathbf{r}$ is the radius vector from the center of the planet to the fluid parcel. The time derivative in the inertial system can be expressed in terms of the time derivative in the rotating system,

$$\left(\frac{d\mathbf{V}_I}{dt}\right)_I = \frac{d\mathbf{V}_I}{dt} + \mathbf{\Omega} \times \mathbf{V}_I. \qquad (9.2.3)$$

Using Eqs. (9.2.2) and (9.2.3), and explicitly displaying the pressure gradient, gravitational, and frictional terms contributing to **G**, Eq. (9.2.1) becomes

$$\frac{d\mathbf{V}}{dt} + 2\mathbf{\Omega} \times \mathbf{V} = -\frac{1}{\rho}\nabla p + \mathbf{g} + \mathbf{M}, \qquad (9.2.4)$$

where p is pressure, ρ is density, **g** is the effective gravitational acceleration, and **M** is the total frictional force or momentum damping per unit mass. The operator d/dt here represents the time derivative following the motion of a fluid particle and is called the advective or material derivative. The effective gravitational acceleration is defined as the sum of the true gravitational acceleration and the centrifugal acceleration, $\mathbf{\Omega} \times (\mathbf{\Omega} \times \mathbf{r})$. The second term on the left side of Eq. (9.2.4), $2\mathbf{\Omega} \times \mathbf{V}$, is the Coriolis acceleration, which can profoundly affect the behavior of motion in the atmosphere of a rapidly rotating planet. This gyroscopic effect results in phenomena not encountered in nonrotating fluids, such as the occurrence of large-scale flow at right angles to the pressure gradient, ∇p.

The mass continuity equation,

$$\frac{\partial \rho}{\partial t} + \nabla \cdot (\rho \mathbf{V}) = 0, \qquad (9.2.5)$$

is completely analogous to the electric continuity equation, discussed in Section 1.1, and implies the absence of sources and sinks of mass within the fluid. Any change of mass within a volume must be associated with a net flow of fluid into or out of the volume. Equation (9.2.5) can be rewritten in the form

$$\frac{1}{\rho}\frac{d\rho}{dt} + \nabla \cdot \mathbf{V} = 0. \qquad (9.2.6)$$

If the density remains constant following the flow, the fluid is said to be incompressible, and $\nabla \cdot \mathbf{V} = 0$. Although the gases comprising planetary atmospheres are highly compressible, the horizontal components of flow in a shallow atmosphere can sometimes be treated as if the flow were incompressible to a first approximation.

An equation of state for the atmospheric gas must be specified, and is generally of the form $p = p(\rho, T)$. For the present applications, the ideal gas law is an adequate approximation,

$$p = R\rho T, \qquad (9.2.7)$$

where R is the universal gas constant divided by the mean molecular weight of the atmosphere. The use of a mean molecular weight is valid for the well-mixed portions of the atmospheres that we consider.

9.2 Atmospheric motion

The final equation required is the first law of thermodynamics,

$$\frac{dE}{dt} + \frac{dW}{dt} = Q, \tag{9.2.8}$$

where E is the internal energy per unit volume, W is the work per unit volume performed by a gas parcel on its surroundings as the result of expansion, and Q is the rate at which heat is added to the parcel. Assuming the atmosphere can be treated as a perfect gas, the time rate of change of internal energy per unit volume is $\rho c_v dT/dt$ where c_v is the specific heat at constant volume. The rate at which work is performed per unit volume is $p\nabla \cdot \mathbf{V}$. Inserting these relations in Eq. (9.2.8) gives

$$\rho c_v \frac{dT}{dt} + p\nabla \cdot \mathbf{V} = Q. \tag{9.2.9}$$

When phase changes occur within the atmosphere, latent heat and the heat capacity of the condensate must also be included in Eq. (9.2.9). The heating per unit volume, Q, is called the diabatic heating and usually results from a combination of solar energy absorption and infrared radiative transfer. Heating due to frictional dissipation of the flow is generally negligible, and is omitted in Eq. (9.2.9).

Equations (9.2.4), (9.2.5), (9.2.7), and (9.2.9) are sufficient for describing the motions of a planetary atmosphere. The vector notation used in these expressions is convenient for the study of their general properties, but for most applications it is necessary to write the equations in a specific coordinate system. For some calculations, a local rectangular system may suffice. More generally, spherical coordinates are employed with the origin at the center of the planet and the polar axis coincident with $\mathbf{\Omega}$. In this coordinate system, the equations take the form

$$\frac{Du}{Dt} + \frac{uw^*}{r} - \frac{uv}{r}\tan\theta - 2\Omega\sin\theta v + 2\Omega\cos\theta w^* = -\frac{1}{\rho r \cos\theta}\frac{\partial p}{\partial \phi} + M_\phi \tag{9.2.10}$$

$$\frac{Dv}{Dt} + \frac{vw^*}{r} - \frac{u^2 \tan\theta}{r} + 2\Omega\sin\theta u = -\frac{1}{\rho r}\frac{\partial p}{\partial \theta} + M_\theta \tag{9.2.11}$$

$$\frac{Dw^*}{Dt} - \frac{u^2 + v^2}{r} - 2\Omega\cos\theta u = -\frac{1}{\rho}\frac{\partial p}{\partial r} - g + M_r \tag{9.2.12}$$

$$\frac{1}{\rho}\frac{D\rho}{Dt} + \frac{1}{r\cos\theta}\left[\frac{\partial u}{\partial \phi} + \frac{\partial}{\partial \theta}(v\cos\theta)\right] + \frac{\partial w^*}{\partial r} = 0 \tag{9.2.13}$$

$$\frac{D}{Dt}(\ln T) - \frac{R}{c_p}\frac{D}{Dt}(\ln p) = \frac{Q}{\rho c_p T}, \tag{9.2.14}$$

where ϕ is longitude, θ latitude, and r the radial distance from the planetary center. The operator D/Dt is the advective derivative in spherical coordinates,

$$\frac{D}{Dt} = \frac{\partial}{\partial t} + \frac{u}{r\cos\theta}\frac{\partial}{\partial \phi} + \frac{v}{r}\frac{\partial}{\partial \theta} + w^*\frac{\partial}{\partial r}. \qquad (9.2.15)$$

The zonal (eastward), meridional (northward), and vertical velocity components are u, v, and w^*, respectively. Eqs. (9.2.6) and (9.2.7) have been combined with Eq. (9.2.9) to obtain Eq. (9.2.14), which is an alternate form of the first law of thermodynamics valid for an ideal gas.

If only radiative heating and cooling contribute significantly to Q, then it can be written in the form $Q = -\nabla \cdot (\pi \mathbf{F})$ where $\pi \mathbf{F}$ is the spectrally integrated net radiative energy flux. When the abundances and distributions of optically active atmospheric constituents are known, Q can be calculated using the radiative transfer techniques discussed in the preceding section. In general, the infrared cooling part of Q will depend strongly on the atmospheric temperature T, which must be evaluated as part of the solution. If the components of the frictional force can also be specified or expressed in terms of the dependent variables of the problem, then Eqs. (9.2.10)–(9.2.14) along with the equation of state, Eq. (9.2.7), form a closed system, which can be solved for the atmospheric motion if appropriate boundary and initial conditions are given. In practice, it is impractical to work with this complex set of highly nonlinear equations, and approximations must be sought. Much effort has been devoted to the development of meaningful approximations, with the most complex and sophisticated being used in numerical models of the general circulation of the atmosphere. Our goals here are relatively modest; we wish to develop simple approximations, which, either directly or indirectly, provide information on atmospheric motion from remotely sensed data.

We begin our simplification of Eqs. (9.2.10)–(9.2.14) by letting $r = a + z^*$, where a is the mean planetary radius referred either to the solid planetary surface or, in the case of the giant planets, to a reference pressure surface. If $z^* \ll a$ for all atmospheric levels of interest, we can replace r with a everywhere it occurs in the coefficients of the equations. This is a valid approximation for all cases considered here. A further simplification can be achieved if the aspect ratio of the flow is such that $D/L \ll 1$, where D and L are the characteristic scales of the vertical and horizontal motions, respectively. From considerations of mass continuity, it follows that the ratio of the vertical to horizontal velocities must also be of order D/L; therefore, terms involving w^* can be neglected relative to other terms in the horizontal momentum equations (9.2.10) and (9.2.11). For this class of motion, the dominant balance in the vertical momentum equation (9.2.12) is between the vertical component of the pressure gradient and the gravitational acceleration, yielding the

9.2 Atmospheric motion

hydrostatic approximation

$$\frac{\partial p}{\partial z^*} = -g\rho. \tag{9.2.16}$$

The pressure at a given level equals the weight of the atmospheric column above that level. While this relationship is strictly true for completely motionless conditions, it is approximately true for slow, large-scale motion with a small aspect ratio.

The hydrostatic approximation permits definition of a new coordinate system, frequently used in dynamical meteorology, and one that is advantageous for our purposes here. The geometric height, z^*, is replaced by a new height variable

$$z = -H \ln(p/p_r) \tag{9.2.17}$$

where $H = RT_r/g$ is a constant scale height, and p_r and T_r are a reference pressure and temperature, respectively. If the atmosphere were isothermal at temperature T_r, H would be the scale height over which the pressure changes by 1/e, and z would be identical with the geometric height. The coordinate system using z as the height variable is called the 'log-p' system. In this system the density is eliminated, and the equations of motion are significantly simplified. For our purposes, it is especially convenient since remotely sensed temperatures are usually retrieved on constant pressure surfaces. The z-coordinate defined here is equivalent to that previously introduced in Chapter 8 [Eq. (8.2.1)] to within the constant multiplicative factor H.

The transformation of the horizontal momentum equations (9.2.10) and (9.2.11) to the log-p system requires the pressure gradient terms to be expressed in appropriate forms. For example, the derivative of pressure with respect to longitude can be written

$$\left(\frac{\partial p}{\partial \phi}\right)_{z^*} = -\left(\frac{\partial p}{\partial z^*}\right)_\phi \left(\frac{\partial z^*}{\partial \phi}\right)_p = \rho g \left(\frac{\partial z^*}{\partial \phi}\right)_p, \tag{9.2.18}$$

where the subscripts denote the variables held constant, and the second step follows from the hydrostatic approximation. An analogous expression can be obtained for $(\partial p/\partial \theta)_{z^*}$. Because the gravitational acceleration, g, is in general a function of position, it is convenient to replace the geometric height z^* with the gravitational potential, Φ, as a dependent variable. The latter can be defined as the work required to move unit mass from reference level $z^* = 0$ to an arbitrary level,

$$\Phi = \int_0^{z^*} g \, dz^*. \tag{9.2.19}$$

Specifications of pressure gradients on surfaces of constant z^* are thereby replaced by equivalent gradients of Φ on constant pressure surfaces. The first law of thermodynamics, Eq. (9.2.14), can be easily transformed, while the transformation of the continuity equation is straightforward but tedious. Neither derivation will be repeated here; the interested reader is referred to the discussion given by Haltiner & Williams (1980). Applying this transformation to Eqs. (9.2.10)–(9.2.14) along with the small aspect ratio approximations discussed above, we obtain

$$\frac{Du}{Dt} - \left(f + u\frac{\tan\theta}{a}\right)v = -\frac{1}{a\cos\theta}\frac{\partial\Phi}{\partial\phi} + M_\phi, \tag{9.2.20}$$

$$\frac{Dv}{Dt} + \left(f + u\frac{\tan\theta}{a}\right)u = -\frac{1}{a}\frac{\partial\Phi}{\partial\theta} + M_\theta, \tag{9.2.21}$$

$$\frac{RT}{H} = \frac{\partial\Phi}{\partial z}, \tag{9.2.22}$$

$$\frac{1}{a\cos\theta}\left[\frac{\partial u}{\partial\phi} + \frac{\partial}{\partial\theta}(v\cos\theta)\right] + \frac{1}{\rho_0}\frac{\partial}{\partial z}(\rho_0 w) = 0, \tag{9.2.23}$$

$$\frac{DT}{Dt} + \frac{R}{c_p}\frac{w}{H}T = \frac{Q}{\rho c_p}. \tag{9.2.24}$$

Here we have defined a reference density, $\rho_0(z) = p/RT_r$, and $f = 2\Omega\cos\theta$ is called the Coriolis parameter. The 'vertical velocity' is now defined as $w = Dz/Dt$. In general $w \neq w^*$; however, for most purposes w is an adequate approximation to the true vertical velocity. The continuity equation (9.2.23) is linear when written in this coordinate system, and the horizontal momentum equations, (9.2.20) and (9.2.21), are also simplified with the density no longer appearing explicitly.

Equations (9.2.20)–(9.2.24) are sometimes called the 'primitive equations' in the meteorological literature. They serve as the starting point for most studies of large-scale atmospheric motions. Although these relations are much less complex than Eqs. (9.2.10)–(9.2.14), they remain a formidable set of coupled, nonlinear differential equations. To proceed further in achieving a physical understanding of large-scale atmospheric dynamics and in developing relations appropriate to remote sensing applications, additional approximations must be introduced.

The geostrophic relations and the associated thermal wind equations can provide significant insight into the behavior of rotating atmospheres; they are the lowest order approximation in a systematic development of large-scale atmospheric dynamics. In addition, these equations have been used to obtain information on atmospheric winds from remotely sensed measurements for many of the planetary atmospheres considered here. Therefore, we examine the geostrophic approximation from a general point of view. More specialized approximations are introduced in the discussions of individual atmospheres.

To motivate the geostrophic approximation, we invoke a scale analysis approach. Assume frictional forces can be neglected, and the atmospheric motions have a characteristic horizontal length-scale, L, and velocity scale, U. Recalling the definition of the advective derivative operator D/Dt, we find that the magnitude of the acceleration terms, Du/Dt and Dv/Dt in Eqs. (9.2.20) and (9.2.21), is U^2/L, provided the magnitude of the time scale of the motion is greater than or equal to the advective time, L/U. The terms proportional to $\tan\theta$ are of order U^2/a, and the Coriolis terms are of order fU. If $L \leq a$, then the ratio of each of the terms to the Coriolis term is less than or comparable to the Rossby number, defined as

$$Ro = \frac{U}{fL}. \qquad (9.2.25)$$

If the atmospheric motion is sufficiently large scale and sufficiently slow, such that $Ro \ll 1$, then Eqs. (9.2.20) and (9.2.21) reduce to

$$fv = \frac{1}{a\cos\theta} \frac{\partial \Phi}{\partial \phi} \qquad (9.2.26)$$

and

$$fu = -\frac{1}{a} \frac{\partial \Phi}{\partial \theta}. \qquad (9.2.27)$$

The horizontal atmospheric motion is governed by a balance between the pressure gradient and the Coriolis acceleration to lowest order. For example, if the pressure decreases toward the pole (equivalent to a poleward decrease in geopotential, Φ, on a constant pressure surface), then an atmospheric flow in the direction of the planetary rotation experiences a Coriolis acceleration that balances the pressure gradient acceleration. Thus, the flow is parallel to the lines of constant pressure (or Φ) rather than down the pressure gradient as might be expected in a nonrotating fluid. Geostrophic balance cannot hold at the equator since the horizontal component of the Coriolis acceleration vanishes there, and other terms in Eqs. (9.2.20) and (9.2.21) become dominant.

If the pressure field could be measured within a planetary atmosphere, the geostrophic relations could be used to estimate the horizontal wind field. However, for our purposes we would like to relate the wind field to the temperature field, which can be directly obtained by remote sensing techniques. This can be accomplished by combining the geostrophic relations, Eqs. (9.2.26) and (9.2.27), with the hydrostatic equation (9.2.22). If Eqs. (9.2.26) and (9.2.27) are differentiated with respect to z, and the hydrostatic relation is differentiated with respect to ϕ or θ, then Φ can be eliminated, yielding

$$f\frac{\partial v}{\partial z} = \frac{R}{Ha\cos\theta}\frac{\partial T}{\partial \phi} \qquad (9.2.28)$$

and

$$f\frac{\partial u}{\partial z} = -\frac{R}{Ha}\frac{\partial T}{\partial \theta}. \qquad (9.2.29)$$

These are the thermal wind equations, which relate the vertical derivative of the geostrophic wind to the horizontal gradient of temperature on a constant pressure surface. Thus, if the temperature is retrieved on constant pressure surfaces by remote sensing, the vertical shear of the geostrophic wind can be obtained directly. If the wind is known independently on some constant pressure surface, this can then be used as a boundary condition for the integration of Eqs. (9.2.28) and (9.2.29). If the horizontal temperature gradient vanishes, the geostrophic wind is independent of height. In this case the surfaces of constant pressure and constant density coincide, and the atmosphere is said to be barotropic. When the horizontal temperature gradient is nonzero, surfaces of constant density and pressure are inclined with respect to one another, the geostrophic wind changes with height, and the atmosphere is baroclinic. Thus, only the baroclinic component of the geostrophic flow is directly accessible from temperature field measurements alone.

We have developed the basic tools necessary for the application of remotely sensed data to problems in the dynamics of planetary atmospheres. The thermal wind equations are used extensively for this purpose, while the complete set of primitive equations, (9.2.20)–(9.2.24), forms the starting point for most of the other relevant approximations and models. We now turn to selected examples of applications to specific planetary atmospheres.

b. Mars

Martian meteorology exhibits both similarities and differences when compared to that of Earth. Both planets possess thin atmospheres in the sense that significant amounts of solar flux penetrate to their surfaces. The length of the Martian solar day is only slightly longer than the terrestrial day, and the obliquity of Mars is comparable with that of Earth. However, the mass of the Martian atmosphere is substantially less than that of Earth, resulting in an average surface pressure of ∼5 mbar compared to the terrestrial value of ∼1000 mbar. Mars has no oceans to store and transport heat nor are latent heat effects significant in its atmosphere, except in the vicinity of the carbon dioxide polar caps. Because of the absence of oceans and phase change effects throughout much of the atmosphere, the dynamic regime of Mars is expected to be simpler in some respects than that of Earth. The Martian atmosphere does have certain complicating factors, however. The time required for the atmosphere to radiatively adjust to temperature perturbations is of the order of a day, more than an order of magnitude shorter than the radiative response time in the lower atmosphere of the Earth. This is a consequence of the lower Martian atmospheric mass combined with a high infrared opacity. The topographic relief of the Martian surface is large, approaching a pressure scale height in some cases, and this can strongly influence the circulation. Finally, dust raised from the surface and entrained in the atmosphere absorbs solar energy and can provide a strong source of thermal forcing. Global dust storms, with dust optical depths substantially exceeding unity at visible wavelengths, occur quasi-annually. Local and regional dust storms occur often.

Much information on Martian meteorology has been obtained from the Mariner 9 orbiter, from the Viking 1 and 2 orbiters and landers, from the Mars Global Surveyor orbiter, and from Earth-based measurements. An understanding of Martian atmospheric dynamics has emerged from a synthesis of data from many sources. Leovy (1979) has presented a post-Viking review of the subject. Here we will give two examples of the inference of information on the dynamics of the Martian atmosphere using infrared remote sensing. The first example is an analysis of the zonal wind field and the meridional circulation using data taken when the atmosphere was relatively dust-free. The second example consists of an analysis of the thermally driven atmospheric tide under dust storm conditions.

Martian temperature profiles were retrieved from nadir-viewing spectra obtained with the Mariner 9 infrared spectrometer. These retrievals provided the first detailed global characterization of the atmospheric thermal structure. More recently, the Thermal Emission Spectrometer (TES) carried on Mars Global Surveyor has acquired extensive sets of both nadir- and limb-viewing spectra that provide information on temperature, atmospheric dust opacity, and water ice clouds. Since the spectra from both spacecraft include the 15 μm carbon dioxide band, and the Martian atmosphere is predominately carbon dioxide, measurements within that band can be inverted to obtain atmospheric temperature as a function of barometric pressure. Constrained linear algorithms similar to those discussed in Chapter 8 are used for that purpose. TES data from the aerobraking and science-phasing portions of the Mars Global Surveyor mission have been used to study the detailed evolution of the atmospheric thermal structure over about one half of a Martian year; the results are summarized by Conrath et al. (2000). During the subsequent mapping phase of the mission, it has been possible to construct zonal mean (averaged over

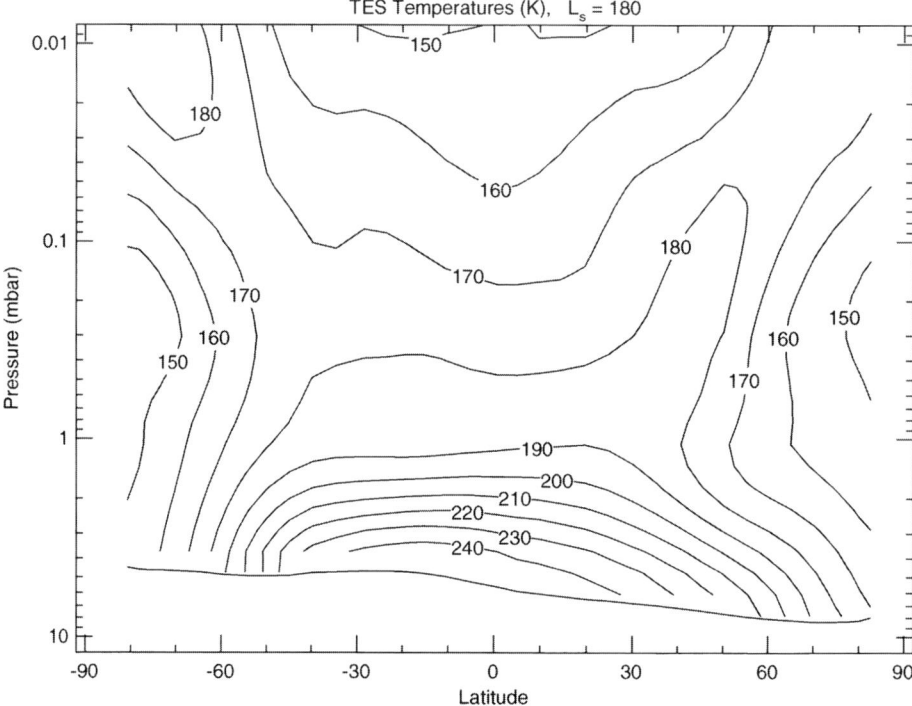

Fig. 9.2.1 Zonal mean meridional temperature cross section of Mars obtained by inversion of Mars Global Surveyor (TES) measurements. The data were acquired at the time of the northern hemisphere fall equinox. The zonal mean atmospheric pressure at the planetary surface is indicated (Smith *et al.*, 2001).

longitude) meridional cross sections of temperature on a daily basis. An example of such a cross section is shown in Fig. 9.2.1 where limb and nadir data have been combined to obtain temperatures from about the 0.008 mbar level down to the planetary surface. The data were taken at a time of the northern hemisphere fall equinox, corresponding to an areocentric solar longitude of $L_s = 180$. Information on the large-scale properties of the atmospheric circulation can be inferred from cross sections such as this. The northward component of the temperature gradient on a constant pressure surface, $\partial T/\partial \theta$, can be estimated by numerical differentiation of the temperature field. To obtain the eastward or zonal component of the wind, Eq. (9.2.29) can be integrated, providing the wind speed can be specified *a priori* at some pressure level as a boundary condition. Figure 9.2.2 shows a thermal wind cross section calculated from the temperature field of Fig 9.2.1 under the assumption that $u = 0$ at the planetary surface. The latter assumption is plausible since surface friction is expected to reduce the wind speed immediately above the ground; however, it must be borne in mind that any nonzero near-surface wind (the barotropic

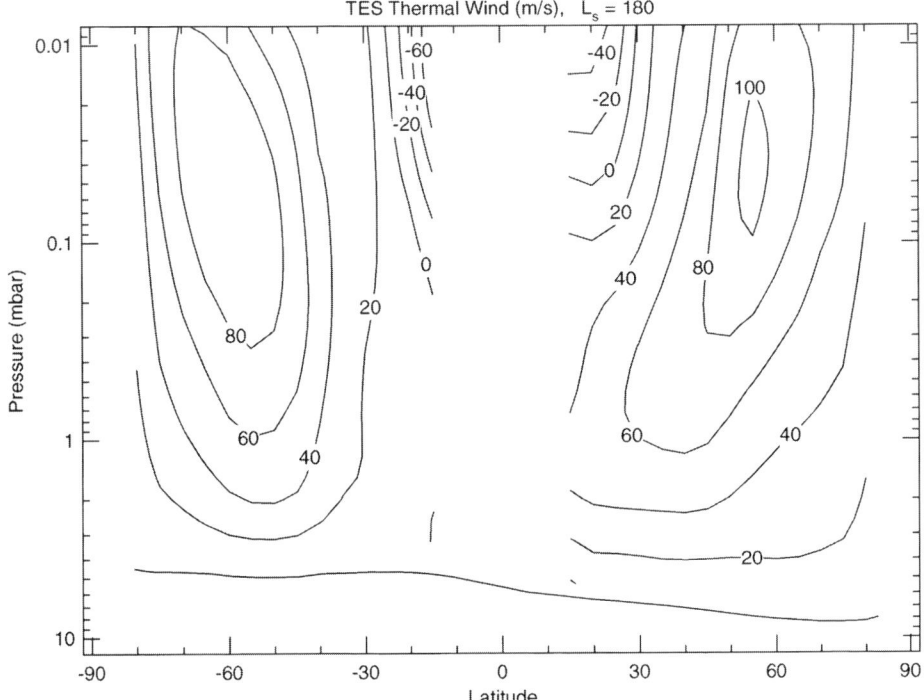

Fig. 9.2.2 Mars zonal mean thermal winds derived from the temperature cross section of Fig. 9.2.1. The thermal wind relation was used to calculate the vertical shear of the zonal wind, and the wind speeds shown were then obtained assuming zero velocity at the planetary surface as a boundary condition (Smith *et al.*, 2001).

component) must be added to the thermal winds (the baroclinic component) to obtain the total wind field. In both hemispheres at high latitudes the circulation is dominated by broad eastward jets. The growth in intensity of these jets with height at the lower atmospheric levels is associated with the decrease in temperature toward each pole. The decay of the jets in the upper levels reflects the reversal of the temperature gradients aloft with temperatures increasing towards the poles. This reversed gradient also accounts for the westward upper level winds at lower latitudes.

It is less straightforward to obtain information on the meridional components of the circulation from temperature fields alone. From Eq. (9.2.28) it can be seen that the meridional component of the thermal wind shear vanishes in the zonal mean, and higher order approximations are required to treat this component of the flow. However, examination of meridional cross sections, such as that shown in Fig. 9.2.1, permits some qualitative statements to be made. For this purpose an approximate form of the thermodynamic energy equations, Eq. (9.2.24), is useful. First, consider the diabatic heating term, which can be written

$$Q = Q_{IR} + Q_S. \qquad (9.2.30)$$

Q_{IR} is the infrared heating and cooling, and Q_S is the heating due to direct absorption of solar energy in the atmosphere. The temperature structure for which $Q = 0$ is the radiative equilibrium profile as discussed in Section 9.1, and is denoted by T_e. In the present example we are interested in the radiative equilibrium profile corresponding to the vanishing of the diurnal mean of Q at each latitude and atmospheric level. In general, for a given atmospheric level Q_{IR} will contain terms representing radiative cooling to space, radiative exchange with the surface, and radiative exchange with all other atmospheric levels. If the sum of the exchange terms with other atmospheric levels and the surface exchange term are small relative to the cooling to space term then, to a first approximation, Q_{IR} depends on the temperature at the given level only. The solar absorption term, Q_S, depends on temperature through the absorption coefficients, but this dependence is weak. Thus, we can regard Q at a given level as a function of T at that level only and expand about T_e to obtain

$$\frac{Q}{\rho c_p} \sim \frac{T_e - T}{t_R}. \qquad (9.2.31)$$

The radiative relaxation time t_R is defined as

$$\frac{1}{t_R} = \frac{1}{\rho c_p} \frac{dQ_{IR}}{dT} \qquad (9.2.32)$$

and is the characteristic time required for a perturbation to the radiative equilibrium temperature profile to decay to 1/e of its initial value.

For large-scale motion with a time scale greater than a Martian day, only the term $w \partial T/\partial z$ contributes significantly to the advective derivative of the temperature, DT/Dt, on the left side of Eq. (9.2.24). With this approximation and Eq. (9.2.31), the thermodynamic energy equation reduces to the relatively simple relation

$$w\left(\frac{\partial T}{\partial z} + \frac{R}{c_p}\frac{T}{H}\right) = \frac{T_e - T}{t_R}. \qquad (9.2.33)$$

The temperature lapse rate is defined as the negative of the vertical temperature gradient, $\partial T/\partial z$. If the vertical temperature profile follows an adiabat, then $T \sim p^{R/c_p}$ and $\partial T/\partial z = -RT/c_p H$; therefore, $RT/c_p H$ is the adiabatic lapse rate in log-p

coordinates. The factor in parentheses on the left side of Eq. (9.2.33) is just the difference between the adiabatic and the actual lapse rate of the atmosphere. Examination of the observed temperature field indicates that the lapse rate is less than adiabatic so this factor is positive. For an upward moving parcel, the left side of Eq. (9.2.33) represents the rate at which the temperature would decrease due to adiabatic expansion, while for a downward moving parcel, this term represents the rate of increase in temperature due to adiabatic compression. These processes are sometimes called adiabatic cooling and heating. Equation (9.2.33) states that adiabatic heating or cooling due to vertical motion is balanced by the radiative relaxation of the temperature field. Thus, if we have sufficient information to calculate the radiative equilibrium temperature field and the radiative relaxation time, the temperature field retrieved from infrared measurements can be used to estimate the large-scale vertical velocity.

Calculation of T_e requires a detailed treatment of the absorption of sunlight by carbon dioxide and atmospheric dust, the absorption of sunlight and the emission of infrared radiation by the surface, and the transfer of infrared radiation in the atmosphere as discussed in Section 9.1. However, the basic characteristics of T_e can be estimated without resorting to a detailed calculation. In the absence of significant dust loading in the thin Martian atmosphere, the distribution of the diurnally integrated sunlight incident on the planetary surface essentially determines the latitude dependence of the diurnal mean radiative equilibrium temperature. For the equinox condition shown in Fig. 9.2.1, we would expect maximum radiative equilibrium temperatures at low latitudes with a monotonic decrease towards either pole. We find that the observed temperatures indeed follow this behavior at the lower atmospheric levels; however, at pressures less than $\sim$0.5 mbar, temperatures increase with latitude in both hemispheres from the equator up to high latitudes. We conclude that the thermal structure has been significantly perturbed away from the radiative equilibrium configuration by atmospheric circulation. If the observed temperature results from the zonal mean meridional circulation, then reference to Eq. (9.2.33) suggests that rising motion must occur at low latitudes resulting in adiabatic cooling at upper levels while descending motion at high latitudes results in adiabatic heating in both hemispheres. From considerations of mass continuity we infer that there must be a meridional flow aloft from low to high latitudes in both hemispheres with a return flow at low levels. With this relatively simple approach we have been able to qualitatively deduce the properties of the postulated meridional circulation. More sophisticated approaches have been taken using general circulation models to calculate the wind and temperature fields, and the latter are compared with the observations (Pollack *et al.*, 1981; Haberle *et al.*, 1982, 1993; Hourdin *et al.*, 1993; Wilson & Hamilton, 1996).

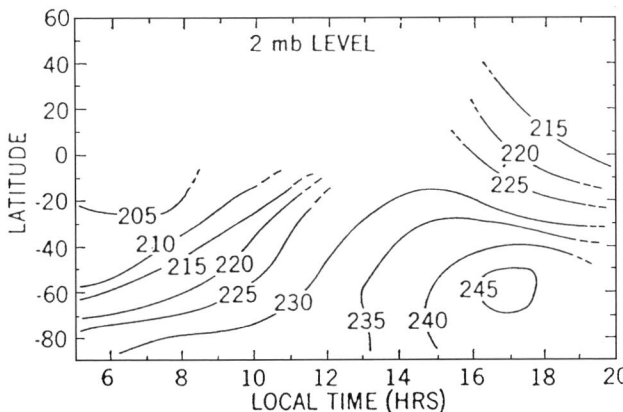

Fig. 9.2.3 Martian atmospheric temperature as a function of local time and latitude for the 2 mbar level. These results were obtained by inversion of spectral measurements acquired with the Michelson interferometer (IRIS) carried on Mariner 9. A strong global dust storm was in progress, resulting in the observed large diurnal amplitude (Hanel et al., 1972a).

We now turn to the second example of the study of a Martian dynamical phenomenon using remotely sensed infrared data. Thermally driven atmospheric tides form a significant component of the Martian meteorology, especially during global dust storms when a substantial amount of solar energy is deposited in the atmosphere due to absorption by the dust. At the time Mariner 9 was injected into orbit around Mars in 1971, a planet-wide dust storm was in progress, and the derived atmospheric thermal structure displayed a strong diurnal variation. Figure 9.2.3 shows temperatures in a layer centered at 2 mbar as a function of latitude and local time. An 'hour' in this case is defined as one twenty-fourth of a Martian solar day. A diurnal temperature fluctuation is observed at all latitudes, reaching a maximum of ~30 K at 60°S.

Because the time scale of the anticipated atmospheric motion associated with the temperature fluctuations is only one day, the thermal wind approximation cannot be used. However, diurnal variations in the pressure and wind fields can be estimated from the observed temperature field using classical tidal theory. The basic concept of the formulation is sketched here; a detailed treatment can be found in Chapman & Lindzen (1970). The theory is based on a linearization of the primative equations. A motionless atmospheric reference state is assumed with temperature profile $T_0(z)$ and a corresponding geopotential surface $\Phi_0(z)$. It is further assumed that the diabatic heating and all other quantities vary as $\exp[i(s\phi - \omega t)]$ where s is a longitudinal wavenumber and ω is $2\pi/$(solar day) or integer multiples thereof. The amplitudes of the time-varying, dependent variables are taken to be sufficiently small so that only terms of first order need be retained. With these assumptions,

Eqs. (9.2.20)–(9.2.24) yield

$$i\omega u - fv + \frac{is}{a\cos\theta}\Phi' = 0, \quad (9.2.34)$$

$$i\omega v + fu + \frac{1}{a}\frac{\partial\Phi'}{\partial\theta} = 0, \quad (9.2.35)$$

$$\frac{\partial\Phi'}{\partial z} - \frac{R}{H}T' = 0, \quad (9.2.36)$$

$$\frac{1}{a\cos\theta}\left[isu + \frac{\partial}{\partial\theta}(v\cos\theta)\right] + \frac{1}{\rho_0}\frac{\partial}{\partial z}(\rho_0 w) = 0, \quad (9.2.37)$$

$$i\omega T' + \left(\frac{dT_0}{dz} + \frac{R}{c_p}T_0\right)w = J. \quad (9.2.38)$$

The variables T' and Φ' are the spatially and temporally varying deviations of temperature and geopotential from $T_0(z)$ and $\Phi_0(z)$ respectively, and $J = Q/\rho c_p$ is called the thermotidal heating. If J is specified, along with boundary conditions, the tidal equations (9.2.34)–(9.2.38) can be solved for u, v, w, Φ', and T'. At the lower boundary, the component of the velocity field normal to the surface must vanish. As $z \to \infty$ solutions with an exponential behavior are required to remain finite; for vertically propagating solutions, the energy flux is required to be directed upward.

In one approach to the analysis of the Martian tidal regime, the thermotidal heating is assumed known and the tidal fields are calculated by solving the system of Eqs. (9.2.34)–(9.2.38). The resulting temperature field is compared with the measured values, and adjustments to J are made until agreement is obtained. Leovy et al. (1973) used this approach in a study of tides during the planet-wide dust storm of 1971. They assumed that the thermotidal heating, J, was confined to a layer between the planetary surface and an upper level, $z = z_c$, and was independent of height within that layer. An acceptable agreement between the calculated diurnal temperature amplitude and that obtained from the Mariner 9 IRIS measurements was obtained for a thermotidal heating having a layer thickness $z_c \geq 4$ scale heights and an absorption of approximately 20% of the available solar flux in the atmospheric column. The calculated and measured temperatures are compared in Fig. 9.2.4. Density weighted vertical averages of the diurnal temperature amplitudes are shown as functions of latitude.

In a second approach to the same atmospheric tidal problem, the measured diurnal variation of the temperature is incorporated directly into the tidal Eqs. (9.2.34)–(9.2.38), which are then solved for the remaining dependent variables. This method was applied to the Mariner 9 IRIS temperature retrievals by Pirraglia & Conrath (1974) who obtained the diurnal surface pressure variation shown in Fig. 9.2.5. The

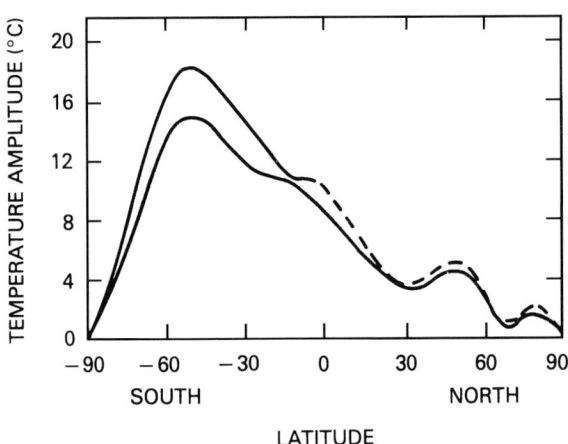

Fig. 9.2.4 Comparison of calculated (lower curve) and observed (upper curve) Martian diurnal temperature amplitudes. The calculated amplitude is based on a classical tidal model with a thermotidal heating function as described in the text. The observed amplitude was obtained from Mariner 9 IRIS measurements with an extrapolation to high northern latitudes denoted by the broken curve. In each case a density weighted, vertical mean amplitude is shown (after Leovy et al., 1973).

maximum amplitude is approximately 6% of the mean surface pressure, which can be compared with a typical terrestrial value less than 0.1%. Both methods yield a maximum diurnally varying surface wind amplitude of ~ 20 m s^{-1}. The large Martian tidal amplitudes are a consequence of the short radiative time constant combined with strong thermal forcing. It has been suggested that the augmentation of the near surface wind field by the tidal winds may be a contributing factor in the initiation of Martian global-scale dust storms. Studies of Martian atmospheric thermal tides have recently been made using Mars Global Surveyor TES results (Banfield et al., 2000).

c. The outer planets

The outer planets provide the opportunity for the study of significantly different meteorological regimes than those of Mars and Earth. Their atmospheres are composed primarily of hydrogen with smaller amounts of helium and traces of other species, including methane and ammonia, and extend to great depths before major phase boundaries are encountered. With the exception of Uranus, the giant planets possess significant internal heat sources, comparable in magnitude to the solar energy absorbed in their atmospheres. A major problem in the dynamics of these atmospheres is to understand how the two energy sources conspire to drive the atmospheric motions.

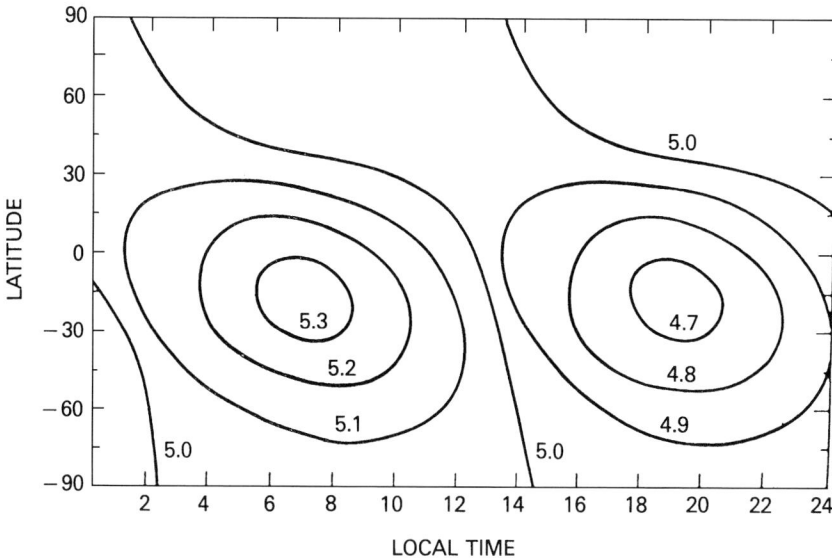

Fig. 9.2.5 Calculated Martian atmospheric surface pressure as a function of latitude and local time during global dust storm conditions. A tidal model with measured atmospheric temperature fields as input was used to obtain the results shown (Pirraglia & Conrath, 1974).

As on Mars, the principal contribution of infrared spectroscopy and radiometry to an understanding of the dynamics of the atmospheres of the outer planets has been the provision of information on the temperature fields. Ground-based measurements, data from the infrared radiometers on Pioneers 10 and 11, and measurements of infrared spectra from Voyager 1 and 2 have all provided significant information on atmospheric temperature structure. The largest set of spatially resolved data has been obtained by the Voyager infrared instruments. We will discuss analyses of those data to illustrate their usefulness.

Temperature retrievals from Voyager infrared spectra for Jupiter, Saturn, Uranus, and Neptune have previously been shown in Fig. 8.2.2. Qualitatively, all giant planets display a similar behavior in that a tropopause or temperature minimum occurs near the 100 mbar level and at deeper levels the profiles appear to approach adiabats. The latter result suggests that atmospheric motion rather than radiative transfer is the primary means of vertical heat transport in the deeper layers. When the atmosphere becomes sufficiently opaque in the infrared, steep vertical temperature gradients are needed to produce the required outward radiative heat flux. If the resulting temperature lapse rate exceeds the adiabatic lapse rate, an atmospheric parcel displaced upward adiabatically will find itself warmer than its surroundings and will therefore be buoyant. A downward moving parcel will be cooler than its surroundings and will continue to sink. Thus, an overturning convective motion

438   *Interpretation of results*

will be initiated, which will transport heat upward and reduce the temperature lapse rate. The resulting lapse rate will be just sufficiently larger than the adiabatic lapse rate to produce the required heat flux. For the heat fluxes in the outer planets, the actual lapse rates are expected to be indistinguishable from the adiabatic values because of the efficiency of convection in transporting heat. In practice, the convective processes can be considerably more complicated than the simple model presented here because of latent heat release and molecular weight stratification due to condensation processes, and because of the possible disequilibrium of the ortho- and para-modifications of molecular hydrogen, which can modify the thermodynamic behavior of the atmosphere.

Sufficiently large quantities of spatially resolved data have been obtained for all four giant planets to permit the meridional, upper tropospheric thermal structure of each to be reasonably well defined. The latitudinal dependence of temperature at selected pressure levels are shown in Fig. 9.2.6 for Jupiter. These results have

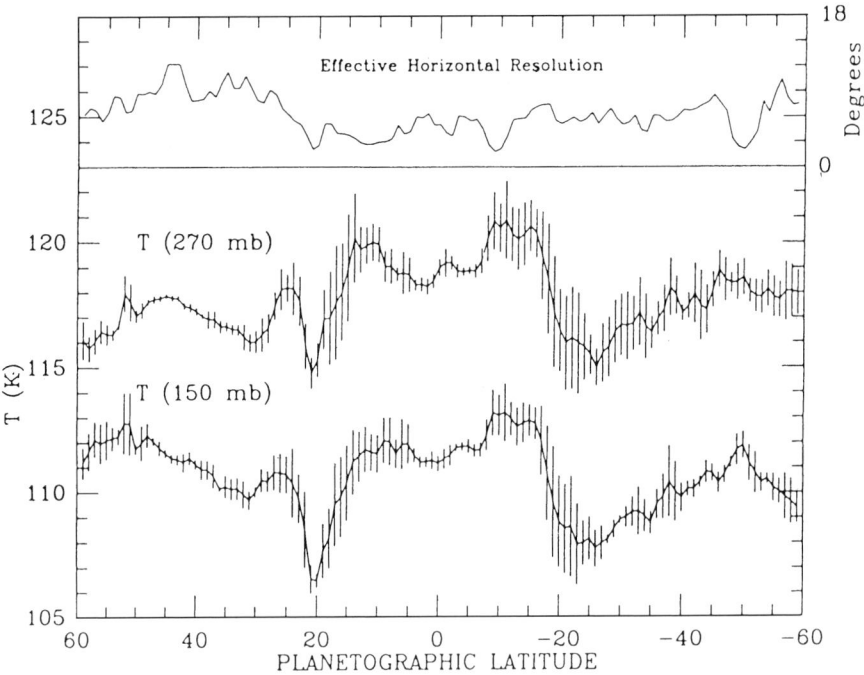

Fig. 9.2.6 Zonal mean temperature as a function of latitude at the 150 and 270 mbar levels in the atmosphere of Jupiter. The temperatures were obtained by inversion of spectral data from the Michelson interferometer (IRIS) carried on Voyager 1. The vertical bars indicate the standard deviation which contains contributions from both instrument noise and actual zonal structure. The effective horizontal resolution in degrees of great circle arc on the planet is shown in the upper panel (Gierasch *et al.*, 1986).

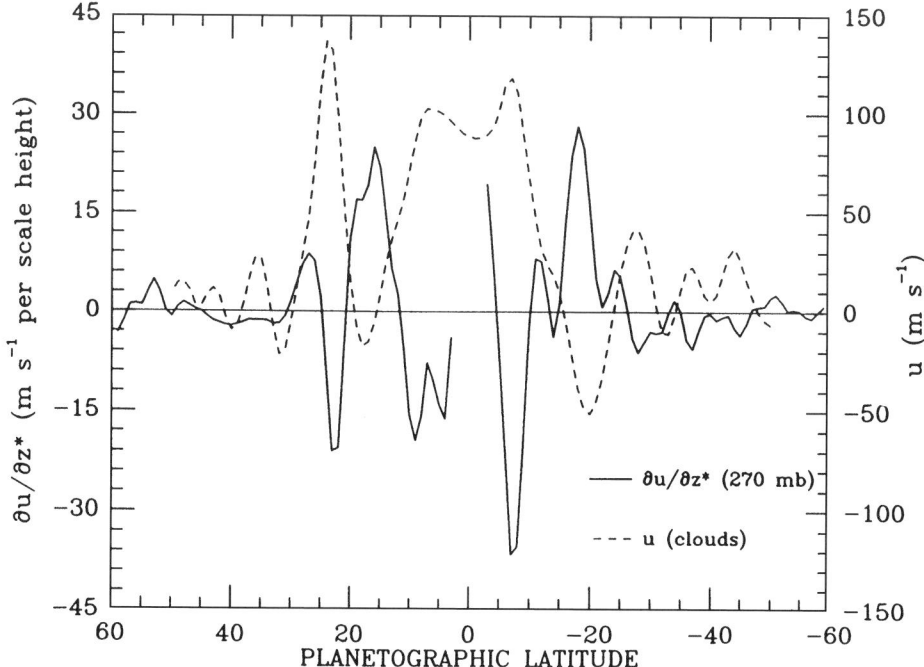

Fig. 9.2.7 Thermal wind shear calculated from the 270 mbar temperatures in Fig. 9.2.6 using Eq. (9.2.29). For comparison, cloud tracer wind velocities from Limaye (1986) are shown as the broken curve (Gierasch et al., 1986).

been used in a number of diagnostic studies of upper tropospheric and stratospheric dynamics. We consider two examples of these analyses.

The first example is a study of the behavior of the zonal wind. By observing the motions of cloud features in images obtained with the Voyager television cameras, zonal wind speed has been derived as a function of latitude. The level of the visible clouds to which the wind speeds pertain is not known with certainty; however, it is believed to be in the 500–700 mbar region. A series of alternating eastward and westward jets is observed. Estimates of the Rossby number are less than unity, which suggests that the zonal thermal wind relation, Eq. (9.2.29), should be valid. Application of this relation to the thermal structure for Jupiter, shown in Fig. 9.2.6, yields the thermal wind shear as a function of latitude, shown in Fig. 9.2.7. Also displayed is the wind speed, obtained from the cloud motions as a function of latitude. Comparison of the wind speed and the vertical shear indicates a tendency toward anticorrelation, which implies that the wind speed decreases with height. Similar results are obtained for Saturn, indicating that the jet systems decay with height in the upper troposphere. In the cases of Uranus and Neptune, comparisons of the cloud-top winds with the thermal wind shear calculated from the temperature data again suggests the zonal winds are decaying with height.

As another example, we consider an approach that extends the analysis of temperature and winds somewhat beyond the use of thermal wind balance alone. A simplified flow model, which describes the balances between averages over longitude of the temperature and the three wind components, will be employed. The zonal thermal wind balance is retained, but the decay of the zonal jets with height suggests that some damping mechanism must be operating, and this mechanism should be incorporated into the model. This is accomplished by using a linearized, time-independent form of the zonal momentum equation (9.2.10), and replacing the frictional term, M_ϕ, by a linear drag term, called Rayleigh friction. The resulting expression is

$$fv = \frac{u}{t_F}, \qquad (9.2.39)$$

where t_F is a characteristic frictional damping time. The drag on the zonal wind, u, is balanced by the Coriolis force acting on the meridional wind, v. It is further assumed that diabatic heating is balanced by adiabatic heating and cooling due to vertical motion. The model is now completely described by Eqs. (9.2.27), (9.2.33), (9.2.39), and an average over longitude of the continuity equation, (9.2.23). If T_e, t_R, and t_F are specified along with appropriate boundary conditions, the equations can be solved for the temperature and wind velocity components. This approach has been applied to an interpretation of the temperature and wind measurements of Jupiter, Saturn, and Uranus (Flasar *et al.*, 1987; Conrath *et al.*, 1990). The radiative equilibrium temperature, T_e, and the radiative damping time, t_R, were calculated using a radiative transfer model as described in detail by Conrath *et al.* Since the frictional damping time, t_F, is not known *a priori*, a strategy was adopted in which the observed wind field was specified as a lower boundary condition, and t_F was varied until the calculated temperature was approximately in agreement with the observed temperature field. The procedure tacitly assumes that the primary source of forcing for the jets lies below the region of observation; no attempt is made to specify the nature of this forcing. Results of this approach applied to Jupiter are shown in Figs. 9.2.8 and 9.2.9 where it was assumed that $t_F = t_R$. The calculated meriodional cross section of temperature, shown in Fig. 9.2.8, is consistent with the measured temperatures in the upper troposphere. The calculated meridional circulation, shown in Fig. 9.2.9, is expressed in terms of the meridional stream function. The stream function is defined such that the meridional flow is parallel to the contour lines shown in the figure; arrows indicate the direction of the flow. The meridional circulation consists of multiple cells associated with the zonal jet

9.2 Atmospheric motion

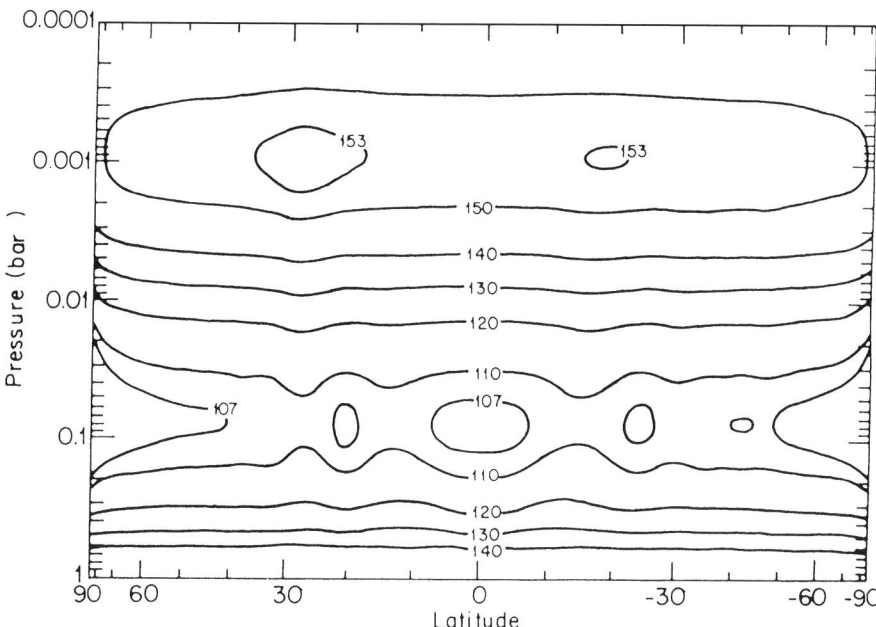

Fig. 9.2.8 Zonal mean meridional temperature cross section of the atmosphere of Jupiter obtained from a linearized atmospheric circulation model. Cloud tracer wind velocities were used as a lower boundary condition (Conrath *et al.*, 1990).

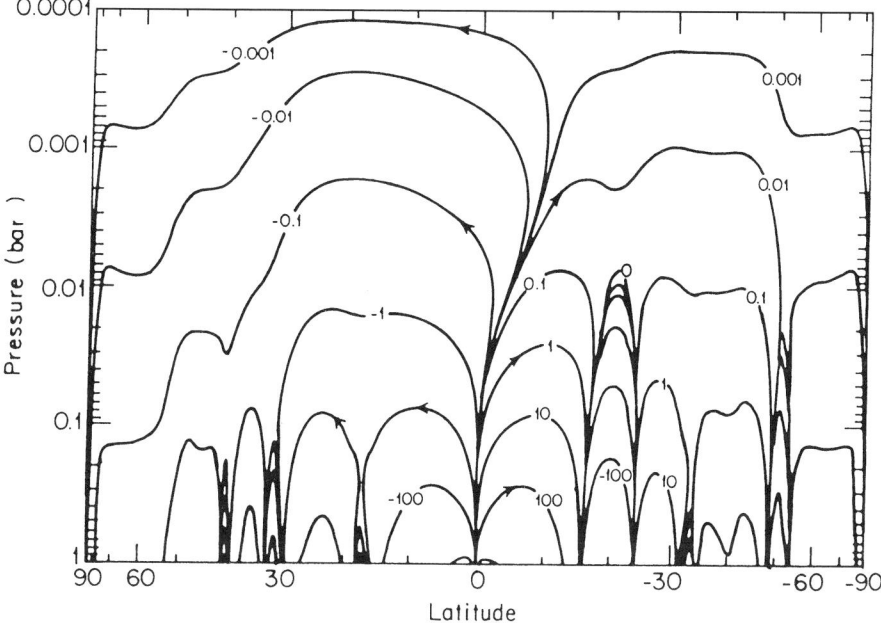

Fig. 9.2.9 Mean meridional circulation of the atmosphere of Jupiter obtained from a linearized atmospheric circulation model. The circulation is expressed in terms of a stream function (g cm^{-1} s^{-1}), and arrows indicate the direction of flow (Conrath *et al.*, 1990).

structure such that rising motion occurs on the equatorward side of eastward jets with sinking motion on the poleward side. For all four planets, a best fit to the measured temperatures is found when the frictional time constant is of the same order of magnitude as the radiative damping time even though t_R varies almost two orders of magnitude from Jupiter and Saturn to Uranus and Neptune. The source of this frictional damping is not yet understood, but may be associated with vertical wave propagation within the atmospheres.

d. Venus

Venus provides an example of a slowly rotating body with a deep atmosphere. The surface pressure is approximately 95 bars and the rotational period in inertial space is 243 days. Infrared remote sensing has been applied to the study of the dynamics of the atmosphere of Venus, and an example of such an application is discussed.

The planetary disk shows distinct ultraviolet markings, which have been observed both from the Earth and from spacecraft. Wind velocities as high as 100 m s^{-1} in a direction the same as the sense of rotation of the planet have been deduced from the motions of these features. This motion has been called the 'four-day atmospheric rotation' corresponding to the approximate time required for a feature to traverse 360° in longitude. In addition to ultraviolet imaging, infrared measurements have also contributed to our understanding of the dynamics of the atmosphere of Venus. As an example we consider an analysis of measurements obtained with the infrared radiometer carried on the Pioneer Venus Orbiter (Taylor *et al.*, 1979*a*). Like Mars, the atmosphere of Venus is composed primarily of carbon dioxide so that measurement within the 15 μm absorption band of that gas can be used for temperature sounding. The radiometer obtained measurements in this spectral region using both a pressure modulator and a grating. Figure 9.2.10 shows zonally averaged brightness temperatures from three channels as functions of latitude. In this case the brightness temperatures have been taken as representative of the atmospheric temperatures at the altitudes associated with the peaks of the weighting functions.

We now relate the measured temperatures to the mean zonal wind. In this case it is not possible to use the geostrophic thermal wind equation (9.2.29) as an examination of Eq. (9.2.21) shows. The ratio of the second term to the first term in the brackets on the left side of the equation is of the order of the ratio of the 243-day planetary rotation period to the four-day atmospheric rotation period or ∼60; hence, the second term dominates. This suggests a first approximation:

$$u^2 \frac{\tan\theta}{a} = -\frac{1}{a}\frac{\partial \Phi}{\partial \theta}. \qquad (9.2.40)$$

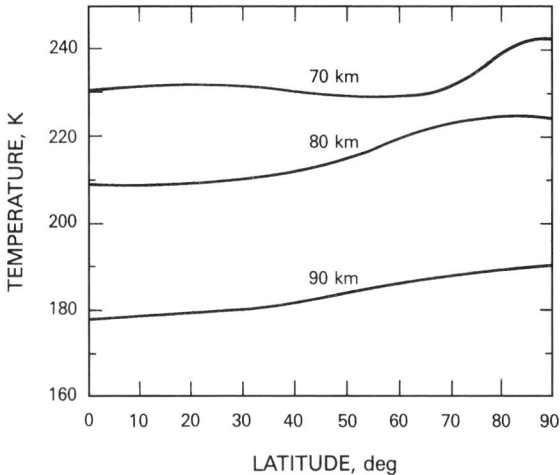

Fig. 9.2.10 Venus atmospheric temperatures. The results shown are brightness temperatures obtained with the infrared radiometer carried on the Pioneer Venus Orbiter. They are representative of temperatures at the indicated atmospheric levels corresponding to weighting function peaks (Elson, 1979).

The relevance of this relation to Venus was first recognized by Leovy (1973); Eq. (9.2.40) is called cyclostrophic balance. It states that the horizontal component of the centrifugal force produced by the zonal atmospheric flow is balanced by the meridional component of the geopotential gradient. Unlike the case of geostrophic balance, the sign of the zonal flow cannot be determined from the geopotential gradient, and the gradient can only be negative. If Eq. (9.2.40) is combined with the hydrostatic relation, Eq. (9.2.22), we obtain

$$u^2 = u^2(z_0) - \text{ctn}\,\theta \frac{R}{H} \int_0^z \frac{\partial T}{\partial \theta} dz. \tag{9.2.41}$$

The lower boundary condition, $u(z_0)$, is the zonal wind speed at level z_0. Elson (1979) has applied this diagnostic relation to the temperatures shown in Fig. 9.2.10. The lower boundary was taken to be at the cloud top height, $z_0 = 65$ km, and the wind speed, $u(z_0)$, was based on observed cloud motion. The resulting cyclostrophic component of the zonal wind in the northern hemisphere is shown in Fig. 9.2.11 as a function of latitude and z. At levels above the zero contour line pure cyclostrophic balance cannot hold since $\partial \Phi / \partial \theta$ becomes positive. Other terms in Eq. (9.2.21) then become important in this part of the atmosphere. This result has led to the application of more complex diagnostic models by Taylor et al. (1980).

Another example of an atmosphere in cyclostrophic balance is that of Titan. This case has been discussed in detail by Flasar et al. (1981), Hunten et al. (1984), and Flasar (1998) and is not pursued further.

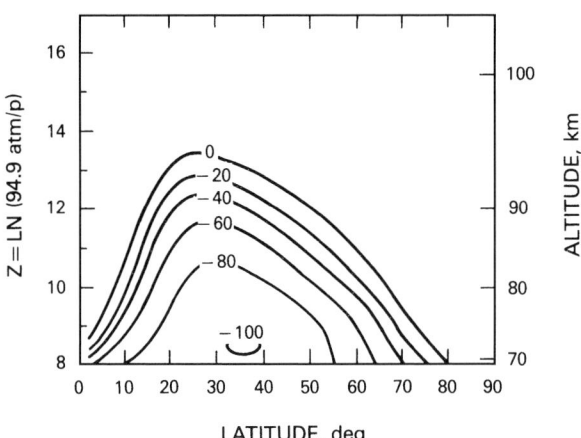

Fig. 9.2.11 Cyclostrophic winds calculated from the temperatures in Fig. 9.2.10 using Eq. (9.2.41). Wind speed based on observed cloud top motion was used as the lower boundary condition (Elson, 1979).

9.3 Evolution and composition of the Solar System

The present composition of planetary atmospheres and interiors is of great interest not only in its own right, but also for studies of the history of the Solar System. All evolution theories must start from an interstellar cloud of matter and end at the present conditions. Precise measurements of the composition of planets, their satellites, asteroids, comets, and meteorites are, therefore, of vital importance for any understanding of the evolution of the Solar System. Many remote sensing investigations have been carried out to gain composition information on planetary objects. Wide segments of the electromagnetic spectrum have been used for that purpose from the extreme ultraviolet to radio frequencies. On the Moon even gamma rays have been employed to study the spatial distribution of several radioactive elements (Adler *et al.*, 1973). As we have seen in Chapters 6 and 8, the infrared part of the spectrum is particularly well suited to discover the presence and determine the abundances of molecules, including important isotopic species. On Venus and Mars additional atmospheric and surface information has been collected *in situ* with mass spectrometers and gas chromatographs. Results from the Galileo probe have already been mentioned. Further *in situ* measurements are planned for the probe of Project Cassini into the atmosphere of Titan in 2004.

Conditions in the Earth's atmosphere and interior are, of course, much better known than conditions on the other planets. At this time, the forefront of composition studies in the Earth's atmosphere concerns the concentrations of man-made and natural pollutants. In particular, photochemical reactions as well as dynamic effects in the stratosphere are the subject of intense investigations. The steady increase of greenhouse gases and disturbances in the ozone balance have potentially an enormous impact on weather patterns and climatic conditions. As important and interesting as this subject is, however, we do not pursue it here. Instead, we direct our attention to the basic composition of the atmospheres and interiors of the planets and what can be deduced from that about the history of the Solar System.

At first thought, the great diversity in atmospheric and surface compositions found among the planets and their satellites is a surprise, considering that all Solar System objects must have formed from the same primordial cloud. An explanation of this diversity must invoke either inhomogeneities in the composition of the circumsolar disk as a function of heliocentric distance, or peculiarities in the individual planetary formation processes, or dissimilarities in the subsequent developments of the planets. Such developments could have altered atmospheric composition in many ways – for example, by escape of light gases to space, by geochemical reactions, or by radial redistributions of certain compounds. As we shall see, all of these processes have played roles in shaping present conditions. The events that lead from the accretion disk to the proto planets and, finally, to the present conditions are very complex and often not well understood in detail, although a number of plausible models have been suggested.

Part a of this section summarizes present theories of Solar System formation. Parts b and c provide brief discussions of evolutionary processes of the terrestrial and the outer planets, respectively. In these discussions we give a few examples where composition measurements have contributed to the understanding of planetary evolution.

a. Formation of the Solar System

It is generally accepted that the Solar System formed about 4.6×10^9 years ago from an interstellar cloud of gases and solid matter, including remnants from a supernova. Material of atomic masses heavier than that of iron, now present in the Solar System, could have formed only in a nova explosion. Our Sun is a second or possibly third generation star. Hydrogen and a large number of organic and inorganic molecules, neutral and ionized, as well as radicals may have been present in the cloud. Similar sets of constituents have been detected in presently existing interstellar clouds (Irvine *et al.*, 1987). Hydrogen is by far the dominant constituent. Much of the helium was formed in the 'big bang'. Helium is presently produced in hydrogen-burning stars. In interstellar clouds solid matter is believed to exist mostly in the form of small grains of refractory materials, such as amorphous carbon, silicates, and many other elements and molecules, including metals, their oxides, and sulfur compounds. Ices and condensed organic matter may have formed coatings around these grains. A small imbalance of the momentum of the individual clumpy segments of the cloud may have given rise to stronger and stronger rotation as parts of the cloud gravitationally contracted. The inner part of the collapsing segment formed a central body simultaneously with a surrounding disk flattened by rotation. The effects of turbulence or of magnetic fields on the ions and indirectly on the neutral matter in the cloud are not well understood. Nor is the mass distribution between central body and circumsolar disk known.

Originally, the cloud may have had a temperature of 10 K. As long as the cloud was optically thin in the far infrared (the intensity distribution of a 10 K blackbody peaks at 20 cm^{-1}), the slow contraction occurred isothermally, but as the density increased so did the opacity and the process became more and more adiabatic. In the center of the newly formed proto Sun temperature and density increased rapidly with the continued influx of matter. Eventually, conditions in the center reached several million degrees and nuclear fusion started; at that point a new star was born.

The dominant nuclear reaction inside the Sun transforms protons into helium nuclei, temporarily generating ^{2}H and ^{3}He in that process. Byproducts of this reaction are positrons, neutrinos, and gamma radiation. The gamma rays are absorbed locally; their energy is converted to heat. The neutrinos escape from the Sun altogether. The positrons annihilate electrons and generate more heat in the process. The energy production by what amounts to a burning of hydrogen into helium was first recognized by Bethe & Critchfield (1938), Weizsäcker (1938), and Bethe (1939). In addition to the dominant process a second set of reactions uses ^{12}C as a catalyst and forms and destroys ^{13}N, ^{13}C, ^{14}N, ^{15}O, and ^{15}N at intermediate steps in the formation of ^{4}He from protons – again, with a net production of energy. This second reaction, sometimes referred to as the Bethe or carbon cycle, contributes about 7% to the present solar luminosity (e.g., Strömgren, 1953).

Numerical models have been relatively successful in describing the physical conditions in the solar interior. The models generally assume that the Sun consists of mass fractions of hydrogen, X, helium, Y, and heavier elements, Z. The fraction Z lumps together all elements from lithium to uranium; Z is often taken to be ~0.02 in the models. Since $X + Y + Z = 1$ only one free parameter is left to characterize solar bulk composition. The models also assume spherical symmetry and hydrostatic equilibrium. Except for a small zone near the visible surface, temperatures inside the Sun are so high that atoms are fully ionized. The equation of state, the radiative opacity, and the energy production rates, all needed in the models as functions of temperature and density, can then be found from stellar interior and nuclear theories. Of course, matters are more complicated; readers interested in stellar models may wish to consult books on that subject, for example, the one by Clayton (1968).

Taking the known mass and radius as boundary conditions permits numerical integration of the appropriate equations from the surface towards the center. Only models that yield the measured solar luminosity are accepted. At the center of the Sun, all of these models yield a temperature of approximately 15×10^6 K and a density of about 1.6×10^5 kg m^{-3}, roughly 160 times that of water. A blackbody of 15×10^6 K has an intensity maximum near 3×10^7 cm^{-1}, that is, at approximately 0.3 nm (3 ångströms). Most recent models require a helium mass fraction, Y, between 0.27 and 0.28 (Van den Berg, 1983; Noels *et al.*, 1984; Lebreton & Maeder, 1986; Cahen, 1986); this is important for the discussions of the helium to hydrogen ratios in the outer planets. The models also show the nuclear reactions in the Sun to be confined to a relatively small, probably convective core. This core is surrounded by a large zone in radiative equilibrium occupying a substantial fraction of the solar volume. Energy transport in the radiative regime is by scattering, absorption, and reemission of electromagnetic radiation. A zone in convective equilibrium, perhaps 0.2×10^5, possibly up to 2×10^5 km thick, surrounds the volume in radiative equilibrium; for comparison, the solar radius is 7×10^5 km. The outer boundary of the convective zone is near the visible surface. The opacity in the convective shell is so high that radiation alone cannot carry the energy flux and convective processes set in. The granular appearance of the solar disk at certain wavelengths is a manifestation of convective circulation cells.

While stellar models have been relatively successful in describing conditions in the Sun (with the possible exception of predicting the neutrino flux), it has been more difficult to model the formation of the circumsolar accretion disk and the subsequent events that, eventually, must have led to the formation of the planets. Matter forming the disk must also have experienced some degree of adiabatic heating in the collapse from the interstellar cloud as well as heating by radiation from the simultaneously forming proto Sun. Large temperature gradients must have existed in the accretion disk. Near the Sun, even the most refractory materials, such as silicates, must have evaporated, while further out temperatures were low enough to maintain water in frozen form (Lewis, 1972, 1974; Prinn & Fegley, 1989). Still further out, ammonia and then even methane could have existed in the form of ices. The main noncondensable components of the disk were molecular hydrogen, helium, and neon. Although the overall composition of the accretion disk must have been very similar to that of the Sun, large gradients in the concentration of condensible matter must have existed. On a small scale, strong inhomogeneities were probably present, as is evident from the great inhomogeneities found in condritic meteorites. The solar composition has recently been studied by Cameron (1982).

Models in which larger solid particles gravitated towards the equatorial disk plane and formed a zone of high concentration of solid matter (Podolak & Cameron,

1974; Mizuno, 1980; Weidenschilling, 1980; Lin, 1981; Cabot *et al.*, 1987) are now favored over models in which a more massive disk with a more homogeneous distribution of gas and solids started to collapse on local volumes of higher density (Cameron, 1978, 1985). In the former type of models small dust particles coagulate and first form clumps of micrometer and millimeter size particles, which soon agglomerate to objects of a few centimeters in size or even larger. The larger objects settle in the midplane of the disk while the smaller dust particles remain with the more uniformly distributed gas. At the midplane clusters of many centimeter size objects orbit the proto Sun at nearly the same speed. According to Goldreich & Ward (1973) these clusters tend to combine and form planetesimals of 100 m radii within a relatively short time of a few years. On a much longer time scale, groups of these planetesimals drift together under the influence of gravity and gas drag and form larger, second generation planetesimals with radii of the order of 5 km. The Goldreich–Ward type theories of accretion from micron to kilometer size objects are not without difficulties. The disruptive role of gas turbulence at the early stage of aggregation, where gravitational forces are minuscule, is just one of these difficulties. After the formation of kilometer size planetesimals the processes are somewhat better understood. Gravitational forces dominate, and accretion can be visualized in terms of the net growth of a few larger objects at the expense of many smaller planetesimals by numerous inelastic collisions.

In this framework the terrestrial planets formed by accumulating such kilometer size planetesimals into larger and larger proto planets. With the increase in size of the proto planets the impacts became more energetic, leading to shock heating and devolatilization of infalling objects. That heat, and energy released by radioactivity and other processes, caused partial melting as well as geologic differentiation of the newly formed proto planets. Mercury, Venus, Earth, and Mars developed iron-rich cores surrounded by lighter, silicate-rich mantles. At that time the Moon may have formed by a very large planetesimal, perhaps almost as big as Mars, impacting the proto Earth, and separating a large portion of the Earth's still partially liquid mantle (Hartmann & Davis, 1975; Cameron & Ward, 1976; Ringwood, 1979; Wänke & Dreibus, 1986; Cameron & Benz, 1991; Halliday & Drake, 1999). The high number of lunar impact craters and the existence of mare indicate that the period of planetesimal impact and wide-spread melting continued after the formation of the Moon.

Heating by impact also liberated large quantities of gases trapped in the mantles of the proto planets and in the arriving projectiles including comets. Water vapor, carbon dioxide, methane, and ammonia, but also smaller quantities of noble and other gases formed primitive atmospheres. However, the terrestrial planets never grew big enough to gravitationally attract large amounts of hydrogen or helium, the lighter gases of the accretion disk.

In contrast to this, the proto planet cores in the outer Solar System became massive enough to attract large quantities of gaseous matter contained in the disk, that is, hydrogen, helium, and smaller solid particles entrained in the gases. The cores consisted mainly of iron, silicates, and ices of water, ammonia, and methane. The ices, some in the form of clathrates, evaporated in the accumulation process and formed primitive atmospheres, which mixed with the gradually increasing lighter gases of the accretion disk. At that time the outer planets may have been substantially larger than they are today. Mimicking the processes in the Solar System as a whole, the outer planets formed rotating accretion disks of their own, in this way giving rise to their regular satellite systems.

At that time Uranus may have collided with a large planetesimal, causing a major tilt of its spin axis and ejection of material from the planet (Cameron, 1975). Some of that material formed the regular satellites of Uranus (see also Pollack *et al.*, 1991).

Spectral observations of Hα lines in T-Tauri stars show Doppler shifted components on the red and blue sides of the line, suggesting strong outflow of material in opposite directions, presumably along the polar axes (Herbig, 1962). A similar conclusion can be reached from Doppler measurements on CO lines of many young, solar-size stars. Apparently, such mass flows are common episodes of many newly formed, low-mass stars just entering the main sequence of stellar evolution. More recent summary articles on Solar System evolution and T-Tauri wind are by Boss *et al.* (1989), Basri (1992), and by Lissauer *et al.* (1995). It must be assumed that our Sun has also gone through one or several such phases. As a consequence of this so called T-Tauri wind, a newly formed star may shed as much as 20% of its mass in a relatively short time of perhaps 10^5 years. Most of that outflow is in the polar directions, possibly in well-formed jets, but the effect would also be quite noticeable in the equatorial plane, that is, where a circumsolar disk would be. The origin and the detailed mechanism of the T-Tauri wind are not well understood; however, there is little doubt that it exists. The effect of the wind may have been a clearing of the disk of gas and grains that had not already been included in the planetary bodies. With the clearing of interplanetary space the planets were exposed directly to the very strong ultraviolet radiation from the Sun, ionizing atmospheric constituents and greatly enhancing atmospheric escape processes. Large parts of the primordial atmospheres were lost from the terrestrial planets. The outer planets also may have lost some fraction of their original atmospheres, but their larger masses and greater distances from the Sun may have provided greater protection against losses. After the T-Tauri phase the Sun became stable again and the planets continued their evolutionary processes. Recent summaries of the planetary formation can be found in a workshop report on the origins of Solar Systems (Nuth & Sylvester, 1988), and in the book by Atreya *et al.* (1989).

b. Evolution of the terrestrial planets

After the Sun had settled as a typical main-sequence star, the solar constant was only about 80% of its present value, according to stellar evolution models. However, the same models suggest that the ultraviolet and X-ray emission of the young Sun was much above present levels. At that time the terrestrial planets had already differentiated, forming iron-rich cores and silicate-rich mantles. Some of the primitive atmospheres, as well as remnants of the accretion disk, had been ejected from the Solar System by the T-Tauri wind permitting the planetary surfaces to cool by radiation directly to space. With progressive cooling the solidified crusts increased in thickness. Volcanic activity must have been widespread with simultaneous emission of gases. If presently exhaled volcanic gases can serve as a guide, the emissions contained H_2O, CO_2, SO_2, N_2, H_2, and probably CH_4 and NH_3 as well. Influx of large meteorites continued providing another source of gaseous matter. More recently, Delsemme (1998) has pointed out that the temperature in the accretion disk at a distance of 1 AU was so high ($\sim$1000 K) that the planetesimals forming the Earth must have been completely degassed. Delsemme suggests that all the oceans and the present atmosphere of Earth are of cometary origin. Either by outgassing or by cometary influx, or by both, secondary atmospheres started to form.

Mercury, however, was too small to permit accumulation of such an atmosphere. Exhaled gases escaped rapidly. Except for minute quantities of hydrogen and helium (possibly supplied by the present solar wind), and sodium and potassium atoms (possibly supplied by the surface), Mercury is without a gaseous envelope; the same is true for the Moon. Venus, on the other hand, must soon have developed a relatively thick atmosphere with increasing quantities of H_2O and CO_2 as the major constituents, and N_2, SO_2, and other gases as minor constituents. It is also possible that the planetesimals forming Venus contained less water of hydration than those forming Earth. Therefore, Venus may have started out with less water than Earth in the degassing process. The relatively close proximity of Venus to the Sun and an increasingly strong greenhouse effect by the dominant atmospheric constituents may have prevented the solid surface from cooling below the condensation temperature of water. This sequence of events is the core of the runaway greenhouse model (Hoyle, 1955; Sagan, 1960; Gold, 1964; Dayhoff *et al.*, 1967; Ingersoll, 1969; Rasool & DeBergh, 1970; Pollack, 1971; Goody & Walker, 1972; Walker, 1975; Watson *et al.*, 1984). In contrast to this, the so-called moist greenhouse model also accounts for convective heat transport in the atmosphere, as well as for condensation and cloud formation (Kasting *et al.* 1984), and eventually for changes of the planetary albedo (Vardavas & Carver, 1985; Kasting, 1988; Kasting & Toon, 1989). In this model, oceans did originally form on Venus, but have evaporated long ago. In either greenhouse model, water vapor was dissociated in the upper atmosphere by strong ultraviolet radiation. Hydrogen is light enough to have escaped readily. The high deuterium/hydrogen (D/H) ratio found on Venus by Pioneer supports the water escape theory (Donahue *et al.*, 1982). The enrichment of D/H by a factor of 100 over the terrestrial value can be explained by the higher escape rate of H in comparison with that of D. The escape of gases from planetary atmospheres is a complex phenomenon, involving several mechanisms. A discussion of such mechanisms can be found in Chamberlain & Hunten (1987) and Hunten *et al.* (1989). Ionized oxygen can also escape readily; some of the remaining oxygen probably was removed from the atmosphere by oxidizing surface minerals. With the destruction of water vapor, the Venus atmosphere was left with large quantities of CO_2 (mole fraction of 0.965), followed in concentration by N_2 (0.035), and a number of trace gases (see Appendix 3).

The Earth evolved differently. Being farther away from the Sun, and possibly also as a consequence of a slower outgassing rate, the surface of the Earth had cooled sufficiently to permit water vapor condensation before the accumulation of a deep H_2O–CO_2 atmosphere. Otherwise, the greenhouse effect would have prevented cooling of the surface below the condensation temperature of water, and the Earth would have followed the evolutionary path of the runaway greenhouse model proposed for Venus. Fortunately, sufficient quantities of water did condense on Earth and did form oceans. Some of the water may have been of cometary origin. Equally important, carbon dioxide dissolves in liquid water and leads, via the Urey reaction, to the formation of sedimentary carbonate rocks. Today, large quantities of the carbon, originally outgassed in the form of CO_2, exist in rock formations on past and present ocean floors as well as in living and buried organic matter.

Nitrogen accumulated in the atmosphere at a relatively slow rate. Because there is no effective sink for N_2, however, its concentration has increased over geologic times enabling N_2 to become the dominant atmospheric constituent. Most of the present atmospheric oxygen is of biological origin. Plants convert CO_2 and H_2O by photosynthesis into hydrocarbon products and O_2. The early atmosphere of the Earth contained much smaller quantities of oxygen, produced mainly by photochemical dissociation of water vapor and subsequent escape of hydrogen. Much of the free oxygen produced in this way was used in various weathering processes, such as oxidizing exposed ferrous iron (FeO) into the ferric form (Fe_2O_3). The third most abundant constituent of the Earth's atmosphere, argon (Ar) 40, forms as the end product in the radioactive decay of potassium. The nonradiogenic primordial isotopes are argon 36 and the less abundant argon 38.

The evolution of the atmosphere of Earth was studied by computer modeling by Hart (1978). He included many known processes in his computer model, and required solutions that reproduce present conditions after evolving for 4.6×10^9

years. Although the solutions obtained may not be unique and some of his assumptions and procedures have been criticized, there is little doubt that the atmosphere–ocean system has changed over geologic time in substantial ways. A more detailed account of our present understanding of the evolution of the terrestrial atmospheres can be found in Walker (1977); Pollack & Yung (1980); Walker (1982); Lewis & Prinn (1984); and Chamberlain & Hunten (1987).

Mars must also have outgassed moderate quantities of H_2O, CO_2, N_2, and other gases, although its smaller size must have caused less devolitalization of infalling planetesimals. Furthermore, escape of light gases must have been more efficient there than on the larger planets. It is not known how much water is presently hidden below the polar caps, in permafrost layers, and in the regolith in the form of water of hydration. Today Mars has a relatively thin atmosphere (~ 5 mbar) of mostly CO_2 (mole fraction of 0.95), N_2 (0.03), Ar (0.016), and traces of CO. Water vapor is now a minor atmospheric constituent that occasionally forms thin ice clouds and frost on the surface. Liquid water is not stable on the surface and would evaporate rapidly. However, fluvious channels photographed by Mariner 9 and Viking suggest the presence of rivers in the geological past. Possibly only episodically, but perhaps over longer periods of time, Mars must have supported a more extensive atmosphere, much richer in H_2O and probably also in CO_2. This, in turn, would have led to a stronger greenhouse effect and higher surface temperatures. Could the conditions that led to the formation of life on Earth also have existed on Mars at that time? Was a biological evolution on Mars interrupted with the disappearance of liquid water? These are intriguing questions that have not yet been resolved.

c. Evolution of the giant planets

As already mentioned, according to presently favored theories the giant planets first started to form cores of high-Z material. In this context high-Z materials are matter composed of elements with atomic number (Z) of three and higher. These planetary cores contained elements in proportions similar to that of the planetesimals and comets responsible for the formation of the terrestrial planets, with the possible exception of a larger fraction of condensables than their counterparts of the inner Solar System. Water ice, clathrates containing ammonia and methane, and other matter of high vapor pressure must have evaporated as a consequence of shock and adiabatic heating resulting from the impact of planetesimals onto the cores of the proto planets. Subsequently, primitive atmospheres formed surrounding solid cores of iron, nickel, silicates, and other refractory compounds of low vapor pressure. In these primitive atmospheres carbon monoxide, nitrogen, and the heavier noble gases also may have increased in abundance as the proto planets grew steadily by the infall of planetesimals and comets.

Table 9.3.1 *Elemental ratios, in solar units, observed in the outer atmospheres of the giant planets**

	Sun	Jupiter / Sun	Saturn / Sun	Uranus / Sun	Neptune / Sun
He/H	0.097	0.807 ± 0.02	0.56 ± 0.82^a	0.92 ± 0.2	0.1 ± 0.2^b
C/H	3.62×10^{-4}	2.9 ± 0.5	~ 6	$\sim 24^c$	$\sim 35^d$
				$\sim 25^d$	$\sim 50 \pm 25^e$
				$\sim 20 \pm 10^e$	
N/H	1.12×10^{-4}	1–1.3 (Global)	$2–4^{f,g}$	$\ll 1?^g$	$\gtrsim 30^{b,h}$
		3.6 ± 0.5 (Hotspot)			
O/H	8.51×10^{-4}	0.35 (Hotspot, 19 bars)			
P/H	3.73×10^{-7}	0.82		4–9	
S/H	1.62×10^{-5}	2.5 ± 0.15			

* References: a: Conrath & Gautier (2000); b: Conrath *et al.* (1993) (assumes N_2 mol fraction = 0.003); c: Lindal *et al.* (1987); d: Lutz *et al.* (1976); e: Pollack *et al.* (1986); f: Marten *et al.* (1980); g: de Pater & Massie (1985); h: Gautier *et al.* (1995). All other values for the Sun, Jupiter, and Saturn adopted from Atreya *et al.* (1999).

As the mass of the proto planets increased steadily, larger and larger amounts of hydrogen and helium mixed with the existing primitive atmosphere (Gautier & Owen, 1983). With the light gases also came very small solid particles and other gases, more or less in solar proportions, which had not been included in the previously formed planetesimals. By this process Jupiter accumulated at least 300 Earth masses of hydrogen and helium, (Fegley & Prinn, 1988a,b) and Saturn approximately 80, while Uranus and Neptune acquired only one or two. For Jupiter and Saturn this stage of hydrogen and helium accumulation occurred within a relatively short time of a few hundred years, while for Uranus and Neptune a slightly longer interval was probably required. The infall of low-Z compounds was limited by the amount of gas available within the zone of influence of the individual proto planets. As a consequence Uranus and Neptune captured much less hydrogen and helium than Jupiter and Saturn.

The newly arrived light gases from the accretion disk readily mixed with the already present core atmospheres of mostly heavier gases. After that phase of accumulation of hydrogen and helium, the influx of matter in the form of planetesimals, meteorites, and comets must have continued, although at a much reduced rate. At the end of the low-Z accretion phase the outer planets had reached nearly their present masses. This sequence of events explains the nearly equal core sizes of the outer planets as well as the strong enrichment of their present atmospheres in heavy elements in comparison with solar composition (Grossman *et al.*, 1980). Moreover, the enrichment seems to be in proportion to the core-to-hydrogen mass ratio of the individual planets. Measured elemental abundances are summarized in Table 9.3.1.

As already mentioned, an alternative theory postulates a gravitational collapse of accretion disk material without first forming planetary cores (Cameron, 1978). In this case one would expect all of the outer planets to contain elements in nearly the same proportions, very similar to that of the Sun. Small increases of heavy elements over solar abundance ratios are possible from a post-accumulation infall of planetesimals, but the large enrichment of some of the heavy elements, as indicated in Table 9.3.1, cannot be explained. In contrast, the theory suggesting core formation to have occurred before the bulk of the low-Z gases arrived readily explains such an enrichment.

Remote sensing measurements from spacecraft, astronomical observations from Earth, and even measurements from future entry probes determine the composition only of the outermost atmospheric layers down to pressures of at best 10 or 20 bars, in most cases to much lower pressure levels. The question arises: how representative are these abundance measurements of the composition of the atmospheres as a whole? On Uranus and Neptune pressures at the lower boundary of the atmospheres are estimated to be several hundred thousand and on Jupiter and Saturn several million bars; large extrapolations are therefore necessary.

To investigate this subject we must study the processes that would maintain as well as destroy a uniform vertical distribution of specific elements. The degree of convective motion, chemical reactions, condensation and phase changes in general, as well as gravitational stratification all can play roles in establishing and preventing compositional homogeneity. If one of these processes prevents an element from being uniformly mixed, the remotely sensed abundance is not representative of the planet as a whole and should not be used for our specific inquiry.

Hydrogen is by far the dominant atmospheric constituent on all giant planets. For that reason, the elemental abundances are often ratioed to the hydrogen abundances. This is fully justified if hydrogen has never participated in chemical reaction inside the planets after their formation. Pollack *et al.* (1986) have pointed out the possibility that significant amounts of carbon monoxide may have been reduced to methane and water. If this did occur, the present hydrogen content would be smaller than the original one. On the other hand if methane dissociates at pressures in excess of 10^5 bar, as Ross & Ree (1980) have suggested, the hydrogen produced in this way could have increased the hydrogen reservoir collected in the late accretion phase. While such reactions are possible, they are expected to account for at most a few percent in the hydrogen abundance, but they cannot be responsible for factors of two or larger shown in Table 9.3.1.

The second most abundant constituent of the atmospheres of the giant planets, helium, does not react chemically and cannot condense even at the cold tropopauses of Uranus and Neptune. For a long time helium was expected to be uniformly mixed in all giant planet atmospheres. Smoluchowski (1967), Salpeter (1973), Hubbard &

Smoluchowski (1973), Stevenson & Salpeter (1976), and others have pointed out the immiscibility of helium in metallic hydrogen at certain temperatures. Pollack *et al.* (1977) suggested that the same process may not only occur on Jupiter but even more so on Saturn. Indeed the helium to hydrogen ratios measured by Voyager are quite different among the giant planets (Conrath *et al.*, 1987; Conrath & Gautier, 2000). Since the depletion of helium from the molecular atmospheres of Jupiter and Saturn is intimately related to the generation of internal heat, we discuss this subject in Subsection 9.4.b devoted to the energy balance of the outer planets. Here we mention only that the apparent depletion of helium measured in the atmospheres of Jupiter and Saturn also implies that vertical mixing by convective motion must be quite efficient in these atmospheres. At high densities the opacity due to collision-induced transitions is very high. Energy transport must then be dominated by convective motion, large scale mass flow, or possibly by the transport of latent heat. This assures relatively good vertical mixing, particularly on Jupiter, Saturn, and Neptune with their large internal heat sources. The efficiency of vertical mixing on Uranus is more problematic because of the observed relatively small internal heat flow from the interior of that planet. Hubbard *et al.* (1995) have discussed the possibility of suppression of convection by compositional gradients, while Smith & Gierash (1995) have suggested episodic convection in the presence of molecular hydrogen ortho–para conversion. In any event, the superadiabaticity is expected to be very small in all four planets, generally not measurable with present techniques.

Acting against these homogenizing forces are chemical reactions studied extensively by Lewis (1973), Lewis & Prinn (1984), and Prinn & Fegley (1989). Of course, formation of clouds by condensation also depletes the colder atmosphere above the clouds of the cloud forming substance. Another constituent may then dissolve into the clouds, removing that compound from the gaseous phase as well. The low N/H ratios measured on Uranus and Neptune may be caused by such a process.

On Jupiter and Saturn atmospheric temperatures are nowhere low enough to permit the formation of CH_4 clouds, therefore, the CH_4 abundances measured in the troposphere and lower stratosphere should be representative of these atmospheres as a whole. Destruction of CH_4 by photodissociation takes place in the upper stratosphere with the simultaneous formation of C_2H_2, C_2H_6, and many other hydrocarbons, but the production rate of these carbon bearing compounds is so low that it would not affect the tropospheric CH_4 concentration significantly. At higher pressures the hydrocarbons are converted back to CH_4. Other carbon bearing constituents are CO and HCN; again, their concentrations are very low and can easily be included in the C/H budget. Destruction of CH_4 and simultaneous production of additional H_2 at very high pressures (Ross & Ree, 1980) would tend to increase the estimate of the primordial C/H ratio and make the present ratio a lower limit of the original ratio. The true excess of C/H compared to the Sun would then be even larger than indicated in Table 9.3.1. In any case a precise measurement of the CH_4 abundance is very important. The composition of Saturn's atmosphere has been studied also by Courtin *et al.* (1984) and Jupiter's C/H ratio by Gautier *et al.* (1982).

On Uranus and Neptune temperatures in the upper troposphere are low enough for the formation of CH_4 clouds. To obtain the CH_4 concentration representative of the atmosphere as a whole the CH_4 abundance below the CH_4 cloud deck must be determined. Fortunately, this was possible for Uranus from data obtained by the Voyager Radio Science Investigation (Lindal *et al.*, 1987) and from ground-based microwave measurements (Lutz *et al.*, 1976). Scattering within the clouds complicates the interpretation of ground-based near infrared measurements.

The N/H and P/H measurements shown in Table 9.3.1 are consistent with an excess of heavy elements in the outer planets except for the low N/H ratios derived from NH_3 abundance measurements at Uranus and Neptune. The low NH_3 abundances may be a consequence of dissolution of NH_3 in water clouds or in a water ocean at great depth below the hydrogen–helium atmospheres (Atreya & Romani, 1985; Gautier & Owen, 1989). Also abnormal is the low O/H ratio on Jupiter derived from water vapor lines apparent in the infrared spectra of the 5 μm hot spots (Bjoracker *et al.*, 1986). Slightly larger amounts of the O/H ratio are indicated from CO measurements (Noll *et al.*, 1988). In contrast, the germanium content of Jupiter derived from the GeH_4 abundance (also measured at 5 μm) is greatly in excess of the abundance derived from the cosmic and solar abundances of that element and chemical equilibrium calculations. Strong deviations from equilibrium chemistry must exist in the deep interior of that planet. In addition to the main elements their isotopic ratios are of interest. For example, a discussion of recent measurements of the HD and CH_3D abundances in the atmospheres of the giant planets by ground- and space-based infrared spectrometers as well as an interpretation of these results have recently been given by Gautier & Owen (1989), by Pollack & Bodenheimer (1989), and by Gautier (1998).

As can be seen from Table 9.3.1 the elemental abundance values of the outer planets are rather incomplete and often not as precisely known as would be desirable. Clearly more precise, high resolution spectroscopy in the infrared and microwave regions as well as *in situ* measurements from descending probes will be required.

After the giant planets had accumulated most of their mass, the interior temperatures reached their maximum values. Evolutionary models indicate central temperatures as high as 50 000 K for Jupiter (Graboske *et al.*, 1975), slightly less for Saturn, and considerably less for Uranus and Neptune. At the same time the planetary diameters may have been 10 to 20 times as large as their present dimensions. The initial luminosity of Jupiter may have been six to seven orders of magnitude higher than present values. The other giant planets must have had similar excesses of luminosity and diameters that included the present orbits of their regular satellites. High rotation rates, acquired in the low-Z accumulation phase, may have been responsible for the formation of disks around the planets. Processes may have been similar to those that caused the formation of the primordial accretion disk around the Sun. Further contraction provided an additional heat source, slowing down the cooling process. The need to shed some of the angular momentum may have helped in the formation of the regular satellites. The irregular satellites are probably captured asteroids or, as in the case of Triton, fragments of planetesimals that were not incorporated into the planets. A recent summary of these events is presented by Pollack & Bodenheimer (1989). The planets eventually contracted and cooled to their present diameters and luminosities. The present heat balance is the subject of the next section.

9.4 Energy balance

From the time of formation of the Solar System nearly 4.6×10^9 years ago to the present epoch the planets have evolved from newly agglomerated, hot bodies to their current shape and form. This evolutionary process will continue until the Sun reaches the end of its life as a main sequence star and turns into a red giant.

As discussed in Section 9.3, a large number of uncertainties exist in the reconstruction of the events that led to the formation of the Solar System as a whole and the subsequent evolution of the individual planets. For example, did the cores of the outer planets form first with subsequent accumulation of the hydrogen–helium envelope, or did the bulk of the material, light gases, ices, and heavier elements, collapse more or less simultaneously onto denser regions of the circumsolar disk? Many of these questions can be addressed using models based on physical and chemical principles and a precise knowledge of the present state of the Solar System, including conditions in the interiors of planets, satellites, asteroids and comets. The present conditions of surfaces and interiors of Solar System objects can be found with different degrees of certainty from observations. Besides obvious parameters such as mass, shape, size, and atmospheric as well as surface composition, additional information is required for the construction of reasonably valid models; this includes data on the moments of inertia, the existence and shape of the magnetic field, the temperature at a certain pressure level, and most importantly for an understanding of the outer planets, the energy balance. In this section we will direct our attention to the energy balance, although composition, especially the helium content of the outer planet atmospheres, will play an important diagnostic role as well. First, however, the energy balance and related terms must be defined.

A planet absorbs sunlight and emits thermal radiation in the infrared. In most cases planets also contain internal heat sources, which are of great interest for the construction of models of the interior and for evolutionary theories. In a steady state, thermal emission must be balanced by absorbed solar radiation and the internally generated heat. For a spherical planet of radius R,

$$4\pi R^2 \sigma T_{\text{eff}}^4 = \pi R^2 (1 - A)\pi S/D^2 + P_{\text{int}}, \qquad (9.4.1)$$

where σ is the Stefan–Boltzmann constant and T_{eff} the effective planetary temperature. This is the temperature of a blackbody with the same *spectrally integrated* power as the planet. The first term on the right represents the absorbed solar power; A is the planetary Bond albedo, πS the solar constant at Earth, and D is the heliocentric distance of the planet in Astronomical Units. The last term in Eq. (9.4.1) is the power released by internal heat sources.

In the discussion of the energy balance a number of processes have to be considered:

(1) Primordial heat left over from the planetary formation,
(2) Gravitational energy converted to heat liberated by subsequent shrinking in the cooling process,
(3) Gravitational energy converted to heat liberated by the vertical redistribution of mass,
(4) Heat generated by absorption of solar energy,
(5) Heat generated by radioactive decay in the rock component, and
(6) Heat generated by tidal processes.

For the terrestrial planets, only items 4 and 5 are of importance. For the outer planets, all items are of interest, except item 6 which plays a dominant role only on Io, a much smaller one on Europa, and possibly a very small one on Ganymede and Callisto. On Jupiter and Saturn items 1, 2, 3, 4, and 5 contribute to the total infrared emission, the only quantity which can be measured by remote sensing techniques [left term in Eq. (9.4.1)]. On Uranus and Neptune a metallic hydrogen core is not expected to exist; therefore, the radial redistribution of helium cannot take place. However, other processes can conceivably affect the measured helium-to-hydrogen ratio as will be discussed further below. Item 4 is important for all planets. The absorbed solar radiation [first term on the right side of Eq. (9.4.1)] is found by a measurement of the planetary Bond albedo and a knowledge of the radius, the solar constant, and the heliocentric distance of the object.

On the outer planets the internal heat is found by measuring the thermal emission and subtracting the term representing absorbed solar power. Consequently, careful measurements of the effective planetary temperatures and the Bond albedos are required. The quantities R and D in Eq. (9.4.1) are relatively well known for each planet, as is S. On Earth the internal power is small in comparison with the other terms of Eq. (9.4.1). Thus it would be difficult to find the internal heat by subtracting two almost equal quantities. The terrestrial internal power can be found much more precisely from measurements of the vertical temperature gradient in the Earth's crust.

The ratio of the total emitted power to the absorbed power is customarily called the energy balance of a planet. First, we consider the determination of this quantity for the Earth and by implication for the other terrestrial bodies. Then we review measurements of the energy balance of the outer planets. In that process we discuss the implications of the internal heat sources for theories of Solar System evolution.

a. The terrestrial planets

The terrestrial planets are almost in energy balance, that is, thermal emission nearly equals absorbed solar power. On Earth only a small internal heat source exists, which manifests itself by a vertical temperature gradient in the outer layers of the crust. Early measurements, mostly from a few deep mines and bore holes, indicated a temperature increase with depth of 10–40 K km^{-1}. With reasonable assumptions on the thermal conductivity of rocks this corresponds to an internal heat source of approximately 2.6×10^{13} W (Bullard, 1954). More recent estimates, including data from deep sea drillings, yield a slightly higher value of $4.3 \pm 0.6 \times 10^{13}$ W (Williams & von Herzen, 1974). In contrast, solar radiation absorbed by the Earth amounts to approximately 1.2×10^{17} W. The internal heat flux is, therefore, only 3.5×10^{-4} of the absorbed solar radiation and, consequently, the energy balance of the Earth is approximately 1.000 35.

If one assumes, as Lord Kelvin did, that the internal heat is still part of the primordial heat, now emerging as a consequence of the cooling process from a once molten state, the age of the Earth would be only 30×10^6 years, which is inconsistent with all other evidence. Now we know the Earth has, for all practical purposes, completely cooled from the primordial state and the internal heat presently originates from radioactive decay of uranium, thorium, and potassium, all elements found in igneous rock such as granite and basalt.

An internal heat source of 4.3×10^{13} W corresponds to a surface flux of about 5×10^{-2} W m^{-2}, which is too small to be measured from space, considering the accuracy of presently available remote sensing techniques and the great variability of meteorological conditions. It is generally assumed that the other terrestrial planets and the Moon have similar internal heat sources, proportional to their share of radioactive material in their interiors.

b. The giant planets

Recent measurements of the energy balance of the outer planets reveal the presence of, in most cases, remarkably large internal heat sources. Several parameters

Table 9.4.1 *Energy balance and helium abundance of the outer planets. See text for the sources of the numerical values.*

	Jupiter	Saturn	Uranus	Neptune
Effective temperature, K	124 ± 0.3	95 ± 0.4	59.1 ± 0.3	59.3 ± 0.8
Emitted power, 10^{16} W	83.65 ± 0.84	19.77 ± 0.32	0.560 ± 0.011	0.534 ± 0.029
Bond albedo	0.343 ± 0.032	0.342 ± 0.030	0.300 ± 0.049	0.290 ± 0.067
Absorbed power, 10^{16} W	50.14 ± 14	11.14 ± 0.50	0.526 ± 0.037	0.204 ± 0.019
Energy balance	1.67 ± 0.09	1.78 ± 0.09	1.06 ± 0.08	2.61 ± 0.28
Helium mass mixing ratio, Y	0.234 ± 0.005	0.18 to 0.25	0.262 ± 0.048	0.32 ± 0.05

pertinent to a discussion of the energy balance of the giant planets are summarized in Table 9.4.1. The numerical values have been derived from the spacecraft and ground-based measurements discussed in Section 8.6. First we discuss Jupiter. This planet is so massive that the heat generated by the contraction process begun 4.6×10^9 years ago is still significant. Primordial heat and heat from the subsequent, more gradual, contraction are nearly sufficient to explain the internal heat source (Graboske *et al.*, 1975). The infrared interferometer, IRIS, on Voyager 1, using spectra between 230 and 2300 cm^{-1} and model calculations below 230 cm^{-1}, derived the thermal emission of Jupiter to be $8.365 \pm 0.084 \times 10^{17}$ W, which corresponds to an equivalent blackbody temperature of 124.4 ± 0.3 K (Hanel *et al.*, 1981*b*). This number is in good agreement with Pioneer results, 125 ± 3 K (Ingersoll *et al.*, 1976) and aircraft measurements 123 ± 2 K (Erickson *et al.*, 1978), but somewhat lower than earlier measurements.

A geometric albedo of 0.274 ± 0.013 was also derived by Hanel *et al.* (1981*b*). Combining this with the Pioneer-derived phase integral of 1.25 (Tomasko *et al.*, 1978) leads to a Bond albedo of 0.343 ± 0.032. However, the error bar should in reality be larger. The geometric albedo was derived by calibrating the IRIS radiometer in flight using a diffuse reflector mounted on the spacecraft. Measurements from Voyager 2, using a similar reflector, yielded a 12% higher albedo value of Jupiter. Only one witness sample generated together with the coating of the diffuse reflector was available for laboratory measurement of the reflectivity of the coating. Obviously, one or other plate (or both) had changed between manufacturing and the first measurements shortly before the Jupiter encounter (see also Pearl & Conrath, 1991). A possible general degradation of the radiometer itself does not enter into this discussion (it possibly appears in an indirect way, e.g., if the spectral response

9.4 Energy balance

of the radiometer has changed substantially) since in the data reduction process the radiometer signal from Jupiter is compared directly to the radiometer signal while viewing the diffusor plate. The Voyager 1 derived geometric albedo of 0.274 agrees well with the ground-based derivation of 0.28 of the same quantity by Taylor (1965). Nevertheless, the in-flight calibration using the diffusor plate on Voyager must be taken with caution.

A problem with the Voyager value of the helium-to-hydrogen ratio derived by combining radio occultation data with infrared spectra recently came to light. The helium-to-hydrogen ratio of Jupiter was measured by instruments on the Galileo probe to be 0.157 ± 0.003 (mass fraction $Y = 0.234 \pm 0.005$; von Zahn et al., 1998) and with consistent results by the mass spectrometer (Niemann et al., 1998). These numbers are larger than the helium-to-hydrogen ratio derived by Voyager, 0.11 ± 0.032 (mass fraction $Y = 0.18 \pm 0.04$; Gautier et al., 1981; Conrath et al., 1984). This discrepancy prompted Conrath & Gautier (2000) to reexamine the Voyager procedure. Excellent agreement among the probe measurements, the infrared spectra, and the radio occultation profile was obtained after the latter was systematically increased by 2 K. This subject is also discussed in Section 8.3.f.

The proto solar value of the helium-to-hydrogen ratio – that is the value pertinent to the Solar Nebula from which the Sun and the planets formed – is somewhat uncertain. However, the most reliable numbers come from models of stellar interiors, yielding $Y = 0.28$ (Profitt, 1994). The difference between 0.28 and 0.234 indicates that Jupiter's outer atmosphere is now to some degree depleted of helium, which seems to confirm the theory, originally proposed by Smoluchowski (1967), that helium may be migrating towards the metallic hydrogen core (see also Salpeter, 1973; Smoluchowski, 1973; Hubbard & Smoluchowski, 1973; Stevenson & Salpeter, 1976). This redistribution of mass is a significant source of heat and contributes to the energy balance of Jupiter. The relative importance of the heat sources 1, 2, and 3 mentioned above must be found from models of the interior of Jupiter, using measurements of the energy balance, the helium abundance, the moments of inertia, and the magnetic field.

On Saturn the thermal emission and albedo measurements are much more complicated then those of Jupiter. Complications arise primarily due to the presence of the rings and the tilt of Saturn's axis of rotation with respect to its orbital plane. The rings cast shadows and reflect sunlight onto Saturn. However, these effects vary over the orbit of Saturn around the Sun. The Voyager measurements occurred at one particular point on that orbit, but only the mean orbital values are pertinent to the energy balance of Saturn. First, it was necessary to remove the effects of the rings from the measurements of the thermal emission and the albedo of Saturn. Then the effect of the rings needed to be introduced over the full orbital motion of Saturn (Hanel et al., 1983). An effective temperature of 95.0 ± 0.4 K was found.

Using the same on-board calibration with the diffuse reflecting plate, a geometric albedo of 0.242 ± 0.012 was derived. Again combining this with a Pioneer-derived phase integral led to a Bond albedo of 0.342 ± 0.030. Because of the uncertainty of the radiometer calibration this number must be taken with the same reservations as the corresponding number for Jupiter.

On Saturn only remote sensing information on the helium-to-hydrogen ratio is available today. The very low number of He/$H_2 = 0.034 \pm 0.024 (Y = 0.06 \pm 0.05$; Conrath *et al.*, 1984) was derived from a combination of radio occultation data and infrared spectra both from Voyager. The possibility exists that a systematic error in the radio occultation data, similar to that suspected for Jupiter, may be present for Saturn as well. An analysis based on only the infrared spectra to retrieve simultaneously the temperature profile, the ortho–para ratio of hydrogen, and the helium-to-hydrogen ratio was carried out by Conrath & Gautier (2000). A most likely helium-to-hydrogen ratio between 0.11 and 0.16 (Y between 0.18 and 0.25) was found. Although not as precise as would be desired, this analysis excludes the low helium abundance inferred previously. Clearly, a careful analysis of the systematic and random errors of the radio occultation data would be desirable.

The helium-to-hydrogen ratio of Saturn obtained by Conrath & Gautier (2000) using only IRIS spectra, although higher than the previous results, is much lower than the proto solar value of $Y = 0.28$ (Profitt, 1994), indicating a helium deficiency in Saturn's outer atmosphere, possibly even slightly larger than that found on Jupiter. The vertical redistribution of helium, first recognized by Smoluchowski (1967), plays an important role on Saturn as well.

The albedo, effective temperature, and energy balance of Uranus was investigated by Pearl *et al.* (1990) using Voyager 2 data. They found an effective temperature of 59.1 ± 0.3 K. Due to the low temperatures of Uranus useful measurements were derived from the infrared spectra only between 180 and 400 cm^{-1}. Pearl *et al.* (1990) give a detailed description of the extrapolation procedure used to obtain the spectral information below 180 cm^{-1}.

For the radiometer measurements, the same procedure, including calibration of the radiometer data with the diffusor plate, was used to derive a geometric albedo of 0.215 ± 0.046. Earlier measurements, also discussed by Pearl *et al.* (1990), yielded somewhat higher values. A limited set of radiometer data, augmented with imaging data (Pollack *et al.*, 1986), permitted the construction of a phase function. Integration of the phase function yielded a phase integral of 1.40 ± 0.14. Combining the geometrical albedo and the phase integral lead to a Bond albedo of 0.300 ± 0.049. Based on the uncertainties of the radiometer calibration, the error bar may be larger. A Bond albedo of 0.319 ± 0.051 was derived from the imaging system of Voyager by Pollack *et al.* (1986), probably also using the same diffusor plate for calibration. Spacecraft and ground-based data pertinent to the energy balance of Uranus have been reviewed by Conrath *et al.* (1991*b*).

The helium-to-hydrogen ratio of Uranus was again investigated using radio occultation data and infrared spectra by Conrath *et al.* (1987). They inferred a mass mixing ratio of $Y = 0.262 \pm 0.048$. Considering the error bar, this value is consistent with the proto solar value of 0.28 of Profitt (1994) and would indicate that the helium-to-hydrogen ratio on Uranus is proto solar. It would also indicate that the radio occultation data in this case are probably without systematic errors and no other atmospheric gas contributes significantly to the mean molecular weight in the applicable pressure range. The helium-to-hydrogen ratio on Uranus is also reviewed by Fegley *et al.* (1991); see also Hubbard (1980, 1984) and Hubbard & MacFarlane (1980).

The albedo, effective temperature, and energy balance of Neptune have been investigated by Pearl & Conrath (1991); see also the review by Gautier *et al.* (1995). In deriving the thermal emission of that planet, Pearl & Conrath found it necessary to account for a small latitudinal variation of that quantity. The effective temperature of Neptune, 59.3 ± 0.8 K, was found to be nearly the same as that derived for Uranus, 59.1 ± 0.3 K, although Neptune (30.1 AU) is much further from the Sun than Uranus (19.2 AU). The geometric albedo of Neptune, 0.215 ± 0.050, also turned out to be identical to that found for Uranus, 0.215 ± 0.046. As for Uranus, it was possible to derive a phase function for Neptune. The phase integral was found to be 1.35 ± 0.16, which led to a Bond albedo of 0.290 ± 0.067, again almost the same as for Uranus. Of course, on Uranus and on Neptune the albedo values may be systematically off by the same amount due to the uncertainty in the absolute calibration of the radiometer using the diffusor plate. However, the ratio of the albedo measurements of Neptune and Uranus is independent of the difficulties with the absolute calibration. Considering their heliocentric distances, the similarity of the albedo values and of the effective temperatures clearly show a fundamental difference in the internal energy balance of these otherwise so similar planets.

The helium-to-hydrogen ratio of Neptune was derived by Conrath *et al.* (1991*a*) to be 0.190 ± 0.032 (corresponding to $Y = 0.32 \pm 0.05$), using radio occultation data and infrared spectra. The helium mass fraction Y appears to be higher than the proto solar value of 0.28 (Profitt, 1994) although just within the error bar. A number of possible causes for the high value of Y can be found. The original derivation of the helium abundance by Conrath *et al.* (1991*a*) assumed that, besides hydrogen, only helium contributes significantly to the mean molecular weight of the atmosphere. A subsequent analysis by Conrath *et al.* (1993) showed that a small mole fraction of molecular nitrogen may easily reduce the measured helium mass fraction to the proto solar value. The induced opacity due to H_2–N_2 collisions (N_2 mole fraction

0.003) and possibly a small effect from methane clouds are sufficient to yield the same helium abundance as was found for Uranus. Clouds and hazes on Neptune have been reviewed by Baines *et al.* (1995).

The values pertinent to the energy balance of all giant planets are summarized in Table 9.4.1. Although Uranus and Neptune are very similar in size, Bond albedo, and other parameters, they are quite different in the energy balance; see also the review paper by Hubbard *et al.* (1995).

The above summary demonstrates that presently available observations, such as those of the energy balance and the helium composition of the atmospheres of the outer planets, provide important constraints on the choice of parameters used in planetary evolutionary models and, in turn, on the reconstruction of the events which led to the formation of the Solar System as a whole.

Closing remarks

We have surveyed the present state of the exploration of the Solar System by remote sensing in the thermal infrared. In the first three chapters we reviewed the physical theories that serve as the basis of remote sensing. Models presented in Chapter 4 were designed to illustrate how the physical state of an atmosphere influences the observed spectrum. Chapter 5 was devoted to instrumental techniques, with emphasis placed on physical principles and measurement methods. Several real instruments served as examples in the discussions. Chapter 6 addressed measurements of planets up to Neptune and their satellites. Wherever possible, data were selected to illustrate how physical attributes of various bodies can often be inferred qualitatively by a visual inspection of the data. Pluto and Charon, as well as comets and asteroids, were the subjects of Chapter 7. In Chapter 8 we treated numerical retrieval methods. Again, examples demonstrated how atmospheric temperature profiles, mole fractions of gases, information on clouds, and surface parameters can be extracted from measurements. Finally, in Chapter 9 four areas were discussed where remotely sensed data have made significant contributions to our present understanding of underlying physical processes. These areas concern radiative equilibrium, the temperature field and dynamical processes, the atmospheric composition in the context of Solar System evolution, and the energy balance of the planets.

It should be clear from the content of this book that much has been achieved in planetary research in the last decades, not only from spacecraft, but also from ground-based and air-borne platforms. It should be equally clear that much more needs to be done to obtain a more nearly complete understanding of the processes that have shaped and are still shaping our Solar System. Both theoretical and observational advances are necessary. In the coming decades new instrumental techniques will evolve, more advanced detectors will become available, and more sophisticated observational methods will be applied, all leading to better experimental results. At the same time theoretical progress will be made. Observations and measurements will test new hypotheses, and new theories will stimulate additional measurements. This interplay between experiment and theory has been fruitful in astronomy and science in general since the days of Galileo and Kepler; undoubtedly it will continue.

Curiosity is one of mankind's strongest motivations. Be it the caves in the next valley, the rest of the Earth, the Solar System, or the universe, exploration of our environment has always been intrinsically compelling. Since the beginning of civilization the desire to solve not only the problems of daily life, but to go one step further and engage in seemingly unnecessary tasks has slowly expanded our awareness of the world around us. It is this inborn human need to explore and to understand that has conquered unknown territories, carried us to the Moon, and led us through the Solar System. Future generations will say 'We would not be here were it not for the constant search for knowledge of past generations'. In the pursuit of this endeavor remote sensing will remain one of the fundamental investigative tools. We hope this book will help to sharpen this implement. We also hope it will encourage students and researchers alike to actively use this tool and pursue the 'Exploration of the Solar System by infrared remote sensing'.

Appendix 1

MATHEMATICAL FORMULAS

A1.1 Vector quantities

In the following, vectors are boldface, scalars are not. Basis vectors are $\hat{\mathbf{i}}, \hat{\mathbf{j}}, \hat{\mathbf{k}}$ in Cartesian coordinates and $\hat{\mathbf{r}}, \hat{\boldsymbol{\theta}}, \hat{\boldsymbol{\phi}}$ in spherical coordinates. Alternative notations associated with the vector $\mathbf{A}$ and scalar A are listed in Table A1.1.

Cartesian coordinates

$$\nabla A = \frac{\partial A}{\partial x}\hat{\mathbf{i}} + \frac{\partial A}{\partial y}\hat{\mathbf{j}} + \frac{\partial A}{\partial z}\hat{\mathbf{k}},$$

$$\nabla \cdot \mathbf{A} = \frac{\partial A_x}{\partial x} + \frac{\partial A_y}{\partial y} + \frac{\partial A_z}{\partial z},$$

$$\nabla \times \mathbf{A} = \left(\frac{\partial A_z}{\partial y} - \frac{\partial A_y}{\partial z}\right)\hat{\mathbf{i}} + \left(\frac{\partial A_x}{\partial z} - \frac{\partial A_z}{\partial x}\right)\hat{\mathbf{j}} + \left(\frac{\partial A_y}{\partial x} - \frac{\partial A_x}{\partial y}\right)\hat{\mathbf{k}}.$$

Spherical coordinates

$$\nabla A = \hat{\mathbf{r}}\frac{\partial A}{\partial r} + \hat{\boldsymbol{\theta}}\frac{1}{r}\frac{\partial A}{\partial \theta} + \hat{\boldsymbol{\phi}}\frac{1}{r\sin\theta}\frac{\partial A}{\partial \phi},$$

$$\nabla \cdot \mathbf{A} = \frac{1}{r^2}\frac{\partial}{\partial r}(r^2 A_r) + \frac{1}{r\sin\theta}\left[\frac{\partial}{\partial \theta}(\sin\theta A_\theta) + \frac{\partial A_\phi}{\partial \phi}\right],$$

$$\nabla \times \mathbf{A} = \hat{\mathbf{r}}\frac{1}{r\sin\theta}\left[\frac{\partial}{\partial \theta}(\sin\theta A_\phi) - \frac{\partial A_\theta}{\partial \phi}\right] + \hat{\boldsymbol{\theta}}\frac{1}{r}\left[\frac{1}{\sin\theta}\frac{\partial A_r}{\partial \theta} - \frac{\partial}{\partial r}(rA_\phi)\right]$$

$$+ \hat{\boldsymbol{\phi}}\frac{1}{r}\left[\frac{\partial}{\partial r}(rA_\theta) - \frac{\partial A_r}{\partial \theta}\right].$$

Appendix 1

Table A1.1 *Components of* **A**

Cartesian coordinates	Spherical coordinates	Gradient	Divergence	Curl
A_x	A_r	grad A	div **A**	curl **A**
A_y	A_θ	∇A	$\nabla \cdot \mathbf{A}$	rot **A**
A_z	A_ϕ			$\nabla \times \mathbf{A}$

The Laplacian $\nabla^2 \mathbf{A}$ in both coordinate systems is defined as

$$\nabla^2 \mathbf{A} = \nabla(\nabla \cdot \mathbf{A}) - \nabla \times (\nabla \times \mathbf{A}).$$

Transformation of coordinates

$$A_r = A_x \sin\theta \cos\phi + A_y \sin\theta \sin\phi + A_z \cos\theta,$$
$$A_\theta = A_x \cos\theta \cos\phi + A_y \cos\theta \sin\phi + A_z \sin\theta,$$
$$A_\phi = -A_x \sin\phi + A_y \cos\phi.$$

A1.2 Spherical Bessel functions

The generalized spherical Bessel functions $z_n(\rho)$ of integer order n are defined by

$$\rho^2 \frac{d^2 z_n}{d\rho^2} + 2\rho \frac{dz_n}{d\rho} + [\rho^2 - n(n+1)] z_n = 0 \quad (n = 0, \pm 1, \pm 2, \ldots),$$

and can be expressed in terms of ordinary half-order cylindrical (Bessel, Neumann, and Hankel) functions $Z_{n+\frac{1}{2}}$ by

$$z_n(\rho) = \left(\frac{\pi}{2\rho}\right)^{\frac{1}{2}} Z_{n+\frac{1}{2}}(\rho).$$

Series representations are as follows.
Spherical Bessel functions of the first kind:

$$j_n(\rho) = \frac{\rho^n}{1 \cdot 3 \cdot 5 \cdot \ldots \cdot (2n+1)} \left[1 - \frac{\frac{1}{2}\rho^2}{1!(2n+3)} + \frac{\left(\frac{1}{2}\rho^2\right)^2}{2!(2n+3)(2n+5)} - \cdots \right];$$

Spherical Bessel (Neumann) functions of the second kind:

$$y_n(\rho) = \frac{1 \cdot 3 \cdot 5 \cdot \ldots \cdot (2n-1)}{\rho^{n+1}} \left[1 - \frac{\frac{1}{2}\rho^2}{1!(1-2n)} + \frac{\left(\frac{1}{2}\rho^2\right)^2}{2!(1-2n)(3-2n)} - \cdots \right].$$

Spherical Bessel (Hankel) functions of the third kind obey the relations:

$$h_n^{(1)}(\rho) = j_n(\rho) + iy_n(\rho),$$
$$h_n^{(2)}(\rho) = j_n(\rho) - iy_n(\rho).$$

Asymptotic values as $\rho \to \infty$ are:

$$h_n^{(1)}(\rho) = \frac{(-i)^{n+1}}{\rho} e^{i\rho},$$

$$h_n^{(2)}(\rho) = \frac{i^{n+1}}{\rho} e^{-i\rho}.$$

A1.3 Legendre polynomials

The Legendre polynomials $P_n(\mu)$, where $-1 \leq \mu \leq 1$, obey the second order differential equation

$$(1-\mu^2)\frac{d^2 P_n(\mu)}{d\mu^2} - 2\mu \frac{dP_n(\mu)}{d\mu} + n(n+1)P_n(\mu) = 0 \quad (n = 0, 1, 2, \ldots).$$

If $f(\mu)$ is a piecewise continuous function in the interval $-1 \leq \mu \leq 1$, the Legendre series corresponding to $f(\mu)$ is

$$f(\mu) \sim \sum_{n=0}^{\infty} \tilde{\omega}_n P_n(\mu),$$

where

$$\tilde{\omega}_n = \frac{2n+1}{2} \int_{-1}^{+1} f(\mu') P_n(\mu') \, d\mu' \quad (n = 0, 1, 2, \ldots).$$

The correspondence is an equality except at each point of discontinuity, where the mean value of $f(\mu)$ is assumed. If $f(\mu)$ is a polynomial the series is finite.

Legendre polynomials obey the orthogonality relation

$$\frac{1}{2}(2n+1)\int_{-1}^{+1} P_m(\mu)P_n(\mu)\,\mathrm{d}\mu = \delta_{m,n},$$

where

$$\delta_{m,n} = \begin{cases} 1 & (m=n) \\ 0 & (m \neq n) \end{cases}.$$

They also obey the recursion relation

$$(2n+1)\,\mu\,P_n(\mu) = (n+1)\,P_{n+1}(\mu) + n\,P_{n-1}(\mu).$$

The first three values are

$$P_0(\mu) = 1,$$
$$P_1(\mu) = \mu,$$
$$P_2(\mu) = \tfrac{3}{2}\mu^2 - \tfrac{1}{2}.$$

Appendix 2

Table A2.1 *Physical constants**

Symbol	Name	Numerical value	Units
ε_0	Permittivity of free space	$10^7/4\pi c^2 = 8.854\,187\,817 \times 10^{-12}$	[F m^{-1}]
μ_0	Permeability of free space	$4\pi \times 10^{-7} = 12.566\,370\,614 \times 10^{-7}$	[N A^{-2}]
c	Velocity of light in free space	299 792 458	[m s^{-1}]
σ	Stefan–Boltzmann constant	$5.670\,51(19) \times 10^{-8}$	[W m^{-2} K^{-4}]
h	Planck constant	$6.626\,075\,5(40) \times 10^{-34}$	[J s]
k	Boltzmann constant	$1.380\,658(12) \times 10^{-23}$	[J K^{-1}]
$2hc^2$	First radiation constant	$1.191\,043\,9(07) \times 10^{-16}$	[J m^2 s^{-1}]
hc/k	Second radiation constant	$1.438\,769(12) \times 10^{-2}$	[m K]
G	Gravitational constant	$6.672\,59(85) \times 10^{-11}$	[m^3 kg^{-1} s^{-2}]
e	Charge of electron	$1.602\,177\,33(49) \times 10^{-19}$	[C]
m_e	Mass of electron	$9.109\,389\,7(54) \times 10^{-31}$	[kg]
m_p	Mass of proton	$1.672\,623\,1(10) \times 10^{-27}$	[kg]
m_n	Mass of neutron	$1.674\,928\,6(10) \times 10^{-27}$	[kg]
a_0	Bohr radius	$0.529\,177\,249(24) \times 10^{-10}$	[m]

* After Cohen & Giacomo, 1987

The digits in parentheses are the one-standard deviation uncertainty in the last digits in the given value

Appendix 3

Table A3.1 *The terrestrial planets*

Parameter	Mercury	Venus	Earth	Mars
Mean distance from Sun, AU*	0.387	0.723 3	1.000	1.523 7
Sidereal period, tropical yr	0.241	0.62	1.000	1.88
Orbital eccentricity	0.206	0.006 82	0.016 7	0.093 33
Inclination to ecliptic, deg	7.00	3.395	0.000	1.850
Max. phase angle from Earth, deg	180	180	–	46.8
Equatorial radius, km	2439.0	6051.0	6378.5	3393.0
Polar radius, km	2439.0	6051.0	6356.0	3375.0
Oblateness	0	0	0.003 527	0.005 305
Volume of planet (Earth = 1)*	0.057	0.869	1.000	0.152
Mass of planet (Earth = 1)*	0.056	0.817	1.000	0.108
Mean density ($\times 10^3$ kg m^{-3})	5.50	5.26	5.52	3.97
Sidereal rotation period	58.6 d	−243.05 d	23.933 h	24.633 h
Obliquity, deg	0	177.34	23.45	25.19

* 1 AU = 149.6 × 10^9 m; volume of Earth = 1.070 × 10^{21} m^3; mass of Earth = 5.98 × 10^{24} kg

Table A3.2 *The giant planets*

Parameters	Jupiter	Saturn	Uranus	Neptune
Mean distance from Sun, AU*	5.202 561	9.554 747	19.218 14	30.109 57
Sidereal orbital period, tropical yr	11.862 3	29.458	84.01	164.79
Orbital eccentricity	0.048 5	0.055 6	0.047 2	0.008 6
Orbital inclination to ecliptic, deg	1.30	2.49	0.77	1.77
Max. phase angle from Earth, deg	11.42	6.33	3.13	1.92
Equatorial radius at 1 bar, km	71 492	60 268	25 559	24 764
Polar radius at 1 bar, km	66 854	54 364	24 950	24 340
Oblateness	0.064 87	0.097 96	0.023 83	0.017 12
Volume of planet (Earth = 1)*	1321.6	763.6	63	56
Mass of planet (Earth = 1)*	317.893	95.147	14.54	17.23
Mean density ($\times 10^3$ kg m^{-3})	1.325	0.693	1.18	1.56
Sidereal body rotation period, h	9.924 9	10.656 2	17.24	16.11
Obliquity, deg	3.08	26.73	97.92	28.8

* 1 AU = 149.6×10^9 m; volume of the Earth = 1.070×10^{21} m^3; mass of Earth = 5.98×10^{24} kg

Table A3.3 *Atmospheric constituents in the Solar System**

Object	Major species	Minor species
Sun	H, He	CO, C_2, H_2, CH, NH, OH, MgH, CaH, TiO, SiH, ZrO, CoH, NiH, SiO, H_2O
Mercury		Na
Venus	CO_2	Ar, CO, SO, N_2, HCl, HF, H_2O, SO_2, OCS, H_2SO_4
Earth	N_2, O_2	Ar, NO, CO, HF, H_2O, CO_2, O_3, N_2O, NH_3, CH_4, ...
Mars	CO_2	Ar, NO, N_2, O_2, CO, H_2O, HDO, O_3
Jupiter	H_2, He	CO, HD, H_2O, CO_2, H_2S, NH_3, PH_3, AsH_3, CH_3, C_2H_2, CH_4, CH_3D, GeH_4, C_2H_4, C_4H_2, C_3H_4, C_2H_6, C_6H_6
Io		Na, K, SO, S_2, SO_2
Saturn	H_2, He	CO, HD, H_2O, CO_2, NH_3, PH_3, CH_3, C_2H_2, CH_4, CH_3D, C_2H_4, C_4H_2, C_3H_4, C_2H_6, C_6H_6
Titan	N_2	Ar, H_2, CO, HCN, H_2O, CO_2, C_2H_2, C_2N_2, CH_4, CH_3D, HC_3N, C_2H_4, C_4H_2, CH_3CN, C_3H_4, C_2H_6, C_3H_8
Uranus	H_2, He	HD, H_2O, C_2H_2, CH_4
Neptune	H_2, He	HD, H_2O, CO_2, C_2H_2, CH_3, CH_4, C_2H_4, C_2H_6, HCN, CO
Triton	[N_2, CO, CO_2, H_2O, CH_4]**	
Pluto	[N_2, CO, H_2O, CH_4]**	
Comets		He, Na, K, Ar, O^+, OH, C_2, CO, CO^+, N_2, S_2, SO, CN, CS, NH, H_2O, HDO, H_2O^+, HCN, DCN, HNC, OCS, NH_2, H_2S, SO_2, C_3, HCO^+, CO_2, H_3O^+, C_2H_2, NH_3, H_2CO, H_2CS, CH_4, HC_3N, HCOOH, CH_3OH, CH_3CN, NH_2CHO, C_2H_6, CH_3OCHO

* Updated from Jennings (1988)
** Surface ices. Volatile N_2 and CH_4 may form a tenuous atmosphere

References

Abbas, M. M., Mumma, M. J., Kostiuk, T., & Buhl, D. (1976). Sensitivity limits of an infrared heterodyne spectrometer for astrophysical applications. *Applied Optics*, **15**, 427–36.

Ade, P. A. R., Costley, A. E., Cunningham, C. T., Mok, C. L., Neill, G. F., & Parker, T. J. (1979). Free-standing grids for spectroscopy at far-infrared wavelength. *Infrared Physics*, **19**, 599–601.

Adler, I., Trombka, J. L., Lowman, P., Schmadebeck, R., Blodget, H., Eller, E., Yin, L., Lamothe, R., Osswald, G., Gerard, J., Gorenstein, P., Bjorkholm, P., Gursky, H., Harris, B., Arnold, J., Metzger, A., & Reedy, R. (1973). Apollo 15 and 16 results of the integrated geochemical experiment. *The Moon*, **7**, 487–504.

Allen, H. C. & Cross, P. C. (1963). *Molecular Vibration – Rotors – The Theory and Interpretation of High Resolution Infrared Spectra*. New York: John Wiley & Sons, Inc.

Andrews, D. G., Holton, J. R., & Leovy, C. B. (1987). *Middle Atmosphere Dynamics*. New York: Academic Press.

Annable, R. V. (1970). Radiant cooling. *Applied Optics*, **9**, 185–93.

Aptaker, I. M. (1987). A Near Infrared Mapping Spectrometer (NIMS) for investigation of Jupiter and its satellites. In *Imaging Spectroscopy 11*, ed. G. Vane. *SPIE Proceedings*, **834**, 196, Bellingham, Washington.

Aronson, J. R. & Emslie, A. G. (1973). Spectral reflectance and emittance of particulate materials, 2, application and results. *Applied Optics*, **12**, 2573–84.

Asphaug, E. (2000). The small planets. *Scientific American*, **282**, 46–55.

Astheimer, R. W., DeWaard, R., & Jackson, E. A. (1961). Infrared radiometric instruments on Tiros II. *Journal of the Optical Society of America*, **51**, 1386–93.

Atreya, S. K., Pollack, J. B., & Matthews, M. S. (1989). *Origin and Evolution of Planetary and Satellite Atmospheres*. Tucson: The University of Arizona Press.

Atreya, S. K. & Romani, P. N. (1985). Photochemistry and clouds on Jupiter, Saturn, and Uranus. In *Recent Advances in Planetary Meteorology*, ed. G. E. Hunt. Cambridge: Cambridge University Press.

Atreya, S. K., Wong, M. H., Owen, T. C., Mahaffy, P. R., Niemann, H. B., de Pater, I., Drossart, P., & Encrenaz, Th. (1999). A comparison of the atmospheres of Jupiter and Saturn: deep atmospheric composition, cloud structure, vertical mixing, and origin. *Planetary and Space Science*, **47**, 1243–62.

Atreya, S., Wong, M. H., Owen, T., Niemann, H., & Mahaffy, P. (1997). Chemistry and clouds of the atmosphere of Jupiter: A Galileo perspective. In *Three Galileos: The*

Man, The Spacecraft, The Telescope, 249–60, ed. C. Barbieri, J. Rahe, T. Johnson, & A. Sohus. Dordrecht: Kulwer Academic Publishers.

Bachet, G., Cohen, E. R., Dore, P., & Birnbaum, G. (1983). The translational–rotational absorption spectrum of hydrogen. *Canadian Journal of Physics*, **61**, 591–603.

Backus, G. & Gilbert, F. (1970). Uniqueness in the inversion of inaccurate gross earth data. *Philosophical Transactions Royal Society London*, **A266**, 123–92.

Baines, K. H., Hammel, H. B., Rages, K. A., Romani, P. N., & Samuelson, R. E. (1995). Clouds and hazes in the atmosphere of Neptune. In *Neptune and Triton*, 489–546, ed. D. Cruikshank. Tucson: University of Arizona Press.

Bandeen, W., Kunde, V., Nordberg, W., & Thompson, H. (1964). Tiros III meteorological satellite, radiation observations of a tropical hurricane. *Tellus*, **16**, 481–502.

Banfield, D., Conrath, B. J., Pearl, J. C., Smith, M. D., & Christensen, P. R. (2000). Thermal tides and stationary waves on Mars as revealed by MGS TES. *Journal of Geophysical Research*, **105**, 9521–38.

Barker, J. L. ed. (1985*a*). Landsat 4 science characterization, early results. *NASA Conference Publication 2355*, Goddard Space Flight Center, Greenbelt, Maryland, 20771.

Barker, J. L. ed. (1985*b*). Landsat 4 science investigation summary. *NASA Conference Publication 2326*, Goddard Space Flight Center, Greenbelt, Maryland, 20771.

Barkstrom, B. R. & Smith, G. L. (1986). The Earth radiation budget experiment: science and implementation. *Review of Geophysics*, **24**, 379–90.

Basri, G. (1992). Stars, T-Tauri. In *Astronomy and Astrophysics Encyclopedia*, 797–800, ed. S. Maran. New York and Cambridge: Van Nostrand Reinhard and Cambridge University Press.

Beer, R. & Marjaniemi, D. (1966). Wavefronts and construction tolerances for the cat's-eye retroreflector. *Applied Optics*, **5**, 1191–7.

Beer, R. & Norton, R. H. (1988). Analysis of spectra using correlation functions. *Applied Optics*, **27**, 1255–61.

Beer, R. & Taylor, F. W. (1973). The abundance of CH_3D and the D/H ratio in Jupiter. *Astrophysical Journal*, **179**, 309–27.

Beerman, H. P. (1975). Investigation of pyroelectric material characteristics for improved infrared detector performance. *Infrared Physics*, **15**, 225–31.

Bell, E. E. (1971). Amplitude spectroscopy. *Aspen International Conference on Fourier Spectroscopy*, ed. G. A. Vanasse, A. T. Stair, & D. J. Baker, US Air Force Cambridge Research Laboratory, 71-0019.

Bell, R. J. (1972). *Introductory Fourier Transform Spectroscopy*. New York: Academic Press.

Bergstralh, J. T., Miner, E. D., & Matthews, M. S. ed. (1991). *Uranus*. Tucson: University of Arizona Press.

Bethe, H. A. (1939). Energy production in stars. *Physical Review*, **55**, 434–56.

Bethe, H. A. & Critchfield, C. L. (1938). The formation of deuterons by proton combination. *Physical Review*, **54**, 248–54.

Betz, A. L., Johnson, M. A., McLaren, R. A., & Sutton, E. C. (1976). Heterodyne detection of CO_2 emission lines and wind velocities in the atmosphere of Venus. *Astrophysical Journal*, **208**, L141–4.

Bézard, B., Encrenaz, Th., Lellouch, E., & Feuchtgruber, H. (1999*a*). A new look at the Jovian planets. *Science*, **283**, 800–1.

Bézard, B., Feuchtgruber, H., Moses, J. I., & Encrenaz, T. (1998). Detection of methyl radicals (CH_3) on Saturn. *Astronomy and Astrophysics*, **334**, L41–4.

Bézard, B., Romani, P. N., Fechtgruber, H., & Encrenaz, T. (1999b). Detection of methyl radical on Neptune. *Astrophysical Journal*, **515**, 868–72.

Bibring, J., Combes, M., Langevin, Y., Soufflot, A., Cara, C., Drossart, P., Encrenaz, T., Erard, S., Forni, O., Gondet, B., Ksanformality, L., Lellouche, E., Masson, P., Moroz, V., Rocard, F., Rosenqvist, J., & Sortin, C. (1989). Results from the ISM experiment. *Nature*, **341**, 591–2.

Binzel, R. P., Gehrels, T., & Matthews, M. S. ed. (1989). *Asteroids II*. Tucson: University of Arizona Press.

Binzel, R. P. & Hubbard, W. B. (1997). Mutual events and stellar occultations. In *Pluto and Charon*, 85–102, ed. S. A. Stern & D. J. Tholen. Tucson: University of Arizona Press.

Birnbaum, G. (1978). Far infrared absorption in H_2–H_2 and H_2–He mixtures. *Journal of Quantitative Spectroscopy and Radiative Transfer*, **19**, 51–62.

Birnbaum, G. (1979). The shape of collision broadened lines from resonance to the far wings. *Journal of Quantitative Spectroscopy and Radiative Transfer*, **21**, 597–607.

Birnbaum, G. ed. (1985). *Phenomena Induced by Intermolecular Interactions*. New York: Plenum Press.

Birnbaum, G. & Cohen, E. R. (1976). Theory of the line shape in pressure induced absorption. *Canadian Journal of Physics*, **54**, 593–602.

Bjoraker, G. L., Larson, H. P., & Kunde, V. G. (1986). The gas composition of Jupiter derived from 5-micron airborn spectroscopic observations. *Icarus*, **66**, 579–609.

Blackman, R. B. & Tukey, J. W. (1958). *The Measurement of Power Spectra*. New York: Dover Publications, Inc.

Blamont, J. E. & Luton, J. M. (1972). Geomagnetic effect on the neutral temperature of the F region during the magnetic storm of September 1969. *Journal of Geophysical Research*, **77**, 3534–56.

Bockelee-Morvan, D., Gautier, D., Lis, D. C., Young, K., Keene, J., Phillips, T., Owen, T., Crovisier, J., Goldsmith, P. F., Bergin, E. A., Despois, D., & Wootten, A. (1998). Deuterated water in Comet C/1996 B2 (Hyakutake) and its implications for the origin of comets. *Icarus*, **133**, 147–62.

Bohr, N. (1913). On the constitution of atoms and molecules. *Philosophical Magazine*, (6) **26**, 476 and 857.

Boltzmann, L. (1884). Ableitung des Stefanschen Gesetzes betreffend die Abhängigkeit der Wärmestrahlung von der Temperatur aus der electromagnetischen Lichttheorie. *Wiedemanns Annalen*, **22**, 291–4.

Born, M. (1926). Quantenmechanik der Stossvorgänge. *Zeitschrift für Physik*, **38**, 803–27.

Born, M. (1927). Das Adiabatenprinzip in der Quantenmechanik. *Zeitschrift für Physik*, **40**, 167–92.

Born, M. & Wolf, E. (1959). *Principles of Optics* (5th edn, revised, 1975). Oxford: Pergamon Press.

Borysow, A. & Tang, C. (1993). Far infrared CIA spectra of N_2–CH_4 pairs for modeling of Titan's atmosphere. *Icarus*, **105**, 175–83.

Boss, A. P., Morfill, G. E., & Tscharnuter, W. M. (1989). Models of the formation and evolution of the Solar nebula. In *Origin and Evolution of Planetary and Satellite Atmospheres*, 35–77, ed. S. K. Atreya, J. B. Pollak, & M. S. Matthews. Tucson: University of Arizona Press.

Boyle, W. S. & Rodgers, K. F. (1959). Performance characteristic of a new low-temperature bolometer. *Journal of the Optical Society of America*, **49**, 66–9.

Brattain, W. H., & Becker, J. A. (1946). Thermistor bolometers. *Journal of the Optical Society of America*, **36**, 354.
Broadfoot, A. L., Atreya, S. K., Bertaux, J. I., Blamont, J. E., Dessler, A. J., Donahue, T. M., Forrester, W. T., Hall, D. T., Herbert, F., Holberg, J. B., Hunten, D. M., Krasnopolsky, V. A., Linick, S., Lunine, J. I., McConnell, J. C., Moos, H. W., Sandel, B. R., Schneider, N. M., Shemansky, D. E., Smith, G. R., Strobel, D. F., & Yelle, R. V. (1989). Ultraviolet spectrometer observations of Neptune and Triton. *Science*, **246**, 1459–66.
de Broglie, L. (1924). Recherches sur la théorie des quanta. Thesis, University of Paris; see also (1925) *Annals de physique*, Série 10, **3**, 22–128.
Brown, M. E. & Calvin W. M. (2000). Evidence of crystalline water and ammonia ices on Pluto's satellite Charon. *Science*, **287**, 107–9.
Brown, R. (1828). Mikroskopische Beobachtungen. *Poggendorfs Annalen*, **14**, 294.
Brown, R. H. & Matson, D. L. (1987). Thermal effects of insolation propagation into the regoliths of airless bodies. *Icarus*, **72**, 84–94.
Buie, M. W. (1992). Pluto and its moon. In *Astronomy and Astrophysics Encyclopedia*, 552–4, ed. S. Maran. New York and Cambridge: Van Nostrand Reinhold and Cambridge University Press.
Buijs, H. L. (1979). A class of high resolution ruggedized Fourier transform spectrometers. *Society of Photo-Optical Instrumentation Engineers (SPIE)*, 191, *Multiplex and/or High-Throughput Spectroscopy*, 116–19.
Bullard, E. (1954). The interior of the Earth. In *The Earth as a Planet*, ed. G. P. Kuiper, Chapter 2. Chicago: University of Chicago Press.
Bunsen, R. & Kirchhoff, G. (1861, 1863). Untersuchungen über das Sonnenspektrum und die Spectren der Chemischen Elemente. Berlin: *Abhandlungen der königlichen Akademie der Wissenschaften*.
Burgdorf, M. J., Caux, E., Clegg, P. E., Davis, G. R., Emery, R. J., Ewart, J. D., Griffin, M. J., Gry, C., Harwood, A. S., King, K. J., Lim, T., Lord, S., Molinari, S., Swinyard, B. M., Texier, D., Tommasi, E., Trams, N., & Unger, S. J. (1998). Calibration and performance of the LWS. *Space Based Astronomy: ISO, AGN, Radiopulsars and the Sun*, **21**, 5–9.
Burkert, P., Fergg, F., & Fischer, H. (1983). A compact interferometer for passive atmospheric sounding. *Institute of Electrical and Electronic Engineers. Transactions for Geoscience and Remote Sensing* GE-21, 345–9.
Cabot, W., Canuto, V. M., Hubickyi, O., & Pollack, J. B. (1987). The role of turbulent convection in the primitive solar nebula. *Icarus*, **69**, 387–457.
Cahen, S. (1986). About the standard solar model. In *Proceedings of the 2nd Institut d'Astrophysique de Paris Workshop: Advances in Nuclear Astrophysics*, ed. E. Vangioni-Flam *et al.*, pp. 97–104.
Caldwell, J. J. (1977). Thermal radiation from Titan's atmosphere. In *Planetary Satellites*, 436–50, ed. J. A. Burns. Tucson: University of Arizona Press.
Cameron, A. G. W. (1975). Cosmogonial considerations regarding Uranus. *Icarus*, **24**, 280–4.
Cameron, A. G. W. (1978). Physics of the primitive solar accretion disk. *Moon and Planets*, **18**, 5–40.
Cameron, A. G. W. (1982). Elemental and nuclidic abundances in the solar system. In *Essays in Nuclear Astrophysics*, ed. C. A. Barnes, D. D. Clyton, & D. N. Schramm. Cambridge: Cambridge University Press.
Cameron, A. G. W. (1985). Formation and evolution of the primitive solar nebula. In *Protostars and Planets II*, ed. D. C. Black & M. S. Matthews, pp. 1073–99. Tucson: The University of Arizona Press.

Cameron, A. G. W. & Benz, W. (1991). The origin of the Moon and single impact hypothesis–IV. *Icarus*, **92**, 204–16.

Cameron, A. G. W. & Ward, W. R. (1976). The origin of the Moon. *Lunar and Planetary Science, VII*, 120–2 (abstract).

Carli, B., Mencaraglia, F., & Bonetti, A. (1984). Submillimeter high-resolution FT spectrometer for atmospheric studies. *Applied Optics*, **23**, 2594–603.

Carlson, R. W., Anderson, M. S., Johnson, R. E., Smythe, W. D., Hendrix, A. R., Barth, C. A., Soderblom, L. A., Hansen, G. B., McCord, T. B., Dalton, J. B., Clark, R. N., Scirley, J. H., Ocampo, A. C., & Matson, D. L. (1999). Hydrogen peroxide on the surface of Europa. *Science*, **283**, 2062–4.

Carlson, R. W., Weissman, P. R., Smythe, W. D., Mahoney, J. C., & the NIMS Science and Engineering Teams (1992). Near-Infrared Mapping Spectrometer experiment on Galileo. *Space Science Review*, **60**, 457–502.

Carr, M. H., Belton, M. J. S., Chapman, C. R., Davis, A. S., Geissler, P., Greenberg, R., McEwen, A. S., Tufts, B. R., Greeley, R., Sullivan, R., Head, J. W., Pappalardo, R. T., Klassen, K. P., Johnson, T. V., Kaufman, J., Senske, D., Moore, J., Neukum, G., Schubert, G., Burns, J. A., Thomas, P., & Veverka, J. (1998). Evidence for a subsurface ocean on Europa. *Nature*, **391**, 363–5.

Cartwright, C. H. (1933). Radiation thermopiles for use at liquid air temperature. *Review of Scientific Instruments*, **4**, 382–4.

Castles, S. (1980). Design of an adiabatic demagnetization refrigerator for studies in astrophysics. *Refrigeration for Cryogenic Sensors NASA Conference Publication 2287*, 389–404.

Chahine, M. T. (1968). Determination of the temperature profile in an atmosphere from its outgoing radiance. *Journal of the Optical Society of America*, **58**, 1634–7.

Challener, W. A., Richards, P. L., Zilio, S. C., & Garvin, H. L. (1980). Grid polarizers for infrared Fourier spectrometers. *Infrared Physics*, **20**, 215–22.

Chamberlain, J. (1971). Phase modulation in far infrared (submillimetre-wave) interferometers. I. Mathematical formulation. *Infrared Physics*, **11**, 25–55.

Chamberlain, J. W. & Hunten, D. M. (1987). *Theory of Planetary Atmospheres*. New York: Academic Press, Inc., Harcourt Brace Jovanovich, Publishers.

Chandrasekhar, S. (1950). *Radiative Transfer*. Oxford: Oxford University Press. (Reprinted and slightly revised by Dover Publications, Inc. in 1960.)

Chanin, G. (1972). Liquid helium-cooled bolometers. In *Infrared Detection Techniques for Space Research*, ed. V. Manno & J. Ring. Dordrecht, Holland: D. Reidel Publ. Co.

Chapman, C. R. (1992). Asteroids. In *Astronomy and Astrophysics Encyclopedia*, 30–2, ed. S. P. Maran. New York and Cambridge: Van Nostrand Reinhold and Cambridge University Press.

Chapman, S. & Lindzen, R. S. (1970). *Atmospheric Tides*. Dordrecht, Holland: D. Reidel Publ. Co.

Chase, S. C., Jr., Engel, J. L., Eyerly, H. W., Kieffer, H. H., Palluconi, F. D., & Schofield, D. (1978). Viking Infrared thermal mapper. *Applied Optics*, **17**, 1243–51.

Chase, S. C., Jr., Hatzenbeler, H., Kieffer, H. H., Miner, E., Münch, G., & Neugebauer, G. (1972). Infrared radiometry experiment on Mariner 9. *Science*, **175**, 308–9.

Chase, S. C., Jr., Miner, E. D., Morrison, D., Münch, G., & Neugebauer, G. (1976). Mariner 10 infrared radiometer results: temperatures and thermal properties of Mercury. *Icarus*, **28**, 565–78.

Chatanier, M. & Gauffre, G. (1971). An infrared transducer for space applications. In *Infrared Detection Techniques*, ed. V. Manno & J. Ring. Dordrecht, Holland: D. Reidel Publ. Co.

Christensen, P. R. (1998). Variation of the Martian surface composition and cloud occurences determined from thermal emission spectroscopy: Analysis of Viking and Mariner 9 data. *Journal of Geophysical Research* **103**, 1733–46.

Christensen, P. R., Anderson, D. L., Chase, S. C., Jr., Clancy, R. T., Clark, R. N., Conrath, B. J., Kieffer, H. H., Kuzmin, R. O., Malin, M. C., Pearl, J. C., Roush, T. L., & Smith, M. D. (1998). Results from the Mars Global Surveyor Thermal Emission Spectrometer. *Science*, **279**, 1692–8.

Christensen, P. R., Anderson, D. L., Chase, S. C., Jr., Clark, R. N., Kieffer, H. H., Malin, M. C., Pearl, J. C., Carpenter, J., Bandiera, N., Brown, F. G., & Silverman, S. (1992). Thermal Emission Spectrometer experiment: Mars Observer Mission. *Journal of Geophysical Research*, **97**, 7719–34.

Christensen, P. R., Bandfield, J. L., Clark, R. N., Edgett, K. S., Hamilton, V. E., Hoefen, T., Kieffer, H. H., Kuzmin, R. O., Lane, M. D., Malin, M. C., Morris, R. V., Pearl, J. C., Pearson, R., Roush, T. L., Ruff, S. W., & Smith, M. D. (2000*a*). Detection of crystalline hematite mineralization on Mars by the Thermal Emission Spectrometer: Evidence for near-surface water. *Journal of Geophysical Research*, **105**, 9623–42.

Christensen, P. R., Bandfield, J. L., Smith, M. D., Hamilton, V. E., & Clark, R. N. (2000*b*). Identification of a basaltic component on the Martian surface from Thermal Emission Spectrometer data. *Journal of Geophysical Research*, **105**, 9609–21.

Christensen, P. R. & Pearl, J. C. (1997). Initial data from the Mars Global Surveyor thermal emission spectrometer experiment: Observations of the Earth. *Journal of Geophysical Research*, **102**, 10875–10880.

Christy, J. W. & Harrington, R. S. (1978). The satellite of Pluto. *Astronomical Journal*, **83**, 1005–8.

Chynoweth, A. G. (1956*a*). Dynamic method for measuring the pyroelectric effect with special reference to barium titanate. *Journal of Applied Physics*, **27**, 78–84.

Chynoweth, A. G. (1956*b*). Surface space-charge layers in barium titanate. *Physical Review*, **102**, 705–14.

Clancy, R. T., Lee, S. W., Gladstone, G. R., McMillan, W. W., & Roush, T. (1995). A new model for Mars atmospheric dust based upon analysis of ultraviolet through infrared observations. *Journal of Geophysical Research*, **100**, 5251–63.

Clegg, P. E. (1992). The long-wavelength spectrometer in ISO. In *Infrared Astronomy with ISO*, 87–102, ed. Th. Encrenaz & M. F. Kessler. New York: Nova Science Publishers, Inc.

Clegg, P. E., Ade, P. A. R., Armand, C., Baluteau, J. P., Barlow, M. J., Buckley, M. A., Berges, J. C., Burgdorf, M., Caux, E., Ceccarelli, C., Cerulli, R., Church, S. E., Cotin, F., Cox, P., Cruvellier, P., Culhane, J. L., Davis, G. R., Di Giorgio, A., Diplock, B. R., Drummond, D. L., Emery, R. J., Ewart, J. D., Fischer, J., Furniss, I., Glencross, W. M., Greenhouse, M. A., Griffin, M. J., Gry, C., Harwood, A. S., Hazell, A. S., Joubert, M., King, K. J., Lim, T., Liseau, R., Long, J. A., Lorenzetti, D., Molinari, S., Murray, A. G., Naylor, D. A., Nasini, B., Norman, K., Omont, A., Orfei, R., Patrick, T. J., Pequignot, D., Pouliquen, D., Price, M. C., Nguyen-Q-Rieu, Rogers, A. J., Robinson, F. D., Saisse, M., Saraceno, P., Serra, G., Sidher, S. D., Smith, A. F., Smith, H. A., Spinoglio, L., Swinyard, B. M., Texier, D., Towlson, W. A., Trams, N. R., Unger, S. J., & White, G. J. (1996). The ISO long-wavelength spectrometer. *Astronomy and Astrophysics*, **315**, L38–42.

Clayton, D. D. (1968). *Principles of Stellar Evolution and Nucleosynthesis*. New York: McGraw-Hill. (Reprinted in 1983 by The University of Chicago Press.)

Cohen, C. J. & Hubbard, E. C. (1965). Liberations of the close approaches of Pluto to Neptune. *Astronomical Journal*, **70**, 10–13.

Cohen, E. R. & Giacomo, P. (1987). Symbols, units, nomenclature, and fundamental constants in physics. *Physica*, **146A**, 1–68.

Combes, M., Vapillon, L., Gendron, E., Coustenis, A., Lai, O., Wittenberg, R., & Sirday, R. (1997). Spatially resolved images of Titan by means of adaptive optics. *Icarus*, **129**, 482–97.

Connes, J. (1961). Recherches sur la spectroscopie par transformation de Fourier. *Revue d'Optique*, **40**, 45–79, 116–40, 171–90, 231–64.

Connes, J. & Connes, P. (1966). Near-infrared planetary spectra by Fourier spectroscopy. 1. Instruments and results. *Journal of the Optical Society of America*, **56**, 896–910.

Connes, J., Connes, P., & Maillard, J. (1967*a*). Spectroscopie astronomique par transformation de Fourier. *Journal de Physique*, **28**, C2–120.

Connes, J., Connes, P., & Maillard, J. (1967*b*). *Atlas des Spectres Planétaires Infrarouges*. Paris: Editions du CNRS.

Connes, P., Connes, J., Benedict, W. S., & Kaplan, L. D. (1967*c*). Traces of HCl and HF in the atmosphere of Venus. *Astrophysical Journal*, **147**, 1230–7.

Connes, P., Connes, J., Kaplan, L., & Benedict, W. (1968). Carbon monoxide in the Venus atmosphere. *Astrophysical Journal*, **152**, 731–43.

Conrady, A. E. (1957). *Applied Optics and Optical Design*, Vols. I and II. New York: Dover Publications, Inc.

Conrath, B., Curran, R., Hanel, R., Kunde, V., Maguire, W., Pearl, J., Pirraglia, J., Welker, J., & Burke, T. (1973). Atmospheric and surface properties of Mars obtained by infrared spectroscopy on Mariner 9. *Journal of Geophysical Research*, **78**, 4267–78.

Conrath, B., Flasar, F. M., Hanel, R., Kunde, V., Maguire, W., Pearl, J., Pirraglia, J., Samuelson, P., Gierasch, P., Weir, A., Bézard, B., Gautier, D., Cruikshank, D., Horn, L., Springer, R., & Shaffer, W. (1989*a*). Infrared observations of the Neptunian system. *Science*, **246**, 1454–9.

Conrath, B. & Gautier, D. (1979). Thermal structure of Jupiter's atmosphere obtained by inversion of Voyager 1 infrared measurements. In *Remote Sensing of Atmospheres and Oceans*, ed. A. Deepak, pp. 611–30. New York: Academic Press.

Conrath, B. J. & Gautier, D. (2000). Saturn helium abundance: A reanalysis of Voyager measurements. *Icarus*, **144**, 124–34.

Conrath, B. J., Gautier, D., Hanel, R. A., & Hornstein, J. S. (1984). The helium abundance of Saturn from Voyager measurements. *Astrophysical Journal*, **282**, 807–15.

Conrath, B. J., Gautier, D., Hanel, R., Lindal, G., & Marten, A. (1987). The helium abundance of Uranus from Voyager measurements. *Journal of Geophysical Research*, **92**, 15003–10.

Conrath, B. J., Gautier, D., Lindal, G. F., Samuelson, R. E., & Shaffer, W. A. (1991*a*). The helium abundance of Neptune from Voyager measurements. *Journal of Geophysical Research*, **96**, 18921–39.

Conrath, B. J., Gautier, D., Owen, T., & Samuelson, R. E. (1993). Constraints on N_2 in Neptune's atmosphere from Voyager measurements. *Icarus*, **101**, 168–72.

Conrath, B. J., Gierasch, P. J., & Leroy, S. S. (1990). Temperature and circulation in the stratospheres of the outer planets. *Icarus*, **83**, 255–81.

Conrath, B. J., Gierasch, P. J., & Ustinov, E. A. (1998). Thermal structure and para hydrogen fraction on the outer planets from Voyager IRIS measurements. *Icarus*, **135**, 501–17.

Conrath, B. J., Hanel, R. A., Prabhakara, C., Kunde, V. G., Revah, I., & Salomonson, V. V. (1971). Vertical sounding of the atmosphere with the Nimbus IV infrared spectrometer experiment. *Proceedings of the Twenty-first Astronautical Congress*. Amsterdam: North-Holland Publ. Co.

Conrath, B. J., Hanel, R. A., & Samuelson, R. E. (1989*b*). Thermal structure and heat balance of the outer planets. In *Origin and Evolution of Planetary and Satellite Atmospheres*, ed. S. K. Atreya, J. B. Pollack, & M. S. Matthews. Tucson: The University of Arizona Press.

Conrath, B. J., Pearl, J. C., Appleby, J. F., Lindal, G. F., Orton, G. S., & Bézard, B. (1991*b*). Thermal structure and energy balance of Uranus. In *Uranus*, 204–52, ed. J. T. Bergstralh, E. D. Miner, & M. S. Matthews. Tucson: University of Arizona Press.

Conrath, B. J., Pearl, J. C., Smith, M. D., Maguire, W. C., Christensen, P. R., Dason, S., & Kaelberer, M. S. (2000). Mars Global Surveyor Thermal Emission Spectrometer (TES) observations: Atmospheric temperatures during aerobraking and science phasing. *Journal of Geophysical Research*, **105**, 9509–19.

Cooley, J. W. & Tuckey, J. W. (1965). An algorithm for the machine calculation of complex Fourier series. *Mathematics of Computation*, **19**, 297–301.

Cooper, J. (1962). Minimum detectable power of a pyroelectric thermal receiver. *Review of Scientific Instruments*, **33**, 92–5.

Coron, N., Dambier, G., & Leblanc, J. (1972). A new type of helium-cooled bolometer. In *Infrared Detection Techniques for Space Research*, ed. V. Manno & G. Ring. Dordrecht, Holland: D. Reidel Publ. Co.

Courant, R. & Hilbert, D. (1931). *Methoden der Mathematischen Physik*. Berlin: Springer-Verlag.

Courtin, R. (1988). Pressure-induced absorption coefficients for radiative transfer calculations in Titan's atmosphere. *Icarus*, **75**, 245–54.

Courtin, R., Gautier, D., Marten, A., Bézard, B., & Hanel, R. (1984). The composition of Saturn's atmosphere at northern temperate latitudes from Voyager IRIS spectra: NH_3, PH_3, C_2H_2, C_2H_4, CH_3D, CH_4 and the Saturnian D/H isotopic ratio. *Astrophysical Journal*, **287**, 899–916.

Courtin, R., Gautier, D., Marten, A., & Kunde, V. (1983). The $^{12}C/^{13}C$ ratio in Jupiter from the Voyager infrared investigation. *Icarus*, **53**, 121–32.

Courtin, R., Gautier, D., & McKay, C. P. (1995). Titan's thermal emission spectrum: reanalysis of the Voyager infrared measurements. *Icarus*, **114**, 144–62.

Coustenis, A., Bézard, B., & Gautier, D. (1989*a*). Titan's atmosphere from Voyager infrared observations I. The gas composition of Titan's equatorial region. *Icarus*, **80**, 54–76.

Coustenis, A., Bézard, B., & Gautier, D. (1989*b*). Titan's atmosphere from Voyager infrared observations II. The CH_3D abundance and D/H ratio from the 900–1200 cm^{-1} spectral region. *Icarus*, **82**, 67–80.

Coustenis, A., Bézard, B., Gautier, D., Marten, A., & Samuelson, R. (1991). Titan's atmosphere from Voyager infrared observations III. Vertical distributions of hydrocarbons and nitriles near Titan's north pole. *Icarus*, **89**, 152–67.

Coustenis, A., Encrenaz, Th., Salama, A., Lellouch, E., Gautier, D., Kessler, M. F., de Graauw, Th., Samuelson, R. E., Bjoraker, G., & Orton, G. (1997). ISO Observations of Titan with SWS/Grating. In *Proceedings of First ISO Workshop on Analytical Spectroscopy*. (ESA SP-419.)

Coustenis, A., Salama, A., Lellouch, E., Encrenaz, Th., Bjoraker, G. L., Samuelson, R. E., de Graauw, Th., Feuchtgruber, H., & Kessler, M. F. (1998). Evidence for water

vapor in Titan's atmosphere from ISO/SWS data. *Astronomy and Astrophysics*, **336**, L85–9.
Coustenis, A. & Taylor, F. (1999). *Titan: The Earth-Like Moon*. Singapore, New Jersey, London, and Hong Kong: World Scientific Publishing Co. Pte. Ltd.
Craig, I. J. D. & Brown, J. C. (1986). *Inverse Problems in Astronomy*. Boston: Adam Hilger Ltd.
Cruikshank, D. P. ed. (1995). *Neptune and Triton*. Tucson: University of Arizona Press.
Cruikshank, D. P., Pilcher, C. B., & Morrison, D. (1976). Pluto: Evidence of methane frost. *Science*, **194**, 835–7.
Cruikshank, D. P., Roush, T. L., Moore, J. M., Sykes, M., Owen, T. C., Bartholomew, M. J., Brown, R. H., & Tryka, K. A. (1997). The surfaces of Pluto and Charon. In *Pluto and Charon*, 221–67, ed. S. A. Stern & D. J. Tholen. Tucson: University of Arizona Press.
Cruikshank, D. P., Roush, T. L., Owen, T. C., Quirico, E., & de Bergh, C. (1998). Pluto and Kuiper Disk. In *Solar System Ices*, 655–84, ed. B. Schmitt, C. de Bergh, & M. Festou. Dordrecht: Kluwer Academic Publishers.
Curran, R. J., Conrath, B., Hanel, R., Kunde, V., & Pearl, J. (1973). Mars: Mariner 9 spectroscopic evidence for H_2O ice clouds. *Science*, **182**, 381–3.
Czerny, M. (1930). Messungen am Steinsalz im Ultraroten zur Prüfung der Dispersionstheorie. *Zeitschrift für Physik*, **65**, 600–31.
Dall'Oglio, G., Melchiorri, B., & Natale, V. (1974). Comparison between carbon, silicon, and germanium bolometers and Golay cell in the far infrared. *Infrared Physics*, **14**, 347–50.
Danielson, R. E., Caldwell, J. J., & Larach, D. R. (1973). An inversion in the atmosphere of Titan. *Icarus*, **20**, 437–43.
Davis, G. R., Furniss, I., Towlson, W. A., Ade, P. A. R., Emery, R. J., Glencross, W. M., Naylor, D. A., Patrick, T. J., Sidey, R. J., & Swinyard, B. M. (1995). Design and performance of cryogenic, scanning Fabry–Perot interferometers for the Long-Wavelength Spectrometer on the Infrared Space Observatory. *Applied Optics*, **34**, 92–106.
Davis, G. R., Griffin, M. J., Naylor, D. A., Oldham, P. G., Swinyard, B. M., Ade, P. A. R., Calcutt, S. B., Encrenaz, Th., De Graauw, Th., Gautier, D., Irwin, P. G. J., Lellouch, E., Orton, G. S., Armand, C., Burgdorf, M., Di Giorgio, A., Ewart, D., Gry, C., King, K. J., Lim, T., Molinari, S., Price, M., Sidher, S., Smith, A., Texier, D., Trams, N., & Unger, S. J. (1996). ISO LWS measurement of the far-infrared spectrum of Saturn. *Astronomy and Astrophysics*, **315**, L393–6.
Dayhoff, M. O., Eck, R., Lippincott, E. R., & Sagan, C. (1967). Venus: atmospheric evolution. *Science*, **155**, 556–7.
De Graauw, Th., Feuchtgruber, H., Bézard, B., Drossart, P., Encrenaz, Th., Beintema, D. A., Griffin, M., Heras, A., Kessler, M., Leech, K., Lellouch, E., Morris, P., Roelfsema, P. R., Roos-Serote, M., Salama, A., Vandenbussche, B., Valentijn, E. A., Davis, G. R., & Naylor, D. A. (1997). First results of ISO–SWS observations of Saturn: detection of CO_2, CH_3C_2H, C_4H_2, and tropospheric H_2O. *Astronomy and Astrophysics*, **321**, L13–16.
De Graauw, Th., Haser, L. N., Beintema, D. A., Roelfsema, P. R., van Agthoven, H., Barl, L., Bauer, O. H., Bekenkamp, H. E. G., Boonstra, A. J., Boxhoorn, D. R., Cote, J., de Groe, P., van Dijkhuizen, C., Drapatz, S., Evers, J., Feuchtgruber, H., Frericks, M., Genzel, R., Haerendel, G., Heras, A. M., van der Hucht, K. A., van der Hulst, Th., Huygen, R., Jacobs, H., Jakob, G., Kamperman, Th., Katterloher, R. O., Kester, D. J.

M., Kunze, D., Kussendrager, D., Lahuia, F., Lamers, H. J. G. L. M., Leech, K., van der Lei, S., van der Linden, R., Luinge, W., Lutz, D., Melzner, F., Morris, P. W., van Nguyen, D., Ploeger, G., Price, S., Salama, A., Schaidt, S. G., Sijm, N., Smoorenburg, C., Spakman, J., Spoon, H., Steinmayer, M., Stoecker, E. A., Valentijn, E. A., Vandenbussche, B., Visser, H., Waelkens, C., Waters, L. B. F. M., Wensink, J., Wesselius, P. R., Wiezorrek, E., Wieprecht, E., Wijnbergen, J. J., Wildeman, K. J., & Young, E. (1996). Observing with the ISO Short-Wavelength Spectrometer. *Astronomy and Astrophysics*, **315**, L49–54.

Delsemme, A. H. (1992). Comets. In *Astronomy and Astrophysics Encyclopedia*, 104–7, ed. S. P. Maran. New York and Cambridge: Van Nostrand Reinhold and Cambridge University Press.

Delsemme, A. H. (1997). Organic chemistry in comets from remote and *in situ* observations. In *Comets and the Origin and Evolution of Life*, 29–68, ed. P. Thomas *et al*. New York: Springer Verlag.

Delsemme, A. H. (1998). Recollections of a cometary scientist. *Planetary and Space Science*, **46**, 111–24.

Delsemme, A. H. (1999). The deuterium enrichment observed in recent comets is consistent with the cometary origin of seawater. *Planetary and Space Science*, **47**, 125–31.

Deming, D., Espenak, F., Jennings, D., Kostiuk, T., & Mumma, M. J. (1982). Evidence for high altitude haze thickening on the dark side of Venus from 10-micron heterodyne spectroscopy of CO_2. *Icarus*, **49**, 35–48.

Dereniak, E. L. & Boreman, G. D. (1996). *Infrared Detectors and Systems*. New York: John Wiley & Sons, Inc.

Dereniak, E. L. & Crowe, D. G. (1984). *Optical Radiation Detectors*. New York: John Wiley & Sons.

Dereniak, E. L. & Sampson, R. E. eds. (1998). *Infrared detectors and focal plane arrays V. SPIE Proceedings*, **3379**, Bellingham, Washington.

De Waard, R. & Wormser, E. M. (1959). Description and properties of various thermal detectors. *Proceedings of the Institute of Radio Engineers*, **47**, No. 9, 1508–13.

Dobrovolskis, A. R., Peale, S. J., & Harris, A. W. (1997). Dynamics of the Pluto–Charon binary. In *Pluto and Charon*, 159–90, ed. S. A. Stern & D. J. Tholen. Tucson: University of Arizona Press.

Donahue, T. M., Hoffman, J. H., Hodges, R. R., Jr., & Watson, A. J. (1982). Venus was wet: a measurement of the ratio of deuterium to hydrogen. *Science*, **216**, 630–3.

Dragovan, M. & Moseley, S. H. (1984). Gold absorbing film for a composite bolometer. *Applied Optics*, **23**, 654–6.

Dreyfus, M. G. & Hilleary, D. T. (1962). Satellite infrared spectrometer. *Aerospace Engineering*, **21**, 42–5.

Drossart, P., Rosenqvist, J., Erard, S., Langevin, Y., Bibring, J. P., & Combes, M. (1991). Martian aerosol properties from the Phobos–ISM experiment. *Annals of Geophysics*, **9**, 754–60.

Drummond, J. R., Houghton, J. T., Pescett, G. D., Rodgers, D. C., Wale, M. J., Whitney, J., & Williamson, E. J. (1980). The stratospheric and mesopheric sounder on Nimbus 7. *Philosophical Transactions Royal Society London*, **A296**, 219–41.

Edgeworth, K. E. (1949). The origin and evolution of the solar system. *Monthly Notices of the Royal Astronomical Society*, **109**, 600–9.

Eichhorn, W. L. & Magner, T. J. (1986). Large aperture, freestanding wire grid polarizers for the far-infrared: fabrication, characterization, and testing. *Optical Engineering*, **25**, 541–4.

Einstein, A. (1905*a*). A heuristic standpoint concerning the production and transformation of light. *Annalen der Physik*, (4), **17**, 132–48.

Einstein, A. (1905*b*). Investigations on the theory of the Brownian motion. *Annalen der Physik*, (4), **17**, 549.

Einstein, A. (1906*a*). Zur Theorie der Brownschen Bewegung. *Annalen der Physik*, (4), **19**, 371–81.

Einstein, A. (1906*b*). Zur Theorie der Lichterzeugung und Licht absorption. *Annalen der Physik*, (4), **20**, 199–206.

Eisberg, R. & Resnick, R. (1974). *Quantum Physics* (2nd edn 1985). New York: John Wiley & Sons, Inc.

Elson, L. S. (1979). Preliminary results from the Pioneer Venus infrared radiometer: temperature and dynamics in the upper atmosphere. *Geophysical Research Letters*, **6**, 720–2.

Emslie, A. G. & Aronson, J. R. (1973). Spectral reflectance and emittance of particulate material, 1. Theory. *Applied Optics*, **12**, 2563–72.

Encrenaz, Th., de Graauw, Th., Schaeidt, S., Lellouch, E., Feuchtgruber, H., Beintema, D. A., Bézard, B., Drossart, P., Griffin, M., Heras, A., Kessler, M., Leech, K., Morris, P., Roelfsema, P. R., Roos-Serote, M., Salama, A., Vandenbussche, B., Valentijn, E. A., Davis, G. R., & Naylor, D. A. (1996). First results of ISO–SWS observations of Jupiter. *Astronomy and Astrophysics*, **315**, L397–400.

Encrenaz, Th., Drossart, P., Carlson, R. W., & Bjoraker, G. (1997*a*). Detection of H_2O in the splash phase of G- and R-impacts from NIMS–Galileo. *Planetary and Space Science*, **45**, 1189–96.

Encrenaz, Th., Lellouch, E., Feuchtgruber, H., Altieri, B., Bézard, B., Davis, M. G., de Graauw, Th., Drossart, P., Griffin, M. J., Kessler, M. F., & Oldham, P. G. (1997*b*). The giant planets as seen by ISO. *Proceedings First Workshop on Analytical Spectroscopy*, 6–8 October 1997, 125–30. (ESA SP 419, December 1997.)

Erard, S. & Celvin, W. (1997). New composite spectra of Mars from 0.4 to 3.14 μm. *Icarus*, **130**, 449–60.

Erickson, E. F., Goorvitch, D., Simpson, J. P., & Stecker, D. W. (1978). Far infrared brightness temperature of Jupiter and Saturn. *Icarus*, **35**, 61–73.

Fabry, C. & Perot, A. (1899). Théorie et application d'une nouvelle méthode de spectroscopie interférentielle. *Annals de Chimie et de Physique*, **16**, 115–44.

Farmer, C. B., Davis, D. W., Holland, A. L., LaPorte, D. D., & Doms, P. E. (1977). Mars: Water vapor observations from the Viking Orbiters. *Journal of Geophysical Research*, **82**, 4225–48.

Farmer, C. B. & LaPorte, D. D. (1972). The detection and mapping of water vapor in the Martian atmosphere. The Viking Mars Orbiter. *Icarus*, **16**, 34–46.

Farmer, C. B. & Norton, R. H. (1989). *A High-Resolution Atlas of the Infrared Spectrum of the Sun and the Earth Atmosphere from Space. NASA Reference Publication 1224*, Vol. II. Washington, DC: NASA, Scientific and Technical Information Division.

Farmer, C. B. & Raper, O. F. (1986). High resolution infrared spectroscopy from space: A preliminary report on the results of the atmospheric trace molecule (ATMOS) experiment on Spacelab 3. *NASA Publication CP-2429*.

Fegley, B., Gautier, D., Owen, T., & Prinn, R. G. (1991). Spectroscopy and chemistry of the atmosphere of Uranus. In *Uranus*, 147–203, ed. J. Bergstralh, E. Miner, & M. S. Matthews. Tucson: University of Arizona Press.

Fegley, M. B., Jr. & Prinn, R. G. (1988*a*). Chemical constraints on the water and total oxygen abundances in the deep atmosphere of Jupiter. *Astrophysical Journal*, **324**, 621–5.

Fegley, M. B., Jr. & Prinn, R. G. (1988b). The predicted abundances of deuterium-bearing gases in the atmosphere of Jupiter and Saturn. *Astrophysical Journal*, **326**, 490–508.

Fellgett, P. B. (1951). Ph.D. Thesis, University of Cambridge (unpublished).

Fellgett, P. B. (1958). Spectrométre interférentiel multiplex pour mesures infra-rouges sur les étoiles. *Journal de Physique et le Radium*, **19**, 237–40.

Fellgett, P. B. (1971). The origins and logic of multiplex, Fourier, and interferometric methods in spectrometry. *Aspen International Conference on Fourier Spectroscopy*, ed. G. A. Vanasse, A. T. Stair, & D. J. Baker, pp. 139–42. US Air Force Cambridge Research Laboratory 71-0019.

Feuchtgruber, H., Lellouch, E., Bézard, B., Encrenaz, Th., de Graauw, Th., & Davis, G. R. (1999). Detection of HD in the atmospheres of Uranus and Neptune: a new determination of the D/H ratio. *Astronomy and Astrophysics*, **341**, L17–21.

Feuchtgruber, H., Lellouch, E., de Graauw, Th., Bézard, B., Encrenaz, Th., & Griffin, M. (1997). External supply of oxygen to the atmospheres of the giant planets. *Nature*, **389**, 159–62.

Fimmel, R. O., Colin, L., & Burgess, E. (1983). *Pioneer Venus. NASA Special Publication SP-461*.

Fink, U., Larson, H. P., & Treffers, R. R. (1978). Germane in the atmosphere of Jupiter. *Icarus*, **34**, 344–54.

Flasar, F. M. (1998). The dynamic meteorology of Titan. *Planetary and Space Science*, **46**, 1125–47.

Flasar, F. M., Conrath, B. J., Gierasch, P. J., & Pirraglia, J. A. (1987). Voyager infrared observations of Uranus' atmosphere: thermal structure and dynamics. *Journal of Geophysical Research*, **92**, 15011–18.

Flasar, F. M., Samuelson, R. E., & Conrath, B. J. (1981). Titan's atmosphere: temperature and dynamics. *Nature*, **292**, 693–98.

Forman, M. L. (1966). Fast Fourier transform technique and its application to Fourier spectroscopy. *Journal of Optical Society of America*, **56**, 978–9.

Formisano, V. (1999). Private information.

Formisano, V., Grassi, D., Piccioni, G., Pearl, J., Hanel, R., Bjoraker, G., & Conrath, B. (2000). IRIS Mariner 9 data revisited: An instrumental effect. *Planetary and Space Science*, **48**, 569.

Foster, M. (1961). An application of Wiener–Kolmogorov smoothing theory to matrix inversion. *Journal of the Society of Industrial Applied Mathematics*, **9**, 387–92.

Fowler, A. M. ed. (1993). *Infrared detectors and instrumentation. SPIE Proceedings*, **1946**. Bellingham, Washington.

Fraunhofer, J. (1817). *Gilberts Annalen*, **56**, 264.

Fresnel, A. (1816). Sur la diffraction de la lumière, où l'on examine particulièrement de phénomène des franges colorées que présentent les ombres des corps éclairés par une point lumineux. *Annals de Chimie et de Physique*, **1**, 239–81.

Fresnel, A. (1819). Du mémoire sur la diffraction de la lumière. *Annals de Chimie et de Physique*, **11**, 337–8.

Frommhold, L., Samuelson, R., & Birnbaum, G. (1984). Hydrogen dimer structure in the far infrared spectra of Jupiter and Saturn. *Astrophysical Journal Letters*, **283**, L79–82.

Fung, A. K. & Eom, H. J. (1981). Emission from a Rayleigh layer with irregular boundaries. *Journal of Quantitative Spectroscopy and Radiative Transfer*, **26**, 397–409.

Gaidos, E. J., Nealson, K. N., & Kirschvink, J. L. (1999). Life in ice-covered oceans. *Science*, **284**, 1631–3.

Gallinaro, G. & Varone, R. (1975). Construction and calibration of a fast superconducting bolometer. *Cryogenics*, **15**, 292–3.

Gautier, D. (1998). Deuterium in the Solar System and cosmogonical implications. In *Planetary Systems: The Long View, 9th Rencontres de Blois*, 17–24, ed. L. M. Celnikier & J. Tran Thanh Van. Paris: Editions Frontieres.

Gautier, D., Bézard, B., Marten, A., Baluteau, J. P., Scott, N., Chedin, A., Kunde, V., & Hanel, R. (1982). The C/H ratio in Jupiter from the Voyager infrared investigation. *Astrophysical Journal*, **257**, 901–12.

Gautier, D., Conrath, B., Flasar, M., Hanel, R., Kunde, V., Chedin, A., & Scott, N. (1981). The helium abundance of Jupiter from Voyager. *Journal of Geophysical Research*, **86**, 8713–20.

Gautier, D., Conrath, B., Owen, T., de Pater, I., & Atreya, S. (1995). The troposphere of Neptune. In *Neptune and Triton*, 547–611, ed. D. Cruikshank. Tucson: University of Arizona Press.

Gautier, D. & Grossman, K. (1972). A new method for the determination of the mixing ratio hydrogen to helium in the giant planets. *Journal of Atmospheric Science*, **29**, 788–92.

Gautier, D. & Owen, T. (1983). Cosmological implications of helium and deuterium abundances on Jupiter and Saturn. *Nature*, **302**, 215–18.

Gautier, D. & Owen, T. (1989). The composition of the outer planets' atmospheres. In *Origins and Evolution of Planetary and Satellite Atmospheres*, ed. S. K Atreya, J. B. Pollack, & M. S. Matthews. Tucson: The University of Arizona Press.

Gehrels, T. (1976). The results of the imaging photopolarimeter on Pioneers 10 and 11. In *Jupiter*, ed. T. Gehrels. Tucson: The University of Arizona Press.

Gehrels, T. ed. (1994). *Hazards due to Comets and Asteroids*. Tucson: University of Arizona Press.

Gehrels, T., Coffeen, D., Tomasko, M., Doose, L., Swindell, W., Castillo, N., Kendall, J., Clements, A., Hämeen-Antilla, J., Knight, C. K., Blenman, C., Baker, R., Best, G., & Baker, L. (1974). The imaging photopolarimeter experiment on Pioneer 10. *Science*, **183**, 318–20.

Gierasch, P. J., Conrath, B. J., & Magalhães, J. A. (1986). Zonal mean properties of Jupiter's upper troposphere from Voyager infrared observations. *Icarus*, **67**, 456–83.

Gille, J. C., Russell, J. M., Bailey, P. L., Gordley, L. L., Remsberg, E. E., Liesesch, J. H., Planet, W. G., House, F. B., Lyjak, L. V., & Beck, S. A. (1984). Validation of temperature retrievals obtained by the Limb Infrared Monitor of the Stratosphere (LIMS) experiment on Nimbus 7. *Journal of Geophysical Research*, **89**, 5147–60.

Goguen, J. D. & Sinton, W. M. (1985). Characterization of Io's volcanic activity by infrared polarimetry. *Science*, **230**, 65–9.

Golay, M. J. E. (1947). A pneumatic infrared detector. *Review of Scientific Instruments*, **18**, 357–62.

Gold, T. (1964). Outgassing processes on the Moon and Venus. In *The Origin and Evolution of Atmospheres and Oceans*, ed. P. J. Brancazio & A. J. W. Cameron, pp. 249–56. New York: Wiley.

Goldberg, I. L. & McCulloch, A. W. (1969). Annular aperture diffracted energy distribution for an extended source. *Applied Optics*, **8**, 1451–8.

Goldman, S. (1953). *Information Theory*. New York: Prentice-Hall, Inc. (Reprinted by Dover Publications, Inc. in 1968.)

Goldreich, P., & Ward, W. R. (1973). The formation of planetesimals. *Astrophysical Journal*, **183**, 1051–61.

Golombek, M. P. (1998). The Mars Pathfinder Mission. *Scientific American*, **279**, 40–9.

Golombek, M. P., Cook, R. A., Economou, T., Folkner, W. M., Haldemann, A. F. C., Kallemeyn, P. H., Knudsen, J. M., Manning, R. M., Moore, H. J., Parker, T. J., Rieder, R., Schofield, P. H., Smith, P. H., Vaughan, R. M. (1997). Overview of the Mars Pathfinder Mission and assessment of landing site predictions. *Science*, **278**, 1734–74.

Gomes, R. (1999). On the edge of the Solar System. *Science*, **286**, 1487–8.

Goody, R. M. (1964). *Atmospheric Radiation*. Oxford: Clarendon Press.

Goody, R. M. & Walker, J. C. G. (1972). *Atmospheres*. Englewood Cliffs, NJ: Prentice-Hall.

Goody, R. M. & Yung, Y. L. (1989). *Atmospheric Radiation Theoretical Basis*. Oxford: Clarendon Press (second edition).

Graboske, H. C., Jr., Pollack, J. B., Grossman, A. S., & Olness, R. J. (1975). The structure and evolution of Jupiter, the fluid contraction phase. *Astrophysical Journal*, **199**, 265–81.

Griffin, M. J., Naylor, D. A., Davis, G. R., Ade, P. A. R., Oldham, P. G., Swinyard, B. M., Gautier, D., Lellouch, E., Orton, G. S., Encrenaz, Th., de Graauw, Th., Furniss, I., Smith, H., Armand, C., Burgdorf, M., Di Giorgio, A., Ewart, D., Gry, C., King, K. J., Lim, T., Molinari, S., Price, M., Sidher, S., Smith, A., Texier, D., Trams, N., Unger, S. J., & Salama, A. (1996). First detection of the 56-μm rotational line of HD in Saturn's atmosphere. *Astronomy and Astrophysics*, **315**, L389–92.

Griffith, C. A., Owen, T., Miller, G. A., & Geballe, T. (1998). Transient clouds in Titan's lower atmosphere. *Nature*, **395**, 575–8.

Groebner, W. & Hofreiter, N. (1957). *Integraltafeln*. Vienna, Innsbruck: Springer-Verlag.

Grossman, A. S., Pollack, J. B., Reynolds, R. T., & Summers, A. L. (1980). The effect of dense cores on the structure and evolution of Jupiter and Saturn. *Icarus*, **42**, 358–79.

Guillot, T. (1999). Interiors of giant planets inside and outside the Solar System. *Science*, **286**, 72–7.

de Haas-Lorentz, G. L. (1913). *Die Brownsche Bewegung*. Braunschweig: Vieweg Verlag.

Haberle, R. M., Leovy, C. B., & Pollack, J. B. (1982). Some effects of global dust storms on the atmospheric circulation of Mars. *Icarus*, **50**, 322–67.

Haberle, R. M., Pollack, J. B., Barnes, J. R., Zurek, R. W., Leovy, C. B., Murphy, J. R., Lee, H., & Schaeffer, J. (1993). Mars atmospheric dynamics as simulated by the NASA Ames general circulation model 1. The zonal-mean circulation. *Journal of Geophysical Research*, **98**, 3093–123.

Hagen, E. & Rubens, H. (1903). Über Beziehungen des Reflexions- und Emissionsvermögens der Metalle zu ihrem elektrischen Leitvermögen. *Annalen der Physik* (4), **11**, 873–901.

Halliday, A. N., & Drake, M. J. (1999). Colliding theories. *Science*, **283**, 1861–3.

Haltiner, G. J. & Williams, R. T. (1980). *Numerical Prediction and Dynamic Meteorology* (2nd edn). New York: John Wiley & Sons.

Hanel, R. A. (1961*a*). Low resolution radiometer for satellites. *American Rocket Society Journal*, 246–50.

Hanel, R. A. (1961*b*). An instrument to measure the solar constant from satellites. *NASA Technical note D-1152*.

Hanel, R. A. (1961*c*). Dielectric bolometer: a new type of thermal radiation detector. *Journal of the Optical Society of America*, **51**, 220–4.

Hanel, R. A. (1983). Planetary exploration with spaceborne Michelson interferometers in the thermal infrared. In *Spectrometric Techniques, III*. ed. G. A. Vanasse, Chapter 2, New York and London: Academic Press.

Hanel, R. A., Conrath, B., Flasar, F. M., Kunde, V., Maguire, W., Pearl, J., Pirraglia, J., Samuelson, R., Herath, L., Allison, M., Cruikshank, D., Gautier, D., Gierasch, P., Horn, L., Koppany, R., & Ponnamperuma, C. (1981*a*). Infrared observations of the Saturnian system from Voyager 1. *Science*, **212**, 192–200.

Hanel, R. A. & Conrath, B. (1969). Interferometer experiment on Nimbus 3: preliminary results. *Science*, **165**, 1258–60.

Hanel, R. A., Conrath, B., Flasar, M., Kunde, V., Lowman, P., Maguire, W., Pearl, J., Pirraglia, J., & Samuelson, R. (1979). Infrared observations of the Jovian system from Voyager 1. *Science*, **204**, 972–6.

Hanel, R. A., Conrath, B., Flasar, F. M., Kunde, V., Maguire, W., Pearl, J., Pirraglia, J., Samuelson, R., Cruikshank, D., Gautier, D., Gierasch, P., Horn, L., & Schulte, P. (1986). Infrared observations of the Uranian system. *Science*, **233**, 70–4.

Hanel, R. A., Conrath, B. J., Herath, L. W., Kunde, V. G., & Pirraglia, J. A. (1981*b*). Albedo, internal heat, and energy balance of Jupiter: preliminary results of the Voyager infrared investigation. *Journal of Geophysical Research*, **86**, 8705–12.

Hanel, R. A., Conrath, B., Hovis, W., Kunde, V., Lowman, P., Maguire, W., Pearl, J., Pirraglia, J., Prabhakara, C., Schlachman, B., Levin, G., Straat, P., & Burke, T. (1972*a*). Investigation of the Martian environment by infrared spectroscopy on Mariner 9. *Icarus*, **17**, 423–42.

Hanel, R. A., Conrath, B. J., Hovis, W. A., Kunde, V. G., Lowman, P. D., Pearl, J. C., Prabhakara, C., & Schlachman, B. L. (1972*b*). Infrared spectroscopy experiment on the Mariner 9 mission: preliminary results. *Science*, **175**, 305–8.

Hanel, R. A., Conrath, B. J., Kunde, V. G., Pearl, J. C., & Pirraglia, J. A. (1983). Albedo, internal heat flux, and energy balance of Saturn. *Icarus*, **53**, 262–85.

Hanel, R. A., Conrath, B. J., Kunde, V. G., Prabhakara, C., Revah, I., Salomonson, V. V., & Wolford, G. (1972*c*). The Nimbus 4 infrared spectroscopy experiment, 1. Calibrated thermal emission spectra. *Journal of Geophysical Research*, **77**, 2639–41.

Hanel, R. A., Crosby, D., Herath, L., Vanous, D., Collins, D., Creswick, H., Harris, C., & Rhodes, M. (1980). Infrared spectrometer for Voyager. *Applied Optics*, **19**, 1391–400.

Hanel, R. A., Forman, M., Meilleur, T., Westcott, R., & Pritchard, J. (1969). A double beam interferometer for the middle infrared. *Applied Optics*, **8**, 2059–65.

Hanel, R. A., Forman, M., Stambach, G., & Meilleur, T. (1968). Preliminary results of Venus observations between 8 and 13 microns. *Journal of Atmospheric Science*, **25**, 586–93.

Hanel, R. A., Schlachman, B., Breihan, E., Bywaters, R., Chapman, F., Rhodes, M., Rogers, D., & Vanous, D. (1972*d*). The Mariner 9 Michelson interferometer. *Applied Optics*, **11**, 2625–34.

Hanel, R. A., Schlachman, B., Clark, F. D., Prokesh, C. H., Taylor, J. B., Wilson, W. M., & Chaney, L. (1970). The Nimbus 3 Michelson interferometer. *Applied Optics*, **9**, 1767–74.

Hanel, R. A., Schlachman, B., Rogers, D., & Vanous, D. (1971). The Nimbus 4 Michelson interferometer. *Applied Optics*, **10**, 1376–81.

Hansen, J. E. (1969). Absorption line formation in a scattering planetary atmosphere: a test of van de Hulst's similarity relations. *Astrophysical Journal*, **158**, 337–49.

Hansen, N. P. & Strong, J. (1971). Performance of a simple spherical lamellar grating beamsplitter by wavefront division. *Aspen International Conference on Fourier Spectroscopy*, ed. G. A. Vanasse, A. T. Stair, Jr., & D. J. Baker, US Air Force Cambridge Research Laboratory 71-0019.

Hapke, B. (1981). Bidirectional reflectance spectroscopy. 1. Theory. *Journal of Geophysical Research*, **86**, 3039–54.
Hapke, B. (1984). Bidirectional reflectance spectroscopy. 3. Correction for macroscopic roughness. *Icarus*, **59**, 41–59.
Hapke, B. (1986). Bidirectional reflectance spectroscopy. 4. The extinction coefficient and the opposition effect. *Icarus*, **67**, 264–80.
Hapke, B. & Wells, E. (1981). Bidirectional reflectance spectroscopy. 2. Experiments and observations. *Journal of Geophysical Research*, **86**, 3055–60.
Harrington, J., Deming, D., & Ressler, M. (2000). Private communication.
Hart, M. H. (1978). The evolution of the atmosphere of the Earth. *Icarus*, **33**, 23–9.
Hartman, W. K. & Davis, D. R. (1975). Satellite-sized planetesimals and lunar origin. *Icarus*, **24**, 504–15.
Harvey, J. E. & Ftaclas, C. (1988). Diffraction effects of secondary mirror spiders upon telescope imaging quality. *Current Developments in Optical Engineering III, Proceedings of the Society of Photo-Optical Instrumentation Engineers*, **965**, 7–17.
Hass, G., ed. (1963–1971). *Physics of Thin Films*, Vol. 1–6. New York: Academic Press.
Hayes, P. B., Killeen, T. C., & Kennedy, B. C. (1981). The Fabry–Perot interferometer on dynamics explorer. *Space Science Instrumentation*, **5**, 395–416.
Heavens, O. S. (1955). *Optical Properties of Thin Solid Films*. London: Butterworth's Scientific Publications, Ltd. (Reprinted by Dover Publications, Inc. in 1965.)
Hegyi, D. J., Traub, W. A., & Carlton, N. P. (1972). Cosmic background radiation at 1.32 mm. *Physical Review Letters*, **28**, 1541–4.
Heisenberg, W. (1925). Über quantentheoretische Umdeutung kinematischer und mechanischer Beziehungen. *Zeitschrift für Physik*, **33**, 879–93.
Herbig, G. H. (1962). The properties of T-Tauri stars and related objects. *Advances in Astronomy and Astrophysics*, **1**, 47–103.
Herr, K. C., Forney, P. B., & Pimentel, G. C. (1972). Mariner Mars 1969 infrared spectrometer. *Applied Optics*, **11**, 493–501.
Herr, K. C. & Pimentel, G. C. (1969). Infrared absorptions near three microns recorded over the polar cap of Mars. *Science*, **166**, 496–9.
Herr, K. C. & Pimentel, G. C. (1970). Evidence for solid carbon dioxide in the upper atmosphere of Mars. *Science*, **167**, 46–9.
Herschel, W. (1800). Experiments on the refrangibility of the invisible rays of the Sun. *Philosophical Transactions Royal Society London*, **90**, 284.
Hertz, H. (1888). *Sitzungsberichte der Berliner Akademie der Wissenschaften*, 2 Feb. 1888; *Wiedemanns Annalen*, 34. (English translation in *Electric Waves*. London: Macmillan, 1893.)
Herzberg, G. (1939). *Spectra of Diatomic Molecules*. New York: Van Nostrand Reinhold.
Herzberg, G. (1945). *Molecular Spectra and Molecular Structure. II. Infrared and Raman Spectra of Polyatomic Molecules*. New York: Van Nostrand Reinhold.
Herzberg, G. (1950). *Molecular Spectra and Molecular Structure. I. Spectra of Diatomic Molecules*. New York: Van Nostrand Reinhold.
Hess, S. L., Henry, R. M., Leovy, C. B., Ryan, J. A., & Tillman, J. E. (1977). Meteorological results from the surface of Mars: Viking 1 and 2. *Journal of Geophysical Research*, **82**, 4559–74.
Hess, S. L., Henry, R. M., & Tillman, J. E. (1979). The seasonal variation of atmospheric pressure on Mars as affected by the south polar cap. *Journal of Geophysical Research*, **84**, 2923–7.
Hickey, J. R. (1970). Laboratory methods of experimental radiometry including data analysis. In *Advances in Geophysics Vol. 14, Precision Radiometry*, ed. A. J. Drummond, Chapter 8. New York and London: Academic Press.

Hickey, J. R. & Daniels, D. B. (1969). Modified optical system for the Golay detector. *Review of Scientific Instruments*, **40**, 732–3.

Hofmann, R., Drapatz, S., & Michel, K. W. (1977). Lamellar grating Fourier spectrometer for balloon-borne telescope. *Infrared Physics*, **17**, 451–6.

Hofstadter, M. D., & Muhlman, D. O. (1989). Latitude variation of ammonia in the atmosphere of Uranus: an analysis of microwave observations. *Icarus*, **87**, 396–412.

Holton, J. R. (1992). *An Introduction to Dynamic Meteorology*. San Diego: Academic Press.

Hoppa, G. V., Tufts, B. R., Greenberg, R., & Geissler, P. E. (1999). Formation of cycloidal features on Europa. *Science*, **285**, 1899–902.

Horn, D., McAfee, J. M., Winer, A. M., Herr, K. C., & Pimentel, G. C. (1972). The composition of the Martian atmosphere: minor constituents. *Icarus*, **16**, 543–56.

Houghton, J. T. & Smith, S. D. (1970). Remote sounding of atmospheric temperature from satellites. *Proceedings of the Royal Society, London*, **A320**, 23–33.

Houghton, J. T., Taylor, F. W., & Rodgers, C. D. (1984). *Remote Sensing of Atmospheres*. Cambridge: Cambridge University Press.

Hourdin, F., Van, P. L., Forget, F., & Talagrand, O. (1993). Meteorological variability and the annual surface pressure cycle on Mars. *Journal of the Atmospheric Sciences*, **50**, 3625–40.

House, F. B., Gruber, A., Hunt, G. E., & Mecherikunnel, A. T. (1986). History of satellite missions and measurements of the Earth radiation budget (1957–1984). *Journal of Geophysical Research*, **24**, 357–77.

Hovis, W. A., Jr., Blaine, L. R., & Callahan, W. R. (1968). Infrared aircraft spectra over desert: 8.5–16 μm. *Applied Optics*, **7**, 1137–40.

Hovis, W. A., Jr., & Callahan, W. R. (1966). Infrared reflectance spectra of igneous rocks, tuffs, and red sandstone from 0.5 μm to 22 μm. *Journal of the Optical Society of America*, **56**, 639–43.

Hovis, W. A., Jr., Kley, W. A., & Strange, M. G. (1967). Filter wedge spectrometer for field use. *Applied Optics*, **6**, 1057–8.

Hovis, W. A., Jr. & Tobin, M. (1967). Spectral measurements from 1.6 μm to 5.4 μm of natural surfaces and clouds. *Applied Optics*, **6**, 1399–402.

Hoyle, F. (1955). *Frontiers in Astronomy*, pp. 68–72. London: William Heinemann.

Hubbard, W. B. (1980). Intrinsic luminosity of the Jovian planets. *Review of Geophysics and Space Physics*, **18**(1), 1–9.

Hubbard, W. B. (1984). Interior structure of Uranus. In *Uranus and Neptune*, ed. J. T. Bergstralh. *NASA Conference Publication 2330*, pp. 291–325.

Hubbard, W. B. & MacFarlane, J. J. (1980). Structure and evolution of Uranus and Neptune. *Journal of Geophysical Research*, **85**, 225–34.

Hubbard, W. B., Podolak, M., & Stevenson, D. J. (1995). The interior of Neptune. In *Neptune and Triton*, 109–38, ed. D. Chruikshank. Tucson: University of Arizona Press.

Hubbard, W. B. & Smoluchowski, R. (1973). Structure of Jupiter and Saturn. *Space Science Review*, **14**, 599–662.

van de Hulst, H. C. (1957). *Light Scattering by Small Particles*. New York: John Wiley & Sons. (Reprinted in 1981 by Dover Publications, Inc.)

Hunt, G., Kandel, R., & Mecherikunnel, A. (1986). A history of presatellite investigations of the Earth's radiation budget. *Reviews of Geophysics*, **24**, 351–6.

Hunt, G. R. & Logan, L. M. (1972). Variation of single particle mid-infrared emission spectrum with particle size. *Applied Optics*, **11**, 142–7.

Hunt, G. R., Logan, L. M., & Salisbury, J. W. (1973). Mars: Components of the infrared spectra and the composition of the dust cloud. *Icarus*, **18**, 459–69.

Hunten, D. M., Donahue, T. M., Walker, J. C. G., & Kasting, J. F. (1989). Escape of atmospheres and loss of water. In *Origin and Evolution of Planetary and Satellite Atmospheres*, ed. S. K. Atreya, J. B. Pollack, & M. S. Matthews. Tucson: The University of Arizona Press.

Hunten, D. M., Tomasko, M. G., Flasar, F. M., Samuelson, R. E., Strobel, D. F., & Stevenson, D. J. (1984). Titan. In *Saturn*, ed. T. Gehrels & M. S. Matthews, pp. 671–759. Tucson: The University of Arizona Press.

Huygens, Christian (1690). *Traite de la lumière*, published in Leyden. (English translation: Treatise on Light by S. P. Thompson, London: Macmillan & Co., 1912.)

Ingersoll, A. P. (1969). The runaway greenhouse: A history of water on Venus. *Journal of Atmospheric Science*, **26**, 1191–8.

Ingersoll, A. P., Muench, G., Neugebauer, G., & Orten, G. S. (1976). Results of the infrared radiometer experiment on Pioneer 10 and 11. In *Jupiter*, 197–215, ed. T. Gehrels. Tucson: University of Arizona Press.

Ingersoll, A. P., Summers, M. E., & Schlipf, S. G. (1985). Supersonic meteorology of Io: sublimation-driven flow of SO_2. *Icarus*, **64**, 375–90.

Irvine, W. M., Goldsmith, P. F., Hjalmarson, A. (1987). Chemical abundances in molecular clouds. In *Interstellar Processes*, ed. D. Hollenbach, & H. Thomson, pp. 343–437. Dordrecht, Holland: D. Reidel Publ. Co.

Irvine, W. M. & Knacke, R. F. (1989). The chemistry of interstellar gas and grains. In *Origin and Evolution of Planetary and Satellite Atmospheres*, 3–34, ed. S. K. Atreya, J. B. Pollack, & M. S. Matthews. Tucson: University of Arizona Press.

Jaacks, R. G. & Rippel, H. (1989). Double pendulum Michelson interferometer with extended spectral resolution. *Applied Optics*, **28**, 29–30.

Jacquinot, P. & Dufour, C. (1948). Condition optique d'emploi des cellules photoélectrique dans les spectrographes et les interféromètres. *Journal Recherche du Centre National Recherche Scientifique Laboratorie Bellevue* (Paris), **6**, 91–103.

Jacquinot, P. & Roizer-Dossier, B. (1964). Apodisation. *Progress in Optics*, **3**, 29–186.

Jahnke, E. & Emde, F. (1933). *Tables of Functions*. Leipzig: Teubner. (Reprinted 1943, New York: Dover Publications, Inc.)

James, J. F. & Sternberg, R. S. (1969). *The Design of Optical Spectrometers*. London: Chapman and Hall.

Jeans, J. H. (1905). Partition of energy between matter and ether. *Philosophical Magazine*, **10**, 91–8.

Jennings, D. E. (1988). Laboratory diode laser spectroscopy in molecular planetary astronomy. *Journal of Quantitative Spectroscopy and Radiative Transfer*, **40**, 221–38.

Jessenberger, E. K., Kissel, J., & Rahe, J. (1989). The composition of comets. In *Origin and Evolution of Planetary and Satellite Atmospheres*, 167–91, ed. S. K. Atreya, J. B. Pollack, & M. S. Matthews. Tucson: University of Arizona Press.

Johnson, J. B. (1925). The Schottky effect in low frequency circuits. *Physical Review*, **26**, 71–8.

Johnson, T. (2000). The Galileo Mission to Jupiter and its moons. *Scientific American*, **282**, 40–9.

Jones, R. C. (1952). Detectivity, the reciprocal of noise equivalent input of radiation. *Nature*, **170**, 937–8, see also *Journal of the Optical Society of America*, **42**, 286.

Jones, R. C. (1959). Phenomenological description of response and detecting ability of radiation detectors. *Proceedings of the Institute of Radio Engineers*, **17**, 1495–502.

Jurgens, D. W., Duval, J. E., Lockhart, R. F., Langevin, Y., Formisano, V., & Belucci, G. (1990). Visible and infrared mapping spectrometer for exploration of comets, asteroids, and the Saturnian system of rings and moons. *International Journal of Imaging Systems and Technology*, **3**, 108–20.

Kaplan, L. D. (1959). Inference of atmospheric structure from remote radiation measurement. *Journal of the Optical Society of America*, **49**, 1004–7.

Kasting, J. F. (1988). Runaway and moist greenhouse atmospheres and the evolution of Earth and Venus. *Icarus*, **74**, 472–94.

Kasting, J. F., Pollack, J. B., & Ackerman, T. P. (1984). Response of Earth's surface temperature to increases in solar flux and implications of loss of water from Venus. *Icarus*, **57**, 335–55.

Kasting, J. F. & Toon, O. B. (1989). Climate evolution on the terrestrial planets. In *Origins and Evolution of Planetary and Satellite Atmospheres*, ed. S. K. Atreya, J. B. Pollack, & M. S. Matthews. Tucson: The University of Arizona Press.

Käufl, U., Rothermel, H., & Drapatz, S. (1984). Center to limb variations in intensity of CO_2 laser emission in Mars. *Astronomy and Astrophysics*, **136**, 319–25.

Keihm, S. J. & Langseth, M. G., Jr. (1973). Surface brightness temperatures at the Apollo 17 heat flow site: thermal conductivity of the upper 15 cm of regolith. *Proceedings of the Fourth Lunar Science Conference*, **3**, 2503–13.

Kerr, R. A. (1999). Another ocean for a Jovian satellite? *Science*, **286**, 1827–8.

Kessler, M. F., Steinz, J. A., Andreegg, M. E., Clavel, J., Drechsel, G., Estaria, P., Faelker, J., Riedinger, J. R., Robson, A., Taylor, B. G., & Ximenez de Ferran, S. (1996). The Infrared Space Observatory (ISO) mission. *Astronomy and Astrophysics*, **315**, L27–31.

Keszthelyi, L., McEven, A. S., & Taylor, G. J. (1999). Revisiting the hypothesis of a mushy global magma ocean in Io. *Icarus*, **141**, 415–19.

Keyes, R. J. (1977). *Optical and Infrared Detectors. Topics in Applied Physics*, Vol. 19. Berlin, Heidelberg, New York: Springer-Verlag.

Khare, B. N., Sagan, C., Arakawa, E. T., Suits, F., Callcott, T. A., & Williams, M. W. (1984). Optical constants of organic tholins produced in a simulated Titanian atmosphere: From soft X-ray to microwave frequencies. *Icarus*, **60**, 127–37.

Kieffer, H. H., Chase, S. C., Miner, E., Münch, G., & Neugebauer, G. (1973). Preliminary report on infrared radiometric measurements from Mariner 9. *Journal of Geophysical Research*, **78**, 4291–312.

Kieffer, H. H., Martin, T. Z., Peterfreund, A. R., Jakossky, B. M., Miner, E. D., & Palluconi, F. D. (1977). Thermal and albedo mapping of Mars during the Viking primary mission. *Journal of Geophysical Research*, **82**, 4249–92.

Kieffer, S. W., Lopes-Gautier, R., McEwen, A., Smythe, W., Keszthelyi, L., & Carlson, R. (2000). Prometheus: Io's wandering plume. *Science*, **288**, 1204–8.

Kinch, M. A. (1971). Compensated silicone-impurity conduction bolometer. *Journal of Applied Physics*, **42**, 5861–3.

Kingston, R. H. (1978). *Detection of Optical and Infrared Radiation*. New York: Springer-Verlag.

Kirchhoff, G. (1883). Zur Theorie des Lichtstrahlers. *Annalen der Physik (2)*, **18**, 663–95.

Kirkland, L. (1999). *Infrared Spectroscopy of Mars*. Ph.D. thesis, Rice University.

Kivelson, M. G., Khurana, K. K., Russel, C. T., Volwerk, M., Walker, R. J., & Zimmer, C. (2000). Galileo magnetometer measurements: A stronger case for a subsurface ocean at Europa. *Science*, **289**, 1340–3.

Kopia, L. P. (1986). Earth radiation budget experiment scanner instrument. *Review of Geophysics*, **24**, 400–6.

Kostiuk, T., Espenak, F., Mumma, M., Deming, D., & Zipoy, D. (1987). Variability of ethane on Jupiter. *Icarus*, **29**, 199–204.

Kostiuk, T. & Mumma, M. J. (1983). Remote sensing by IR heterodyne spectroscopy. *Applied Optics*, **22**, 2644–54.

Kostiuk, T., Mumma, M. J., Espenak, F., Deming, D., Jennings, D. E., Maguire, W., & Zipoy, D. (1983). Measurements of stratospheric ethane in the Jovian South Polar Region from infrared heterodyne spectroscopy of the ν_9 band near 12 μm. *Astrophysical Journal*, **265**, 564–9.

Kourganoff, V. (1952). *Basic Methods in Transfer Problems*. Oxford: Oxford University Press. (Reprinted by Dover Publications, Inc., New York, in 1963.)

Kuiper, G. P. ed. (1949). *The Atmospheres of the Earth and Planets* (2nd edn, 1951), Chapter 12. Chicago: The University of Chicago Press.

Kuiper, G. (1951). On the origin of the Solar System. In *Astrophysics, a Topical Symposium*, 357–424, ed. J. A. Hynek. New York: McGraw Hill.

Kunde, V., Ade, P., Barney, R., Bergman, D., Bonnal, J. F., Borelli, R., Boyd, D., Beasunas, J., Brown, G., Calcutt, S., Carroll, F., Courtin, R., Cretolle, J., Crooke, J., Davis, M., Edberg, S., Fettig, R., Flasar, M., Glenar, D., Graham, S., Hagopian, J., Hakun, C., Hayes, P., Herath, L., Horn, L., Jennings, D., Karpati, G., Kellebenz, C., Lakew, B., Lindsey, J., Lohr, J., Lyons, J., Martineau, R., Martino, A., Matsumura, M., McCloskey, J., Malek, T., Michel, G., Morell, A., Mosier, C., Pack, L., Plants, L., Robinson, D., Rodriguez, L., Romani, P., Schaefer, W., Schmidt, S., Trujillo, C., Vellacott, T., Wagner, K., & Yun, D. (1996). Cassini infrared Fourier spectroscopic investigation. In *Cassini/Huygens: A Mission to the Saturnian Systems*, SPIE Proceedings, **2803**, Denver, CO.

Kunde, V. G., Aikin, A. C., Hanel, R. A., Jennings, D. E., Maguire, W. C., & Samuelson, R. E. (1981). C_4H_2, HC_3N, and C_2N_2 in Titan's atmosphere. *Nature*, **292**, 686–8.

Kunde, V. G., Brasunas, J. C., Conrath, B. J., Hanel, R. A., Herman, J. R., Jennings, D. E., Maguire, W. C., Walser, D. W., Annen, J. N., Silverstein, M. J., Abbas, M. M., Herath, L. W., Buijs, H. J., Berube, J. N., & McKinnon, J. (1987). Infrared spectroscopy of the lower stratosphere with a balloon-borne cryogenic Fourier spectrometer. *Applied Optics*, **26**, 545–53.

Kunde, V. G., Conrath, B., Hanel, R., Maguire, W., Prabhakara, C., & Salomonson, V. (1974). The Nimbus IV infrared spectroscopy experiment: 2. Comparison of observed and theoretical radiances from 425–1450 cm^{-1}. *Journal of Geophysical Research*, **79**, 777–84.

Kunde, V. G., Hanel, R. A., & Herath, L. W. (1977). High spectral resolution ground based observations of Venus in the 450–1250 cm^{-1} region. *Icarus*, **32**, 210–24.

Kunde, V. G., Hanel, R., Maguire, W., Gautier, D., Baluteau, J. P., Marten, A., Chedin, A., Husson, N., & Scott, N. (1982). The tropospheric gas composition of Jupiter's North Equatorial Belt (NH_3, PH_3, CH_3D, GeH_4, H_2O) and the Jovian D/H ratio. *Astrophysical Journal*, **263**, 443–67.

Kurucz, R. L., Traub, N. P., Carlton, N. P., & Lester, J. B. (1977). The rotational velocity and barium abundance of Sirius. *Astrophysical Journal*, **217**, 771–4.

Landau, L. D. & Lifshitz, E. M. (1960). *Electrodynamics of Continuous Media*. New York: Pergamon Press.

Lane, A. L., Hord, C. W., West, R. A., Esposito, L. W., Coffeen, D. L., Sato, M., Simmons, K. E., Pomphrey, R. B., & Morris, R. B. (1982). Photopolarimetry from Voyager 2: preliminary results on Saturn, Titan, and the rings. *Science*, **215**, 537–43.

Langley, S. P. (1881). The bolometer. *Nature*, **25**, 14–16, also *Proceedings of the American Academy of Arts and Sciences*, **16**, 342.

Lara, L. M., Lellouch, E., Lopez-Moreno, J. J., & Rodrigo, R. (1996). Vertical distribution of Titan's atmospheric neutral constituents. *Journal of Geophysical Research*, **101**, 23262–83.

Larson, H. P., Fink, U., Treffers, R., & Gautier, T. N. (1975). Detection of water vapor on Jupiter. *Astrophysical Journal*, **197**, L137–40.

Lebofsky, L. A., Jones, T. D., & Herbert, F. (1989). Asteroid volatile inventories. In *Origin and Evolution of Planetary Atmospheres*, 192–229, ed. S. K. Atreya, J. B. Pollack, & M. S. Matthews. Tucson: University of Arizona Press.

Lebreton, Y. & Maeder, J. (1986). The evolution and helium content of the sun. *Astronomy and Astrophysics*, **161**, 119–24.

Lellouch, E., Coustenis, A., Gautier, D., Raulin, F., Dubouloz, N., & Frère, C. (1989). Titan's atmosphere and hypothesized ocean: A reanalysis of the Voyager 1 radio-occultation and IRIS 7.7 μm data. *Icarus*, **79**, 328–49.

Leovy, C. B. (1973). Rotation of the upper atmosphere of Venus. *Journal of Atmospheric Science*, **30**, 1218–20.

Leovy, C. B. (1979). Martian meterology. *Annual Review of Astronomy and Astrophysics*, **17**, 387–413.

Leovy, C. B., Zurek, R. W., & Pollack, J. B. (1973). Mechanisms for Mars dust storms. *Journal of Atmospheric Science*, **30**, 749–62.

Lewis, J. S. (1972). Low temperature condensation from the solar nebula. *Icarus*, **16**, 241–52.

Lewis, J. S. (1973). Chemistry of the outer solar system. *Space Science Review*, **14**, 401–10.

Lewis, J. S. (1974). The temperature gradient in the solar nebula. *Science*, **186**, 440–2.

Lewis, J. S. & Prinn, R. G. (1984). *Planets and Their Atmospheres: Origin and Evolution*. New York: Academic Press.

Lewis, W. B. (1947). Fluctuations in streams of thermal radiation. *Proceedings of the Physical Society* (*London*), **59**, 34–40.

Lillie, C. F., Hord, C. W., Pang, K., Coffeen, D. L., & Hansen, J. E. (1977). The Voyager mission photopolarimeter experiment. *Space Science Review*, **21**, 159–81.

Limaye, S. S. (1986). Jupiter: new estimates of the mean zonal flow at the cloud level. *Icarus*, **65**, 335–52.

Lin, D. N. C. (1981). Convective accretion disk model for the primordial solar nebula. *Astrophysical Journal*, **246**, 972–84.

Lindal, G. F., Lyons, J. R., Sweetnam, D. N., Eshleman, V. R., Hinson, D. P., & Tyler, G. L. (1987). The atmosphere of Uranus: results of radio occultation measurements with Voyager 2. *Journal of Geophysical Research*, **92**, 14987–5001.

Lindal, G. F., Wood, G. E., Hotz, H. B., Sweetnam, D. N., Eshleman, V. R., & Tyler, G. L. (1983). The atmosphere of Titan: an analysis of the Voyager 1 radio occultation measurements. *Icarus*, **53**, 348–63.

Linsky, J. L. (1966). Models of the lunar surface including temperature dependent thermal properties. *Icarus*, **5**, 606–34.

Lissauer, J. J., Pollack, J. B., Wetherill, G. W., & Stevenson, D. J. (1995). Formation of the Neptune system. In *Neptune and Triton*, 37–108, ed. D. Cruikshank. Tucson: University of Arizona Press.

Loewenstein, E. V. (1966). History and current status of Fourier transform spectroscopy. *Applied Optics*, **5**, 845–54.

Loewenstein, E. V. (1971). Fourier spectroscopy: an introduction. In *Proceedings of the Aspen International Conference on Fourier Spectroscopy*, ed. G. A. Vanasse, A. T. Stair, & D. J. Baker. US Air Force Cambridge Research Laboratory 71-0019.

Logan, L. M. (1979). Signal-to-noise enhancement of Fourier transform spectroscopy (FTS) by electrical filter compensation of slide velocity errors. *Proceedings of the Society of Photo-Optical Instrumentation Engineers, 191, Multiple and/or High-Throughput Spectroscopy*, ed. G. A. Vanasse, pp. 110–13. Bellingham, Washington: Society of Photo-Optical Instrumentation Engineers.

Logan, L. M. & Hunt, G. R. (1970). Emission spectra of particulate silicates under simulated lunar conditions. *Journal of Geophysical Research*, **75**, 6539–48.

Lopes-Gautier, R., Doute, S., Smythe, W. D., Kamp, L. W., Carlson, R. W., Dacis, A. G., Leader, F. E., McWwen, A. S., Geissler, P. E., Kieffer, S. W., Keszthelyi, L., Barbinis, E., Mehlman, R., Segura, M., Shirley, J., & Soderblum, L. A. (2000). A close-up look at Io from Galileo's Near-Infrared Mapping Spectrometer. *Science*, **288**, 1201–4.

Low, F. J. (1961). Low-temperature germanium bolometer. *Journal of the Optical Society of America*, **51**, 1300–4.

Low, F. J. & Hoffman, A. R. (1963). The detectivity of cryogenic bolometers. *Applied Optics*, **2**, 649–50.

Low, F. J. & Rieke, G. H. (1974). The instrumentation and techniques of infrared photometry. In *Methods of Experimental Physics*, Vol. 12, ed. N. Carlton, pp. 415–62. New York: Academic Press.

Lummer, O. & Pringsheim, E. (1897). Die Strahlung eines schwarzen Körpers zwischen 100 und 1300 C. *Wiedemanns Annalen*, **63**, 395–410.

Lummer, O. & Pringsheim, E. (1899). Kritisches zur schwarzen Körper Strahlung. *Annalen der Physik*, **6**, 192–210 (see also *Verhandlungen der Deutschen Physikalischen Gesellschaft*, 1901, **1**, 23).

Lunine, J. I., Atreya, S. K., & Pollack, J. B. (1989). Present state and chemical evolution of the atmospheres of Titan, Triton, and Pluto. In *Origin and Evolution of Planetary and Satellite Atmospheres*, 605–65, ed. S. K. Atreya, J. B. Pollack, & M. S. Matthews. Tucson: University of Arizona Press.

Luther, M. R., Cooper, J. E., & Taylor, G. R. (1986). The Earth radiation budget experiment nonscanner instrument. *Journal Geophysical Research*, **24**, 391–9.

Lutz, B. L., de Bergh, C., & Owen, T. (1983). Titan: discovery of carbon monoxide in its atmosphere. *Science*, **220**, 1374–5.

Lutz, B. L., Owen, T., & Cess, R. D. (1976). Laboratory band strengths of methane and their application to the atmospheres of Jupiter, Saturn, Uranus, Neptune, and Titan. *Astrophysical Journal*, **203**, 541–51.

Maguire, W. C. (1977). Martian isotopic ratios and limits for possible minor constituents as derived from Mariner 9 infrared spectrometer data. *Icarus*, **32**, 85–97.

Maguire, W. C., Hanel, R. A., Jennings, D. E., Kunde, V. G., & Samuelson, R. E. (1981). C_3H_8 and C_3H_4 in Titan's atmosphere. *Nature*, **292**, 683–6.

Malhotra, R. (1993). The origin of Pluto's peculiar orbit. *Nature*, **365**, 819–21.

Malhotra, R. (1999). Migrating planets. *Scientific American*, **281**, 56–63.

Malhotra, R. & Williams, J. G. (1997). Pluto's heliocentric orbit. In *Pluto and Charon*, 127–159, ed. S. A. Stern & D. J. Tholen. Tucson: University of Arizona Press.

Malin, M. C. & Edgett, K. S. (2000). Sedimentary rocks of early Mars. *Science*, **290**, 1927–37.

Manno, V. & Ring, J., eds. (1972). *Infrared Detection Techniques for Space Research*. Dordrecht, Holland: D. Reidel, Publ. Co.

Marcialis, R. L. (1997). The first 50 years of Pluto–Charon research. In *Pluto and Charon*, 27–83, ed. S. A. Stern & D. J. Tholen. Tucson: University of Arizona Press.

Marten, A., Courtin, R., Gautier, D., & Lacombe, A. (1980). Ammonia vertical density profiles on Jupiter and Saturn from their radioelectric and infrared emissivities. *Icarus*, **41**, 410–22.

Marten, A., Gautier, D., Tanguy, L., Lecacheux, A., Rosolen, C., & Paubert, G. (1988). Abundance of carbon monoxide in the stratosphere of Titan from millimeter heterodyne observations. *Icarus*, **76**, 558–62.

Martin, D. H. (1982). Polarizing (Martin–Puplett) interferometric spectrometers for the near- and submillimeter spectra. In *Infrared and Millimeter Waves, Vol. 6, Systems and Components*, ed. K. J. Button, Chapter 2. New York: Academic Press.

Martin, D. H. & Bloor, D. (1961). The application of superconductivity to the detection of radiant energy. *Cryogenics*, **1**, 159–65.

Martin, D. H. & Puplett, E. (1969). Polarized interferometric spectrometry for the millimeter and submillimeter spectrum. *Infrared Physics*, **10**, 105–9.

Maserjian, J. (1970). A thin-film capacitive bolometer. *Applied Optics*, **9**, 307–15.

Mather, J. C. & Boslough, J. I. (1996). *The Very First Light*. New York: BasicBooks, A Division of Harper-Collins Publishers, Inc.

Mather, J. & Kelsall, T. (1980). The Cosmic Background Explorer Satellite. *Physica Scripta*, **21**, 671–7.

Matson, D. L. ed. (1986). IRAS Asteroid and Comet Survey. *JPL Report*, D 3698.

Maxwell, J. C. (1873). *A Treatise on Electricity and Magnetism*, 2 Vols. Oxford: Clarendon Press.

Maymon, P., Dittman, M., Pasquale, B., Jennings, D. E., Mehalick, K., & Trout, C. (1993). Optical design of the Composite Infrared Spectrometer (CIRS) for the Cassini Mission. In *Space Astronomical Telescopes and Instruments II*, SPIE *Proceedings*, **1945**, 100–11, Orlando, FL.

McCord, T. B., Carlson, R. W., Smythe, W. D., Hansen, G. B., Clark, R. N., Hibbitts, C. A., Fanale, F. P., Granahan, J. C., Segura, M., Matson, D. L., Johnson, T. V., & Martin, P. D. (1997*a*). Organics and other molecules in the surface of Callisto and Ganymede. *Science*, **278**, 271–5.

McCord, T. B., Hansen, G. B., Fanale, F. P., Carlson, R. W., Matson, D. L., Johnson, T. V., Smythe, W. D., Crowley, J. K., Martin, P. D., Ocampo, A., Hibbitts, C. A., Granahan, J. C., & the NIMS team. (1998). Salts on Europa's surface detected by Galileos's Near Infrared Mapping Spectrometer. *Science*, **280**, 1242–5.

McCord, T. B. Hansen, G. H., Hibbitts, C., Carlson, R. W., Smythe, W., Matson, D. L., Johnson, T. V., Clark, R. M., Granahan, J., Fanale, F., Segura, M., & the NIMS team (1997*b*). Analysis of Near-Infrared Mapping Spectrometer reflectance spectra for the icy Galilean satellites. *Meteoritics & Planetary Science*, **32**(4) Supplement, A86–7.

McEwen, A. S., Belton, M. J. S., Breneman, H. H., Fagents, S. A., Geissler, P., Greeley, R., Head, J. W., Hoppa, G., Jaeger, W. L., Johnson, T. V., Keszthelyi, L., Klaasen, K. P., Lopes-Gautier, R., Magee, K. P., Milazzo, M. P., Moore, J. M., Pappalardo, R. T., Phillips, C. B., Radebaugh, J., Schubert, G., Schuster, P., Simonelli, D. P., Sullivan, R., Thomas, P. C., Turtle, E. P., & Williams, D. A. (2000). Galileo at Io: Results from high-resolution imaging. *Science*, **288**, 1193–8.

McEwen, A. S. & Soderblom, L. A. (1983). Two classes of volcanic plumes on Io. *Icarus*, **55**, 191–217.

McKellar, A. R. W. (1984). Possible identification of sharp features in the Voyager far-infrared spectra of Jupiter and Saturn. *Canadian Journal of Physics*, **62**, 760–3.

McKinnon, W. B., Lunine, J. I., & Banfield, D. (1995). Origin and evolution of Triton. In *Neptune and Triton*, 807–78, ed. D. P. Cruikshank. Tucson: University of Arizona Press.

Meier, R., Owen, T. C., Jewitt, D. C., Matthews, E. H., Senay, M., Biver, M., Bockelee-Morvan, D., Crovisier, J., & Gautier, D. (1998). Deuterium in Comet C 1995 01 (Hale Bopp): Detection of DCN. *Science*, **279**, 1707–10.

Melloni, M. (1833). Durchgang der Wärmestrahlung durch verschiedene Körper. *Poggendorfs Annalen*, **28**, 240 and 371.
Mertz, L. (1965). *Transformations in Optics*. New York: John Wiley & Sons.
Michelson, A. A. (1881). The relative motion of the Earth and the luminiferous ether. *American Journal of Science* (3), **22**, 120–9.
Michelson, A. A. (1898). Radiation in a magnetic field. *Astrophysical Journal*, **7**, 131–8.
Michelson, A. A. (1927). *Studies in Optics*. Chicago and London: Phoenix Books, The University of Chicago Press.
Michelson, A. A. & Morley, E. W. (1887). On the relative motion of the Earth and the luminiferous ether. *American Journal of Science* (3), **34**, 333–45.
Michelson, A. A. & Stratton, S. W. (1898). A new harmonic analyzer. *American Journal of Science* (4), **5**, 1–13.
Mie, G. (1908). Beiträge zur Optik trüber Medien, speziell kolloidaler Metallösungen. *Annalen der Physik* (4), **25**, 377–445.
Millis, R. L., Wassermann, L. H., Franz, O. G., Nye, R. A., Elliot, J. L., Dunham, E. W., Bosh, A. S., Young, L. A., Slivan, S. M., Gilmore, A. C., Kilmartin, P. M., Allen, P. M., Watson, W. H., Dieters, R. D., Hill, S. W., Giles, A. B., Blow, G., Priestley, J., Kissling, W. M., Walker, W. S. G., Marino, B. F., Dix, D. G., Page, A., Ross, J. E., Kennedy, H. D., & Klemola, R. H. (1993). Pluto's radius and atmosphere: Results from the entire 9 June 1988 occultation data set. *Icarus*, **105**, 282–97.
Minnaert, M. (1941). The reciprocity principle in lunar photometry. *Astrophysical Journal*, **93**, 403–10.
Mizuno, H. (1980). Formation of the giant planets. *Progress in Theoretical Physics*, **64**, 544–57.
Mok, C. L., Chambers, W. G., Parker, T. J., & Costley, A. E. (1979). The far-infrared performance and application of free-standing grids wound from 5 μm diameter tungsten wire. *Infrared Physics*, **19**, 437–42.
Moon, P. & Steinhardt, L. R. (1938). The dielectric bolometer. *Journal of the Optical Society of America*, **28**, 148–62.
Morabito, L. A., Synnott, S. P., Kupferman, P. N., & Collins, S. A. (1979). Discovery of currently active extraterrestrial volcanism. *Science*, **204**, 972.
Moroz, V. I., Spänkuch, D., Linkin, V. M., Döhler, W., Matsygorin, I. A., Schäfer, K., Zasova, D., Oertel, A. V., Dyachkov, A. V., Schuster, R., Kerzhanovich, V. V., Becker-Ross, H., Ustinov, E. A., & Stadthaus, W. (1986). Venus spacecraft infrared radiance spectra and some aspects of their interpretation. *Applied Optics*, **25**, 1710–18.
Morrison, D. M. ed. (1982). *Satellites of Jupiter*. Tucson: University of Arizona Press.
Morrison, D. M. & Lebofsky, L. A. (1979). Radiometry of asteroids. In *Asteroids*, 184–205, ed. T. Gehrels. Tucson: University of Arizona Press.
Moses, J. I. (1996). SL9 impact chemistry: Long-term photochemical evolution. In *The Collision of Comet Shoemaker–Levy 9 and Jupiter*, 243–68, ed. K. S. Knoll, H. A. Weaver, & P. D. Feldman. Cambridge: Cambridge University Press.
Moses, J. I. (1997). Dust ablation during the Shoemaker–Levy 9 impacts. *Journal of Geophysical Research – Planets*, **102**, 21619–43 and 28727.
Mumma, M. J. (1992). Comets, atmospheres. In *Astronomy and Astrophysics Encyclopedia*, 107–9, ed. S. Maran. New York and Cambridge: Van Nostrand Reinhold and Cambridge University Press.
Mumma, M. J., Buhl, D., Chin, G., Deming, D., Espenak, F., Kostiuk, T., & Zipoy, D. (1981). Discovery of natural gain amplification in the 10 μm CO_2 laser bands on Mars. *Science*, **212**, 45–9.

Mumma, M. J., Weissmann, P. R., & Stern, S. A. (1993). Comets and the origin of the solar system: Reading the Rosetta stone. In *Protostars and Planets III*, ed. E. H. Levy & J. I. Lunine. Tucson: University of Arizona Press.

Murchie, S., Mustard, J., Bishop, J., Head, J., Peaters, C., & Erard, S. (1993). Spatial variations in the spectral properties of bright regions on Mars. *Icarus*, **105**, 454–68.

Murcray, D. G. & Goldman, A. (1981). *CRC Handbook of High Resolution Infrared Laboratory Spectra of Atmospheric Interest*. Boca Raton, Florida: CRC Press.

Murcray, F. H., Murcray, F. J., Murcray, D. G., Pritchard, J., Vanasse, G., & Sakai, H. (1984). Liquid nitrogen-cooled Fourier transform spectrometer system for measuring atmospheric emission at high altitudes. *Journal of Atmospheric and Oceanic Technology*, **1**, 351–7.

Murray, B. C., Wildey, R. L., & Westphal, J. A. (1963). Infrared photometric mapping of Venus through the 8–14 micron window. *Journal of Geophysical Research*, **58**, 4813–18.

Nelson, R. M., Buratti, B. J., Wallis, B. D., Lane, A. L., West, R. A., Simmons, K. E., Hord, C. W., & Esposito, L. W. (1987). Voyager 2 photopolarimeter observations of the Uranian satellites. *Journal of Geophysical Research*, **92**, No. A13, 14905–10.

Neugebauer, G., Habing, H. J., van Dunen, R., Aumann, H. H., Baud, B., Beichman, C. A., Beinthema, D. A., Bogess, N., Clegg, P. E., de Young, T., Emerson, J. P., Gautier, T. N., Gillett, F. C., Harris, S., Hauser, M. G., Houck, J. R., Jennings, R. E., Low, F. J., Marsden, P. L., Miley, G., Olnon, F. M., Pottasch, S. R., Raimond, E., Rowan-Robinson, M., Soifer, B. T., Walker, R. G., Wesselius, P. R., & Young, E. (1984). The Infrared Astronomical Satellite (IRAS) mission. *Astrophysical Journal*, **278**, L1–6.

Neugebauer, G., Münch, G., Kieffer, H., Chase, S. C., & Miner, E. (1971). Temperature and thermal properties of the Martian surface. *Astronomical Journal*, **76**, 719–28.

Nicholson, P. D., Gierasch, P. J., Hayward, T. L., McGhee, C. A., Moersch, J. E., Squires, S. W., Vancleve, J., Matthews, K., Neugebauer, G., Shupe, D., Weinberger, A., Miles, J. W., & Conrath, B. J. (1995). Palomar observations of the impact of comet Shoemaker–Levy-9. 2. Spectra. *Geophysical Research Letters*, **22**, 1617–20.

Niemann, H. B., Atreya, S. K., Carigan, G. R., Donahue, T. M., Haberman, J. A., Harpold, D. N., Hartle, R. E., Hunten, D. M., Kasprzak, W. T., Mahaffy, P. R., Owen, T. C., Spencer, N. W., & Way, S. H. (1998). The composition of the Jovian atmosphere determined by the Galileo mass spectrometer. *Journal of Geophysical Research*, **103**, 22831–46.

Niemann, H. B., Atreya, S. K., Carigan, G., Donahue, T. M., Hartle, R., Haberman, J., Harpold, D., Hartle, R. E., Hunten, D. M., Kasprzak, W. T., Mahaffy, P. R., Owen, T. C., Spencer, N. W., & Way, S. H. (1996). The Galileo Probe mass spectrometer: Composition of Jupiter's atmosphere. *Science*, **272**, 846–8.

Nishioka, N. S., Richards, P. L., & Woody, D. P. (1978). Composite bolometers for submillimeter wavelength. *Applied Optics*, **17**, 1562–7.

Nobili, L. (1831). *Poggendorfs Annalen*, **24**, 640.

Noels, A., Soulfaire, R., & Gabriel, M. (1984). Influence of the equation of state on the solar five-minute oscillation. *Astronomy and Astrophysics*, **130**, 389–96.

Noll, K. S., Knacke, R. F., Geballe, T. R., & Tokunaga, A. T. (1988). The origin and vertical distribution of carbon monoxide on Jupiter. *Astrophysical Journal*, **324**, 1210–18.

Nordberg, W., Bandeen, W. R., Conrath, B. J., Kunde, V. G., & Persano, I. (1962). Preliminary results of radiation measurements from the TIROS III meteorological satellite. *Journal of Atmospheric Science*, **19**, 20–30.

Nuth, J. A. & Sylvester, P., ed. (1988). Workshop on the origins of solar systems. *Lunar and Planetary Institute Technical Report Number 88–04*.

Nyquist, H. (1928). Thermal agitation of electric charge in conductors. *Physical Review*, **32**, 110–13.

Oort, J. H. (1950). The structure of the cloud of comets surrounding the solar system and a hypothesis concerning its origin. *Bulletin of the Astronomical Institute of the Netherlands*, **11**, 91–110.

Öpik, E. J. (1963). The stray bodies in the Solar System. I. Survival of cometary nuclei and the asteroids. *Advances in Astronomy and Astrophysics*, **2**, 219–62.

Öpik, E. J. (1966). The dynamical aspects of the origin of comets. *Memoirs of the Royal Astronomical Society* Liege, Ser. 5, **12**, 523–74.

Pappalardo, R. T., Head, J. W., & Greeley, R. (1999). The hidden ocean of Europa. *Scientific American*, **281**, 54–63.

Park, J. H., Zander, R., Farmer, C. B., Rinsland, C. P., Russell, J. M. III, Norton, R. H., & Raper, O. F. (1986). Spectroscopic detection of CH_3Cl in the upper troposphere and lower stratosphere. *Geophysical Research Letters*, **13**, 765–8.

de Pater, I. & Massie, S. T. (1985). Models of the millimeter–centimeter spectra of the giant planets. *Icarus*, **62**, 143–71.

de Pater, I., Romani, P. N., & Atreya, S. K. (1989). Uranus' deep atmosphere revealed. *Icarus*, **82**, 288–313.

Pauling, L. & Wilson, E. B., Jr. (1935). *Introduction to Quantum Mechanics with Applications to Chemistry*. New York and London: McGraw Hill.

Pearl, J. (1988). A review of Voyager IRIS results on Io. *EOS Transaction, American Geophysical Union*. Abstract 32–05, 394.

Pearl, J. P. & Conrath, B. J. (1991). The albedo, effective temperature, and energy balance of Neptune, as determined from Voyager data. *Journal of Geophysical Research*, **96**, 18921–30.

Pearl, J. C., Conrath, B. J., Hanel, R. A., Pirraglia, J. A., & Coustenis, A. (1990). The albedo, effective temperature, and energy balance of Uranus, as determined from Voyager IRIS data. *Icarus*, **84**, 12–28.

Pearl, J. C., Conrath, B. J., Smith, M. D., Bandfield, J. L., & Christensen, P. R. (2001). Observations of Martian ice clouds by the Mars Global Surveyor Thermal Emission Spectrometer. The first Martian year. *Journal of Geophysical Research – Planets*, **106**, 12325–38.

Pearl, J. C., Hanel, R., Kunde, V., Maguire, W., Fox, K., Gupta, S., Ponnamperuma, C., & Raulin, F. (1979). Identification of gaseous SO_2 and new upper limits of other gases on Io. *Nature*, **280**, 755–8.

Pearl, J. C. & Sinton, W. M. (1982). Hot Spots of Io. In *Satellites of Jupiter*, ed. D. Morrison, pp. 724–55. Tucson: The University of Arizona Press.

Pearson, W. B. (1954). Thermopiles suitable for use at low temperature. *Journal Scientific Instruments*, **31**, 444.

Pedloskey, J. (1979). *Geophysical Fluid Dynamics*. New York: Springer-Verlag.

Peltier, M. (1834). Nouvelles expériences sur la coloricité des courans électriques. *Annals de Chimie*, **56**, 371–86.

Petroff, M. D. & Stapelbroek, M. G. (1980). Blocked impurity band detectors, radiation hard high performance LWIR detectors. In *Proceedings, IRIS Specialty Group on Infrared Detectors*, Menlo Park, California.

Petroff, M. D., Stapelbroek, M. G., & Kleinhans, W. A. (1987). Detection of individual 0.4–28 μm wavelength photons via impurity-impact ionization in a solid-state photomultiplier. *Applied Physics Letters*, **51**, 406–8.

Pinard, J. (1969). Réalisation d'un spectrometre par transformation de Fourier á très haut pouvoir de résolution. *Annals de la Physique*, **4**, 147–96.

Pirraglia, J. A. (1984). Meridional heat balance of Jupiter. *Icarus*, **59**, 169–74.

Pirraglia, J. A. & Conrath, B. J. (1974). Martian tidal pressure and wind fields obtained from the Mariner 9 infrared spectroscopy experiment. *Journal of Atmospheric Science*, **31**, 318–29.

Planck, M. (1900). Zur Theorie des Gesetzes der Energieverteilung im Normalspektrum. *Verhandlungen der Deutschen Physikalischen Gesellschaft*, **2**, 237–45.

Planck, M. (1901). Über des Gesetz der Energieverteilung im Normalspectrum. *Annalen der Physik*, (4), **4**, 553–63.

Planck, M. (1913). *Waermestrahlung*, 2nd edn. (English translation by Morton Masius (1914): *The Theory of Heat Radiation*. Reprinted by Dover Publications, Inc., New York in 1959.)

Podolak, M. & Cameron, A. G. W. (1974). Models of the giant planets. *Icarus*, **22**, 123–48.

Pollack, J. B. (1971). A nongray calculation of the runaway greenhouse: Implication of Venus past and present. *Icarus*, **14**, 295–306.

Pollack, J. B. & Bodenheimer, P. (1989). Theories of the origin and evolution of the giant planets. In *Origin and Evolution of Planetary and Satellite Atmospheres*, ed. S. K. Atreya, J. B. Pollack, & M. S. Matthews. Tucson: The University of Arizona Press.

Pollack, J. B., Grossman, A. S., Moore, R., & Graboske, H. C. (1977). A calculation of Saturn's contraction history. *Icarus*, **30**, 111–28.

Pollack, J. B., Leovy, C. B., Greiman, P. W., & Mintz, Y. (1981). A Martian general circulation model with large topography. *Journal of Atmospheric Science*, **38**, 3–29.

Pollack, J. B., Lunine, J. I., & Tittemore, W. C. (1991). Origin of the Uranian satellites. In *Uranus*, 469–512, ed. J. T. Bergstralh, E. D. Miner, & M. S. Matthews. Tucson: University of Arizona Press.

Pollack, J. B., Rages, K., Baines, K. H., Bergstralh, J. T., Wenkert, D., & Danielson, G. E. (1986). Estimates of the bolometric albedos and radiation balance of Uranus and Neptune. *Icarus*, **65**, 442–66.

Pollack, J. B. & Yung, Y. L. (1980). Origin and evolution of planetary atmospheres. *Annual Review of Earth and Planetary Sciences*, **8**, 425–87.

Potter, A. E. & Morgan, T. H. (1981). Observations of silicate reststrahlen bands in lunar infrared spectra. *Lunar and Planetary Science Conference*, **12B**, 703–13.

Prabhakara, C. & Dalu, G. (1976). Remote sensing of the surface emissivity at 9 μm over the globe. *Journal of Geophysical Research*, **81**, 3719–24.

Prabhakara, C., Fraser, R. S., Dalu, G., Man-Li, W. C., Curran, R. J., & Styles, T. (1988). Thin cirrus clouds: seasonal distribution over oceans deduced from Nimbus-4 IRIS. *Applied Meteorology*, **27**, 379–99.

Press, W. H., Teukolsky, S. A., Vetterling, W. T., & Flannerly, B. P. (1992). *Numerical Recipes, the Art of Scientific Computing* (2nd edn). Cambridge: Cambridge University Press.

Prinn, R. G. & Fegley, B., Jr. (1989). Solar nebula chemistry: origin of planetary, satellite, and cometary volatiles. In *Planetary and Satellite Atmospheres: Origin and Evolution*, ed. S. K. Atreya, J. B. Pollack, & M. S. Matthews. Tucson: The University of Arizona Press.

Profitt, C. R. (1994). Effects on heavy-element settling on solar neutrinos fluxes and interior structure. *Astrophysics Journal*, **425**, 849–55.

Puget, P., Cazes, S., Soufflot, A., Bibring, J., & Combes, M. (1987). Near-infrared mapping spectrometer of the Phobos space mission to the planet Mars. *Proc. SPIE*, **865**, 136–41. Bellingham, Washington.

Putley, E. H. (1977). Thermal detectors. In *Optical and Infrared Detectors*, ed. R. J. Keyes, Chapter 3. Vol. 19 of *Topics in Applied Physics*. Berlin, Heidelberg, New York: Springer-Verlag.

Quirico, E., Doute, S., Schmitt, B., de Bergh, C., Cruikshank, D. P., Owen, T. C., Geballe, T. R., & Roush, T. L. (1999). Composition, physical state, and distribution of ices at the surface of Triton. *Icarus*, **139**, 159–78.

Rasool, S. I. & de Bergh, C. (1970). The runaway greenhouse and the accumulation of CO_2 in the Venus atmosphere. *Nature*, **226**, 1037–9.

Ray, B. S. (1932). Über die Eigenwerte des asymmetrischen Kreisels. *Zeitschrift für Physik*, **78**, 74–91.

Rayleigh, Lord (1881). *Philosophical Magazine*, **12**, 81–101.

Rayleigh, Lord (1900). The law of complete radiation. *Philosophical Magazine*. **49**, 539–40.

Reaves, G. (1997). The prediction and discoveries of Pluto and Charon. *In Pluto and Charon*, 3–25, ed. S. A. Stern & D. J. Tholen. Tucson: University of Arizona Press.

Reber, C. A. (1990). The UARS. *EOS Transactions* AGU 71, 1867–78.

Remsberg, E. E., Russell, J. M. III, Gille, J. C., Gordley, L. L., Bailey, P. L., Planet, W. G., & Harries, J. E. (1984). The validation of Nimbus 7 LIMS measurements of ozone. *Journal of Geophysical Research*, **89**, 5161–78.

Richards, P. L. & Greenberg, L. T. (1982). Infrared detectors for low-background astronomy: incoherent and coherent devices from one micrometer to one millimeter. In *Infrared and Millimeter Waves*, Vol. 6, pp. 149–207. New York: Academic Press.

Ringwood, A. E. (1979). *On the Origin of the Moon*. New York: Springer-Verlag.

Robie, R. A., Hemingway, B. S., & Wilson, W. H. (1970). Specific heats of lunar surface materials from 90 to 350 degrees Kelvin. *Science*, **167**, 749–50.

Rodgers, C. D. (1970). Remote sounding of the atmospheric temperature profile in the presence of cloud. *Quarterly Journal of the Royal Meteorological Society*, **96**, 654–66.

Rodgers, C. D. (2000). *Inverse Methods for Atmospheric Sounding: Theory and Practice*. London: World Scientific Publishing.

Rosenqvist J., Drossart, P., Combes, M., Encrenaz, T., Lellouch, E., Bibring, J. P., Erard, S., Langevin, Y., & Chassefiere, E. (1992). Minor constituents in the Martian atmosphere from the IMS–Phobos experiment. *Icarus*, **98**, 254–70.

Ross, M. & Ree, F. H. (1980). Repulsive forces of simple molecules and mixtures at high density and temperature. *Journal of Chemical Physics*, **73**, 6146–52.

Roush, T. L., Cruikshank, D. P., Pollack, J. B., Young, F. E., & Bartholomew, M. J. (1996). Near-infrared spectral geometric albedos of Charon and Pluto: Constraints on Charon's surface composition. *Icarus*, **119**, 214–18.

Rubens, H. & Wood, R. W. (1911). *Philosophical Magazine*, **21**, 249.

Russell, C. T. ed. (1992). *The Galileo Mission*. Reprinted from Space Science Reviews, Vol. 60/1–4, Dordrecht: Kluwer Academic Publishers.

Saari, J. M. & Shorthill, R. W. (1967). Isothermal and isophotic atlas of the Moon. *NASA Contract Report CR-885*. Washington, DC: National Aeronautics and Space Administration.

Saari, J. M. & Shorthill, R. W. (1972). The sunlit lunar surface I. Albedo studies and full Moon temperature distribution. *The Moon*, **5**, 161–76.

Saari, J. M., Shorthill, R. W., & Winter, D. F. (1972). II. A study of far infrared brightness temperatures. *The Moon*, **5**, 179–99.

Sadtler Research Laboratories (1972). *The Sadtler Standard Spectra*. Gases and Vapors, GS1—GS150. Philadelphia: Sadtler Research Laboratories, Inc.

Safranov, V. S. (1972). Ejection of bodies from the Solar System in the course of the accumulation of the giant planets and the formation of the cometary cloud. In *IAU Symposium* 45, 329–34, ed. G. A. Chebotarev, E. I. Kazimirchak-Polonskaya, & B. G. Marsden. Holland: Reidel-Dordrecht.

Safranov, V. S. (1977). Oort's cometary cloud in the light of modern cosmogony. In *Comets, Asteroids, Meteorites*, 483–4, ed. A. H. Delsemme. Dordrecht, Holland: University of Toledo Press.

Sagan, C. (1960). The radiation balance of Venus. *Jet Propulsion Laboratory Technical Report, No. 32–34*.

Sagan, C. (1979). Sulfur flows on Io. *Nature*, **280**, 750–3.

Sagan, C. & Thompson, W. R. (1984). Production and condensation of organic gases in the atmosphere of Titan. *Icarus*, **59**, 133–61.

Salby, M. L. (1996). *Fundamentals of Atmospheric Physics*. San Diego: Academic Press.

Salpeter, E. (1973). On convection and gravitational layering in Jupiter and in stars of low mass. *Astrophysical Journal*, **181**, L83–L86.

Samuelson, R. E. (1983). Radiative equilibrium model of Titan's atmosphere. *Icarus*, **53**, 364–87.

Samuelson, R. E. (1985). Clouds and aerosols of Titan's atmosphere. In *The Atmospheres of Saturn and Titan*, pp. 99–107. European Space Agency: SP-241.

Samuelson, R. E., Hanel, R. A., Kunde, V. G., & Maguire, W. C. (1981). Mean molecular weight and hydrogen abundance of Titan's atmosphere. *Nature*, **292**, 688–93.

Samuelson, R. E., Maguire, W. C., Hanel, R. A., Kunde, V. G., Jennings, D. E., Yung, Y. L., & Aiken, A. C. (1983). CO_2 on Titan. *Journal of Geophysical Research*, **88**, 8709–15.

Samuelson, R. E. & Mayo, L. A. (1991). Thermal infrared properties of Titan's stratospheric aerosol. *Icarus*, **91**, 207–19.

Samuelson, R. E. & Mayo, L. A. (1997). Steady-state model for methane condensation in Titan's troposphere. *Planetary and Space Science*, **45**, 949–58.

Samuelson, R. E., Nath, N. R., & Borysow, A. (1997). Gaseous abundance and methane supersaturation in Titan's atmosphere. *Planetary and Space Science*, **45**, 959–80.

Schaeidt, S. G., Morris, P. W., Salama, A., Vandenbussche, B., Beinthema, D. A., Boxhoorn, D. R., Feuchtgruber, H., Heras, A. M., Lahuis, F., Leech, K., Roelfsema, P. R., Valentijn, E. A., Bauer, O. H., van der Bliek, N. S., Cohen, M., de Graauw, Th., Haser, L. N., van der Hucht, K. A., Hygen, E., Katterloher, R. O., Kessler, M. F., Koornneef, J., Luinge, W., Luty, D., Planck, M., Spoon, H., Waelkens, C., Waters, L. B. F. M., Wieprecht, E., Wildeman, K. J., Young, E., & Zaal, P. (1996). The photometric calibration of the ISO Short Wavelength Spectrometer. *Astronomy and Astrophysics*, **315**, L55–9.

Schmitt, B., de Bergh, C., & Feston, M. eds. (1998). *Solar System ices*. Astrophysics and Space Science Library, Vol. 227. Dordrecht: Kluver Academic Publishers.

Schrödinger, E. (1926). Quantisierung als Eigenwertproblem. *Annalen der Physik*, **79**, 361–76, 489–527; **80**, 437–90; **81**, 109–39.

Seebeck, T. J. (1826). Magnetische Polarität der Metalle durch Temperaturdifferenz. *Poggendorfs Annalen*, (6), 133 and 263.

Serlemitsos, A. T., Warner, B. A., Castles, S., Breon, S. R., San Sebastian, M., & Hait, T. (1990). Adiabatic demagnetization refrigerator for space use. In *Advances in Cryogenic Engineering*, Vol. 35. New York: Plenum Press.

Showman, A. P. & Malhotra, R. (1999). The Galilean satellites. *Science*, **286**, 77–84.

Sill, G. T. & Clark, R. N. (1982). Composition of the surfaces of the Galilean satellites. In *Satellites of Jupiter*, ed. D. Morrison, pp. 174–212. Tucson: The University of Arizona Press.

Sinton, W. M. (1962). Temperatures of the lunar surface. In *Physics and Astronomy of the Moon*, ed. Z. Kopal, pp. 407–28. New York: Academic Press.

Sinton, W. M., Goguen, J. D., Nagata, T., Ellis, H. B., Jr., & Werner, M. (1988). Infrared polarization measurements of Io in 1986. *Astronomical Journal*, **96**, 1095–105.

Smith, B. A., Soderblom, L. A., Johnson, T. V., Ingersoll, A. P., Collins, S. A., Shoemaker, E. M., Hunt, G. E., Masursky, H., Carr, M. H., Davies, M. E., Cook, A. F. II, Boyce, J., Danielson, G. E., Owen, T., Sagan, C., Beebe, R. F., Veverka, J., Strom, R. G., McCauley, J. F., Morrison, D., Briggs, G. A., & Suomi, V. E. (1979a). The Jupiter system through the eyes of Voyager 1. *Science*, **204**, 951–72.

Smith, B. A., Soderblom, L. A., Beebe, R., Boyce, J., Briggs, G., Carr, M., Collins, S. A., Cook, A. F., Danielson, G. E., Davis, M. E., Hunt, G. E., Ingersoll, A., Johnson, T. V., Masursky, H., McCauley, J., Morrison, D., Owen, T., Sagan, C., Shoemaker, E. M., Strom, R., Suomi, V., & Veverka, J. (1979b). The Galilean satellites and Jupiter: Voyager 2 imaging science results. *Science*, **206**, 927–50.

Smith, B. A., Soderblom, L., Beebe, R., Boyce, J., Briggs, G., Bunker, A., Collins, S. A., Hansen, C. J., Johnson, T. V., Mitchell, J. L., Terrile, R. J., Carr, M., Cook, A. F. II, Cuzzi, J., Pollack, J. B., Danielson, G. E., Ingersoll, A., Davis, M. E., Hunt, G. E., Masursky, H., Shoemaker, E., Morrison, D., Owen, T., Sagan, C., Veverka, J., Strom, R., & Suomi, V. E. (1981). Encounter with Saturn: Voyager 1 imaging results. *Science*, **212**, 163–91.

Smith, B. A., Soderblom, L. A., Beebe, R., Bliss, D., Boyce, A., Brahic, A., Briggs, G. A., Brown, R. H., Collins, S. A., Cook, A. F. II, Croft, S. K., Cuzzi, J. N., Danielson, G. E., Davis, M. E., Dowling, T. E., Godfrey, D., Hansen, C. J., Harris, C., Hunt, G. E., Ingersoll, A. P., Johnson, T. V., Krauss, R. J., Masursky, H., Morrison, D., Owen, T., Plescia, J. B., Pollack, J. B., Porco, C. C., Rages, K., Sagan, C., Shoemaker, E. M., Sromovsky, L. A., Stoker, C., Storm, R. G., Suomi, V. E., Synnott, S. P., Terrile, R. J., Thomas, P., Thompson, W. R., & Veverka, J. (1986). Voyager 2 in the Uranian system: imaging science results. *Science*, **233**, 43–64.

Smith, D. E., Zuber, M. T., Solomon, S. C., Phillips, R. J., Head, J. W., Garvin, J. B., Banerdt, W. B., Muhlman, D. O., Pettengill, G. H., Neumann, G. A., Lemoine, F. G., Abshire, J. B., Aharonson, O., Brown, C. D., Houck, S. A., Ivanov, A. B., McGovern, P. J., Zwally, H. J., & Duxbury, T. C. (1999). The global topography of Mars and implications for surface evolution. *Science*, **284**, 1495–503.

Smith, M. D. & Gierasch, P. J., (1995). Convection in the outer planet atmospheres including ortho–para hydrogen conversion. *Icarus*, **116**, 159–79.

Smith, M. D., Pearl, J. C., Conrath, B. J., & Christensen, P. R. (2000). Mars Global Surveyor Thermal Emission Spectrometer (TES) observations of dust opacity during aerobraking and science phasing. *Journal of Geophysical Research*, **105**, 9539–52.

Smith, M. D., Pearl, J. C., Conrath, B. J., & Christensen, P. R. (2001). Thermal Emission Spectrometer results: Mars atmospheric thermal structure and aerosol distribution. *Journal of Geophysical Research*, **106**(E10), 23925–45.

Smith, P. H., Lemmon, M. T., Lorenz, R. D., Sromovsky, L. A., Caldwell, J. J., & Allison, M. D. (1996). Titan's surface revealed by HST imagery. *Icarus*, **119**, 336–49.

Smith, R. A., Jones, F. E., & Chasmar, R. P. (1957). *The Detection and Measurement of Infra-Red Radiation*. Oxford: Clarendon Press.

Smith, S. D., Holah, G. D., Seeley, J. S., Evans, C., & Hunneman, R. (1972). Survey of the present state of art of infrared filters. In *Infrared Detection Techniques for Space Research*, ed. V. Manno & J. Ring. Dordrecht, Holland: D. Reidel Publ. Co.

Smith, W. (1970). Iterative solution of the radiative transfer equation for the temperature and absorbing gas profile of an atmosphere. *Applied Optics*, **9**, 1993–9.

von Smoluchowski, M. (1906). Zur kinetischen Theorie der Brownschen Molekular-bewegung und der Suspensionen. *Annalen der Physik* (4), **21**, 756–80.

Smoluchowski, R. (1967). Internal structure and energy emission of Jupiter. *Nature*, **215**, 691–5.

Smoluchowski, R. (1973). Dynamics of the Jovian interior. *Astrophysical Journal*, **185**, L95–L99.

Soderblom, L., Johnson, T., Morrison, D., Danielson, E., Smith, B., Veverka, J., Cook, A., Sagan, C., Kupferman, P., Pieri, D., Mosher, J., Avis, C., Gradie, J., & Clancy, T. (1980). Spectrophotometry of Io: preliminary Voyager results. *Geophysical Research Letters*, **7**, 963–6.

Sommerfeld, A. (1896). *Mathematical Annalen*, **47**, 317.

Sommerfeld, A. (1952). *Electrodynamics* (Lectures on Theoretical Physics, Vol. III), English translation. New York: Academic Press.

Sommerfeld, A. (1954). *Optics* (Lectures on Theoretical Physics, Vol. IV), English translation. New York: Academic Press.

Sommerfeld, A. (1964). *Thermodynamics and Statistical Mechanics* (Lectures on Theoretical Physics, Vol. V), English translation. New York: Academic Press.

Spänkuch, D., Oertel, D., Moroz, V. I., Döhler, W., Linkin, V. M., Schäfer, K., Zasova, L. V., Güldner, J., Mazygorin, I. A., Ustinov, E. A., & Dubois, R. (1984). Venus spectra obtained from Venera spacecrafts 15 and 16. *Proceedings of the International Radiation symposium, Perugia, Italy* (21–28 August, 1984).

Spencer, J. R., Jessup, K. L., McGrath, M., Ballester, G. E., & Yelle, R. (2000*a*). Discovery of gaseous S_2 in Io's Pele Plume. *Science*, **288**, 1208–10.

Spencer, J. R., Rathbun, J. A., Travis, L. D., Tammpari, L. K., Bernard, L., Martin, T. Z., & McEwen, A. S. (2000*b*). Io's thermal emission from the Galileo Photopolarimeter–Radiometer. *Science*, **288**, 1198–201.

Spencer, J. R., Stansberry, J. A., Trafton, L. M., & Young, E. F. (1997). Volatile transport, seasonal cycles, and atmospheric dynamics on Pluto. In *Pluto and Charon*, 435–74, ed. S. A. Stern & D. J. Tholen. Tucson: University of Arizona Press.

Stair, A. T., Jr., Pritchard, J., Coleman, I., Bohne, C., Williamsen, W., Rogers, J., & Rawlins, W. T. (1983). Rocketborne cryogenic (10 K) high-resolution interferometer spectrometer flight HIRIS: auroral and atmospheric IR emission spectra. *Applied Optics*, **22**, 1056–69.

Stefan, J. (1879). *Wiener Berichte*, **79**, 391.

Steinfeld, J. I. (1974). *Molecules and Radiation* (2nd edn, 1985). Cambridge, Massachusetts: The MIT Press.

Stern, S. A., McKinnon, W. B., & Lunine, J. I. (1997). On the origin of Pluto, Charon, and the Pluto–Charon binary. In *Pluto and Charon*, 605–63, ed. S. A. Stern & D. J. Tholen. Tucson: University of Arizona Press.

Stevenson, D. J. (1982). Interiors of the giant planets. *Annual Review of Earth and Planetary Sciences*, **10**, 257–95.

Stevenson, D. J. & Salpeter, E. E. (1976). Interior models of Jupiter. In *Jupiter*, ed. T. Gehrels, pp. 85–112. Tucson: The University of Arizona Press.

Strand, O. N. & Westwater, E. R. (1968). Statistical estimation of the numerical solution of a Fredholm integral equation of the first kind. *Journal of the Association for Computing Machinery*, **15**, 100–14.

Strömgren, B. (1953). The Sun as a star. In *The Sun*, ed. G. Kuiper, Chapter 2. Chicago: The University of Chicago Press.

Strong, J. (1954). Interferometric modulator. *Journal of the Optical Society of America*, **44**, 352.

Strong, J. (1957). Interferometry for the far infrared. *Journal of the Optical Society of America*, **47**, 354–7.

Strong, J. (1958). *Concepts of Classical Optics*. San Francisco and London: W. H. Freeman and Company.

Strong, J. & Vanasse, G. (1958). Modulation interférentielle et calculateur analogique pour une spectromètre interférential. *Journal de Physique et le Radium*, **19**, 192–6.

Strong, J. & Vanasse, G. (1960). Lamellar grating far-infrared interferometer. *Journal of the Optical Society of America*, **50**, 113–18.

Summers, M. E., Strobel, D. F., & Gladstone, G. R. (1997). Chemical models of Pluto's atmosphere. In *Pluto and Charon*, 391–434, ed. S. A. Stern & D. J. Tholen. Tucson: University of Arizona Press.

Swinyard, B. M., Clegg, P. E., Ade, P. A. R., Armand, C., Balteau, J. P., Barlow, M. J., Berges, J. C., Burgdorf, M., Caux, E., Ceccarelli, C., Cerulli, R., Church, S. E., Colgan, S., Cotin, F., Cox, P., Cruvellier, P., Davis, G. R., DiGiorgio, A., Emery, R. J., Ewart, D., Fischer, J., Furniss, I., Glencrossi, W. M., Greenhouse, M., Griffin, M. J., Gry, C., Haas, M. R., Joubert, M., King, K. J., Lim, T., Liseau, R., Lord, S., Lorenzetti, D., Molinari, S., Naylor, D. A., Nisini, B., Omont, A., Orfei, R., Patrick, T., Pequinot, D., Price, M. C., NguyenQRieu, Robinson, F. D., Saisse, M., Saraceno, P., Serra, G., Sidher, S. D., Smith, H. A., Spinoglio, L., Texier, D., Towlson, W. A., Trams, N., Unger, S. J., & White, G. J. (1996). Calibration and performance of the ISO Long-Wavelength Spectrometer. *Astronomy and Astrophysics*, **315**, L43–8.

Taylor, D. J. (1965). Spectrophotometry of Jupiter's 34100–1000 Å spectrum and bolometric albedo for Jupiter. *Icarus*, **4**, 362–73.

Taylor, F. W. (1983). The pressure modulator radiometer. In *Spectrometric Techniques Vol. III*, ed. G. Vanasse, Chapter 3. New York: Academic Press.

Taylor, F. W., Beer, R., Chahine, M. T., Diner, D. J., Elson, L. S., Haskins, R. D., McCleese, D. J., Martonchik, J. V., Reichley, P. E., Bradley, S. P., Delderfield, J., Schofield, J. T., Farmer, C. B., Froidevaux, L., Leung, J., Coffey, M. T., & Gille, J. C. (1980). Structure and meterology of the middle atmosphere of Venus: infrared remote sensing from the Pioneer orbiter. *Journal of Geophysical Research*, **85**, 7963–8006.

Taylor, F. W., Diner, D. J., Elson, L. S., McCleese, D. J., Martonchik, J. V., Delderfield, J., Bradley, S. P., Schofield, J. T., Gille, J. C., & Coffey, M. T. (1979*a*). Temperature, cloud structure, and dynamics of the Venus middle atmosphere by infrared remote sensing. *Science*, **205**, 65–7.

Taylor, F. W., Vescelus, F. E., Locke, J. R., Foster, G. T., Forney, P. B., Beer, R., Houghton, J. T., Delderfield, J., & Schofield, J. T. (1979*b*). Infrared radiometer for the Pioneer Venus Orbiter: instrument description. *Applied Optics*, **18**, 3893–900.

Terrile, R. J. & Beebe, R. F. (1979). Summary of historical data: interpretation of the Pioneer and Voyager cloud configurations in a time-dependent framework. *Science*, **204**, 948–51.

Tholen, D. J. & Buie, M. W. (1997). Bulk properties of Pluto and Charon. In *Pluto and Charon*, 193–219, ed. S. A. Stern & D. J. Tholen. Tucson: University of Arizona Press.

Thomson, W. (Lord Kelvin) (1843). On the uniform motion of heat in homogeneous solid bodies and its connection with the mathematical theory of electricity. *Cambridge Mathematical Journal*, **3**, 11–84.

Tikhonov, A. N. (1963). On the solution of incorrectly stated problems and a method of regularization. *Dokl. Acad. Nauk, USSR*, **151**, 501–4.

Toon, O. B., Pollack, J. B., & Sagan, C. (1977). Physical properties of the particles comprising the Martian dust storm of 1971–1972. *Icarus*, **30**, 663–96.

Tomasko, M. G., McMillan, R. S., Doose, L. R., Castillo, N., & Dilley, J. P. (1980). Photometry of Saturn at large phase angles. *Journal of Geophysical Research*, **85**, 5891–903.

Tomasko, M. G., West, R. A., & Castillo, N. D. (1978). Photometry and polarimetry of Jupiter at large phase angles. *Icarus*, **33**, 558–92.

Toublanc, D., Parisot, I. P., Brillet, J., Gautier, D., Raulin, F., & McKay, C. P. (1995). Photochemical modeling of Titan's atmosphere. *Icarus*, **113**, 2–16.

Townes, C. H. & Schawlow, A. L. (1955). *Microwave Spectroscopy*. New York: McGraw-Hill.

Trafton, L. M. (1966). The pressure-induced monochromatic translational absorption coefficients for homopolar and non-polar gases and gas mixtures with particular application to H_2. *Astrophysical Journal*, **146**, 558–71.

Trafton, L. M., Hunten, D. M., Zahnle, K. J., & McNutt, R. L., Jr. (1997). Escape processes at Pluto and Charon. In *Pluto and Charon*, 475–522, ed. S. A. Stern & D. L. Tholen. Tucson: University of Arizona Press.

Trombka, J. I., Squyres, S. W., Brückner, J., Boynton, W. V., Reedy, R. C., McCoy, T. J., Gorenstein, P., Evans, L. G., Arnold, J. R., Starr, R. D., Nittler, L. R., Murphy, M. E., Mikheeva, I., McNutt, Jr., R. L., McClanahan, T. P., McCartney, E., Goldsten, J. O., Gold, R. E., Floyd, S. R., Clark, P. E., Burbine, T. H., Bhangoo, J. S., Bailey, S. H., & Pataev, M. (2000). The elemental composition of Asteroid 433 Eros: Results of the NEAR–Schoemaker X-ray Spectrometer. *Science*, **289**, 2101–5.

Twomey, S. (1963). On the numerical solution of Fredholm integral equations of the first kind by inversion of the linear system produced by quadrature. *Journal of the Association for Computing Machinery*, **10**, 79–101.

Twomey, S. (1977). *Introduction to the Mathematics of Inversion in Remote Sensing and Indirect Measurements*. Amsterdam: Elsevier.

Tyler, G. L., Eshleman, V. R., Anderson, J. D., Levy, G. S., Lindal, G. F., Wood, G. E., & Croft, T. A. (1981). Radio science investigation of the Saturn system with Voyager 1: preliminary results. *Science*, **212**, 201–6.

Ulaby, F. T., Moore, R. K., & Fung, A. K. (1986). *Microwave Remote Sensing: Active and Passive, Vol. III, From Theory to Application*. Norwood, MA: Artech House, Inc.

Valentijn, E. A., Feuchtgruber, H., Kester, D. J. M., Roelfsema, P. R., Barr, P., Bauer, O. H., Beinthema, D. A., Boxhoorn, D. R., de Graaw, Th., Haser, L. N., Haske, A., Heras, A. M., Katterloher, R. O., Lahuis, F., Luty, D., Leech, K. J., Morris, P. W., Salama, A., Schaidt, S. G., Spoon, H. W. W., Vandenbussche, B., Wieprecht, E., Luinge, W., & Wildeman, K. J. (1996). The wavelength calibration and resolution of the SWS. *Astronomy and Astrophysics*, **315**, L60–3.

Vanasse, G. A. & Sakai, H. (1967). Fourier spectroscopy. In *Progress in Optics VI*, ed. E. Wolf, pp. 260–330. Amsterdam: North-Holland Publishing Company.

Vanasse, G. A., Stair, A. T., & Baker, D. J., eds. (1971). *Proceedings of the Aspen International Conference on Fourier Spectroscopy*, US Air Force Cambridge Research laboratory 71-0019.

Van den Berg, D. A. (1983). Star clusters and stellar evolution, I, Improved synthetic color-magnitude diagrams for the oldest clusters. *Astrophysical Journal, Supplement*, **51**, 29–66.

Vardavas, I. M. & Carver, J. H. (1985). Atmospheric temperature response to variations in CO_2 concentration and the solar-constant. *Planetary and Space Science*, **33**, 1187–207.

Vasicek, A. (1960). *Optics of Thin Films*. Amsterdam: North-Holland Publishing Company.

Veverka, J., Robinson, M., Thomas, P., Murchie, S., Bell, J. F., III, Izenberg, N., Chapman, C., Harch, A., Bell, M., Carcich, B., Cheng, A., Clark, B., Domingue, D., Farquhar, R., Gaffey, M. G., Hawkins, E., Joseph, J., Kirk, R., Li, H., Lucey, P., Malin, M., McFadden, L., Merline, W. J., Miller, J. K., Owen Jr., W. M., Peterson, C., Prockter, L., Warren, J., Wellnitz, D., Williams, B. J., & Yeomans, D. K. (2000). NEAR at Eros: imaging and spectral results. *Science*, **289**, 2088–97.

Veverka, J., Thomas, P. J., Bell III, M., Carcich, B., Clark, B., Harch, A., Joseph, J., Martin, P., Robinson, M., Murchie, S., Izenberg, N., Hawkins, E., Warren, J., Farquar, R., Chang, A., Dunham, D., Chapman, C., Merline, W. J., McFadden, L., Wellnitz, D., Malin, M., Owen, W. M., Jr., Miller, J. K., Williams, B. G., & Yeomans, D. K. (1999). Imaging of Asteroid 433 Eros During NEAR's flyby reconnaissance. *Science*, **285**, 562–4.

Vogel, P. & Genzel, L. (1964). Transmission and reflection of metallic mesh in the far infrared. *Infrared Physics*, **4**, 257–62.

Voigt, W. (1912). *Sitzungsberichte der Akademie der Wissenschaften, Mathematisch – Physikalische Klasse*, **42**, 603. Munich.

von Zahn, U. & Hunten, D. M. (1996). The helium mass fraction in Jupiter's atmosphere. *Science*, **272**, 849–951.

von Zahn, U., Hunten, D. M., & Lehmacher, G. (1998). Helium in Jupiter's atmosphere: Results from the Galileo probe helium interferometer experiment. *Journal of Geophysical Research*, **103**, 22815–29.

Walker, J. C. G. (1975). Evolution of the atmosphere of Venus. *Journal of Atmospheric Science*, **32**, 1248–56.

Walker, J. C. G. (1977). *Evolution of the Atmosphere*. New York: Macmillan.

Walker, J. C. G. (1982). The earliest atmosphere of the Earth. *Precambrian Research*, **17**, 147–71.

Wang, S. C. (1929). On the asymmetrical top in quantum mechanics. *Physical Review*, **34**, 243–52.

Wänke, H. & Dreibus, G. (1986). Geochemical evidence for the formation of the Moon by impact-induced fission of the proto-Earth. In *Origin of the Moon*, ed. W. K. Hartmann, R. J. Phillips, & G. J. Taylor, pp. 649–72. Houston: Lunar and Planetary Institute.

Ward, P. D., Montgomery, D. R., & Smith, R. (2000). Altered river morphology in South Africa related to the Permian–Triassic extinction. *Science*, **289**, 1740–3.

Wark, D. Q. & Hilleary, D. T. (1969). Atmospheric temperature: successful test of remote probing. *Science*, **165**, 1256–8.

Watson, A. J., Donahue, T. M., & Kuhn, W. R. (1984). Temperatures in a runaway greenhouse on the evolving Venus: implication for water loss. *Earth and Planetary Science Letters*, **68**, 1–6.

Weidenschilling, S. J. (1980). Dust to planetesimals: settling and coagulation in the solar nebula. *Icarus*, **44**, 172–89.

Weinstein, M. & Suomi, V. E. (1961). Meteorological applications of Explorer VII infrared radiation measurements. *Transactions of the American Geophysical Union*, **42**, 492–8.

Weissman, P. R. (1998). The Oort cloud. *Scientific American*, **279**, 84–9.

Weissman, P. R. & Levison, H. F. (1997). The population of the Trans-Neptunian region: The Pluto–Charon environment. In *Pluto and Charon*, 559–604, ed. S. A. Stern & D. J. Tholen. Tucson: University of Arizona Press.

von Weizsäcker, C. F. (1938). *Physikalische Zeitschrift*, **39**, 633.

Welford, W. T., & Winston, R. (1978). *The Optics of Nonimaging Concentrators of Light and Solar Energy*. New York: Academic Press.

Welsh, H. L. (1972). Pressure induced absorption spectra of hydrogen. *MTP International Review of Science – Physical Chemistry. Series one, vol. III: Spectroscopy*, pp. 33–71. London: Butterworths.

Wesselink, A. J. (1946). Heat conductivity and nature of the lunar surface material. *Bulletin of the Astronomical Institutes of the Netherlands*, **10**, 351–63.

West, R. A., Strobel, D. F., & Tomasko, M. G. (1986). Clouds, aerosols, and photochemistry in the Jovian atmosphere. *Icarus*, **65**, 161–217.

Westphal, J. A. (1969). Observations of localized 5-micron radiation from Jupiter. *Astrophysical Journal*, **157**, L63.

Whipple, F. (1950). A comet model. I. The acceleration of Comet Encke. *Astrophysical Journal*, **111**, 375–94.

Whipple, F. (1951). A comet model. II. Physical relations for comets and meteors. *Astrophysical Journal*, **113**, 464–74.

Whipple, F. (1963). On the structure of the cometary nucleus. In *The Moon, Meteorites, and Comets*. Vol. 4, 639–64, ed. B. M. Middlehurst & G. P. Kuiper. Chicago: University of Chicago Press.

Whiting, E. E. (1968). An Empirical approximation to the Voigt profile. *Journal of Quantitative Spectroscopy and Radiative Transfer*, **8**, 1379–84.

Wiedemann, G., Jennings, D. E., Hanel, R. A., Kunde, V. G., Mosley, S. H., Lamb, G., Petroff, M. D., & Stapelbroek, M. G. (1989). Postdispersion system for astronomical observations with Fourier transform spectrometers in the thermal infrared. *Applied Optics*, **28**, 139–45.

Wien, W. (1893). Temperatur und Entropie der Strahlung. *Sitzungsberichte der Akademie der Wissenschaften*, Berlin, Feb. 9, 1893 (also *Wiedemanns Annalen*, **52**, 132–65, 1894).

Wien, W. (1896). Über die Energieverteilung in Emissionsspectrum eines schwarzan Körpers. *Wiedemanns Annalen*, **58**, 662–9.

Williams, D. L. & von Herzen, R. P. (1974). Heat loss from the Earth: New estimate. *Geology*, **2**, 327–8.

Wilson, J. R. & Hamilton, K. (1996). Comprehensive model simulation of thermal tides in the Martian atmosphere. *Journal of the Atmospheric Sciences*, **53**, 1290–326.

Wolfe, W. L. & Zissis, G. J. (1978; third printing, 1989). *The Infrared Handbook*. The Infrared Information and Analysis (IRIA) Center, Environmental Research Inst. of Michigan.

Wolter, H. (1956). Optik dünner Schichten. *Handbuch der Physik, Vol. XXIV, Grundlagen der Optik*. Springer-Verlag.

Woody, D. P., Mather, J. C., Nishioka, N. S., & Richards, P. L. (1975). Measurement of the spectrum of the submillimeter cosmic background. *Physical Review Letters*, **34**, 1036–9.

Woolley, R. v. d. R. & Stibbs, D. W. N. (1953). *The Outer Layers of a Star*. Oxford: Clarendon Press.

Wyatt, C. L. (1975). Infrared spectrometer: liquid-helium-cooled rocketborn circular-variable filter. *Applied Optics*, **14**, 3086–91.

Yelle, R. V. & Elliot, J. L. (1997). Atmospheric structure and composition: Pluto and Charon. In *Pluto and Charon*, 347–90, ed. S. A. Stern & D. J. Tholen. Tucson: University of Arizona Press.

Yeomans, D. K., Antreasian, P. G., Barriot, J. P., Chesley, S. R., Dunham, D. W., Farquhar, R. W., Georgini, J. D., Helfrich, C. E., Konopliv, A. S., McAdams, J. V., Miller, J. K., Owen, W. M., Jr., Scheeres, D. J., Thomas, P. C., Veverka, J., & Williams, B. G. (2000). Radio science results during the Near–Schoemaker spacecraft rendezvous with Eros. *Science*, **289**, 2085–8.

Yeomans, D. K., Antreasian, P. G., Cheeng, A. Dunham, D. W., Farquhar, R. W., Gaskell, R. W., Georgini, J. D., Helfrich, C. E., Konoplif, A. S., McAdams, J. V., Miller, J. K., Owen, W. M., Jr., Thomas, P. C., Veverka, J., & Williams, B. G. (1999). Estimating the mass of Asteroid 433 Eros during the NEAR spacecraft flyby. *Science*, **285**, 560–1.

Yung, Y. L., Allen, M., & Pinto, J. P. (1984). Photochemistry of the atmosphere of Titan: comparison between model and observations. *Astrophysical Journal Supplement*, **55**, 465–506.

Zuber, M. T., Smith, D. E., Cheng, A. F., Garvin, J. B., Aharonson, O., Cole, T. D., Dunn, P. J., Guo, Y., Lemoine, F. G., Neumann, G. A., Rowlands, D. D., & Torrence, M. H. (2000). The shape of 433 Eros from the NEAR–Shoemaker laser rangefinder. *Science*, **289**, 2097–101.

Zwerdling, S., Smith, R. A., & Theriault, J. P. (1968). A fast, high-responsivity bolometer detector for the very-far infrared. *Infrared Physics*, **8**, 271–336.

Abbreviations

a.c.	alternating current
ASA	American Standards Association
ATMOS	Atmospheric Trace Molecule Spectroscopy
ATS	Application Technology Satellite
$A\Omega$	area times solid angle (étendue)
BIB	blocked-impurity band detectors
BLIP	background limited performance
CCD	Charge Coupled Device
CIA	collision-induced absorption
CIRS	Composite Infrared Spectrometer
CLAES	Cryogenic Limb Array Etalon Spectrometer
COBE	Cosmic Background Explorer
CVF	Continuously Variable Filter
d.c.	direct current
DIRBE	Diffuse Infrared Background Experiment
DTGS	deuterated triglycine sulphate
ERBE	Earth Radiation Budget Experiment
FET	field-effect transistor
FIRAS	Far Infrared Absolute Spectrometer
FTS	Fourier transform spectrometer
GOES	Geostationary Operational Environmental Satellite
G–R	generation–recombination (noise)
HRIR	High Resolution Infrared Radiometer
IF	intermediate frequency
IRAS	Infra Red Astronomical Satellite
IRIS	Infrared Interferometer Spectrometer
IRTM	Infrared Thermal Mapper
ISM	Imaging Spectrometer for Mars
ISO	Infrared Space Observatory
ITOS	Improved TIROS Operational Satellite
JFET	Johnson field-effect transistor
LO	local oscillator
LTE	local thermodynamic equilibrium
MGS	Mars Global Surveyor
MOC	Mars Orbiter Camera

MOLA	Mars Orbiter Laser Altimeter
NASA	National Aeronautics and Space Administration
NEAR	Near Earth Asteroid Rendezvous Mission
NEF	Noise Equivalent Flux
NEP	Noise-Equivalent-Power
NESR	Noise-Equivalent-Spectral-Radiance
NIMS	Near Infrared Mapping Spectrometer
NOAA	National Oceanographic and Atmospheric Administration
OGO	Orbiting Geophysical Observatory
PC	photoconductive
PFS	Planetary Fourier Spectrometer
PV	photovoltaic
RF	radio frequency
RSS	Radio Science System
SAMS	Stratospheric and Mesospheric Sounder
SI	International System
SIRS	Satellite Infrared Spectrometer
S/N	signal-to-noise (ratio)
SSCC	Spin Scan Cloud Camera
SWS	Short Wavelength Spectrometer
TE	transverse electric wave
TES	Thermal Emission Spectrometer
TGS	triglycine phosphate
TIA	trans-impedance amplifier
TIROS	Television and Infrared Observational Satellite
TM	transverse magnetic wave
UARS	Upper Atmosphere Research Satellite
VHRR	Very High Resolution Radiometer
VIMS	Visible Infrared Mapping Spectrometer
VISSR	Visible Infrared Spin Scan Radiometer
VORTEX	Venus Orbiter Radiometer Temperature Experiment

Index

A band, 92
a.c. radiometer, 170, 179, 182, 225
$A\Omega$ advantage, *see* étendue
absorption, 31, 34, 36, 64, 102
abundance determination, 371
accuracy, 281, 282, 283
acetylene (C_2H_2), 87, 91, 321, 326, 330, 419
adiabatic heating and cooling, 433, 440
adiabatic lapse rate, 438
advective derivative, 422, 424, 432
aerosol, 110, 380, 382, 420
Airy disk, 167, 169, 231, 251
albedo
   bolometric, 388
   Bond, 395, 396, 401, 456, 460–3
   geometric, 392, 400, 401, 456–63
   single scattering, 29, 38, 53, 137
   surface, 409, 415
ammonia (NH_3), 81, 87, 89, 320–2, 375–6, 436, 447–8, 450, 456
angular momentum, 59, 66, 72, 73, 83, 87, 88, 457
anharmonic effects, 70, 72, 76, 80, 82, 85
anomalous dispersion, 108
antireflection coating, 160, 198–200
aperture stop, 161–2
apodization, 230, 232, 247
Apollo, manned mission to the Moon, 338, 339, 391
Apollo, Amor, and Atem group of asteroids, 349
Application Technology Satellite (ATS), 175, 177
argon (Ar), 306, 451
arsenic-doped silicon detector, 273
Asteroid Belt, 349
asteroids, 349–51
asymmetric rotor, 87, 89, 92
asymmetry factor, 135, 409
asymptotic scattering function, 110
atmospheric composition, 368–79, 474
atmospheric emissivity, 136
atmospheric escape, 449
atmospheric motion, 321–43
Atmospheric Trace Molecule Spectroscopy (ATMOS), 225, 235–6, 299, 371, 379

aureole, 126
autocorrelation, 370

B band, 92, 93
background limited performance (BLIP), 277, 278
background noise, 211, 221, 240, 242, 263, 264, 278
background radiation, 278
Backus–Gilbert formulation, 361–5
band gap, 273–4
barium fluoride (BaF_2), 189
beamdividers *or* beamsplitters, 200–4
Beer's law, 41
benzene (C_6H_6), 85
bidirectional reflectivity, 397
blackbody, 21–5, 181, 260, 284, 285–9, 312–13, 334
blocked-impurity band detectors (BIB), 275
bolometer, 266–8, 271–2
Boltzmann factor, 95, 96, 98
boundary condition, 14–16
boundary layer, 412
brightness temperature, 147, 385
Brownian motion, 260

C band, 92–3
calcium fluoride (CaF_2), 239
calibration,
   intensity, 281–93
   wavenumber, 293–4
Callisto, satellite of Jupiter, 335
carbon dioxide (CO_2), 81–2, 87, 148–50, 193, 250, 303–17, 320, 324–30, 348, 352, 358–9, 364, 419, 442, 450–2
carbon monoxide (CO), 69–70, 75, 78, 96, 99, 310, 325–6, 344, 348, 452, 454–6
Cassegrain telescope, 162–4, 166, 170
Cassini mission to Saturn, 227, 239, 318–20, 444
cat's eye, 234–5
Centaurs, 348
centrifugal distortion, 74–6, 80, 88–90
cesium bromide (CsBr), 239
cesium iodide (CsI), 189, 192, 320
channel spectrum, 206
charge coupled devices (CCD), 280

Charon, satellite of Pluto, 342–5
Christiansen frequency, 387
circumsolar disk, 447
clathrates, 452
clouds, 110–29, 311, 314–15, 380–5
  water, 311, 314–15
  water ice, 317
collision broadening, 99–102, 129
collision-induced absorption, 305–6, 320, 415, 420
collision-induced transitions, 78–80, 294, 328–9, 375, 377, 455
comets, 346–8
complex refractive index, 21, 107
Composite Infrared Spectrometer (CIRS), instrument on Cassini, 239, 318–19
conductivity,
  electric, 3
  thermal, 258
constrained linear inversion, 356–60
continuity equation, 3, 422
continuous mode, 223
contribution function, 143–8
convection, 455
convective equilibrium, 447
copper-doped germanium detector, 273–4, 276
core formation, 448, 449
Coriolis force, 422, 440
corner cubes, 233–4, 239
correlation analysis, 370–1
Cosmic Background Explorer (COBE), 165, 244, 247
covariance function, 370
Cryogenic Limb Array Etalon Spectrometer (CLAES), instrument on UARS, 205
cryogenically cooled detectors, 243, 271–2, 276
cryolite, 207
Curie point, 269
current noise, 264
cyanoacetylene (HC_3N), 327, 329, 330
cyanogen (C_2N_2), 87, 328–30
cyclostrophic balance, 443–4
Czerny–Turner spectrometer, 216

d.c. radiometer, 170, 182
degenerate state, 81–3, 88, 95–6
degrees of freedom, 80–1
detectivity, 254, 276
detector array, 210–11, 213, 217, 218, 240–1, 243, 280–1
detector noise, 260–4, 277–8
deuterated methane (CH_3D), 87, 321, 330
deuterium to hydrogen ratio (D/H ratio), 325, 451
diabatic heating, 423, 432, 434, 440
diacetylene (C_4H_2), 329, 330
diamond, 190
diatomic molecules, 66–80
dielectric bolometer, 269
dielectric constant, 3, 106–7
diffraction, 166–70, 209, 251
diffraction grating, 211
Diffuse Infrared Background Experiment (DIRBE), 149

dimer, 78–80
dipole moment, 63, 69–70, 74, 97
dipole moment, induced, 78
Doppler broadening, 99–102
Doppler shift, 250, 294
Dynamic Explorer, 209

Earth, 193, 301, 305–8, 310–12, 337, 360, 364, 399, 413–14, 419–20, 444, 448, 450–1, 459
Earth Radiation Budget Experiment (ERBE), 395
Ebert–Fastie spectrometer, 215–16
echelle grating, 215
effective temperature, 284–5, 458
efficiency factors, 122–3, 127–8, 232, 296
eigenvalues, 84, 260
Einstein coefficients, 94, 99
elemental abundances, 453
emission, 32, 36–7, 64
emissivity, 19, 20, 52, 136, 139–40, 256, 258–9, 263, 284, 289, 337, 385–7, 390
energy balance, 395, 457–64
energy levels, 62, 68–9, 88–9
Enke, comet, 347
epsonite ($MgSO_4*6H_2O$), 341
equation
  of heat conduction, 388
  of state, 422, 446
  of transfer, 27–57, 354–5, 410
equations of fluid motion, 421–8, 435
equipartition law, 22–3
Eros, asteroid, 350
étendue, 156, 221–2, 232, 251, 282, 292, 297
ethane (C_2H_6), 87, 93, 250, 318–19, 321, 326–30, 344, 419, 455
Europa, 341, 458
evolution of the Solar System, 444–57
Explorer, 154
exponential integral, 403
extinction coefficient, 384, 415
extinction cross section, 32, 38, 381, 411
extrinsic and intrinsic semiconductors, 273

f-number, 156
Fabry–Perot interferometer, 162, 204–11, 218–20, 240, 294
far-field phase function, 122
Far Infrared Absolute Spectrometer (FIRAS), 225, 244, 247
far-wing absorption, 102–3
field-effect transistor (FET), 270, 279–80
field lens, 159, 161–2
field stop, 159, 161, 164–5
flux, 25–6, 40, 111, 388, 405–19
Fourier Transform Spectroscopy (FTS), 211, 220–48, 294, 360
Fraunhofer diffraction, 111–12, 166, 212
Fresnel equations, 17–20, 197
frictional damping time, 440

Galileo, mission to Jupiter, 323, 340–1, 350
Ganymede, 325, 458
gas filter, 192–4

Gaussian line shape, 101
generation–recombination noise, 277
Geostationary Operational Environmental Satellite (GOES), 175
geostrophic balance, 426–7, 443
geosynchronous altitude, 175–7
germane (GeH_4), 87, 321–2, 456
germanium (Ge), 189, 204, 206–7, 271–2, 456
Giotto, 348
glory, 126
Golay cell, 269
gold black, 266
grating spectrometers, 209–20, 240–3
gravitational potential, 425
greenhouse effect, 415–16, 418, 445, 450, 452

Halley, comet, 325, 346–8
Hamming function, 230
harmonic oscillator, 23, 66, 69–70, 109
heat capacity, 255–8, 391
helium, 295, 318, 377, 400, 445–9, 454
helium to hydrogen ratio, 377–8, 453–63
heterodyne detection, 249–52
High Resolution Infrared Radiometer (HRIR), instrument on Nimbus, 184–6
homogeneous line broadening, 101
hot spot
  on Io, 333–7
  on Jupiter, 321–2
hydrogen,
  atomic, 59, 320
  deuterated, 85, 456
  dimer, 79
  metallic, 105, 455
  molecular, 66, 78, 318–21, 328, 448, 453, 455
  ortho and para forms, 359, 377, 438, 455, 462
hydrogen bond, 104
hydrogen chloride (HCl), 65, 69, 310
hydrogen cyanide (HCN), 81, 84, 326–30, 332, 348, 455
hydrogen fluoride (HF), 310
hydrogen to helium ratio, see helium to hydrogen ratio
hydrostatic equation, 426, 443
hydrostatic equilibrium, 140

image motion compensation, 185, 188
Imaging Spectrometer, Mars (ISM), 218, 316
immersion lens, 160
Improved TIROS Operational Satellite (ITOS), 175–6
indium antimonide detector, 273
induced dipole moment, 78–9
infrared detectors, 253–81
Infrared Interferometer Spectrometer (IRIS), 224–7, 314–5, 318, 322, 329, 337, 359, 373, 377, 382, 404, 434–8
Infrared Space Observatory (ISO), 218–20, 320, 322–9, 332
Infrared Thermal Mapper (IRTM), instrument on Viking, 315
inhomogeneous line broadening, 101
initial state population, 95

intensity calibration, 281–93
interference filters, 204–9
interferogram, 222, 228–9
internal heat source, 457–63
intrinsic and extrinsic semiconductors, 273–4
invariance of $n^2 A\Omega$, 158, 160
invariance principle, 44–50
inversion methods, 352–5
Io, satellite of Jupiter, 295, 333, 335, 338–40
ionic bond, 104
isotopes, 85, 303–4, 372–3, 451, 456
isotopic ratio, 304, 333, 451

Johnson noise, 261, 264, 266–8, 271, 277
Jupiter, 317–25, 347, 359, 377, 413, 416, 421, 433, 439–41, 454–62

kernel, 361–3
Kramers–Kronig relation, 110
Kuiper Belt, also known as Edgeworth–Kuiper Belt, 347

Lambert surface, 400–1
lamellar grating interferometer, 247–8
lattice modes, 103
lead telluride, 207
limb-tangent measurements, 378–80
line shape, 99–103
line strength, 93–8
linear inversion technique, 356–60
liquid helium, 278
lithium fluoride, 192
lithium tantalum oxide detector, 240
Littrow configuration of a grating spectrometer, 212–13
local thermodynamic equilibrium, 37, 352
Loki, area on Io, 333–5, 339–40, 394
Lorentz line shape, 99–102

magnesium carbonate ($MgCO_3$), 190
magnesium oxide (MgO), 190
magnetic permeability, 3
Mariner 6 and 7, missions to Mars, 209, 284, 387
Mariner 9, mission to Mars, 227, 284, 289, 295, 303, 305
Mariner 10, mission to Mercury, 387
Mars, 284, 295, 303, 305–6, 312–17, 413, 421, 428–37, 448
Mars Global Surveyor (MGS), 232, 316–17, 429
Mars Orbital Laser Altimeter (MOLA), instrument on MGS, 316
Martin–Puplett interferometer, 243–7
Maxwell's equations, 1–3, 5–6, 107, 113, 196
Mercury, 301, 325, 337, 391, 448, 450
mercury cadmium telluride detector, 250, 273, 276
methane (CH_4), 321, 344, 348, 450, 456
  deuterated, 321, 456
methyl acetylene (C_3H_4), 85, 87, 89, 326–30, 332
Michelson Interferometer, 162, 208, 220–43, 299
Mie theory, 113, 386
Miranda, satellite of Uranus, 104, 185, 187, 295
moment of inertia, 66, 72, 86, 457

momentum equation, 425–6
Moon, 301, 325, 338–9, 343, 391, 444, 448
multiplex advantage, 211, 221, 240, 299
Mylar, 190, 201–2

nadir viewing, 317, 369
natron ($Na_2CO_3*10H_2O$), 341
natural line broadening, 99
Near Earth Asteroid Rendezvous Mission (NEAR), 350
near infrared calibration, 291–3
Near Infrared Mapping Spectrometer (NIMS), 216–18, 340–1
Neptune, 317, 324, 347, 359, 413, 416, 421, 439, 453–5, 460, 463
net flux, 26
neutrino flux, 447
Nimbus, weather satellite, 216, 224–5, 227, 289, 305, 307–8
nitrogen (N, N_2), 66, 78, 333, 344
noise, 260–4, 277–8
Noise Equivalent Flux (NEF), 252
Noise-Equivalent-Power (NEP), 232, 241–2, 254, 261–3, 266, 268, 271
Noise-Equivalent-Spectral-Radiance (NESR), 232, 290, 296–7, 302
normal modes of vibration, 80–3
nuclear reactions in the Sun, 446–7
numerical filtering, 361
Nyquist formula, 261

oblate symmetric top, 88–9
observing efficiency, 182, 223
off-axis configuration, 163, 165
one-over-f noise, 179, 225, 264, 277
onion peeling method, 379–80
Oort cloud, 346–8
optical depth, 39, 353, 384
optical retardation, 214
Orbiting Geophysical Observatory (OGO), 209
oxygen (O, O_2), 69–70, 78, 209, 451
ozone (O_3), 87, 305–6, 308, 310, 419, 445

P-branch, 77–8, 91–4, 98, 149
parallel and perpendicular bands, 92
paraxial approximation, 156
partition function, 95
path difference modulation, 234–5
Pathfinder Mission to Mars, 316
Pele, area on Io, 334, 336
Peltier effect, 264
perturbation theory, 62
phase characteristic, 175, 227
phase function, 29, 37–8, 52, 126, 135, 139, 400
phase integral, 400–1
phase locked loop, 227–8
phased degenerate vibration, 83
Phobos 2, mission to Mars, 218, 316
phonon, 262, 264
phosphene (PH_3), 318, 321
photoconductors, 253, 274–6

photodiodes, 274–5
photometric investigations, 394–404
photomultipliers, 275
photon detectors, 272–81
photon noise, 277–8
Photopolarimeter–Radiometer, instrument on Galileo, 340
photovoltaic detectors, 253, 275–6
Pioneer 10 and 11, 295, 395, 437
Pioneer Venus, 194, 309, 443
Planck function, 21–5, 36, 260, 288, 334, 380, 403
Planetary Fourier Spectrometer (PFS), 239–40
plasma frequency, 110
Pluto, 333, 342–5
pneumatic detector, 268–9
Pointing–Robertson Effect, 350
polarization,
    electrical, 106–7, 269–70
    optical, 13–14, 252, 386–7, 392–4
polarizing interferometer, 243–7
polyatomic molecules, 80–93
polyethylene, 190
post-dispersion, 240–3
potassium bromide, 109, 189, 203–4
potential energy, 60
Poynting vector, 4–5, 25
precision, 281–2
pressure gradient, 427
pressure modulation, 192–4, 442
primitive equations, 426
primordial atmospheres, 448–50
principal moments of inertia, 86–7
prism spectrometer, 190–2
probability distribution, 61–3
prolate symmetric top, 88–9
propane (C_3H_8), 85–7, 94, 326–30
proto planets, 448, 452
proto Sun, 446
pyroelectric detector, 269–71

Q-branch, 91–4, 149, 307–8, 321
quadruple transition, 64, 70, 74
quantum efficiency, 276
quantum mechanics, 59–64
quantum number, 73, 83–4, 87, 98
quantum theory of Bohr, 59–60
quartz, 189, 312, 337

R-branch, 77–8, 91–4, 98, 149
Ra Patera, area on Io, 394
radiative equilibrium, 405–20, 432, 440, 447
radiative relaxation time, 432
radiative transfer, 27–57, 353, 385, 437
radioactive decay, 407, 458–9
rainbow, 126
Rayleigh friction, 440
Rayleigh scattering, 111
Rayleigh–Jeans law, 22–4
reduced mass, 68, 72
reference interferometer, 225, 227, 294
reflectivity, 52, 386, 397

refractive index, 8, 18, 191, 382, 392–4
   real and imaginary, 10, 106
relaxation algorithm, 360–1
relaxation phenomena, 257
relaxation time, 110
residual ray effect, 190
resolution,
   spatial, 230, 296, 298
   spectral, 213–14, 231, 296–8, 302, 369
resolving power, 192, 214
responsivity, 253–5, 270, 275–7, 284, 288, 290
rigid rotor, 72, 74, 80, 88–9
Rossby number, 427, 439
rotation of diatomic molecules, 64, 72–5
rotation of polyatomic molecules, 64, 86–9
rotational temperature, 97
runaway greenhouse, 450
rutile, 189

sapphire, 189
Satellite Infrared Spectrometer (SIRS), instrument on Nimbus, 216–17
Saturn, 317–21, 324, 347, 359, 377, 413, 416, 419, 421, 439, 453, 460, 462
scale height, 382
scanning function, 172–4
scattering, 31–5
scattering function, 110
Schrödinger equation, 60–4, 83
   in polar coordinates, 73
   time independent, 62
selection rule, 69–70, 72, 74, 77, 85, 91, 98
selective chopper, 192–4
selenium, 160, 199–200
Shoemaker–Levy 9, comet, 325, 347
signal-to-noise ratio, 252, 267, 271–3, 302
silicates, 311–13, 346
silicon, 198, 207, 272
silicon oxide (SiO), 69
silver chloride (AgCl), 189
simultaneous retrieval of temperature and abundance, 376–8
sinc function, 230
sodium, 59, 209
sodium chloride, 104, 192
solar constant, 450
solar wind, 346, 350, 450
solid and liquid surfaces, 103–10
solid surface parameters, 385–94
spatial coherence, 251
specific intensity, 26
spherical Bessel and Hankel functions, 314–15
spherical lamellar grating, 248
spherical top, 88
Spin Scan Cloud Camera (SSCC), 175, 177
spinel, 189
standard deviation, 282
statistical estimation 365–7
stepping mode, 223
Stratospheric and Mesospheric Sounder (SAMS), 194
stray light, 163

sulfur, 334
sulfur dioxide (SO_2), 309, 333–4, 339–40, 450
sulfur oxide (SO), 69
sulfuric acid clouds, 305, 309, 415
Sun, nuclear reactions in the, 446–7
surface emissivity, 52, 256–9, 289, 385–7
surface temperature, 385, 391
surface texture, 392–4
symmetric top, 88
synchronous rectifier, 181–2

TE wave, 196, 201
TM wave, 196, 201
T-Tauri stars, 449–50
Television and Infrared Observational Satellite (TIROS), 158
temperature inversion, 406, 414, 418
temperature lapse rate, 432, 437–8
temperature noise, 261–2
temperature profile retrieval, 355–68
tenuous atmospheres, 333–4
Thematic Mapper, 175, 178–80, 299
thermal capacity, 256, 258, 271–2
thermal conduction, 255, 256, 258, 388
thermal detectors, 255–72
thermal–electric analogies, 258
Thermal Emission Spectrometer (TES), instrument on MGS, 232–3, 316–17, 429–30, 436
thermal inertia, 388, 390
thermal resistance, 255–8
thermal time constant, 257–8
thermal wind equation, 420, 428, 431
thermocouple, 264–6
thermoelectric effect, 265
thermopile, 265
thin film theory, 195–204
thorium fluoride, 207
tidal theory, 435
Titan, 295, 324–33, 420–1, 443–4
trans-impedance amplifier (TIA), 278–80
transition moment, 63
transmission function, 45, 50, 355
Trans-Neptunian Objects, 342–51
triglycine phosphate (TGS), 269
triple point, 105
Triton, satellite of Neptune, 295, 333, 345
turbulent convection, 412, 419
two-stream approximation, 52–7, 132, 411

Upper Atmosphere Research Satellite (UARS), 205
Uranus, 317–18, 347, 359, 421, 434, 449, 453–4, 460–3
Urey reaction, 451

van der Waal forces, 79, 104
Vanguard program, 154
variable thickness filter, 208–9
Venera, mission to Venus, 305–6, 309
Venus, 305–6, 308–9, 320, 413, 419, 442–4, 448
Venus Orbiter Radiometer Temperature Experiment (VORTEX), 194

vertical resolution, 364
vertical temperature sounding, 367
Very High Resolution Radiometer (VHRR), 175
vibration of molecules, 64–72, 75–8, 80–6, 90–3
vibration–rotation interaction of molecules, 75–8, 90–3
vibrational quantum number, 68
vignetting, 161
Viking, mission to Mars, 216, 296, 387, 429, 452
visibility curve, 220–1
Visible Infrared Mapping Spectrometer (VIMS), instrument on Cassini 217–18
Visible Infrared Spin Scan Radiometer (VISSR), 175
Voigt line shape, 100, 102
Voyager 1 and 2, 226–7, 229, 289, 295, 318–23, 359, 421

water (H_2O), 83, 87, 305–6, 308, 310, 317, 448–52
water clouds, 172, 456
water ice, 104–5, 314–15, 324, 337, 341, 344–5, 347, 452
water ice clouds, 315–17
wave equation, 61
wave impedance, 12, 260
wave mechanics, 60
wavenumber calibration, 293–4
weighting function, 143–8, 353, 358, 360, 368
Wien's displacement law, 22
Wien's radiation law, 22-3
Winston cone, 160

zinc sulfide, 200
zonal winds, 439–41